Soil Use and Management

Soil Use and Management

Editor: Lester Bane

R CALLISTO REFERENCE

www.callistoreference.com

Callisto Reference,
118-35 Queens Blvd., Suite 400,
Forest Hills, NY 11375, USA

Visit us on the World Wide Web at:
www.callistoreference.com

ISBN: 978-1-63239-789-8 (Hardback)

The publisher's policy is to use permanent paper from mills that operate a sustainable forestry policy. Furthermore, the publisher ensures that the text paper and cover boards used have met acceptable environmental accreditation standards.

Trademark Notice: Registered trademark of products or corporate names are used only for explanation and identification without intent to infringe.

Printed in the United States of America.

Cataloging-in-publication Data

Soil use and management / edited by Lester Bane.
  p. cm.
Includes bibliographical references and index.
ISBN 978-1-63239-789-8
1. Soil science. 2. Soil management. 3. Land use. 4. Agriculture. 5. Soil chemistry. 6. Soil fertility.
7. Soil productivity. I. Bane, Lester.
S591 .S65 2017
631.4--dc23

Table of Contents

Preface

Soil Use and management refers to the practices used for protecting and managing soil. It incorporates all the functions and treatments required for soil enhancement and nutrient management. Some of the major aspects of soil management are to maintain soil fertility, increase crop yield, improve soil structure and function, reduce tillage, etc. This book provides significant information of this discipline to help develop a good understanding of the area and related fields. Most of the topics introduced in it cover new techniques and the applications within this area. Researchers and students interested in the subject will be assisted by this text as it comprises contributions of experts from across the globe.

This book unites the global concepts and researches in an organized manner for a comprehensive understanding of the subject. It is a ripe text for all researchers, students, scientists or anyone else who is interested in acquiring a better knowledge of this dynamic field.

I extend my sincere thanks to the contributors for such eloquent research chapters. Finally, I thank my family for being a source of support and help.

Editor

Configuration and Specifications of an Unmanned Aerial Vehicle (UAV) for Early Site Specific Weed Management

Jorge Torres-Sánchez*, Francisca López-Granados, Ana Isabel De Castro, José Manuel Peña-Barragán

Department of Crop Protection, Institute for Sustainable Agriculture (IAS) Spanish National Research Council (CSIC), Córdoba, Spain

Abstract

A new aerial platform has risen recently for image acquisition, the Unmanned Aerial Vehicle (UAV). This article describes the technical specifications and configuration of a UAV used to capture remote images for early season site- specific weed management (ESSWM). Image spatial and spectral properties required for weed seedling discrimination were also evaluated. Two different sensors, a still visible camera and a six-band multispectral camera, and three flight altitudes (30, 60 and 100 m) were tested over a naturally infested sunflower field. The main phases of the UAV workflow were the following: 1) mission planning, 2) UAV flight and image acquisition, and 3) image pre-processing. Three different aspects were needed to plan the route: flight area, camera specifications and UAV tasks. The pre-processing phase included the correct alignment of the six bands of the multispectral imagery and the orthorectification and mosaicking of the individual images captured in each flight. The image pixel size, area covered by each image and flight timing were very sensitive to flight altitude. At a lower altitude, the UAV captured images of finer spatial resolution, although the number of images needed to cover the whole field may be a limiting factor due to the energy required for a greater flight length and computational requirements for the further mosaicking process. Spectral differences between weeds, crop and bare soil were significant in the vegetation indices studied (Excess Green Index, Normalised Green-Red Difference Index and Normalised Difference Vegetation Index), mainly at a 30 m altitude. However, greater spectral separability was obtained between vegetation and bare soil with the index NDVI. These results suggest that an agreement among spectral and spatial resolutions is needed to optimise the flight mission according to every agronomical objective as affected by the size of the smaller object to be discriminated (weed plants or weed patches).

Editor: Derek Abbott, University of Adelaide, Australia

Funding: This research was partly financed by the TOAS Project (Marie Curie Program, ref.: FP7-PEOPLE-2011-CIG-293991, EU-7th Frame Program) and the AGL2011-30442-CO2-01 project (Spanish Ministry of Economy and Competition, FEDER Funds). Research of Dr. Peña-Barragán, Ms. De Castro and Mr. Torres-Sánchez was financed by JAEDoc, JAEPre and FPI Programs, respectively. The funders had no role in study design, data collection and analysis, decision to publish, or preparation of the manuscript.

Competing Interests: The authors have declared that no competing interests exist.

* E-mail: jtorres@ias.csic.es

Introduction

Precision agriculture (PA) is defined as "*a management strategy that uses information technology to bring data from multiple sources to bear on decisions associated with crop production*" [1]. PA encompasses all the techniques and methods for crop and field management by taking into account their local and site-specific heterogeneity and variability [2]. Within the context of PA, early season site-specific weed management (ESSWM) involves the development of techniques to detect the weeds growing in a crop and the application of new technologies embedded in specific agricultural machinery or equipment to control them successfully, taking action to maximise economic factors and reduce the environmental impact of the control measurements applied [3]. The efficient development of these practices somehow relies on the use of remote sensing technology for collecting and processing spatial data from sensors mounted in satellite or aerial platforms. This technology has been widely applied in agricultural studies, allowing the mapping of a variety of factors [4], including crop conditions [5], soil properties [6], water content [7] and weed distribution [8], among others. Piloted aircraft and satellites are traditionally the primary platforms used to obtain remote images

for local to global data acquisition. However, these platforms present problems for many aspects of precision agriculture because they are limited in their ability to provide imagery of adequate spatial and temporal resolutions and are strongly affected by weather conditions [9]. In the case of ESSWM, good results have been obtained in late growth stages (normally at the flowering stage) using aerial [10–11] and satellite [12] images, with herbicide savings of more than 50% reported. Nevertheless, in most weed-crop scenarios, the optimal weed treatment is recommended at an early growth stage of the crop, just a few weeks after crop emergence. In this stage, mapping weeds using remote sensing presents much greater difficulties than in the case of the late-stage season for three main reasons [13]: 1) weeds are generally distributed in small patches, which makes it necessary to work with remote images at very small pixel sizes, often on the order of centimetres [14]; 2) grass weeds and monocotyledonous crops (e.g., *Avena* spp. in wheat) or broad-leaved weeds and many dicotyledonous crops (e.g., *Chenopodium* spp. in sunflower) generally have similar reflectance properties early in the season, which decreases the possibility of discriminating between vegetation classes using only spectral information; and 3) soil background reflectance may interfere with detection [15].

Today, difficulties related to spatial and temporal resolutions can be overcome using an Unmanned Aerial Vehicle (UAV) based remote sensing system, which has progressed in recent years as a new aerial platform for image acquisition. UAVs can fly at low altitudes, allowing them to take ultra-high spatial resolution imagery and to observe small individual plants and patches, which has not previously been possible [16]. Moreover, UAVs can supply images even on cloudy days, and the time needed to prepare and initiate the flight is reduced, which allows greater flexibility in scheduling the imagery acquisition. Other advantages of UAVs are their lower cost, and the lower probability of serious accidents compared with piloted aircraft.

Examples of applications of UAVs in agricultural studies are becoming more noticeable in the literature. For instance, Hunt *et al.* (2005) [17] evaluated an aerobatic model aircraft for acquiring high-resolution digital photography to be used in estimating the nutrient status of corn and crop biomass of corn, alfalfa, and soybeans. In other cases, an unmanned helicopter was tested to monitor turf grass glyphosate application [16], demonstrating its ability to obtain multispectral imaging. Other UAV models have been developed, such as the six-rotor aerial platform used by Primicerio *et al.* (2012) [18] to map vineyard vigour with a multi-spectral camera. Recently, Zhang and Kovacs (2012) [19] reviewed the advances in UAV platforms for PA applications. In this review, they indicated the phases in the production of the remote images (including acquisition, georeferencing and mosaicking) and the general workflow for information extraction. Generally, all these authors concluded that these systems provide very promising results for PA and identified some key factors for equipment and system selection, such as maximum UAV payload capacity, platform reliability and stability, sensor capability, flight length and UAV manoeuvrability, among others [20–22].

To our knowledge, however, no detailed investigation has been conducted regarding the application of this technology in the field of ESSWM, in which remote images at centimetre-scale spatial resolution and a narrow temporal window for image acquisition are required [23]. Therefore, this paper defines the technical specifications and configuration of a quadrocopter UAV and evaluates the spatial and spectral requirements of the images captured by two different sensors (a commercial scale camera and a multispectral 6-channel camera) with the ultimate aim of discriminating weed infestations in a sunflower crop-field in the early growing season for post-emergence treatments. Moreover, the steps for preparing and performing UAV flights with both cameras are described as well as the relationships amongst flight altitude, pixel size, sensor properties and image spectral information.

Materials and Methods

1. UAV Description

A quadrocopter platform with vertical take-off and landing (VTOL), model md4-1000 (microdrones GmbH, Siegen, Germany), was used to collect a set of aerial images at several flight altitudes over an experimental crop-field (Figure 1). This UAV is equipped with four brushless motors powered by a battery and can fly by remote control or autonomously with the aid of its Global Position System (GPS) receiver and its waypoint navigation system. The VTOL system makes the UAV independent of a runway, so it can be used in a wide range of different situations and flight altitudes. The UAV's technical specifications and operational conditions, provided by the manufacturer, are shown in Table 1.

The whole system consists of the vehicle, the radio control transmitter, a ground station with the software for mission planning and flight control, and a telemetry system. The radio control transmitter is a handheld device whose main tasks are to start the vehicle's engines, manage take-off and landing, control the complete flight in the manual mode, and activate the autonomous navigation system. The control switchboard consists of several triggers, pushbuttons, scroll bars, a display, and an antenna, and it is equipped with a RF-module synthesiser, which enables the selection of any channel in the 35 MHz band. The ground station works as an interface between the operator and the vehicle and includes the support software mdCockpit (MDC). MDC allow the UAV settings to be configured, implements the flight route plan with the Waypoint Editor (WPE) module, and monitors the flight. The telemetry system collects relevant flight data and retrieves a stream of information in a plain text scheme that includes GPS position data, attitude, altitude, flight time, battery level, and motor power output, among many others. All sensors and control devices for flight and navigation purposes are embedded on-board the vehicle and are managed by a computer system, which can listen telemetry data and make decisions according to the momentary flight situation and machine status, thus avoiding that occasional loss of critical communication between the UAV and the ground station resulting in the vehicle crashing.

Three persons were employed for the secure use of the UAV: a radio control pilot, a ground station operator and a visual observer. The radio control pilot manually takes off and lands the UAV and activates the programmed route during the flight operation. The ground station operator controls the information provided by the telemetry system, i.e., UAV position, flight altitude, flight speed, battery level, radio control signal quality and wind speed. The visual observer is on the lookout for potential collision threats with other air traffic.

2. Sensors Description

The md4-1000 UAV can carry any sensor weighing less than 1.25 kg mounted under its belly, although the maximum recommended payload is 0.80 kg. Two sensors with different spectral and spatial resolutions were separately mounted on the UAV to be tested in this experiment: a still point-and-shoot camera, model Olympus PEN E-PM1 (Olympus Corporation, Tokyo, Japan), and a six-band multispectral camera, model Tetracam mini-MCA-6 (Tetracam Inc., Chatsworth, CA, USA). The Olympus camera acquires 12-megapixel images in true colour (Red, R; Green, G; and Blue, B, bands) with 8-bit radiometric resolution and is equipped with a 14–42 mm zoom lens. The camera's sensor is $4{,}032 \times 3{,}024$ pixels, and the images are stored in a secure digital SD-card. The mini-MCA-6 is a lightweight (700 g) multispectral sensor composed of six individual digital channels arranged in a 2×3 array. The slave channels are labelled from "1" to "5", while the sixth "master" channel is used to define the global settings used by the camera (e.g., integration time). Each channel has a focal length of 9.6 mm and a 1.3 megapixel ($1{,}280 \times 1{,}024$ pixels) CMOS sensor that stores the images on a compact flash CF-card. The images can be acquired with 8-bit or 10-bit radiometric resolution. The camera has user configurable band pass filters (Andover Corporation, Salem, NH, USA) of 10-nm full-width at half-maximum and centre wavelengths at B (450 nm), G (530 nm), R (670 and 700 nm), R edge (740 nm) and near-infrared (NIR, 780 nm). These bandwidth filters were selected across the visible and NIR regions with regard to well-known biophysical indices developed for vegetation monitoring [24]. Image triggering is activated by the UAV according to the

Figure 1. The quadrocopter UAV, model md4-1000, flying over the experimental crop-field.

programmed flight route. At the moment of each shoot, the on-board computer system records a timestamp, the GPS location, the flight altitude, and vehicle principal axes (pitch, roll and heading).

3. Study Site and Field Sampling

The UAV system was tested in a sunflower field situated at the private farm La Monclova, in La Luisiana (Seville, southern Spain, coordinates 37.527N, 5.302W, datum WGS84). The flights were authorized by a written agreement between the farm owners and our research group. We selected sunflower because this is the major oil-seed crop grown in Spain, with a total surface of 850,000 ha in 2012 [25], and because weed control operations (either chemical or physical) with large agricultural machinery represent a significant proportion of production costs, create various agronomic problems (soil compaction and erosion) and represent a risk for environmental pollution. The sunflower seeds were planted at the end of March 2012 at 6 kg ha^{-1} in rows 0.7 m apart. The set of aerial images were collected on May 15th, just when post-emergence herbicide or other control techniques are recommended in this crop. Several visits were periodically made to the field from crop sowing to monitor crop growth and weed emergence and, finally, to select the best moment to take the set of remote images. The sunflower was at the stage of 4–6 leaves

unfolded. The weed plants had a similar size or, in some cases, were smaller than the crop plants (Figure 1).

An experimental plot of 100×100 m was delimited within the crop-field to perform the flights. The coordinates of each corner of the flight area were collected using GPS to prepare the flight route in the mission-planning task. A systematic on-ground sampling procedure was carried out the day of the UAV flights. The procedure consisted of placing 49 square white frames of 1×1 m distributed regularly throughout the studied surface (Figure 2A). Every frame was georeferenced with a GPS and photographed in order to compare on-ground weed infestation (observed weed density) and outputs from image classification (estimated weed density). These numbered cards were also utilised as artificial terrestrial targets (ATTs) to perform the imagery orthorectification and mosaicking process. In the course of the UAV flights, a barium sulphate standard spectralon® panel (Labsphere Inc., North Sutton, NH, USA) of 1×1 m was also placed in the middle of the field to calibrate the spectral data (Figure 2B).

4. UAV Flight and Sensors Tests

4.1. Mission planning. The flight mission was planned with the WPE module of the MDC software installed at the ground station. The flight route was designed over the orthoimages and the digital elevation model (DEM) of the flight area previously

Table 1. Technical specifications and operational conditions of the UAV, model md4-1000.

UAV specification	Value
Technical specifications	
Climb rate	7.5 m/s
Cruising speed	15.0 m/s
Peak thrust	118 N
Vehicle mass	2.65 Kg approx. (depends on configuration)
Recommended payload mass	0.80 Kg
Maximum payload mass	1.25 Kg
Maximum take-off weight	5.55 Kg
Dimensions	1.03 m between opposite rotor shafts
Flight time	Up to 45 min (depends on payload and wind)
Operational conditions	
Temperature	$-10°C$ to $50°C$
Humidity	Maximum 90%
Wind tolerance	Steady pictures up to 6 m/s
Flight radius	Minimum 500 m using radiocontrol, with waypoints up to 40 km
Ceiling altitude	Up to 1,000 m
Take-off altitude	Up to 4,000 m about sea level

Source: UAV manufacturer (microdrones GmbH, Siegen, Germany).

imported from the application Google Earth™ (Keyhole Inc., Mountain View, CA, USA). Three different parameters were needed to plan the route: flight area, camera specifications and UAV tasks (Table 2). The flight area information includes width and length, the direction angle of the main side, and the desired overlap in the imagery. The images were acquired at 60% forward-lap and 30% side-lap. The camera specifications are the focal length and the sensor size. The UAV tasks refer to the actions that the UAV has to perform once it arrives at each point for image acquisition, and it includes the number of photos and dwell time in each point. Once both, this information and the flight altitude were introduced in the WPE module, it automatically generated the flight route and estimated the flight duration according to the total number of images planned (Figure 3). The route file was exported to a memory card embedded in the UAV via a standard serial link.

4.2. UAV flight and image acquisition. The preliminary steps before starting the flight were to upload the flight route to the UAV computer system, attach the camera to the vehicle and check the connectivity and the proper functioning of the whole system. After these steps, the pilot manually launches the UAV with the radio control transmitter and next activates the automatic flight route, making the vehicle go to the first waypoint and then fly along the flight lines until the entire study area is completely covered. Once all the images are taken, the pilot manually lands the UAV, and the ground station operator prepares the vehicle for the next route. During the flight, the ground station operator watches the UAV telemetry data using the downlink decoder,

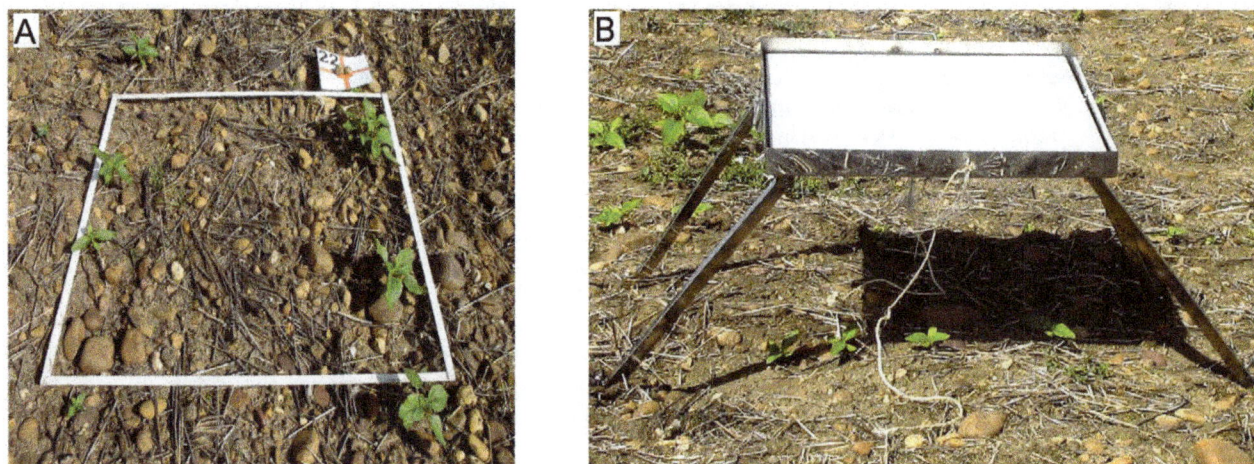

Figure 2. Details of the experimental set. a) 1×1 m frame used in the ground-truth field sampling, and b) reference panel for image spectral calibration.

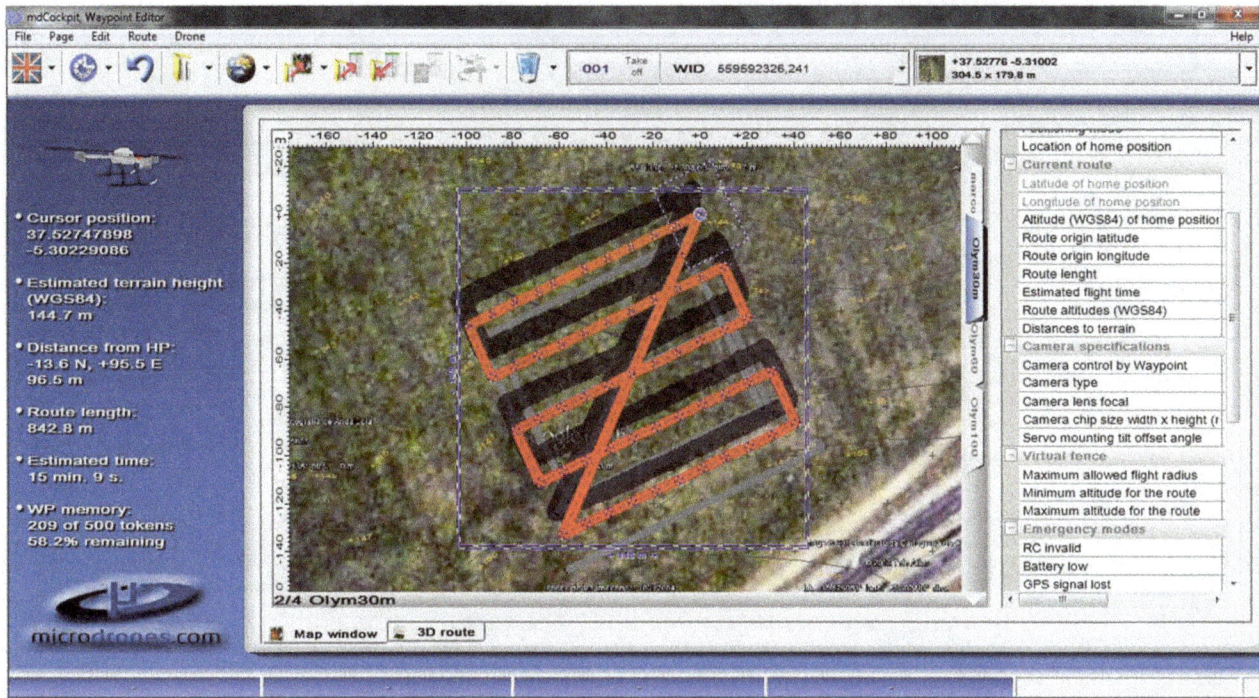

Figure 3. Screen shot of the Waypoint Editor module showing the flight planning.

another component of the MDC software (Figure 4). This program gives information about: 1) operating time of the UAV, 2) current flight time, 3) distance from take-off point to the UAV, 4) quality of the remote control signal received by the UAV, 5) downlink quality, 6) battery status, and 7) GPS accuracy.

Table 2. Data required by the Waypoint Editor software and the route settings used in the experimental field.

Data type	Setting value*
Flight area	
Width	100 m
Length	100 m
Direction angle	65°
Horizontal overlapping	60%
Vertical overlapping	30%
Camera specifications	
Focal length	
RGB camera	14 mm
Multispectral camera	9.6 mm
Sensor size (width ×length)	
RGB camera	17.3×13 mm
Multispectral camera	6.66×5.32 mm
UAV tasks	
Dwell	5 s
Number of images	1

*Values used in the experimental field.

In addition to this information, the downlink decoder supports several important dialog pages, as follows:

– Flight and video. This page shows the video stream captured by the sensor attached to the UAV, making it easier to control the UAV when it is manually driven. Additional data displayed in this page are: 1) distance to the UAV, 2) flight altitude above the take-off position, 3) speed of the UAV, 4) artificial horizon, 5) compass, and 6) roll and tilt angles.

– Technical. This page supplies information about: 1) UAV position (GPS latitude and longitude), 2) UAV altitude (GPS altitude above sea level), 3) current navigation mode, 4) magnetometer status, 5) barometer status, 6) motor power, 7) momentary status of all the radio control channels, and 8) limit values of flight altitude, distance and speed.

– Route. This page shows a tridimensional display of the flight path.

– Waypoint. This section shows information about: 1) the flying route followed by the UAV, 2) the UAV GPS position, and 3) the waypoint command that is being executed at each moment.

– Sensor-payload. This page displays a diagram with sensor data received from the payload.

– Recordings. Three diagrams are displayed in this section: 1) comprising motor power and battery voltage over time, 2) comprising flight attitude (roll, pitch and yaw angles) with GPS data, and 3) comprising velocity, distance, wind profile, flight altitude and radio-control link quality.

4.3. Multispectral band alignment. The images taken by the still camera (Olympus model) can be used directly after downloading to the computer, but those taken by the multispectral camera (mini-MCA-6 Tetracam model) require some pre-processing. This camera takes the images of each channel in raw format and stores them separately on six individual CF cards

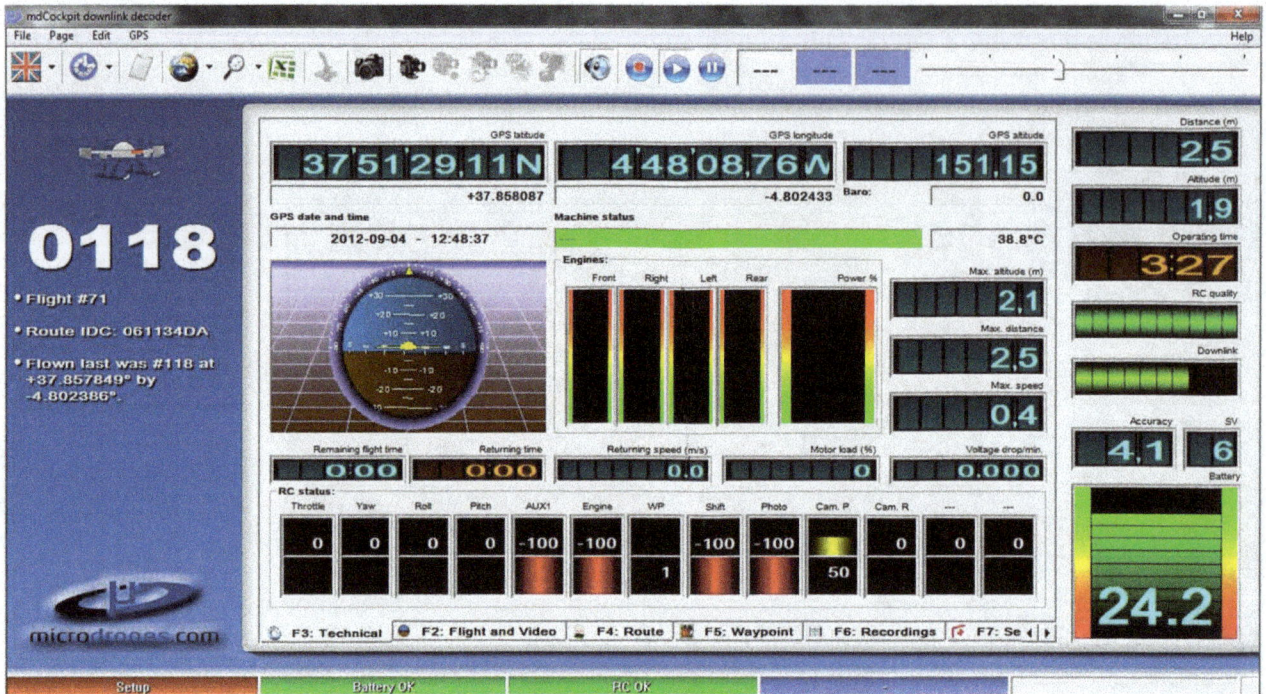

Figure 4. Screen shot of the Downlink Decoder module showing the information displayed during a programmed flight.

embedded in the camera. Therefore, an alignment process is needed to group the six images taken in each waypoint. The Tetracam PixelWrench 2 (PW2) software (Tetracam Inc., Chatsworth, CA, USA), supplied with the multispectral camera, was used to perform the alignment process. The PW2 software provides a band-to-band registration file that contains information about the translation, rotation and scaling between the master and slave channels. Two different options were tested: 1) basic configuration of the PW2 software, as applied by Laliberte et al. (2011) [26], and 2) advanced configuration of PW2, which includes the newest field of view (FOV) optical calculator, which calculates additional offsets to compensate the alignment for closer distances [27]. The quality of the alignment process was evaluated with the help of the spectralon® panel data captured in the images at a 30 m altitude. Spatial profiles were taken across the reference panel for each method and compared with the non-aligned image. The spatial profiles consisted of graphics representing the spectral values for each band along a line 45 pixels long drawn in the multi-band images using the ENVI image processing software (Research System Inc., Boulder, CO, USA).

4.4. Spatial resolution and flight length as affected by flight altitude. Three independent flight routes were programmed for each type of camera to cover the whole experimental field at 30, 60 and 100 m altitude above ground level. The effects of flight altitude and camera resolution with respect to pixel size, area coverage (number of images per hectare) and flight duration were studied, and their implications for weed discrimination in the early season were discussed.

4.5. Spectral resolution as affected by flight altitude. To perform weed mapping based on UAV images, two consecutive phases are usually required [13]: 1) bare soil and vegetation discrimination, which would allow obtaining a two-classes image with vegetal cover (crop and weeds together) and bare soil, 2) crop and weeds discrimination, in which the zones corresponding to crop are identified and masked, and finally, the detection and

location of weeds are obtained. To determine the limitations of each sensor with regard to both phases, spectral values of the three covers present in the field (bare soil, crop and weeds) were extracted. These spectral values were collected in 15 randomly selected sampling areas for each soil use from the images taken during all the flight missions (i.e., both sensors at 30, 60 and 100 m altitudes).

Three well-known vegetation indices (VIs) were derived from these values:

– Normalised Difference Vegetation Index (NDVI, [28])

$$NDVI = (NIR - R)/(NIR + R) \qquad (1)$$

- Normalised Green-Red Difference Index (NGRDI, [29]),

$$NGRDI = (G - R)/(G + R) \qquad (2)$$

- Excess Green Index (ExG, [30], [31]).

$$ExG = 2g - r - b \qquad (3)$$

The potential of the VIs for spectral discrimination was evaluated by performing a least significant difference (LSD) test at $p \leq 0.01$ through a one-way analysis of variance (ANOVA), and applying the M-statistic (equation 4) presented by Kaufman and Remer (1994) [32] in order to quantify the histograms separation of vegetation indices. JMP software (SAS, Cary, NC, USA) was employed to perform the statistical analysis.

$$M = (\mathrm{MEAN_{class1}} - \mathrm{MEAN_{class2}})/(\sigma_{class1} - \sigma_{class2}) \quad (4)$$

M expresses the difference in the means of the class 1 and class 2 histograms normalized by the sum of their standard deviations (σ). Following the research strategy and steps mentioned before, class 1 and class 2 were either, vegetation and bare soil, where vegetation was weeds and crop studied together, or weeds and crop. M values are indicative of the separability or discriminatory power of classes 1 and 2 considered in every step. Two classes exhibit moderate separability when M exceeds 1, showing easier separation for larger M values which will provide a reasonable discrimination [33]. According to Kaufman and Remer (1994) [32], the same difference in means can give different measures of separability depending on the spread of the histograms. Wider histograms (larger σ) will cause more overlap and less separability than narrow histograms (smaller σ) for the same difference in means.

Results and Discussion

1. Image Pre-processing

1.1. Band alignment of multispectral imagery. The images acquired by both cameras were downloaded to a computer by inserting their memory cards into a card reader and copying the data. An alignment process was performed on the multispectral images to match the six bands into a single readable file. The alignment results were examined visually and evaluated using spatial profiles (Figure 5).

The displacement among the curves for each channel in the spatial profiles makes evident the band misalignment of the original non-aligned images. The non-aligned images showed halos around the reference objects (Spectralon and vegetation) and noise in the soil background (Figure 5A). These halos and noise were still recognisable in the image aligned using the basic configuration of the PW2 software (Figure 5B), although they were lesser than in the non-aligned image. These results are similar to those obtained by Laliberte et al. (2011) [26], who reported poor alignment results using PW2 software with the mini-MCA imagery. To solve this problem, they developed the local weighted mean transform (LMWT) method and obtained a satisfactory alignment. However, the latest version of the PW2 software, launched in 2012, which includes the FOV optical calculator, performed a good alignment and allowed elimination of the halos and a high reduction of the background noise (Figure 5C). In fact, these results seem to be quite similar to those achieved using the LMWT method. A good alignment of all the individual bands is crucial for subsequent image analysis, especially when spectral values of different objects of the image are extracted. The vegetation objects present in a weed-crop scenario in the early season are very small, as a consequence a poor alignment might include pixels not belonging to the objects of interest, drastically reducing the success of the image analysis and classification.

Next to the alignment process, the PW2 software generated a unique multi-band image file that is incompatible with the mosaicking software. Therefore, the last step was to convert this multi-band file to a TIFF-readable format using the ENVI software.

1.2. Image orthorectification and mosaicking. A sequence of images was collected in each flight mission to cover the whole experimental crop-field. An important task prior to image analysis was the combination of all these individual and overlapped images by applying two consecutive processes of orthorectification and mosaicking. The Agisoft Photoscan Profes-

sional Edition (Agisoft LLC, St. Petersburg, Russia) software was employed in this task. In the first step, the software asks for the geographical position and principal axes (roll, pitch and yaw) of the vehicle in each acquired image. Next, the software automatically aligns the photos. Finally, some ATT's coordinates are added to assign geographical coordinates to the image. Then, the software automatically performs the orthorectification and mosaicking of the imagery set into a single image of the whole experimental field (Figure 6). The resultant ortho-mosaic shows a high-quality landscape metric and accurate crop row matching between consecutive images, which guarantees good performance of the subsequent image classification.

2. Effect of Flight Altitude on Image Spatial Resolution and Flight Time

The image spatial resolution and the area covered by each image as affected by the UAV flight altitude and the type of camera are shown in Figure 7. The imagery pixel size was directly proportional to the flight altitude. The still RGB camera captured images with pixel sizes of 1.14 cm and 3.81 cm, while the multispectral camera captured images with pixel sizes of 1.63 cm and 5.42 cm at flight altitudes of 30 and 100 m, respectively (Figure 8). At these altitudes, the area covered by each image of the still RGB camera increased from 0.16 ha (46×35 m) to 1.76 ha (153×115 m) and of the multispectral camera from 0.04 (21×17 m) to 0.38 ha (69×55 m), respectively. The differences between both types of images were due to the cameras' technical specifications (Table 2). The camera focal length affects both the pixel size and the area covered by each image, while the camera sensor size only influences the imagery pixel size.

A crucial feature of the remote images for weed mapping in the early season is their high spatial resolution, which can be achieved with low-altitude flights. Of great importance is defining the optimum pixel size needed according to each specific objective, which is calculated from the size of the weed seedlings to be discriminated, the distance between crop rows and the crop type. In general, at least four pixels are required to detect the smallest objects within an image [34]. Accordingly, if the objective is the discrimination of individual weed plants, the pixel size should be approximately 1–4 cm, which corresponds to flight altitudes of 27 to 105 m in the case of the still RGB camera and from 19 to 74 m in the case of the multispectral camera. However, when weed patch detection is aimed, the remote images could have a pixel size of 5 cm or even greater, which corresponds to a flight altitude higher than 100 m in both cameras.

The UAV acquired imagery with 60% forward lap and 30% side lap. From this overlapping and the camera sensor size, the WPE module calculated the number of images needed to capture the whole experimental field and, consequently, the time taken by the UAV to collect them at each flight altitude (Figure 9). The number of images per ha and the flight length were greater when using the multispectral camera, decreasing from 117 images ha^{-1} and 27 min at a 30 m altitude to 12 images ha^{-1} and 6 min at a 100 m altitude. For the still RGB camera, these variables ranged from 42 images ha^{-1} and 12 min at 30 m altitude to 6 images ha^{-1} 5 min at 100 m. A very large number of images can limit the mosaicking process because the number of images per hectare strongly increased at very low altitudes following an asymptotic curve. In addition, the operation timing is limited by the UAV battery duration. All these variables have strong implications in the configuration of the optimum flight mission for weed mapping in the early season, which involves two main conditions: 1) to provide remote images with a fine spatial resolution to guarantee weed discrimination, and 2) to minimise the operating time and the

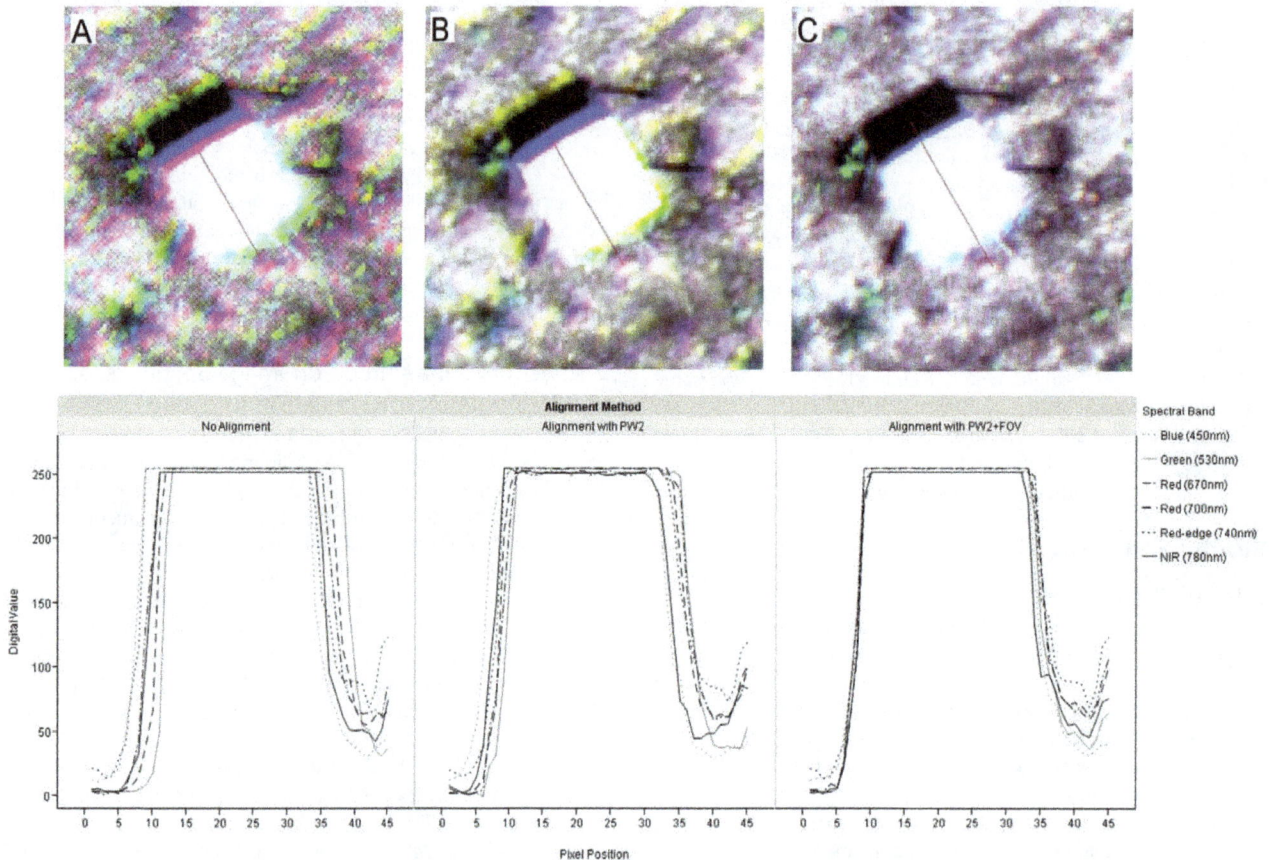

Figure 5. Images captured by the multispectral camera and spatial profiles depicting comparison of band-to-band alignment. a) No alignment, b) Alignment by using the basic configuration of the PW2 software, and c) Alignment by using the PW2 software plus the field of view (FOV) optical calculator.

number of images to reduce the limitation of flight duration and image mosaicking, respectively.

3. Effect of Flight Altitude on Image Spectral Resolution

Spectral information captured by each camera at three flight altitudes was studied to determine significant differences at the pixel level between class 1 and class 2 in the two phases previously mentioned, i.e. between vegetation cover and bare soil, and between weeds and crop. The range and average spectral pixel values of the VIs, and M-statistics are shown in Table 3.

First of all, it was crucial to explore the spectral differences between vegetation and bare soil to identify the potential to perform the first step of our research scheme, such an approach should point out the significant variations in spectral data of both classes, indicating which set of VIs, cameras and altitudes were able for their discrimination. All the indices showed significant differences between vegetation and soil and, in most cases, M-statistics performed reasonably well exceeding 2, being NDVI the index that achieved the highest spectral separability at the three flight altitudes. This is due to NDVI emphasises the spectral response of the NIR band which characterises vegetation vigour and it is less sensitive to soil background effects than the other two indices. The magnitude of M-statistic, usually higher than 2.5 (excepting for ExG at 30 m and 60 altitudes and multispectral camera), offer satisfactory results for a high robustness of vegetation discrimination in all the scenarios. Kaufman and Remer (1994) [32] reported M values ranging from 1.5 to 0.5 for

mapping dense vegetation in forests, whereas Smith *et al.* (2007) [33] obtained M values between 0.24 and 2.18 for mapping burned areas. According to our findings, the M achieved a much higher value ($M = 8.9$ for multispectral camera and NDVI index) suggesting robust separability of classes. NDVI could be the best index to perform the first phase of the proposed classification strategy, although NGRDI and ExG also showed an overall good capacity for distinguishing vegetal cover, which would be very relevant due to RGB camera is much cheaper and easier to use than the multispectral camera.

In order to perform the second proposed phase, it is necessary to test if weeds and crop can be discriminated using either RGB camera or the multispectral sensor. As a general statement, the multispectral camera showed much higher capacity to discriminate crop and weeds than the RGB camera. The better performance of the multispectral camera may be caused by its narrow sensor bandwidth. This camera uses filters with a 10 nm bandwidth, which reduces the interferences caused by other wavelengths, while the RGB camera acquires information in three wider spectral wavebands from the entire visible spectrum. Thus, means of NGRDI and ExG were not significantly different for crop and weeds at any flight altitude and M-statistic values were the lowest ones, excepting for ExG at 30 m altitude where $M = 1.61$. However, even at this altitude, M-statistic value is quite lower than the obtained for ExG and the multispectral camera ($M = 3.02$). A preliminary conclusion could be that the RGB camera is able to discriminate weeds and crop using images from

Figure 6. Ortho-mosaic of the whole experimental field. Composed from six individual images taken by the still RGB camera at 100 meters altitude.

ExG at 30 m altitude. However, one of the key question to elucidate at this point is to determine if $M = 1.61$ provides enough robustness for mapping weeds and crop. That doubt could be clarified going to Figure 10 which shows the significant spectral differences among soil, weeds and crop in all the scenarios. Note that spectral differences among soil, and weeds and crop at 30 m altitude for ExG and RGB camera are clearly significant; however, the range of the standard deviation (see points in Fig. 10) of weeds and crop causes an overlapping which could produce a deficient discrimination between weeds and crop. Therefore, Table 3 offers an overall overview of separation between vegetation and soil, and

weeds and crop; however these results must be deeply studied observing the ranges of minimum and maximum spectral values of every VI (Table 3) and ranges of standard deviation (Figure 10).

In the multispectral camera, NGRDI and ExG were significantly different for weeds and crop in all the flight altitudes tested. However, despite these significant differences observed and as stated before, the M-statistic and Figure 10 must be taken into account since both help to quantify the risk of misclassification due to the overlapping between value ranges of the vegetation indices studied. For instance, at 60 m altitude, NGRDI showed a significant spectral difference for weeds and crop; however M-

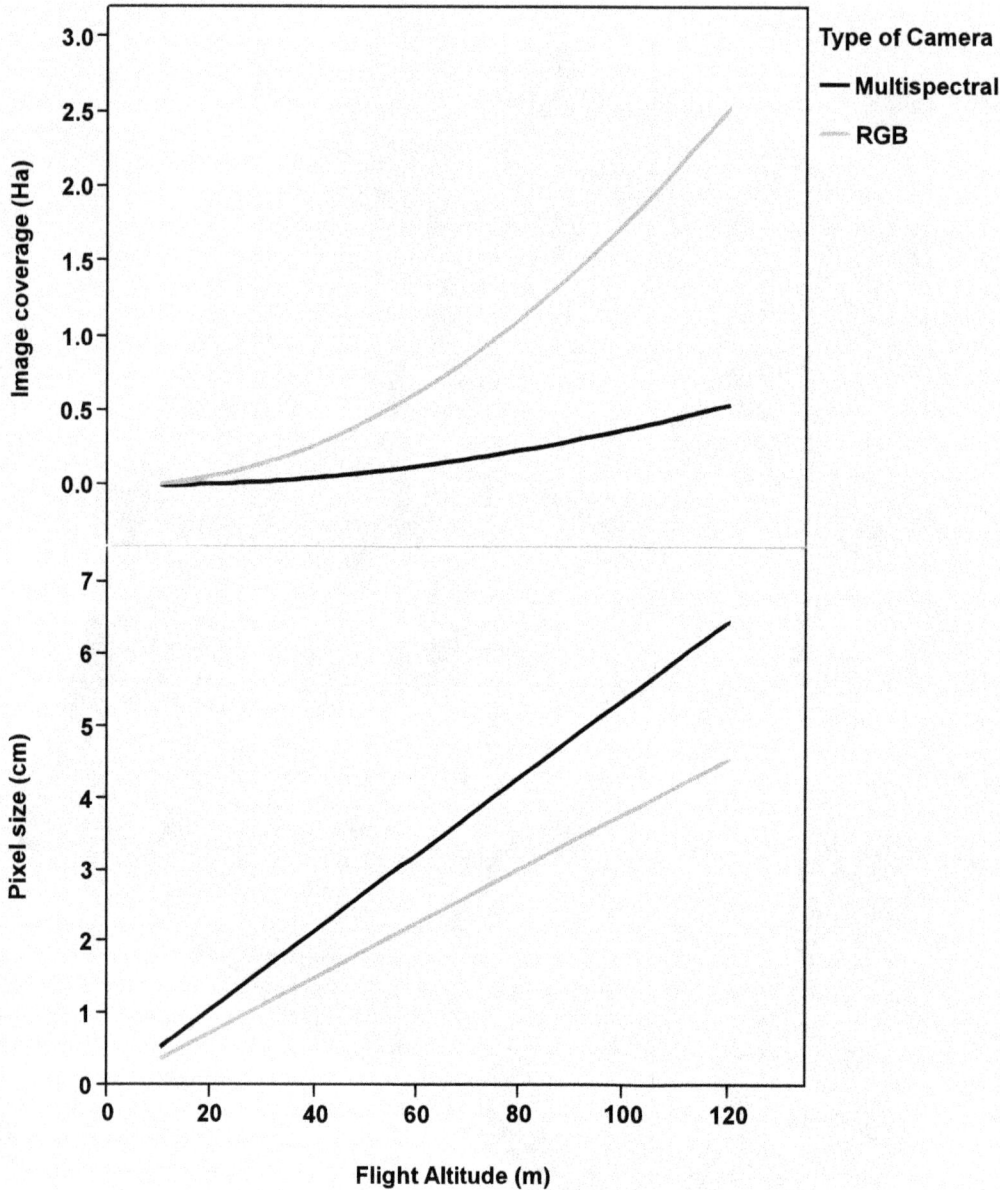

Figure 7. Image spatial resolution and coverage as affected by flight altitude and type of camera.

statistic was lower than 1 ($M = 0.81$). This indicates that, apart from a significant spectral difference, a poor separation is expected between pixels from weeds and crop. This can be clearly appreciated in Figure 10 where the range of the standard deviation between weeds and crop involves an overlapping of values and this is the reason for which having a significant spectral discrimination this is not sufficient to achieve a satisfactory separability (M higher than 1).

The case of ExG is different since this vegetation index showed significant spectral differences and M values higher than 1 at any flight altitude, although M was only slightly superior than 1 ($M = 1.19$) at 60 m altitude. This points out that a good separation would be expected at 30 m and probably at 100 m; however, have the significant spectral differences and $M = 1.19$ obtained in Table 3 sufficient discriminatory power to properly separate crop and weeds at 60 m altitude?. Figure 10 again shows that this magnitude of M probably is not as much as required to successfully

reach this objective due to the apparent overlapping of box-plots of weeds and crop and, consequently, a much more difficult separation would be expected at 60 m altitude. The only index studied using the NIR band was NDVI and it was not able to discriminate between crop and weeds at any flight altitude; in fact, NDVI showed the lowest M-statistic values among the indices calculated from the multispectral camera.

As mentioned in the previous section and according to the objective of minimising the operating time and the number of images taken to reduce the limitation of UAV flight duration and image mosaicking, the optimum flight mission may be to capture images at the highest altitude possible. However, the highest spectral differences and M values of pixels were obtained at the lowest altitudes, i.e., pixel-based methods may be unsuccessful in weeds and crop discrimination in seedling stages at altitudes higher than 30 m due to the spectral similarity among these vegetation classes. Currently, spectral limitations may be solved by imple-

Figure 8. UAV images collected by the two cameras. Still RGB camera (a, b) and multispectral camera (c, d) at 30 m (a, c) and 100 m (b, d) flight altitude.

menting advanced algorithms such as the object-based image analysis (OBIA) methodology [35]. The OBIA methodology identifies spatially and spectrally homogenous units named *objects* created by grouping adjacent pixels according to a procedure known as segmentation. Afterwards, multiple features of localisation, texture, proximity and hierarchical relationships are used that drastically increase the success of image classification [36], [37]. In crop fields at an early stage, the relative position of the plants in the crop rows, rather than their spectral information, may be the key feature to distinguishing them. Consequently, every plant that is not located in the crop row can be assumed to be a weed. Therefore, according our results a strategy for a robust classification of UAV images could be developed involving two steps: 1) discriminating vegetation (weeds and crop) from bare soil by using spectral information, and 2) discriminating weeds from crop-rows using the OBIA methodology. Therefore, future

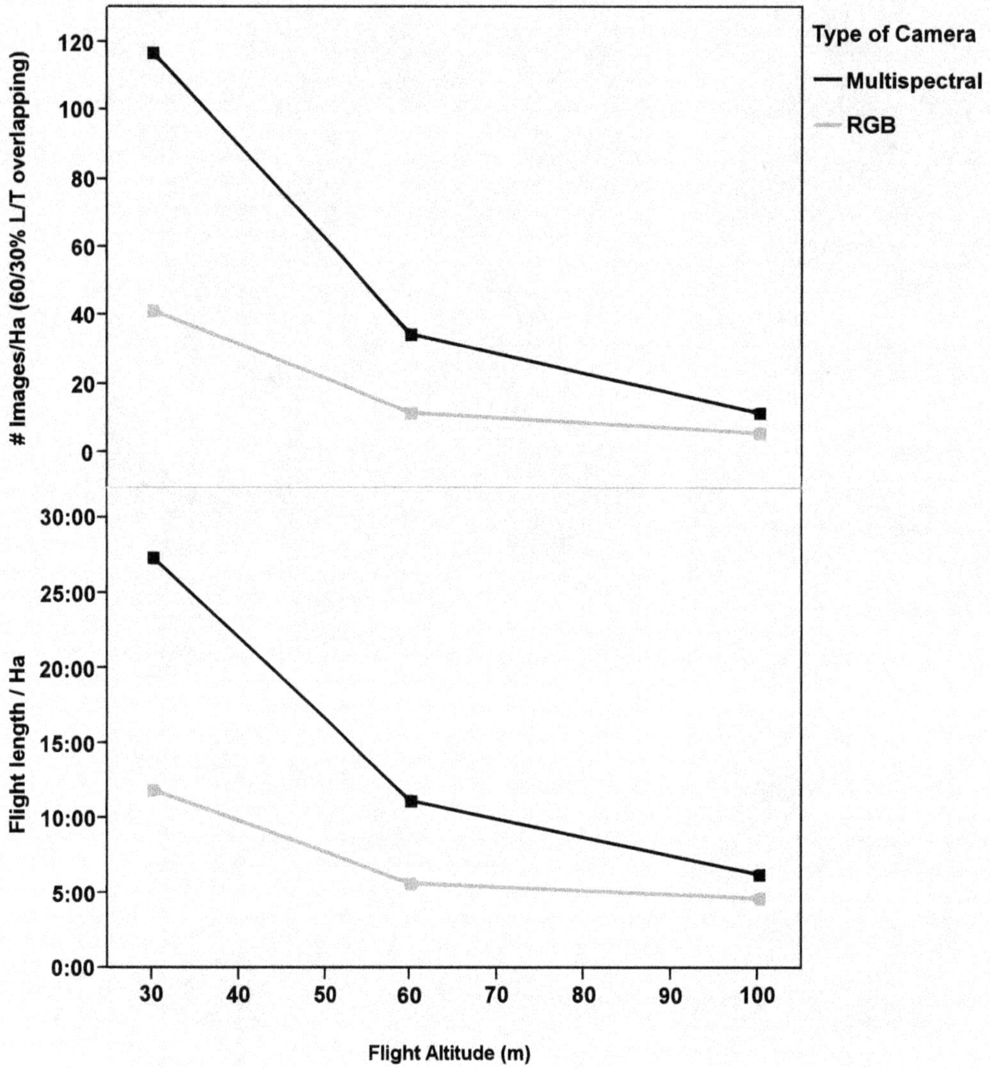

Figure 9. Flight length and number of images per ha as affected by flight altitude and camera.

investigations will be essential to determine the potential of OBIA techniques to distinguish and map weeds and crop using UAV imagery at higher flight altitudes and taken when weeds and crop are at the early phenological stages. Our recent research using OBIA methodology has shown the improvement of using satellite imagery for mapping crops [38] [37] or weeds at late phenological stages in winter wheat [12]. Our hypothesis for further work is based on the idea that the OBIA methodology has confirmed to be a powerful and flexible algorithm adaptable in a number of agricultural situations. The main aim would be to discriminate and map early weeds to enhance the decision making process for developing in-season ESSWM at high altitudes using RGB and ExG index compared to multispectral camera and the pixel-based image analysis. This would allow reducing the number of UAV imagery to improve the performance of the UAV (flight length and efficiency of energy supply) and the mosaicking process. This approach could be a more profitable method for mapping early weed infestations due to both, the covering of larger crop surface area and RGB cameras are cheaper and economically more affordable than multispectral cameras. Considering that the UAV

development is a substantial investment, the possibility of using RGB cameras would reduce significantly the additional costs.

Conclusions

Weeds are distributed in patches within crops and this spatial structure allows mapping infested-uninfested areas and herbicide treatments can be developed according to weed presence. The main objectives of this research were to deploy an UAV equipped with either, RBG or multispectral cameras, and to analyze the technical specifications and configuration of the UAV to generate images at different altitudes with the high spectral resolution required for the detection and location of weed seedlings in a sunflower field for further applications of ESSWM. Due to its flexibility and low flight altitude, the UAV showed ability to take ultra-high spatial resolution imagery and to operate on demand according to the flight mission planned.

The image spatial resolution, the area covered by each image and the flight timing varied according to the camera specifications and the flight altitude. The proper spatial resolution was defined according to each specific objective. A pixel lower than 4 cm was recommended to discriminate individual weed plants, which

Table 3. Least Significant Differences (LSD) test at P≤0.01 and Spectral Separability according to the M-statistic between crop and weed plants and vegetation and bare soil as affected by vegetation index, type of camera and flight altitude.

Flight altitude	Type of camera	VIs[a]	Vegetation vs. Bare soil discrimination										Crop vs. Weed discrimination										
			Vegetation				Bare soil				LSD test (Prob>F)	M[b]	Crop				Weed				LSD test (Prob>F)	M[b]
			Max	Min	Mean	±SD	Max	Min	mean	±SD			Max	Min	mean	±SD	Max	Min	Mean	±SD		
30-m	RGB	NGRDI	0.11	−0.02	0.04	±0.03	−0.08	−0.11	−0.09	±0.01	<0.01	3.61	0.09	0.00	0.05	±0.02	0.11	−0.02	0.03	±0.03	0.10	0.32
		ExG	0.34	0.10	0.21	±0.06	0.02	−0.01	0.00	±0.01	<0.01	2.93	0.34	0.20	0.27	±0.04	0.20	0.10	0.16	±0.01	<0.01	1.61
	Multispectral	NDVI	0.73	0.45	0.58	±0.07	−0.15	−0.19	−0.16	±0.01	<0.01	8.90	0.73	0.52	0.61	±0.06	0.68	0.45	0.55	±0.07	0.03	0.42
		NGRDI	0.35	−0.04	0.14	±0.12	−0.20	−0.27	−0.23	±0.02	<0.01	2.75	0.35	0.15	0.24	±0.06	0.17	−0.04	0.05	±0.06	<0.01	1.59
		ExG	0.18	−0.01	0.08	±0.07	−0.05	−0.09	−0.06	±0.01	<0.01	1.94	0.18	0.11	0.15	±0.02	0.06	−0.01	0.02	±0.02	<0.01	3.02
60-m	RGB	NGRDI	0.06	−0.03	0.01	±0.02	−0.08	−0.10	−0.09	±0.01	<0.01	3.53	0.06	−0.03	0.01	±0.02	0.06	−0.02	0.01	±0.02	0.85	0.03
		ExG	0.26	0.11	0.18	±0.04	0.03	−0.01	0.01	±0.01	<0.01	3.50	0.26	0.14	0.20	±0.04	0.20	0.11	0.15	±0.03	<0.01	0.79
	Multispectral	NDVI	0.51	0.15	0.35	±0.09	−0.08	−0.11	−0.10	±0.01	<0.01	4.52	0.51	0.23	0.38	±0.09	0.46	0.15	0.33	±0.08	0.10	0.31
		NGRDI	0.23	−0.04	0.08	±0.07	−0.11	−0.13	−0.12	±0.01	<0.01	2.77	0.23	0.03	0.12	±0.06	0.11	−0.04	0.04	±0.04	<0.01	0.81
		ExG	0.28	0.10	0.18	±0.05	0.07	0.05	0.06	±0.01	<0.01	1.91	0.28	0.13	0.22	±0.04	0.19	0.10	0.14	±0.02	<0.01	1.19
100-m	RGB	NGRDI	0.04	−0.09	−0.02	±0.04	−0.09	−0.12	−0.10	±0.01	<0.01	1.67	0.04	−0.07	−0.01	±0.04	0.04	−0.09	−0.03	±0.03	0.13	0.28
		ExG	0.23	0.05	0.14	±0.05	0.02	−0.01	0.01	±0.01	<0.01	2.20	0.23	0.12	0.16	±0.05	0.22	0.05	0.12	±0.05	0.02	0.46
	Multispectral	NDVI	0.64	0.23	0.43	±0.10	−0.13	−0.16	−0.14	±0.01	<0.01	5.25	0.64	0.34	0.47	±0.09	0.56	0.23	0.39	±0.09	0.04	0.40
		NGRDI	0.20	−0.12	0.01	±0.09	−0.21	−0.24	−0.21	±0.01	<0.01	2.40	0.20	−0.05	0.08	±0.07	0.02	−0.12	−0.05	±0.04	<0.01	1.16
		ExG	0.23	0.07	0.14	±0.05	0.03	0.00	0.02	±0.01	<0.01	2.28	0.23	0.12	0.18	±0.03	0.12	0.07	0.10	±0.02	<0.01	1.90

[a]Vegetation indices: NGRDI = (G−R)/(G+R); ExG = 2g − r − b; NDVI = (NIR−R)/(NIR+R).

[b]M-statistic = $(MEAN_{class1} − MEAN_{class2})/(\sigma_{class1} + \sigma_{class2})$.

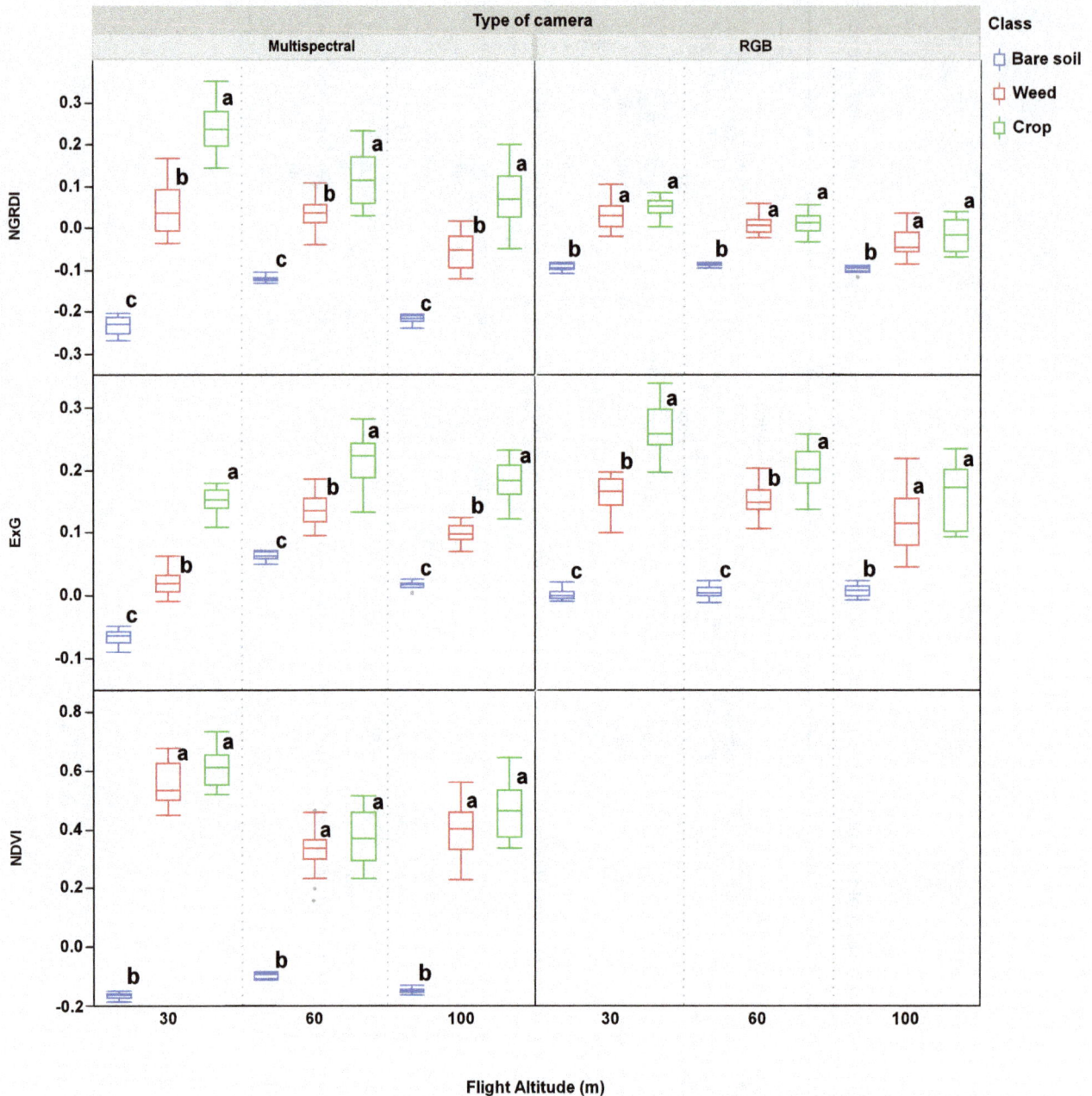

Figure 10. Vegetation index values of each class of soil cover (bare soil, weed and crop). The index values are affected by flight altitude and type of camera. Within a group, box-plots followed by the same letter do not differ significantly according to LSD test at P≤0.01.

corresponded to flight altitudes below 100 m. If the objective was weed patch detection, the UAV can fly to a higher altitude to obtain remote images with pixels of 5 cm or greater. However, the number of images needed to cover the whole field could limit the flight mission at a lower altitude due to the increased flight length, problems with the energy supply, and the computational capacity of the mosaicking software.

Spectral differences between weeds, crop and bare soil were significant for NGRDI and ExG indices, mainly at a 30 m altitude. At higher altitudes, many weed and crop pixels had similar spectral values, which may increase discrimination errors. Greater spectral separability was obtained between vegetation and

bare soil with the index NDVI, suggesting the employment of multispectral images for a more robust discrimination. In this case, the strategy for improving the image mosaicking and classification could be to implement the OBIA methodology to include features of localisation and proximity between weed and crop plants. An agreement among spectral and spatial resolutions is needed to optimise the flight mission according to the size of the smaller objects to be discriminated (weed plants or weed patches).

The information and results herein presented can help in the selection of an adequate sensor and to configure the flight mission for ESSWM in sunflower crops and other similar crop row scenarios (e.g., corn, sugar beet, tomato). Despite the initial complexity of management of the UAV and its components and software, and after a period of training the pilots and operators, the described workflow can be applied recursively.

Acknowledgments

The authors thank Dr. David Gómez-Candón for his very useful help during field work, and Mr. Íñigo de Arteaga y Martín and Mr. Iván de Arteaga del Alcázar for allowing developing our field work in La Monclova farm.

Author Contributions

Interpretation of data: JTS JMPB FLG. Conceived and designed the experiments: JMPB FLG. Performed the experiments: JTS FLG AIDC JMPB. Analyzed the data: JTS AIDC JMPB. Contributed reagents/materials/analysis tools: FLG JMPB. Wrote the paper: JTS FLG JMPB.

References

1. National Research Council (1997) Precision agriculture in the 21st century. Washington, DC: National Academy Press. 149 p.
2. Lelong CCD, Burger P, Jubelin G, Roux B, Labbé S, et al. (2008) Assessment of unmanned aerial vehicles imagery for quantitative monitoring of wheat crop in small plots. Sensors (Basel), 8(5): 3557–3585.
3. Christensen S, Søgaard HT, Kudsk P, Nørremark M, Lund I, et al. (2009) Site-specific weed control technologies. Weed Res 49(3): 233–241.
4. Lee WS, Alchanatis V, Yang C, Hirafuji M, Moshou D, et al. (2010) Sensing technologies for precision specialty crop production. Comput Electro Agr 74(1): 2–33.
5. Houborg R, Anderson M, Daughtry C (2009) Utility of an image-based canopy reflectance modeling tool for remote estimation of LAI and leaf chlorophyll content at the field scale. Remote Sens Environ 113(1): 259–274.
6. López-Granados F, Jurado-Expósito M, Peña-Barragán JM, García-Torres L (2005) Using geostatistical and remote sensing approaches for mapping soil properties. Eur J Agron 23(3): 279–289.
7. Meron M, Tsipris J, Orlov V, Alchanatis V, Cohen Y (2010) Crop water stress mapping for site-specific irrigation by thermal imagery and artificial reference surfaces. Precis Agric 11(2): 148–162.
8. de Castro AI, Jurado-Expósito M, Peña-Barragán JM, López-Granados F (2012) Airborne multi-spectral imagery for mapping cruciferous weeds in cereal and legume crops. Precis Agric 13(3): 302–321.
9. Herwitz S, Johnson L, Dunagan S, Higgins R, Sullivan D, et al. (2004). Imaging from an unmanned aerial vehicle: agricultural surveillance and decision support. Comput Electro Agr 44(1): 49–61.
10. López-Granados F, Jurado-Expósito M, Peña-Barragán JM, García Torres L (2006) Using remote sensing for identification of late-season grass weed patches in wheat. Weed Sci 54: 346–353.
11. Peña-Barragán JM, López-Granados F, Jurado-Expósito M, García-Torres L (2007) Mapping Ridolfia segetum patches in sunflower crop using remote sensing. Weed Res 47: 164–172.
12. de Castro AI, López-Granados F, Jurado-Expósito M (2013) Broad-scale cruciferous weed patch classification in winter wheat using QuickBird imagery for in-season site-specific control. Precis Agric, DOI: 10.1007/s11119-013-9304-y.
13. López-Granados F (2011) Weed detection for site-specific weed management: mapping and real-time approaches. Weed Res 51(1): 1–11.
14. Robert, P. C. 1996. Use of remote sensing imagery for precision farming. Proc of 26th Int. Symposium on Rem. Sens. of Env.: 596–599.
15. Thorp KR, Tian LF (2004) A Review on remote sensing of weeds in agriculture. Precis Agric 5(5): 477–508.
16. Xiang H, Tian L (2011) Development of a low-cost agricultural remote sensing system based on an autonomous unmanned aerial vehicle (UAV). Biosyst Eng 108(2): 174–190.
17. Hunt ER, Cavigelli M, Daughtry CST, McMurtrey JE, Walthall CL (2005) Evaluation of digital photography from model aircraft for remote sensing of crop biomass and nitrogen status. Precis Agric 6(4): 359–378.
18. Primicerio J, Di Gennaro SF, Fiorillo E, Genesio L, Lugato E, et al. (2012) A flexible unmanned aerial vehicle for precision agriculture. Precis Agric 13(4): 517–523.
19. Zhang C, Kovacs J (2012) The application of small unmanned aerial systems for precision agriculture: a review. Prec Agric, 13: 693–712.
20. Laliberte AS, Herrick JE, Rango A, Craig W (2010) Acquisition, orthorectification, and classification of unmanned aerial vehicle (UAV) imagery for rangeland monitoring. Photogramm Eng Rem S 76: 661–672.
21. Hardin PJ, Hardin TJ (2010) Small-scale remotely piloted vehicles in environmental research. Geography Compass 4: 1297–1311.
22. Hardin PJ, Jensen RR (2011) Small-scale unmanned aerial vehicles in environmental remote sensing: Challenges and opportunities. Gisci Remote Sens 48: 99–111.
23. Gray CJ, Shaw DR, Gerard PD, Bruce LM (2008) Utility of multispectral imagery for soybean and weed species differentiation. Weed Technol 22(4): 713–718.
24. Kelcey J, Lucieer A (2012) Sensor Correction of a 6-Band Multispectral imaging sensor for UAV remote sensing. Remote Sens 4(5): 1462–1493.
25. MAGRAMA (2012) http://www.magrama.gob.es/es/agricultura/temas/producciones-agricolas/cultivos-herbaceos/leguminosas-y-oleaginosas/#para3 Accessed 18 August 2012.
26. Laliberte AS, Goforth MA, Steele CM, Rango A (2011) Multispectral remote sensing from unmanned aircraft: Image Processing Workflows and Applications for Rangeland Environments. Remote Sens 3(11): 2529–2551.
27. Tetracam (2012) http://www.tetracam.com/PDFs/PW2%20FAQ.pdf Accessed 12 June 2012.
28. Rouse JW, Haas RH, Schell JA, Deering DW (1973) Monitoring vegetation systems in the Great Plains with ERTS. In: Proceedings of the Earth Resources Technology Satellite Symposium NASA SP-351. Washington, DC, USA. Vol. 1., 309−317.
29. Gitelson AA, Kaufman YJ, Stark R, Rundquist D (2002) Novel algorithms for remote estimation of vegetation fraction. Remote Sens Environ 80(1): 76−87.
30. Woebbecke DM, Meyer GE, Von Bargen K, Mortensen DA (1995) Color indices for weed identification under various soil, residue, and lighting conditions. Transactions of the ASAE. 38(1): 259–269.
31. Ribeiro A, Fernandez-Quintanilla C, Barroso J, Garcia-Alegre MC (2005) Development of an image analysis system for estimation of weed. In: J. Stafford (Ed.). Proceedings of the 5th European Conference on Precision Agriculture (5ECPA), Uppsala, Sweden. The Netherlands: Wageningen Academic Publishers 169–174.
32. Kaufman YJ, Remer LA (1994) Detection of Forests Using Mid-IR Reflectance: An Application for Aerosol Studies. IEEE Trans Geosci Rem Sens, 32(3): 672–683.
33. Smith AMS, Drake NA, Wooster MJ, Hudak AT, Holden ZA, et al. (2007) Production of Landsat ETM+ reference imagery of burned areas within Southern African savannahs: comparison of methods and application to MODIS. Intern J Rem Sens 28(12): 2753–2775.
34. Hengl T (2006) Finding the right pixel size. Comput Geosci-UK 32(9), 1283–1298.
35. Laliberte AS, Rango A (2009) Texture and scale in object-based analysis of sub-decimeter resolution unmanned aerial vehicle (UAV) imagery. IEEE T Geosci Remote 47: 761–770.
36. Blaschke T (2010) Object based image analysis for remote sensing. ISPRS J Photogramm 65: 2–16.
37. Peña-Barragán JM, Ngugi MK, Plant RE, Six J (2011) Object-based crop identification using multiple vegetation indices, textural features and crop phenology. Remote Sens Environ 115(6), 1301–1316.
38. Castillejo-González IL, López-Granados F, García-Ferrer A, Peña-Barragán JM, Jurado-Expósito M, et al. (2009) Object and pixel-based analysis for mapping crops and their agro-environmental associated measures using QuickBird imagery. Comp Electro Agr 68: 207–215.

2

GIS-Based Multi-Criteria Analysis for Arabica Coffee Expansion in Rwanda

Innocent Nzeyimana[1]*, Alfred E. Hartemink[2], Violette Geissen[1]

1 Soil Physics and Land Management Group, Wageningen University, Wageningen, The Netherlands, **2** Department of Soil Science, FD Hole Soils Lab, University of Wisconsin, Madison, Madison, Wisconsin, United States of America

Abstract

The Government of Rwanda is implementing policies to increase the area of Arabica coffee production. Information on the suitable areas for sustainably growing Arabica coffee is still scarce. This study aimed to analyze suitable areas for Arabica coffee production. We analyzed the spatial distribution of actual and potential production zones for Arabica coffee, their productivity levels and predicted potential yields. We used a geographic information system (GIS) for a weighted overlay analysis to assess the major production zones of Arabica coffee and their qualitative productivity indices. Actual coffee yields were measured in the field and were used to assess potential productivity zones and yields using ordinary kriging with ArcGIS software. The production of coffee covers about 32 000 ha, or 2.3% of all cultivated land in the country. The major zones of production are the Kivu Lake Borders, Central Plateau, Eastern Plateau, and Mayaga agro-ecological zones, where coffee is mainly cultivated on moderate slopes. In the highlands, coffee is grown on steep slopes that can exceed 55%. About 21% percent of the country has a moderate yield potential, ranging between 1.0 and 1.6 t coffee ha^{-1}, and 70% has a low yield potential (<1.0 t coffee ha^{-1}). Only 9% of the country has a high yield potential of 1.6–2.4 t coffee ha^{-1}. Those areas are found near Lake Kivu where the dominant soil Orders are Inceptisols and Ultisols. Moderate yield potential is found in the Birunga (volcano), Congo-Nile watershed Divide, Impala and Imbo zones. Low-yield regions (<1 t ha^{-1}) occur in the eastern semi-dry lowlands, Central Plateau, Eastern Plateau, Buberuka Highlands, and Mayaga zones. The weighted overlay analysis and ordinary kriging indicated a large spatial variability of potential productivity indices. Increasing the area and productivity of coffee in Rwanda thus has considerable potential.

Editor: Dafeng Hui, Tennessee State University, United States of America

Funding: The authors acknowledge NUFFIC of the Netherlands for funding the PhD training of Innocent NZEYIMANA at Wageningen University. The funders had no role in study design, data collection and analysis, decision to publish, or preparation of the manuscript.

Competing Interests: The authors have declared that no competing interests exist.

* Email: innocent.nzeyimana@wur.nl

Introduction

Coffee is one of the most important tradable commodities in the world and a major foreign-exchange earner in many developing countries [1]. Arabica coffee accounts for two-thirds of the global coffee market [2]. Coffee is a top export commodity and an important source of revenue in Eastern and Central African countries [3]. In some of these countries, such as Burundi, Uganda, Tanzania, Kenya, and the Democratic Republic of Congo, coffee is occasionally grown in association with agroforestry tree species for nitrogen fixation [4].

Rwanda produces mainly Arabica coffee, largely cultivated by smallholder farmers as mono-crop on plots of less than a hectare scattered on hilly slopes. In South and Central America, coffee is mostly grown on large monoculture plantations or under shade [5]. In Rwanda, coffee is predominantly grown along the shores of Lake Kivu in the west, on the plateau in the central part of Rwanda, and in the Mayaga region in the east [6]. Rwanda has ten agro-ecological zones: Imbo, Impara, Kivu Lake Borders, Birunga (volcano), Congo-Nile Watershed Divide, Buberuka Highlands, Central Plateau, Mayaga-Bugesera, Eastern Plateau,

and Eastern Savanna. Details of the characteristics of the Rwandan agro-ecological zones can be found in [7].

The total area of arable and permanently cropped land in Rwanda is about 1.45 million ha [8], of which about 30 000 ha was under coffee production in 2005 and it increased to 41 762 ha in 2012 [3]. The total area under coffee production in the tropics is about 10.6 million ha [9]. The expansion of land for the production of coffee depends on three main factors: environmental conditions (e.g. topography, soil type, climate, and elevation), practices of agricultural land management, and genetic resources (i.e. coffee varieties) [10]. The growing conditions for Arabica coffee in Rwanda are characterized by an altitude of 1400–1900 m a.s.l., an annual rainfall of 1500–1600 mm, temperatures of 18–22°C, and an average amount of sunlight of 2200–2400 hours per year. Arabica coffee also requires fine-textured soils of at least one-meter with total porosities of 50–60%, a pH of 4.5–6.0, moderate to high sums of basic cations, and 2–5% organic matter [7].

In Rwanda, as in other developing countries, coffee farming is reserved for steep slopes and soils with low fertility [6]. Most of these lands have been degraded by soil erosion and are under

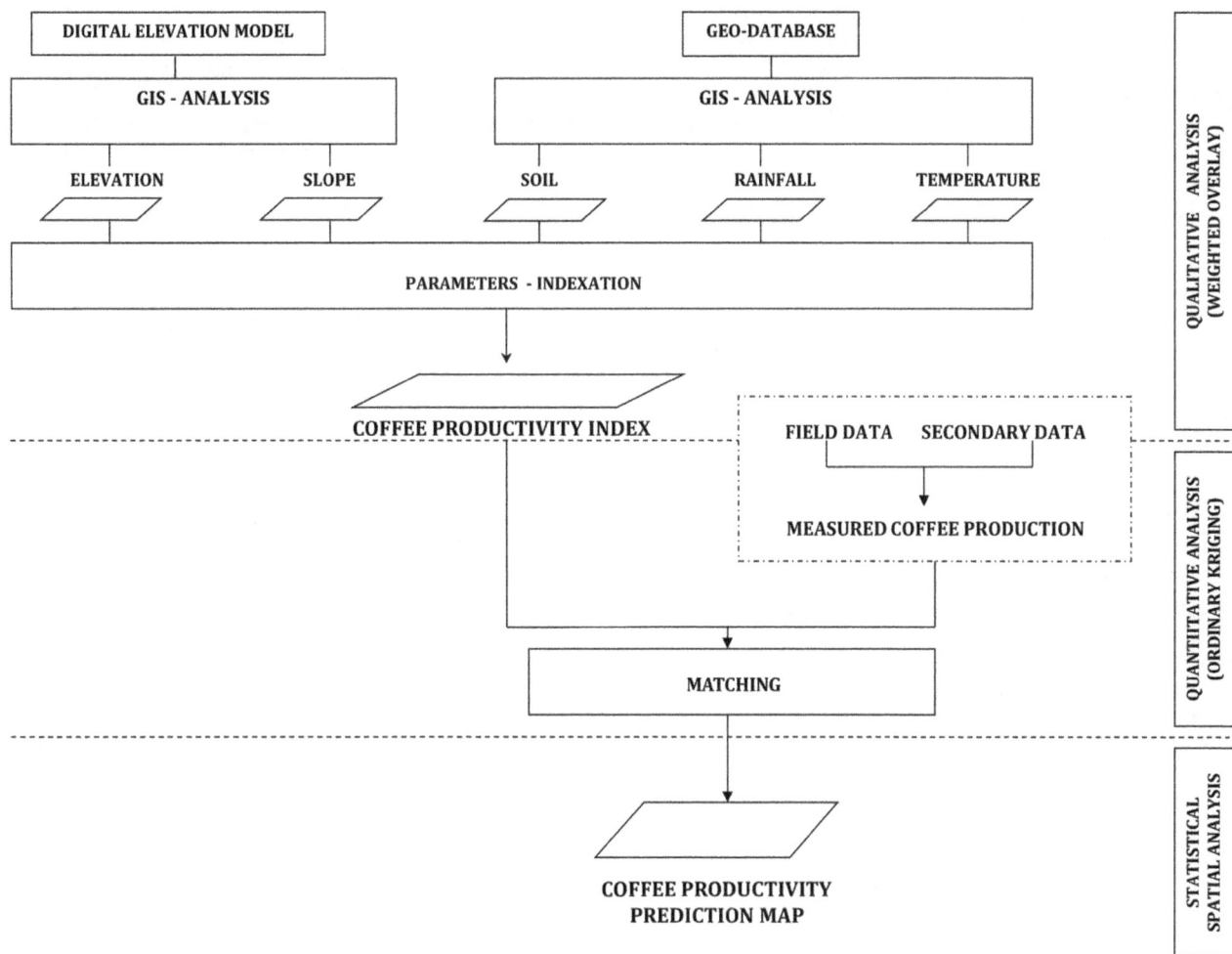

Figure 1. Flow chart of the methodology used to derive coffee productivity indices and predicted Arabica coffee yields.

pressure from intensive cropping by smallholder farmers. Fertile soils are usually reserved for growing staple food crops, which restricts coffee growing to soils of low fertility. In Rwanda, coffee yields above 2.8 t ha^{-1} are rare [6]. The Government of Rwanda has developed a set of policies for improving farmer livelihoods through an increase of sustainable coffee cultivation. A study of agricultural development in Africa has shown that successes are often linked to a cash-crop component and that food crops will profit as a consequence of improved cash income [11]. The identification of potential production zones for expanding coffee production and the prediction of coffee yields are needed to effectively implement these policies.

The evaluation of land is an essential procedure to assess opportunities, potentials, and limitations that a given parcel land can offer for agricultural purposes [12]. Various approaches of land evaluation with specific methodology have been developed to study land-use suitability [12,13]. Geographic information systems (GISs) have been used for mapping and analyzing land-use suitability [14]. Various GIS-based models have been developed by various researchers for land-use planning and suitability analysis [14–21]. The GIS-based models use geo-spatial and

geo-statistical tools to assess the land units and to present the results as suitability maps. The models use multi-criteria evaluative approaches and methods by weights, values, or intensities of preference [14,22]. Weighted overlay analysis is one such approach of GIS modeling using spatial multi-criteria evaluation [23,24]. The objectives of this study were: (1) to analyze the spatial distribution of potential production zones for Arabica coffee production and the current productivity levels in the various zones and (2) to predict potential coffee yields and identify potential productivity zones. To achieve the objectives, we developed a model of land evaluation for expanding the production of Arabica coffee in Rwanda based on standard methodologies for land evaluation and geo-spatial analysis.

Material and Methods

Data acquisition

Digitized and tabulated data were assembled for the entire country, including 43 digital soil maps (scale 1:50 000), a digital elevation model (Shuttle Radar Topography Mission (SRTM) at 90×90 m resolution), the coffee production database for 2005,

Table 1. Weighted environmental criteria for evaluating the qualitative Arabica coffee productivity classes and scores (1 to 5 for worst to best)[a].

Environmental coffee productivity criteria

Elevation (m)	Rainfall (mm)	Temp. (°C)	Soil type[b]	Slope (%)	Qualitative productivity class	Influence value[c] (%)	Score[d]
1600–1800	1400–1600	18–20	MOLL, AND	0–4	Very high	100	5
1400–1600 1800–2000	1200–1400 1600–1800	16–18 20–22	ALF	4–12	High	75	4
1200–1400	1000–1200 1800–2000	15–16 22–24	INCEPT, ULT	12–25	Moderate	50	3
1000–1200>2000	800–1000>2000	14–15 24–26	OX, ENT	25–50	Low	25	2
<1000	<800	<14>26	HIST, VERT	>50	Very low	0	1

[a]The table is a combination matrix that shows the level of productivity if we consider environmental productivity criteria of Arabica coffee production, as guided by [13].
[b]AND, Andisols; ALF, Alfisols; ENT, Entisols; INCEPT, Inceptisols; HIST, Histosols; MOLL, Mollisols; OX, Oxisols; ULT, Ultisols; VERT, Vertisols.
[c]The influence value represents the influence of the raster value compared to the other criteria as a percentage (i.e. 100, 75, 50, 25, or 0%).
[d]Each cell value in each input raster was assigned a new, reclassified score value on an evaluation scale of 1 to 5 (where 5 represented the best score and 1 the worst score 1) The scoring of the environmental productivity criteria of coffee was based on their importance as guided by [13].

general and administrative maps (2006 versions, scale 1:50 000), and a digital agro-climatic database containing temperatures, rainfall, altitudes, and agro-climatic zones. The database contained the amount of coffee produced and the number of trees in each administrative sector. Each sector was divided into cells, which are the lowest administrative units within the Republic of Rwanda.

Data processing: analysis of potential zones for coffee production

The methodology consisted of collecting (on a national scale) data such as soil type and slope gradient and analyzing the spatial distribution of coffee using the spatial-analysis toolset of the ArcGIS software [25]. The methodology aimed at identifying the major production zones based on soil and slope types. The combination of data for coffee distribution and soil was used for identifying the dominant soil types on which coffee is mainly produced and then for estimating the area coverage. The combination of data for coffee distribution and slope was for identifying the dominant slope types on which coffee is mainly produced and then for estimating the area coverage. The spatial distribution of coffee was identified, potential coffee production zones were characterized, and the sizes of the areas of coffee per slope and soil type in the various agro-ecological zones were estimated.

Multi-criteria analysis to estimate a qualitative Arabica coffee productivity index

The assessment of qualitative productivity indices for coffee essentially required the development of a GIS-based database for the optimal use of land resources for coffee. A geo-spatial database of data for elevation, slope, soil parameters, and rainfall and temperature extracted from the digital agro-climatic database, was generated in a GIS multi-criteria model (Figure 1). The landscape characteristics, climatic conditions, and soil parameters of a specific site are the most important determinants of land suitability [17]. The upper part of the flow chart in Figure 1 was thus used to analyze the qualitative productivity indices to indicate the level of productivity in the various agro-ecological zones. The multi-criteria model combined the different layers of data (i.e. elevation, slope, soil parameters, rainfall, and temperature) to identify the major production zones and their current productivity index (CPI). Data for photo synthetically active radiation were not available and so were not included in the model. The multi-criteria analysis used each input raster as a decision variable for sequential GIS interactions between layers. Data were processed using the spatial-analysis tools of ArcGIS [24,25]. The geo-spatial analysis then allowed the combination of the input rasters using weighted overlay analysis in the Model Builder component of ArcGIS to generate output rasters. Each cell value in each input raster was assigned a new, reclassified score value on an evaluation scale of 1 to 5, where 1 represents the lowest suitability and 5 the highest (i.e. scoring of the Arabica coffee requirements over others based on their importance as guided by [13]). Each of the new reclassified score was then weighted by assigning a percentage influence value (i.e. 100, 75, 50, 25, or 0%) (Table 1). This is achieved by multiplying the cell values (i.e. the new reclassified scores) by their percentage influence, and the results are added together to create the output raster. The new output-raster indices were then used as qualitative productivity indices. The weighted Z matrix can have the following form when m input factors and n criteria are considered:

$$Z = \begin{bmatrix} x_{01}..x_{.0j}..x_{0n} \\ .... \\ .... \\ x_{i1}..x_{ij}..x_{in} \\ .... \\ .... \\ x_{m1}..x_{mj}..x_{mn} \end{bmatrix}$$

where Z is the combined-factor weighted matrix, m is the number of input factors, n is the number of criteria describing each input factor, x_{ij} is the score representing the level of importance of input factor i based on criterion set j of the Arabica coffee requirements. The score x_{ij} is assigned a percentage influence value according to the importance of the environmental coffee productivity factor within a single input raster as illustrated in Table 1 [24].

The combination of the output rasters is for understanding the influence of the combined environmental factors and/or each factor separately on coffee productivity. The high-productivity class has no significant limitations for sustainable coffee production. The moderate class is characterized by altitudes of 1200–1400 m, annual rainfall below 900 mm or above 2000 mm, and temperatures varying between 18 and 21°C. The low class is characterized by altitudes below 1200 or above 2000 m, annual rainfall below 800 or above 2000 mm, and temperatures below 10 or above 30°C [7,13].

Multi-criteria analysis to estimate the quantitative Arabica coffee productivity index

The qualitative productivity indices (low, moderate, and high) were then quantified with actual yields to generate quantitative Arabica coffee productivity indices using ordinary kriging. The qualitative indices were extrapolated to 121 sampled sites of actual Arabica coffee yields measured at various sites countrywide.

Actual Arabica coffee yield

Actual yields were collected at 121 farms countrywide. Smallholder coffee fields, particularly those near a coffee-washing station, were selected and monitored for yield. The coffee fields are private farms owned by smallholder farmers, technically supported by the National Agricultural Export Development Board (NAEB). In collaboration with the NAEB, the identified farmers participated voluntary in the selection of sample fields. No specific permissions were required for the field activities. In addition, the field studies did not involve any endangered or protected species.

The yields were measured by sampling three branches of coffee trees (low, middle, and high branches). Experimental plots for data collection were approximately 10×10 m and contained 25 coffee trees (i.e. 2500 trees ha^{-1}), each 2 m apart. The coffee trees were predominantly 20–25 years of age and were cropped as monocultures.

All sample sites were independently selected with equal probability. Five randomly selected trees in each plot were sampled by collecting a composite sample of 500 g of good berries from the three branches weekly from April to September 2009. The coffee berries were cleaned, oven-dried at 60°C for 48 h, adjusted to 12% moisture content, and weighed. Grain yield was determined on each randomly selected tree, and a spatial mean plot yield was calculated as:

$$\bar{y} = \frac{1}{n} \sum_{i=1}^{n} y_i \tag{1}$$

where $\bar{y}$ (t ha^{-1}) is the average yield for 2009, y_i (t ha^{-1}) is the yield at sample site i, and n is the number of sample sites.

Predicted Arabica coffee yield

Ordinary kriging analysis was conducted to predict potential yields and to identify potential productivity zones for Arabica coffee [24,25], based on actual yields measured at the study site.

Table 2. Distribution of Arabica coffee areas (ha) and yields (t ha^{-1}) calculated from the coffee database for 2005 for the ten agro-ecological zones of Rwanda.

AEZ No.	Agro-Ecological Zone (AEZ)	Total area (ha)	Area with scattered coffee trees[a] (ha)	Normalized area[b] (ha)	Range of estimated dry yield[c] (t ha^{-1})	Mean dry coffee yield ± SD[d] (t ha^{-1})
1	Imbo (*Lake Kivu region*)	15 832	15 678	804	0.5–2.6	1.2±0.68
2	Impara (*Lake Kivu region*)	64 954	58 532	3376	0.5–2.6	1.2±0.69
3	Kivu Lake Borders (*Lake Kivu region*)	73 593	70 422	2947	0.3–3.5	1.6±0.79
4	Birunga/Volcano	90 887	1952	65	0.5–2.1	1.0±0.74
5	Congo-Nile Watershed Divide	391 930	136 946	4024	0.3–3.5	1.2±0.91
6	Buberuka Highlands	177 154	81 622	1130	0.3–2.8	0.8±0.53
7	Central Plateau & Granitic Ridge	529 772	461 743	10 155	0.3–2.8	0.8±0.47
8	Mayaga-Bugesera (*eastern region*)	223 573	166 085	3328	0.5–1.8	1.0±0.41
9	Eastern Plateau (*eastern region*)	381 367	350 233	5398	0.3–2.2	0.8±0.65
10	Eastern Savana (*eastern region*)	479 761	134 125	692	0.5–2.2	1.1±0.49
	Total	2 564 255	1 162 338	31 921		1.0±0.65

[a]This area is calculated as the area for each sector in the AEZ; only the sector area in the AEZ is extracted by spatial-analysis tools. The 2005 Rwanda coffee database displays only the number of coffee trees and coffee production per sector. Each sector is an administrative entity divided into cells, which are the lowest administrative units within the Republic of Rwanda.
[b]This area is extracted from the area with scattered coffee trees in each AEZ and is calculated using the standard tree density of 2500 trees ha^{-1}, i.e. b = a/2500.
[c]This yield is calculated by averaging the yields for each part of the sector in the AEZ (i.e. sector yield is calculated as the production of each sector divided by the number of trees using the standard spacing of 2× 2 m, or 2500 trees ha^{-1}).
[d]This yield is calculated using SPSS descriptive statistics; the normality of the data was determined using the Kolmogorov-Smirnov test.

Table 3. Distribution of soil types (ha) and areas of Arabica coffee cultivation in the ten agro-ecological zones of Rwanda.

Agro-Ecological Zone (AEZ)	AEZ label No.	Soil/Coffee coverage	Area per soil type/Area covered by coffee per soil type (ha)[a]								Total	
			ALF	AND	ENT	HIST	INCEPT	MOLL	OX	ULT	ha[b]	%[c]
Buberuka Highlands	6	Soil	2000	99	17 838	10 545	45 835	567	10 492	81 901	167 277	7
		Coffee	21	-	157	19	385	7	73	456	1118	3
Central Plateau & Granitic Ridge	7	Soil	68 244	373	47 504	2 219	164 855	1634	48 054	195 007	527 890	23
		Coffee	1306	5	876	30	3286	34	778	3722	10 036	31
Mayaga-Bugesera (*eastern region*)	8	Soil	14 564	-	13 650	28 515	34 038	5768	75 141	37 048	208 722	9
		Coffee	288	-	206	392	599	86	1182	569	3322	10
Eastern Plateau (*eastern region*)	9	Soil	22 782	-	66 166	10 652	76 506	20 420	99 118	76 238	371 881	16
		Coffee	311	-	910	161	1111	294	1459	1021	5268	16
Eastern Savana (*eastern region*)	10	Soil	10 555	-	48 331	38 421	89 951	20 235	186 596	26 586	420 676	18
		Coffee	21	-	118	58	303	26	287	57	870	3

Data were extracted from the Rwanda soil dataset and analyzed using the geo-spatial tools of ArcGIS.
[a]AND, Andisols; ALF, Alfisols; ENT, Entisols; INCEPT, Inceptisols; HIST, Histosols; MOLL, Mollisols; OX, Oxisols; ULT, Ultisols; VERT, Vertisols.
[b]Total Rwanda soil and Arabica coffee coverage per agro-ecological zone.
[c]Soil and Arabica coffee coverage in percentage per agro-ecological zone over total Rwanda soil area and Arabica coffee area, respectively.

The ordinary kriging is one of the mostly used geo-statistical methods, quite efficient and accurate for spatial prediction and interpolation [26]. The prediction of yield was based on the qualitative productivity indices validated over the actual yields. A variogram was estimated using Matheron's estimator [27,28]:

$$y^{\wedge}(h) = \frac{1}{2Mh} \sum_{i=1}^{Mh} \{Z(x_i) - Z(x_i + h)\}^2 \qquad (2)$$

where $Z(x_i)$ is the actual yield measured at the study site (x_i), h is the lag, i.e. both distance and direction between the sample sites, M_h is the pair of sample sites separated by lag h, and $\gamma^{\wedge}(h)$ is the semi-variance at lag h.

To assess the spatial correlation of the yields, prediction accuracy was calculated by comparing expected yields, $Z^{\wedge}(CYI_j)$, with actual yields measured at the validation sites, (n) - $Z^*(CYI_j)$, and to assess a systematic error, calculated as the mean prediction error (MPE) [29]:

$$MPE = \frac{1}{n} \sum_{j=1}^{1} \left[Z \wedge (CYI_j) - Z(CYI_j) \right] \qquad (3)$$

where CYI is the coffee yield index, $Z^{\wedge}(CYI_j)$ is the expected yield index generated from the qualitative analysis, and $Z^*(CYI_j)$ is the actual yield measured at the validation sites (n). The validation set accounted for 121 sample sites. The accuracy of prediction was calculated as a root mean square error (RMSE) of prediction [29]:

$$RMSE = \sqrt{\frac{1}{n} \sum_{j=1}^{1} \left[Z \wedge (CYI_j) - Z(CYI_j) \right]^2} \qquad (4)$$

The RMSE is a measure of fitness of the prediction curve; the smaller the RMSE, the better the prediction. Ordinary kriging uses and compares different fitting models that perform the analysis, reduce uncertainty, and produce the best prediction map. The RMSE is thus standardized by considering the total variance of the observed values and is then termed the root mean square standardized error (RMSSE) or the mean standardized error (MSE). The RMSSE and the MSE were estimated from the variances between the observed values, i.e. the actual yields measured at the study site [24,29]:

$$RMSSE = \frac{\frac{1}{n} \sum_{j=1}^{1} \left[Z \wedge (CYI_j) - Z * (CYI_j) \right]^2}{s^2} \qquad (5)$$

where s^2 is the total variance of the CYI at the sample site.

A satisfactory accuracy of prediction has an MSE close to zero and an RMSSE close to unity [24,29]. If the RMSSE exceeds unity, the model underestimates the variability at the validation sites, and thus the prediction is unsatisfactory [24,29,30].

The normality of the measured yield data was determined with Kolmogorov-Smirnov test. All data were normally transformed to meet the assumption of normality by comparing different types of model fitting (exponential and Gaussian) for the analysis, and only the model with the smallest Akaike information criterion (AIC) was adopted. The AIC is a measure of how well a model fits the empirical data; the smaller the AIC, the better the fit [24]. In addition, cross-validation, comparing the predicted values with the measured values, checked the quality of the predicted values [31].

Table 4. Distribution of soil types (ha) and areas of Arabica coffee cultivation in the ten agro-ecological zones of Rwanda (Cont'd).

Agro-Ecological Zone (AEZ)	AEZ label No.	Soil/Coffee coverage	Area per soil type/Area covered by coffee per soil type (ha)[a]								Total	
			ALF	AND	ENT	HIST	INCEPT	MOLL	OX	ULT	ha[b]	%[c]
Imbo (*Lake Kivu region*)	1	Soil	3349	-	268	-	6489	-	-	3910	14 017	0.6
		Coffee	197	-	16	-	382	-	-	230	825	3
Impara (*Lake Kivu region*)	2	Soil	11 831	-	138	1269	10 111	-	363	40 436	64 147	3
		Coffee	657	-	7	75	558	-	1	2109	3407	11
Kivu Lake Borders (*Lake Kivu region*)	3	Soil	8526	91	12 708	35	25 248	258	253	25 469	72 590	3
		Coffee	328	-	539	2	1085	11	11	1045	3020	9
Birunga/Volcano	4	Soil	1782	47 176	15 450	79	2182	4222	-	3174	74 065	3
		Coffee	16	23	-	-	5	7	-	15	65	0.2
Congo-Nile Watershed Divide	5	Soil	8879	11 613	23 396	5388	124 675	2107	11 341	203 960	391 359	17
		Coffee	112	-	330	8	1553	-	17	2004	4024	13
		Subtotal[d] (ha)	152 513	59 352	245 450	97 123	579 890	55 211	431 358	693 728	2 314 625	
		Subtotal[d] (%)	7	3	11	4	25	2	19	30		100
		Subtotal[e] (ha)	3257	27	3159	743	9266	465	3809	11 228	31 954	
		Subtotal[e] (%)	10	0.1	10	2	29	1	12	35		100

Data were extracted from the Rwanda soil dataset and analyzed using the geo-spatial tools of ArcGIS (cont'd); See Table 4 for the notes.

[d]Subtotal of soil area per soil type.

[e]Subtotal of Arabica coffee area per soil type.

Table 5. Distribution of slope classes (%) and areas with Arabica coffee in the ten agro-ecological zones of Rwanda.

Agro-Ecological Zone (AEZ)	AEZ label No.	Slope/Coffee coverage	Area per slope category/Area covered by coffee per slope category (ha)			Total	
			<25%	25–55%	>55%	(ha)	(%)[a]
Imbo (*Lake Kivu region*)	1	Slope	10 863	4652	4705	20 220	1
		Coffee	554	238	11	803	3
Impara (*Lake Kivu region*)	2	Slope	45 321	18 165	1407	64 893	3
		Coffee	2417	871	57	3345	10
Kivu Lake Borders (*Lake Kivu region*)	3	Slope	47 780	25 400	413	73 593	3
		Coffee	1911	1016	17	2944	9
Birunga/Volcano	4	Slope	77 170	12 396	1226	90 792	4
		Coffee	59	6	–	65	0.2
Congo-Nile Watershed Divide	5	Slope	229 902	157 739	4258	391 899	16
		Coffee	2059	1922	46	4027	13
Buberuka Highlands	6	Slope	81 303	89 874	5814	176 991	7
		Coffee	435	633	50	1118	3
Central Plateau & Granitic Ridge	7	Slope	369 316	155 345	5070	529 731	22
		Coffee	7212	2944	105	10 261	32
Mayaga-Bugesera (*eastern region*)	8	Slope	214 946	8122	19	223 087	9
		Coffee	3168	152	–	3320	10
Eastern Plateau (*eastern region*)	9	Slope	316 686	63 065	1586	381 337	16
		Coffee	4462	903	24	5389	17
Eastern Savana (*eastern region*)	10	Slope	453 244	25 987	202	479 433	20
		Coffee	641	57	–	698	2
Rwanda - Slope(ha)		Subtotal	1 846 532	560 746	24 701	2 431 979	
(%)			76	23	1		100
Rwanda - Coffee (ha)		Subtotal	22 917	8743	310	31 970	
(%)			72	27	1		100

Data extracted from the digital elevation model (Shuttle Radar Topography Mission at 90×90 m resolution) and analyzed using the geo-spatial tools of ArcGIS.
[a]Slope and coffee coverage in percentages per agro-ecological zone of the total Rwanda slope and coffee areas, respectively.

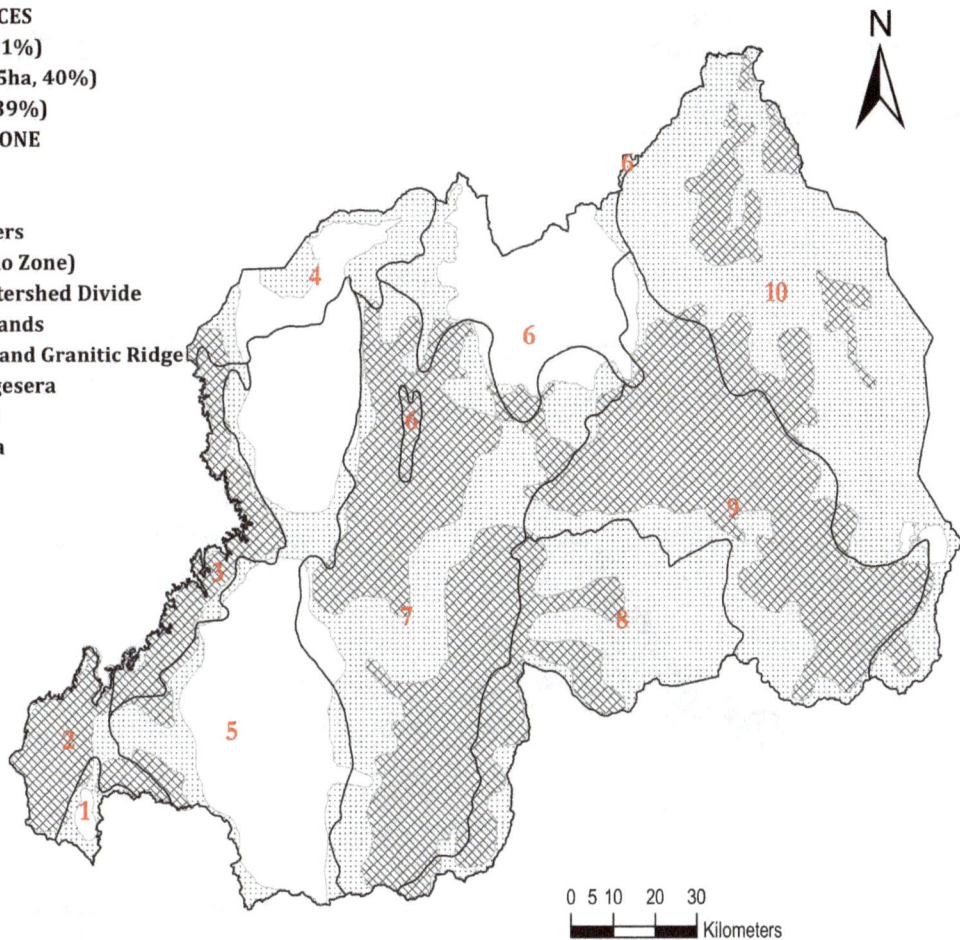

Figure 2. Qualitative Arabica coffee productivity indices (low, moderate, and high) generated by combining factors (elevation, slope, soil type, rainfall, and temperature) using weighted overlay analysis in the ten agro-ecological zones.

Results

Spatial distribution of Arabica coffee and biophysical characterization

The estimated area of Arabica coffee production in Rwanda was about 32 000 ha in 2005, compared to 30 000 ha reported by [3]. This area represents about 2.3% of the total area under agriculture. The Central Plateau had the largest area of coffee production, covering about 32% (10 261 ha) of the total area under coffee cultivation. The Central Plateau was characterized by coffee yields of 0.3–2.8 t ha^{-1} (Table 2). This zone has a wide range of soil types and landscapes. The soils where coffee is cultivated included Alfisols, Inceptisols, and Ultisols, representing 4% (1306 ha), 10% (3286 ha) and 12% (3722 ha), respectively, of the total area under coffee (Tables 3 and 4). Cultivated areas in the Central Plateau and the Granitic Ridge agro-ecological zones are also characterized by moderate (<25%) and steep (25–55%) slopes that cover 23% (7212 ha) and 9% (2944 ha) of the total cultivated area, respectively (Table 5). Coffee productivity is mainly limited by infertile soils derived from schistose and granitic materials. Slopes above 25% affect the productivity of the region due to soil erosion.

The Lake Kivu region (Imbo, Impara, and Kivu Lake Borders zones) in the western province had the highest yields ranging between 0.3 and 3.5 t ha^{-1}, with a mean of 1.6 t ha^{-1}. This

region contained 22% (7127 ha) of the total area cropped with coffee (Table 2). The dominant soil types in the region are Inceptisols and Ultisols, representing ~6% (2025 ha) and 11% (3384 ha), respectively, of the total area devoted to coffee production (Table 4). Arabica coffee in the region is dominantly cultivated on moderate (<25%) and steep slopes (25–55%) that cover 15% (4882 ha) and 6% (2125 ha), respectively, of the areas under coffee (Table 5). The region is characterized by environmental conditions favorable to coffee production.

Yields in the Birunga (volcano) agro-ecological zone ranged between 0.5 and 2.1 t ha^{-1}, with a mean of 1.0 t ha^{-1} (Table 2). The extent of coffee in the zone covered only 65 ha of the land, mainly on Alfisols (16 ha), Andisols (23 ha), and Ultisols (15 ha) (Table 4). Andisols are fertile and productive soils, so the farmers will prefer annual crops over perennial crops such as coffee. Coffee is mainly grown on moderate slopes (Table 5). The effective depth of the soil, dominated by Andisols, is the main factor limiting coffee productivity in the zone.

The Eastern Plateau, Eastern Savanna, Mayaga, and Bugesera zones (i.e. the eastern region) together covered ~30% (9418 ha) of the total area of coffee production. Yields in the eastern region ranged between 0.3 and 2.2 t ha^{-1}, with a mean of 1.0 t ha^{-1} (Table 2). The dominant soil types are Inceptisols, Oxisols, and Ultisols, covering ~6% (2013 ha), 9% (2928 ha), and 5% (1647 ha), respectively, of the area under coffee (Table 3). Coffee

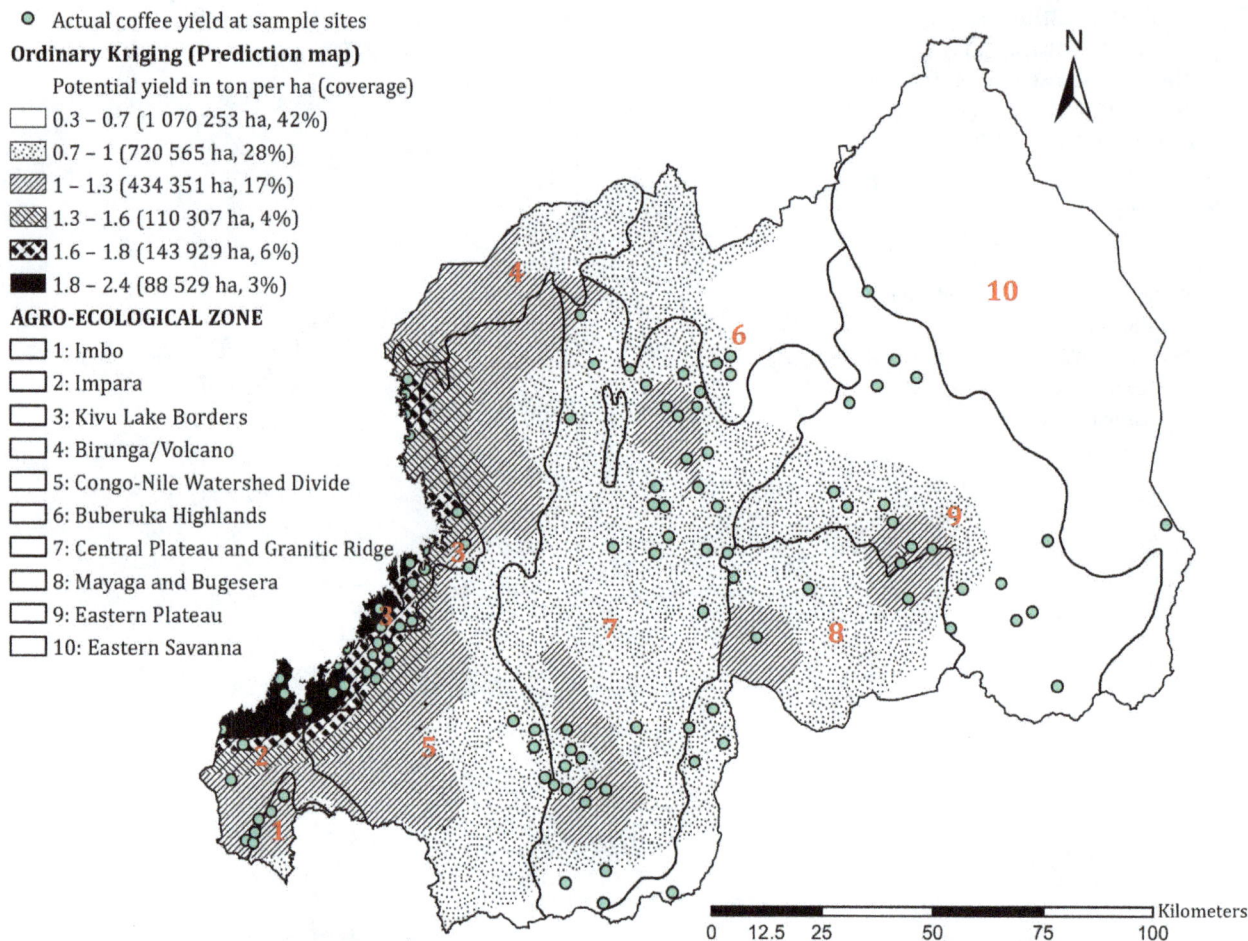

Figure 3. Potential Arabica coffee yield (t ha^{-1}) predicted using ordinary kriging in the ten agro-ecological zones based on actual yields (t ha^{-1}) measured at sample sites.

is cultivated on moderate slopes (<25%) that cover more than 26% (8271 ha) of the area with coffee cultivation (Table 5). The dominant infertile soils of the region, very high temperatures, and low rainfall offer limited conditions for coffee productivity.

The Buberuka Highlands and the Congo-Nile Watershed Divide agro-ecological zones are classified as the highlands of the country (above 2000 m a.s.l.). In both zones, coffee was cultivated on ~16% (5154 ha) of the total coffee area, mainly on moderate (2494 ha, ~8%) and steep (2555 ha, ~8%) (Table 5). Yields in the highlands ranged between 0.3 and 3.5 t ha^{-1}, with a mean of 1.2 t ha^{-1} (Table 2). Inceptisols and Ultisols are the main soil types, representing ~6% (1938 ha) and 8% (2460 ha), respectively, of the total area under coffee cultivation (Table 3). Very low temperatures and heavy rainfall limit the productivity of coffee cultivation in the highlands.

Coffee productivity indices

Qualitative productivity indices were generated based on soil type, elevation, slope, rainfall, and temperature using weighted overlay analysis (Figure 2). The analysis identified three zones with high, moderate, and low productivity indices representing ~930 715 (39%), 949 975 (40%), and 511 945 ha (21%), respectively. Approximately 80% of the total area of the country had moderate to high production potential for Arabica coffee. Zones with high potential productivity indices had fertile soils, moderate slopes and

altitudes, and favorable climates. The zones with low productivity indices were mainly at high altitudes with high rainfall and low temperatures. The semi-dry eastern regions, where Oxisols and Ultisols are the dominant soil types (Figure 5), have zones with low indices.

High predicted yields ranged between 1.6 and 2.4 t ha^{-1} along the shores of Lake Kivu and in the Imbo zone (Figure 3). The calculated yields varied between 0.3 and 3.5 t ha^{-1} (Table 2). The prediction map for the country (Figure 3) shows coffee yields varying between 0.3 and 2.4 t ha^{-1}. Eighty percent of the country had low yield potentials of 0.3–1 t ha^{-1}, whereas 21% of the country had moderate yield potentials of 1.0–1.6 t ha^{-1}. The national average yield was predicted to be 1.12 t ha^{-1}, and the measured yield (n = 121 sampled sites) was 1.1 t ha^{-1} y^{-1}. The correlation between the measured and predicted yields indicated that the prediction model was satisfactory (Coefficient of determination $R^2 = 0.73$) (Figure 4).

Discussion

Some general information on the suitability of coffee cultivation in Rwanda was available. The spatial variation of coffee production can be explained by the growing conditions, which include biophysical factors such as soil type and properties, parent material, altitude, slope, and climatic conditions. For example, the

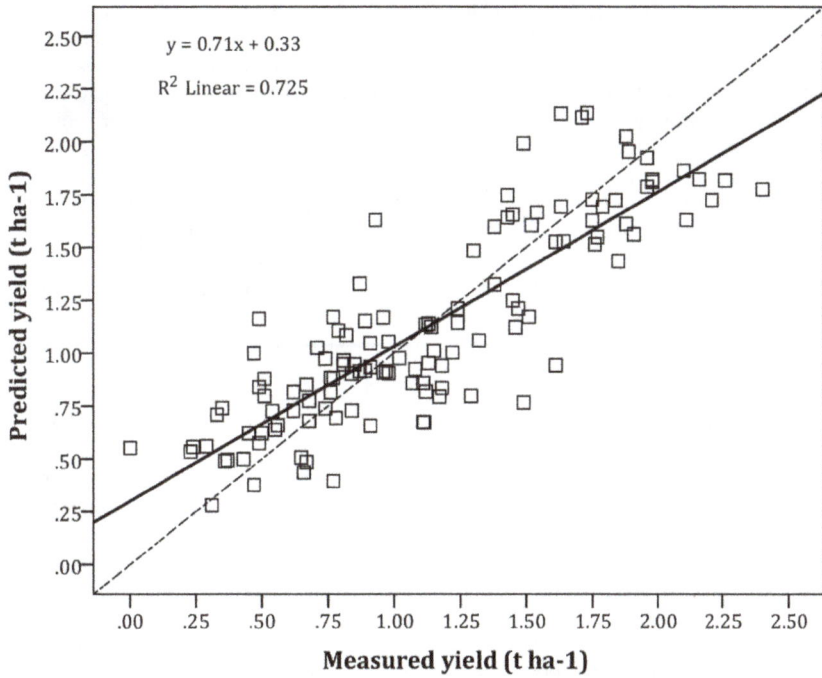

Figure 4. Relationship between measured and predicted Arabica coffee yields – cross validation using ordinary kriging (Predicted Arabica coffee yield index – CYI (t ha^{-1}) = 0.71x + 0.33; Mean Prediction Error – MPE = 0.0187; Root Mean Square Prediction Error – RMSE = 0.278; Root Mean Square Standardized Prediction Error – RMSSE = 0.99; Mean Standardized Prediction Error - MSE = 0.036; Coefficient of determination – R^2 = 0.73; Average Standard Error –Avg. SE = 0.291; Sample points, n = 121).

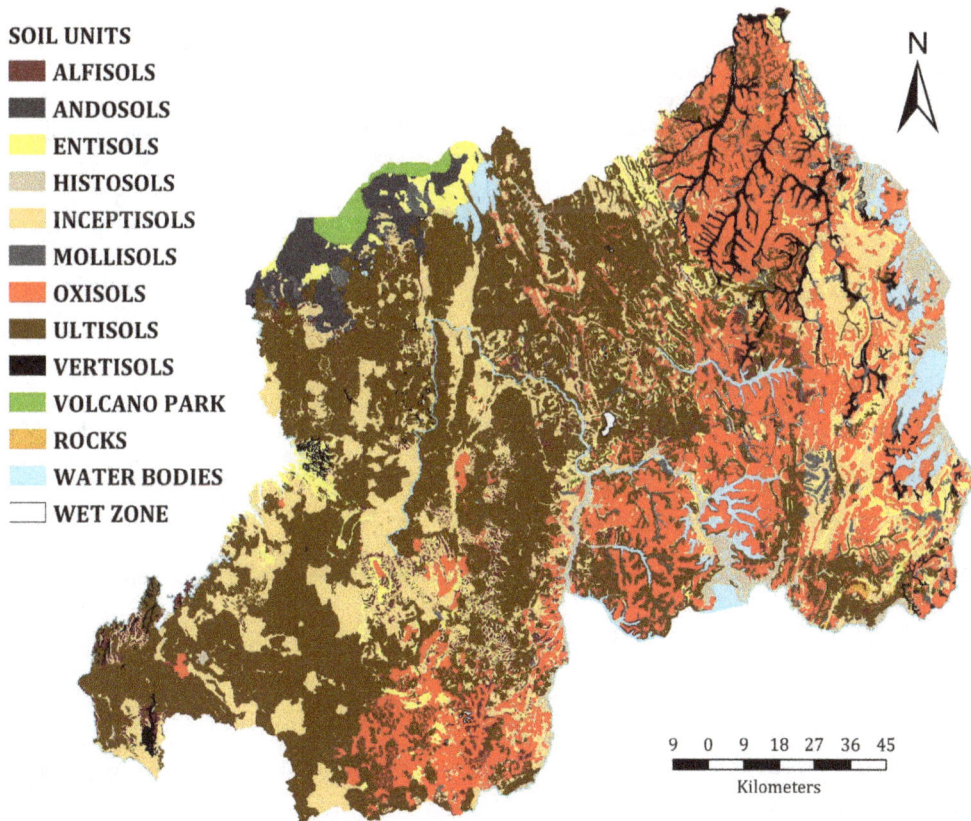

Figure 5. Soil map of Rwanda. Soils are classified using the USDA Soil taxonomy (Source: Data collected from the Ministry of Agriculture and Animal Resources, using the Rwanda soil database) after [32].

yield in the Central Plateau and Granitic Ridge zones was limited by soil acidity and the gravel as these soils developed on granitic and schistose materials. On such soils, erosion on slopes above 25% affects the productivity. Similar observations have been made by [33,34] who indicated that topography can combine with various environmental factors to influence productivity. Topography affects the climate (e.g. variations in temperature and humidity), distribution of soil moisture, soil organic-matter content, soil nutrients, soil textural composition, and soil physical properties, which affect crop growth and yield in a field. The spatial variability of environmental factors can contribute to variability in crop performance, and topography is a vital variable in predicting the spatial variability of crop yields [35].

The multi-criteria overlay analysis used to assess spatial variation in coffee productivity revealed that the Lake Kivu region is an area of high productivity. Similarly, ordinary kriging indicated expected high yields. The Lake Kivu region has favorable soil types and slopes with abundant rainfall and moderate temperatures for the optimal production of coffee. The region has alluvial and very fine clayey soils, developed from basalt, and a high agricultural potential [7].

The Birunga (volcano), Central Plateau, and Buberuka Highlands' zones, the highlands of the Congo-Nile Watershed Divide zone, and the eastern region are areas with both moderate and low productivity indices. The productivity in the Central Plateau zone is mainly limited by infertile soils derived from schistose and granitic materials on the moderately sloped and eroded hillsides. The productivity in the Birunga (volcano) zone is mainly limited by the soil depth (<50 cm) and low temperatures that are suboptimal for coffee production. High yields were expected in this zone based on the high fertile volcanic soils. Instead, low potential yields were predicted by the kriging model, due to the unsuitable climatic conditions that affect the development and maturity of the berries [36]. In the highlands, mainly in the Buberuka Highlands and Congo-Nile Watershed Divide zones, the production of coffee is limited by very low temperatures, heavy rainfall, and steep slopes that could influence the depletion of soil fertility and reduce yields due to water erosion on the hills. Productivity in the eastern regions is limited by infertile soils and the very high temperatures and low rainfall. Our study demonstrated a decrease in yield in very dry conditions that coincided with lower elevations. Coffee is also constrained by very cold temperatures in the highlands that are often cloudy with low solar irradiation and heavy rainfall. Similar trends of the influence of topography on potato yields have also been reported by [37]. Both the ordinary kriging analysis and the multi-criteria factor analysis thus performed well in assessing and predicting potential yields of coffee. The performance of weighted overlay analysis has also been assessed in cotton by [21]. The relationships of soil, elevation, slope, aspect, and curvature with the stability of crop yields were assessed by [38], who deemed ordinary kriging the best method to estimate crop yield as a function of topography and landscape positions.

A wide range of yields of coffee in Rwanda, varying between 0.8 and 2.8 t ha^{-1} of dry parchment coffee, has also been reported by [6]. The low yields were attributed to coffee variety, agroecological conditions, the lack of mineral and organic fertilization, and limited mulching [6]. The national average yield of coffee is estimated at 1.1 t ha^{-1} y^{-1} (this study), but yields above 2.8 t ha^{-1} for dry coffee are rare even with adequate fertilization and sustained crop management [6]. In Uganda, 1.2 t ha^{-1} y^{-1} of dry coffee were recorded for mono-cropped coffee and coffee-banana intercropping systems [39]. The spatial variation in coffee production and productivity are thus mainly influenced by soil properties, soil management, farming practices, and climatic conditions. Similar trends have also been reported by [34,35].

Conclusions

The multi-criteria analysis used to assess spatial variation in potential production zones and the productivity of coffee revealed that agro-ecological factors are largely determined suitable zones of coffee productivity. The spatial variation of coffee productivity in the agro-ecological zones was considerable and was influenced by soil properties, soil management, farming practices, and climatic conditions. High production potentials indicated that smallholder farmers could generate income from coffee and could thus improve their livelihoods. In addition, this may provide an opportunity for farmers to purchase more land and extend the area for the production of coffee.

This study demonstrated that both ordinary kriging analysis and multi-criteria weighted overlay analysis performed well for analyzing the spatial distribution and productivity of coffee and for predicting yield. The depletion of soil fertility due to the lack of erosion control in scattered coffee plots on steep slopes is a major factor limiting coffee productivity in Rwanda. The sustainability of coffee productivity could be ensured by intensifying the use of fertilizers, mainly a well-balanced combination of lime, nitrogen, phosphorus, potassium, zinc, and boron. Limited access to financial resources restricts the purchase of these inputs and the use of different types of mulches that can improve soil properties and reduce the erodibility of the soil is recommended.

Acknowledgments

We thank the Ministry of Agriculture and Animal Resources for providing the digital soil datasets and the National Agricultural Export Development Board (NAEB), former OCIR – CAFÉ, for providing the coffee database and facilitation for field sampling. We also thank the GIS Centre of the National University of Rwanda for providing various spatial datasets and office space. We are very grateful to the NUFFIC for funding the PhD training of Innocent NZEYIMANA at Wageningen University, the Netherlands, and to Mr. Huting of ISRIC and Mrs Mukashema of the GIS Centre for their support in the analysis of the spatial data.

Author Contributions

Conceived and designed the experiments: IN. Performed the experiments: IN. Analyzed the data: IN. Contributed reagents/materials/analysis tools: IN. Wrote the paper: IN AEH VG. Supervision of the research project: AEH VG.

References

1. Ponte S (2002) The "latte revolution"? Regulation, markets and consumption in the global coffee chain. World development 30: 1099–1122.
2. Labouisse JP, Bellachew B, Kotecha S, Bertrand B (2008) Current status of coffee (*coffea Arabica L.*) genetic resources in Ethiopia: implications for conservation. Genetic Resource Crop Evolution 55: 1079–1093.
3. FAO (2014) FAOSTAT - Statistical database. Food and agriculture Organization of the United Nations. Available: http://faostat.fao.org/site/567/DesktopDefault.aspx?PageID=567#ancor. Accessed 2014 Apr 02.
4. Boffa JM, Turyomurugyendo L, Barnekow-Lilleso JP, Kindt R (2009) Enhancing farm tree diversity as a means of conserving landscape-based biodiversity. Mountain Research and Development 25: 212–217.
5. Van Oijen M, Dauzat J, Harmand JM, Gerry L, Vaast P (2010) Coffee agroforestry systems in Central America: II. Development of a simple process-based model and preliminary results. Agroforestry Systems 80.
6. Nzeyimana I, Hartemink AE, de Graaff J (2013) Coffee farming and soil management in Rwanda. Outlook on Agriculture 42 (1): 47–52. Doi 10.5367/oa.2013.0118.

7. Verdoodt A, Van Ranst E (2003) Land evaluation for agricultural production in the tropics: A large – scale land suitability classification for Rwanda. Ghent: Laboratory of soil science, Ghent University. ISBN: 90-76769-89-3.

8. FAO (2014b) FAOSTAT - Statistical database. Food and Agriculture Organization of the United Nations. Available: http://faostat.fao.org/site/666/default.aspx. Accessed 2014 Apr 02.

9. Clay J (2004) Coffee. In: Clay J, editor. World Agriculture and the Environment. Washington: Island press.

10. Bosselmann AS, Dons K, Oberthur T, Olsen CS, Ræbild A, et al. (2009) The influence of shade trees on coffee quality in small holder coffee agroforestry systems in southern Colombia. Agriculture, Ecosystems and Environment 129: 253–260.

11. Gabre-Madhin EZ, Haggblade S (2004) Successes in African Agriculture: results of an expert survey. World development 32: 745–766.

12. Rossiter DG (1996) A theoretical framework for land evaluation. Geoderma 72: 165–190.

13. FAO (1976) A Framework for Land Evaluation. Rome: Soil Resources Development, Land & Water Development Division. FAO Soils Bulletin No 32. 220 p. ISBN: 92-5-100111-1.

14. Malczewski J (2004) GIS-Based land use suitability analysis: a critical overview. Progress in Planning 62: 3–65.

15. De La Rosa D, Moreno JA, Garcia LV, Almorza J (1992) MicroLEIS: a Micro Computer-Based Mediterranean Land Evaluation Information System. Soil Use and Management 8: 89–96.

16. Maji AK, Krishna NDR, Challa O (1998) Geographical Information System in analysis and interpretation of soil resources data for land use planning. Journal of Indian Society of Soil Science 46: 260–263.

17. Coleman AL, Galbraith JM (2000) Using GIS as an Agricultural Land Use Planning Tool. Blacksburg. Department of Crop and Soil Environmental Science, Virginia Tech. Available: http://scholar.lib.vt.edu/ejournals/vaes/00-2.pdf. Accessed 2014 Feb 11.

18. Maji AK, Nayak DC, Krishna NDR, Srinivas CV, Kamble K, et al. (2001) Soil information system of Arunachal Pradesh in a GIS environmental for land use planning. International Journal of Applied Earth Observation and Geo-Information 3 (1): 69–77.

19. Joerin F, Theriault M, Musy A (2001) Using GIS and outranking multi-criteria analysis for land use suitability assessment. International Journal of Geographical Information Science 10 (8): 321–339.

20. Li B, Zhang F, Zhang L-W, Huang J-F, Jin Z-F, et al. (2012) Comprehensive Suitability Evaluation of Tea Crops Using GIS and a Modified Land Ecological Suitability Evaluation Model. Pedosphere 22 (1): 122–130.

21. Walke N, Obi Reddy G, Maji A, Thayalan S (2012) GIS-based multicriteria overlay analysis in soil-suitability evaluation for cotton (Gossypium spp.): A case study in the black soil region of Central India. Computers & Geosciences 41: 108–118.

22. Malczewski J (1996) A GIS-based approach to multiple criteria group decision-making. International Journal of Geographical Information Systems 10 (8): 955–971.

23. Janssen R, Rietveld P (1990) Multi-criteria analysis and geographical information systems. An application to agricultural land use in the Netherlands. In: Scholten HJ, Stillwell JCH, editors. Geographical Information Systems for Urban and Regional Planning. Dordrecht: Kluwer Academic Publishers. pp.129–139. ISBN: 0792307933.

24. ESRI (2011) ArcGIS Desktop: Realease 10. Redlands, CA: Environmental Systems Research Institute.

25. ESRI (2001) ArcGIS Spatial Analyst: Advanced GIS Spatial Analysis Using Raster and Vector Data. Available: http://www.esri.com/software/arcgis/extensions/spatialanalyst/. Accessed 2014 April 02.

26. Diggle PJ, Ribeiro PJJ (2007) Model-Based Geostatistics. New York: Springer.

27. Genton MG (1998) Highly robust variogram estimation. Mathematical Geology 30 (2).

28. Genton MG (2000) The correlation structure of Matheron's classical variogram estimator under elliptically contoured distributions. Mathematical Geology 32 (1).

29. Park SJ, Vlek PLG (2002) Environmental correlation of three-dimensional soil spatial variability: a comparison of three adaptive techniques. Geoderma 109 (1–2): 117–140.

30. Hengl T, Heuvelink GBM, Stein A (2004) A generic framework for spatial prediction of soil variables based on regression-kriging. Geoderma 120 (1–2): 75–93.

31. Mueller TG, Pusuluri NB, Mathias KK, Cornelius PL, Barnhisel RI, et al. (2004) Map Quality for ordinary Kriging and inverse distance weighted interpolation. Soil Science Society of America Journal 68: 2042–2047.

32. Birasa EC, Bizimana I, Bouckaert W, Delflandre A, Chapelle J, et al. (1990) Les Sols du Rwanda: méthodologie, légende et classification. Carte Pédologique du Rwanda. Kigali - Rwanda (Unpublished book): CTB et MINAGRI.

33. Moore ID, Gessler PE, Nielson GA (1993) Soil attribute prediction using terrain analysis. Soil Science Society of American Journal 57: 443–452.

34. Changere A, Lal R (1997) Slope position and erosional effects on soil properties and corn production on a Miamian soil in Central Ohio. Journal of Sustainable Agriculture 11: 5–21.

35. Kravchenko A, Bullock DG (2000) Correlation of corn and soybean grain yield with topography and soil properties. Agronomy Journal 92: 75–83.

36. Descroix F, Snoeck J (2004) Coffee: Growing, Processing, Sustainable Production. In: Wintgens JN, editor. A Guidebook for Growers, Processors, Traders, and Researchers. Switzerland. pp.164–177. ISBN: 352730307311.

37. Soltani A, Stoorvogel JJ, Veldkamp A (2013) Model suitability to assess regional potato yield patterns in northern Ecuador. European Journal of Agronomy 48: 101–108.

38. McKinion JM, Willers JL, Jenkins JN (2010) Spatial analyses to evaluate multi-crop yield stability for a field. Computers and Electronics in Agriculture 70: 187–198.

39. Van Asten PJA, Wairegi LWI, Mukasa D, Uringi NO (2011) Agronomic and economic benefits of coffee–banana intercropping in Uganda's smallholder farming systems. Agricultural Systems 104: 326–334.

3

Abundance, Composition and Activity of Ammonia Oxidizer and Denitrifier Communities in Metal Polluted Rice Paddies from South China

Yuan Liu[1], Yongzhuo Liu[1], Yuanjun Ding[1], Jinwei Zheng[1], Tong Zhou[1], Genxing Pan[1,2]*, David Crowley[3], Lianqing Li[1], Jufeng Zheng[1], Xuhui Zhang[1], Xinyan Yu[1], Jiafang Wang[1]

1 Institute of Resource, Ecosystem and Environment of Agriculture, Nanjing Agricultural University, Nanjing, China, 2 Center of Ecosystem Carbon Sink and Environment Remediation, Zhejiang Agricultural and Forestry University, Linan, Hangzhou, China, 3 Department of Environment Sciences, University of California Riverside, Riverside, California, United States of America

Abstract

While microbial nitrogen transformations in soils had been known to be affected by heavy metal pollution, changes in abundance and community structure of the mediating microbial populations had been not yet well characterized in polluted rice soils. Here, by using the prevailing molecular fingerprinting and enzyme activity assays and comparisons to adjacent non-polluted soils, we examined changes in the abundance and activity of ammonia oxidizing and denitrifying communities of rice paddies in two sites with different metal accumulation situation under long-term pollution from metal mining and smelter activities. Potential nitrifying activity was significantly reduced in polluted paddies in both sites while potential denitrifying activity reduced only in the soils with high Cu accumulation up to 1300 mg kg^{-1}. Copy numbers of *amoA* (AOA and AOB genes) were lower in both polluted paddies, following the trend with the enzyme assays, whereas that of *nirK* was not significantly affected. Analysis of the DGGE profiles revealed a shift in the community structure of AOA, and to a lesser extent, differences in the community structure of AOB and denitrifier between soils from the two sites with different pollution intensity and metal composition. All of the retrieved AOB sequences belonged to the genus *Nitrosospira*, among which species Cluster 4 appeared more sensitive to metal pollution. In contrast, *nirK* genes were widely distributed among different bacterial genera that were represented differentially between the polluted and unpolluted paddies. This could suggest either a possible non-specific target of the primers conventionally used in soil study or complex interactions between soil properties and metal contents on the observed community and activity changes, and thus on the N transformation in the polluted rice soils.

Editor: Manuel Reigosa, University of Vigo, Spain

Funding: The present research was funded by China National Science Foundation under grants number of 40671180 and of 40710019002. The funders had no role in study design, data collection and analysis, decision to publish, or preparation of the manuscript.

Competing Interests: The authors have declared that no competing interests exist.

* Email: gxpan@njau.edu.cn

Introduction

Nitrogen fertilizers are essential for rice production, but their overuse had also caused serious problems from runoff of nitrate and contributes to climate change from production of the greenhouse gas, nitrous oxide [1]. To addressing these problems, many of the environmental and management factors that affect biological transformations of nitrogen had been well studied to determine which soils are most prone to nitrogen losses [2,3]. However, there had been still open questions on the effects of heavy metal pollution on biological nitrogen transformations, mainly owing to the difficulty in determining the extent to which microbial communities adapted to the presence of heavy metals over time when metals were slowly introduced by atmospheric deposition, versus their response to sudden increases in metal concentrations when heavy metals were spiked into soils at high concentrations [4,5]. Another challenge in determining the effects of metals on nitrogen transformations had been the difficulty in

relating changes in soil enzyme activities to changes in the abundance of molecular markers for specific microbial taxa and genes that could be involved in nitrogen transformations [6]. This is especially true for rice paddy soils where wet/dry cycles drive significant changes in redox that simultaneously affected metal bioavailability, microbial community structures, and rates for biological transformations of nitrogen.

Nitrous oxide (N_2O) had been widely accepted as the most radiative greenhouse gas increasing at a year rate of approximate 0.26% per year such that this greenhouse gas had reached with a concentration of 319×10^{-9} mol mol^{-1} in global air by IPCC [7]. However, it had been known also as a product resulted from uncompleted denitrification, in which reduction of nitrite was not completed to form N_2 in soils. Agriculture accounted for about 60% of the global total anthropogenic N_2O emission, of which rice paddies had been considered a major contributor [8–10]. Total N_2O emission from China's rice paddies was estimated at 29.0 Gg

N_2O per year, accounting for 7–11% of annual overall greenhouse gas emission from mainland China croplands [9]. Ammonia oxidization to nitrite had been well known as the initial and rate limiting step in nitrification, being mediated by microorganisms which carry genes encoding for the enzymes AOB [11,12] and/or AOA [13,14]. In contrast to the nitrifier which could be comprised by a few functional taxa, denitrifying bacteria responsible for denitrification could be broadly distributed among many different taxa using nitrate as an alternate electron acceptor for respiration [15].

China had been the largest rice producing country in the world with approximately 20% of global rice production [16]. In the last decade, metal pollution had been widely reported to occur in extensive rice production areas that included the lower Yangtze River delta [17,18], the Pearl River delta [19] and river valleys in the Jiangxi and Guangdong provinces [20]. Much attention had given to the potential health risk through food chain transfer of heavy metals and adverse effects on ecosystem health [21,22]. Recently, there had been observed in metal polluted rice paddies a decline in microbial biomass and fungal to bacterial ratio with an increase in the metabolic quotient that could lead to changes in C cycling [23]. Many laboratory studies had shown that heavy metal contamination in soil could affect the rates of microbial-mediated biogeochemical processes [24,25]. Soil nitrification rates had been known to be suppressed under metal pollution both in spiked soil samples in short term studies, and in polluted fields where metals typically accumulated at a slower rate [26]. In a study on surface wetland sediments [27], total denitrification activity was significantly decreased in multiple metal spiked wetland samples over a period of 7 days incubation with Cd, Cu or Zn ≥ 500 mg kg^{-1}, with Cd being the strongest inhibitor. Whereas, heavy metal pollution could exert effects on ammonia oxidizers [24,25,28,29] and denitrifying bacteria in field soils [30,31].

However, changes in composition and diversity of N-transforming microbial communities with metal pollution had not yet been studied in polluted rice fields. Among these, the potential effect of metal pollution on nitrification and denitrification processes in the rice fields should require primitive study for addressing changes in N cycling in polluted lands. Thus, understanding N loss via soil-air flux of N_2O emission and projecting future N_2O emissions from rice soils would depend on a better understanding of how nitrification and denitrification could contribute to the soil-atmosphere N_2O flux in rice agriculture and how these processes could be affected by heavy metals. In addition, nitrifier and denitrifier could also have different responses to stress of heavy metal pollution in soils owing to their different resource requirements, different metal contamination would exert different effects on the changes in the microbial communities and their biological functions on mediating N transformations in polluted rice fields.

Here we hypothesize that heavy metal contamination could have impact in the abundance, composition and activity of nitrifying and denitrifying communities, which could differ in soils with different soil properties as well as with different metal composition. An experiment was conducted to compare the community composition of nitrifier and denitrifier using two long-term metal polluted soils, which were compared to those in adjacent unpolluted fields at each location. We further studied the possible linkage between gene copy numbers that could serve as molecular markers for nitrogen transformation processes and process rates determined by direct enzyme assays measuring the potential activity of the nitrifying and denitrifying communities.

Ethics statement

The contaminated soils that we collected samples were under rice paddy used for rice production. No specific permission required nor any endangered or protected species involved. The polluted paddies both had a 30 year history of metal pollution due to discharge of mining waste water.

Materials and Methods

Site description

Two locations with soil pollution were selected for this study. Location DX (29°04′N, 117°43′E, Dexing County, Jiangxi) was situated in a copper mining area. The second location, designated DBS (24°26′N, 113°49′E, Wenyuan Municipality, Guangdong Province), was close to a multiple metal mine of zinc, cadmium, lead, and copper. The polluted paddy of at DX site was 1.5 km in distance from the top-hill ore mining and 1 km in distance to an adjacent Cu smelter, being affected by waste water discharge and atmospheric deposition. Whereas, the polluted paddy at DBS was polluted with irrigation by river water discharged from an upstream multiple metal mining ore and the associated smelter 3.5 km in distance. The mining and smelter activity had been taken place since late 1960's in both sites [23]. In each location, unpolluted rice paddies were selected that had no distinct access to polluted irrigation water. Soil samples from the polluted and non-polluted rice paddies at each location were designated as polluted (PS) and background soil (BGS) respectively. The climate at both locations are characterized by a subtropical monsoon climate with a mean annual temperature from 18°C to 25°C, and mean annual rainfall from 1200 mm to 1450 mm. In agriculture, double rice cropping has been traditionally practiced in the rice paddies in the area where these sites situated.

Soil sampling

Soil sampling was conducted at each location before rice planting in spring of 2009. Three composite topsoil samples at a depth of 0–15 cm were randomly collected respectively in 3 subplots from both PS and BGS fields. Each composite sample consisted of 5 sub-samples that were collected following a "Z" shaped sampling pattern with a distance of ~5 m in each subplot. The composite samples were mixed thoroughly and kept on ice until they were transported to the laboratory within two days after sampling. The samples were processed to remove gravels and plant detritus if any, and passed through a 2 mm sieve. Each composite sample was divided into three portions, one of which was used for DNA extraction. A second portion was stored at 4°C for measuring potential nitrifying and denitrifying activities. The third portion was air-dried at room temperature for analysis of soil chemical and physical properties as described below.

Analysis of soil properties and metal contents

Measurements of the basic properties and metal contents of the samples were conducted following the protocols described by Lu [32]. Briefly, soil pH was measured with a glass electrode using a 1/2.5 soil/water ratio. Soil textural class and particle size fractions were determined with a hydrometer method after dispersion with 0.5 mol L^{-1} NaOH solution. Soil organic carbon (SOC) was measured using wet digestion and oxidation with potassium dichromate. Total nitrogen was analyzed using the Kjeldahl method. For total heavy metal content determination, the samples were digested with a solution of HF/HClO$_4$/HNO$_3$ (10/2.5/2.5, v/v/v) and then extracted with 1 mol L^{-1} HCl. Content of Cd was determined with graphite furnace atomic absorption spectrometry (GFAAS, SpectrAA220Z, Varian, USA) while those of

Table 1. Soil physicochemical properties of the studied soil samples.

Site	Plot	pH (H₂O)	SOC (g kg⁻¹)	TN (g kg⁻¹)	Clay (%)	Silt (%)	Sand (%)
DX	BGS	4.87±0.05	22.79±1.58	1.58±0.05	21.0	31.2	47.8
	PS	4.10±0.06	22.25±0.35	1.96±0.07	27.0	32.2	40.8
DBS	BGS	5.58±0.13	15.23±0.60	1.03±0.03	22.2	23.2	54.6
	PS	5.45±0.16	19.11±0.68	1.49±0.08	27.0	28.8	44.2

BGS: Background soil; PG: Polluted soil; SOC: soil organic carbon; TN: total nitrogen.

Pb, Cu and Zn with flame atomic adsorption spectrophotometry (FAAS, TAS-986, China). Soil physic-chemical properties and metal contents determined are presented in Tables 1 and 2, respectively.

The Nemerow pollution index [33] was used to evaluate the overall extent of heavy metal pollution. It was calculated using the following equation:

$$P_n \sqrt{(MaxP_i^2 + AveP_i^2)/2} \qquad (1)$$

Where, P_n was the Nemerow pollution index value, and calculated as the sum of n metal elements analyzed for a soil sample; P_i was a single pollution intensity index of ith metal element with its measured concentration (C_i) divided by the guideline standard of environmental quality (RS_i). $MaxP_i$ and $AveP_i$ were the maximum and average pollution intensity respectively of the analyzed metals in a given soil. A relative metal accumulation degree using content of a single element in PS over in BGS was calculated and then averaged as the overall pollution intensity in a single site. All these calculated values are given in Table 2.

DNA extraction and real-time PCR assay

Total DNA was extracted from 0.25 g fresh soil with a PowerSoil DNA Isolation Kit (Mo Bio Laboratories Inc., CA) according to the manufacturer's protocol. The primers and thermal cycling procedures are listed in Table S1. The primer sets of $nirK$876 and $nirK$1040 were used in this study as they had been considered representing typical denitrifying bacterial community in multiple samples [36]. Each reaction was performed in a 25 μl reaction volume containing 15 ng of DNA, 1 μl of 10 μM of each primer and 12.5 μl of SYBR premix EX Taq TM (Takara Shuzo, Shinga, Japan). The amplified PCR products of $amoA$ (AOB and AOA) and $nirK$ genes were purified using PCR solution purification kit (Takara), ligated into pEASY-T3 cloning vector (Promega, Madison, WI) and cloned into *Escherichia coli* DH5α. Clones containing correct inserts were chosen as the standards for real-time PCR (qPCR). High amplification efficiencies of AOB-$amoA$ (101%), AOA-$amoA$ (97%) and $nirK$ (102%) were obtained for gene quantification, with r^2 values of 0.996, 0.997 and 0.991 respectively.

PCR-DGGE of ammonia oxidizing and denitrifying communities

Profiles of the ammonia oxidizing and denitrifying communities were generated by PCR-DGGE using the primer sets $amoA$-1F-GC and $amoA$-2R, which had been considered specific for the AOB [34], Arch-$amoA$F-GC and Arch-$amoA$R which could be specific for AOA [35] and the $nirK$876-GC and $nirK$1040 set for the denitrifying bacterial communities [36]. The GC clamp described by Muyzer and Smalla [37] was incorporated into the 5′ end of primers. PCR reactions were performed with reaction mixtures that contained 12.5 μl Go TaqH Green Master Mix (Promega, Madison, WI), 1 μl of 10 μM of each primer, and 1 μl of DNA template. For DGGE analysis, PCR products were separated on an 8% (w/v) polyacrylamide gel of acrylamide/bisacrylamide (37.5/1, v/v) containing denaturing gradients of 45–60% for AOB, 25–50% for AOA and 50–65% for $nirK$. DGGE was performed in 1× TAE buffer at 60°C, 200 V for 5 min, then 140 V for 500 min. Gels were silver stained [38] and scanned using a gel document system (Bio-Rad, USA).

Table 2. Total heavy metal contents and Nemerow pollution index (Means ± S.D.) of the studied soils.

Sample	Cd (mg kg^{-1})	Pb (mg kg^{-1})	Cu (mg kg^{-1})	Zn (mg kg^{-1})	Nemerow index	Pollution intensity
DX-B	0.48±0.14b	58.95±0.90b	640.19±2.98b	94.57±17.21b	9.45±0.03b 19.74±1.86a	/
DX-P	1.55±0.14a	95.17±7.07a	1333.68±129.72a	163.90±15.41a		3.23±0.15B
DBS-B	0.29±0.00b	33.37±2.08b	21.87±1.02b	70.40±1.5b	0.76±0.001b 4.04±0.56a	/
DBS-P	1.49±0.24a	133.27±6.67a	224.83±5.68a	248.48±5.87a		5.32±0.23A

Different lowercase characters indicate significant different ($p<0.05$) between polluted soils (PS) and background soils (BGS) in a single site. Different capital letters for pollution intensity indicate significant different ($p<0.05$) between the two sites.

Sequencing and phylogenetic analysis

Prominent bands from the DGGE gels for the AOB and denitrifying communities were numbered and excised from the gels for sequence analysis. The retrieved sequences were compared with GenBank data base sequences using BLAST (Basic Local Alignment Search Tool) (http://www.ncbi.nlm.nih/gov/blast/) to search for best matches. Sequences of the DGGE bands had been deposited in GenBank under the accession numbers JF264803 to JF264813 (AOB) and JF264827 to JF264835 (*nirK*).

Measurement of potential nitrification and denitrification activities

Potential nitrification activity (PNA) was measured using the short-term nitrification assay described by Schmidt and Belser [39]. Briefly, 20 g moist soil was added to a 250 mL cotton-stopped flask with 100 ml of 0.5 mM phosphate buffer (pH 7.2), 0.5 mM $(NH_4)_2SO_4$ and 10 mM $KClO_3$ to stop further oxidation of NO_2^-. Triplicate flasks for each treatment were incubated for 24 h on a shaker at 175 rpm at 25°C. Net rate of NO_2^- production was calculated based on the linear correlation of NO_2^- concentration versus time.

Potential denitrification activity (PDA) assay was analyzed by the acetylene (C_2H_2) inhibition method following the procedure described by Tiedje et al. [40]. N_2O was measured using a gas chromatograph (Agilent 7890D, Santa Clara, CA, USA). Assays were conducted by addition of 20 g moist soil at 60% WHC to 250 mL glass bottles, in soil slurries produced by adding 20 mL of a substrate solution containing 1 mM glucose and 1 mM KNO_3. The linear rates of N_2O production over time were observed within 6 h of initiating the incubations.

Data processing and statistical analysis

The data were processed with Microsoft Excel 2010 and expressed in means plus/minus standard deviations. A paired t-test was used to evaluated differences between the polluted (PS) and background (BGS) samples from a single site with a significance defined at $p<0.05$. Digitized DGGE images were analyzed with Quantity One image analysis software (Version 4.0, Bio-Rad, USA). This software identified the band with the same position in the different lanes of the gel and also measures the intensity of identified bands. Cluster analysis and principal component analysis (PCA) of the DGGE profiles were performed to elucidate the microbial community structures based on relative band intensity and positions using the Minitab v.15 software. Redundancy analysis (RDA) was used to evaluate the relationship among DGGE fingerprints and soil properties (pH, SOC, TN, Nemerow index; total concentration of Cd, Pb, Cu and Zn) using the Canoco 4.5 software.

Results

Soil physicochemical properties and heavy metal pollution

As showed in Table 1, there were generally hardly differences in the basic properties between PS and BGS in a single site. For example, soil pH of the background soil at the DBS site (pH 5.58) was higher than at DX site (pH 4.87), without remarkable difference between PS and BGS in a single site. SOC and TN ranged from 15.23 g kg^{-1} to 22.79 g kg^{-1} and from 1.03 g kg^{-1} to 1.96 g kg^{-1} for the two sites, respectively. Here, N was generally higher in polluted soil than the background soil for less N consumed by rice with reduced production under metal pollution in both sites. However, C/N ratio seemed similar between the two sites both of polluted soils (14.4 and 14.8 respectively for DX and DBS) and of background soils (11.4 and 11.8 respectively for DX and DBS). As listed in Table 2, there were consistent differences in contents of total Cd, Pb, Cu and Zn between PS and BGS in a single site though the contents of a single heavy metal element varied with sites. While the total Cu content in PS was approximately 6 times higher from DX site than from DBS site, though a high Cu content up to 640 mg kg^{-1} owing to the lithological background [23]. As calculated based on environmental quality guideline value of China, Nemerow pollution index value was 4.0 for DBS and 20 for DX. While Cd, Pb and Zn contents of PS were similar between the two sites, the overall metal pollution intensity in PS estimated of the contents of all individual elements in PS over the BGS was significantly higher in DBS site than in DX site.

Abundances of ammonia oxidizer (AOB and AOA) and denitrifier (*nirK*)

The relative abundances of *amoA* and *nirK* gene copy numbers were assessed by qPCR (Table 3). The results showed that copy numbers of AOA *amoA* gene of the two soils ranged from 1.1×10^8 to 4.6×10^8 g^{-1} soil, and approximately 10–100 fold greater than those of AOB *amoA* genes, which ranged from 6.4×10^6 to 5.8×10^7 g^{-1} soil. Compared to background soil, copy numbers of AOB *amoA* gene were apparently reduced in PS by 19% at DX but 80% at DBS, corresponding to their metal pollution intensity difference. However, copy numbers of the AOA *amoA* gene were significantly reduced in polluted soil only at DX. Accordingly, the ratio of AOA to AOB in PS decreased at DX site but increased at DBS site while AOA genes were present in higher copy numbers than AOB genes at both sites, with ratios of AOA to AOB ranging from 2.5 to 59 across all samples.

Copy numbers of the *nirK* gene ranged from 2.32×10^8 to 8.33×10^8 g^{-1} soil in the studied paddies, while there was no significant change between PS and BGS in a single site. The ratios of *nirK* : *amoA* (sum of AOA and AOB gene copy numbers) ranged

Table 3. Ammonia oxidizer and denitrifier gene copy numbers and the relative ratios (Means ± S.D.) of the soils studied.

Sample	AOB ($\times 10^7$)	AOA ($\times 10^7$)	nirK ($\times 10^8$)	Ratio of AOA to AOB	Ratio of nirK to amoA
DX-B	0.79±0.07a	46.40±4.16a	6.25±1.86a	59.02±5.84a	1.32±0.35b
DX-P	0.64±0.05b	10.80±0.31b	8.33±0.25a	16.97±0.61b	7.92±0.20a
DBS-B	5.75±0.52a	14.80±5.78a	2.32±0.48a	2.55±0.89a	1.13±0.03a
DBS-P	1.13±0.06b	12.40±8.41a	3.38±0.68a	10.86±4.36a	3.34±1.03a

Different lowercase characters indicate significant different ($p<0.05$) between polluted soils (PS) and background soils (BGS) in a single site.

from 1.13 to 7.92 across all of the soil samples. Compare to GBSs, the *nirK*: *amoA* ratios increased in polluted soils from both locations (Table 3).

Community structures of ammonia oxidizer (*amoA*) and denitrifier (*nirK*)

Cluster analysis and principal component analysis (PCA) were used to group sampled soils based on similarity in relative band intensity and position of the DGGE profiles. PCA of DGGE profiles of AOB and AOA at the two sites yielded good summaries of data, as 86% (AOB) and 76% (AOA) of the total variability was explained by the first two components (Fig. S1). The AOB community did not show clear shifts between PS and BGS soils at DX site, but was separated between of PS and BGS of DBS site on the basis of PC2. Cluster analysis of AOB community from samples with or without heavy metal pollution revealed similarity of 59% at the DX site and 30% at the DBS site (Fig. 1a). In contrast, the AOA community profiles showed well separation between PS and BGS soils at each site, and separated on the basis of PC1 for DBS samples and on the basis of PC2 for DX samples (Fig. S1b). Cluster analysis clearly showed that AOA community under PS was distinguishable from BGS soils in both sites with an intragroup similarity of 40% for the DX site and 34% for the DBS site (Fig. 1b). Both PCA and cluster analysis of DGGE profiles of *nirK* gene revealed distinct communities between PS and BGS soils at the DX site with 26% similarity; but no clear differences in *nirK* community structure at the DBS site with 48% similarity.

Phylogenetic analysis

Selected DNA bands from the DGGE profiles were sequenced to identify the predominant taxa associated with these bands and the effects of metal pollution on specific taxa (Fig. S2). Some bands were present in the profiles from both BGS and PSs, but their intensity varied between locations. The results showed that all of the gene sequences represented in the DGGE profile of AOB *amoA* gene were associated with the genus *Nitrosospira*. The DGGE profile of AOB revealed several bands (B17, B18, B19 and B20) that were present in both PS and BGS from the DX site, and that had the same relative positions as bands present in BGS from the DBS site. However, sequences identified from the BGS at the DBS site belonged to Cluster 4, whereas bands B17–B20 from the DX site were affiliated with Cluster 12 (Fig. 2). None of the sequences of *nirK* from the different soils were affiliated with either *α-proteobacteria* or *β-proteobacteria* (Fig. 3). Whereas, most of the clones in the PS at the DBS site represented different taxa from those in BGS, sequences representing taxa from the DX site were similar for both BGS and PS.

Soil potential nitrifying and denitrifying activities

Potential nitrifying and denitrifying activities and differences associated with metal pollution varied with location. Transformation of ammonia to nitrate was approximately 3-fold lower in PSs compared to BGS at both locations (Fig. 4a). BGSs from both the DX and DBS sites released 0.008 and 0.023 mg NO_2^- -N kg^{-1} soil h^{-1}, respectively, which compared to 0.003 mg NO_2^- -N kg^{-1} soil h^{-1} at the DX site and 0.007 mg NO_2^- -N kg^{-1} soil h^{-1} at the DBS site in BSs. An 8-fold decrease in denitrification rates was found of PS (0.009 mg N_2O -N kg^{-1} soil h^{-1}) over that of BGS (0.001 mg N_2O -N kg^{-1} soil h^{-1}) at the DX site despite of no significant change between PS and BGS at the DX site. (Fig. 4b).

Correlation of microbial community structure with soil properties

Using redundancy analysis, we identified the factors which can best explained soil microbial community structure at each site (Fig. 5). In the redundancy analysis biplot, axis 1 and axis 2 values explained 80.2% and 79.7% of the variability, respectively, in AOB profiles and AOA profiles. Soil properties most important in explained AOB community composition were soil pH, SOC, TN, Nemerow index and Cu concentration, which were strongly related to the first axis, and soil pH, Nemerow index and Cu, Pb, Zn concentration, which were most important soil factors in determining AOA community composition. In the redundancy analysis biplot of *nirK* gene fragments, axis 1 explained 45.6% and axis 2 explained 29.3% of the variability. Soil pH, TN, Nemerow index and Cu concentration were strongly related to the first axis, and Cd concentration was strongly related to the second axis (Fig. 5c).

Discussion

Changes in ammonia oxidizer and denitrifier populations with metal pollution

Our results showed a strong impact of metal pollutants on ammonia oxidizer with the degree varying with the metal pollution situation in the two different locations. Copy numbers of *amoA* genes for eubacteria (10^6 to 10^7 per g dry soil) and archaea (10^8 per g dry soil) in the BGSs were both within the ranges that had been reported in previous studies of agricultural soils. In contrast to genes encoding enzymes for ammonia oxidation, the abundances of *nirK* genes representing the denitrifying bacteria seemed not affected by pollution in a single site. The *nirK* gene abundances ranged from 1.5×10^8 to 8.3×10^8 copies g^{-1} dry soil, which seemed within the range of those reported by Dandie et al. [41]. Decreases in copy numbers of *nirK* and *nosZ* genes were reported for estuary sediments with spiked copper, though recovered after prolonged incubation [31]. Likewise, in a study by Holtan-Hartwig et al. [42], a mixture of Cd, Cu and Zn caused a temporary reduction in N_2O production

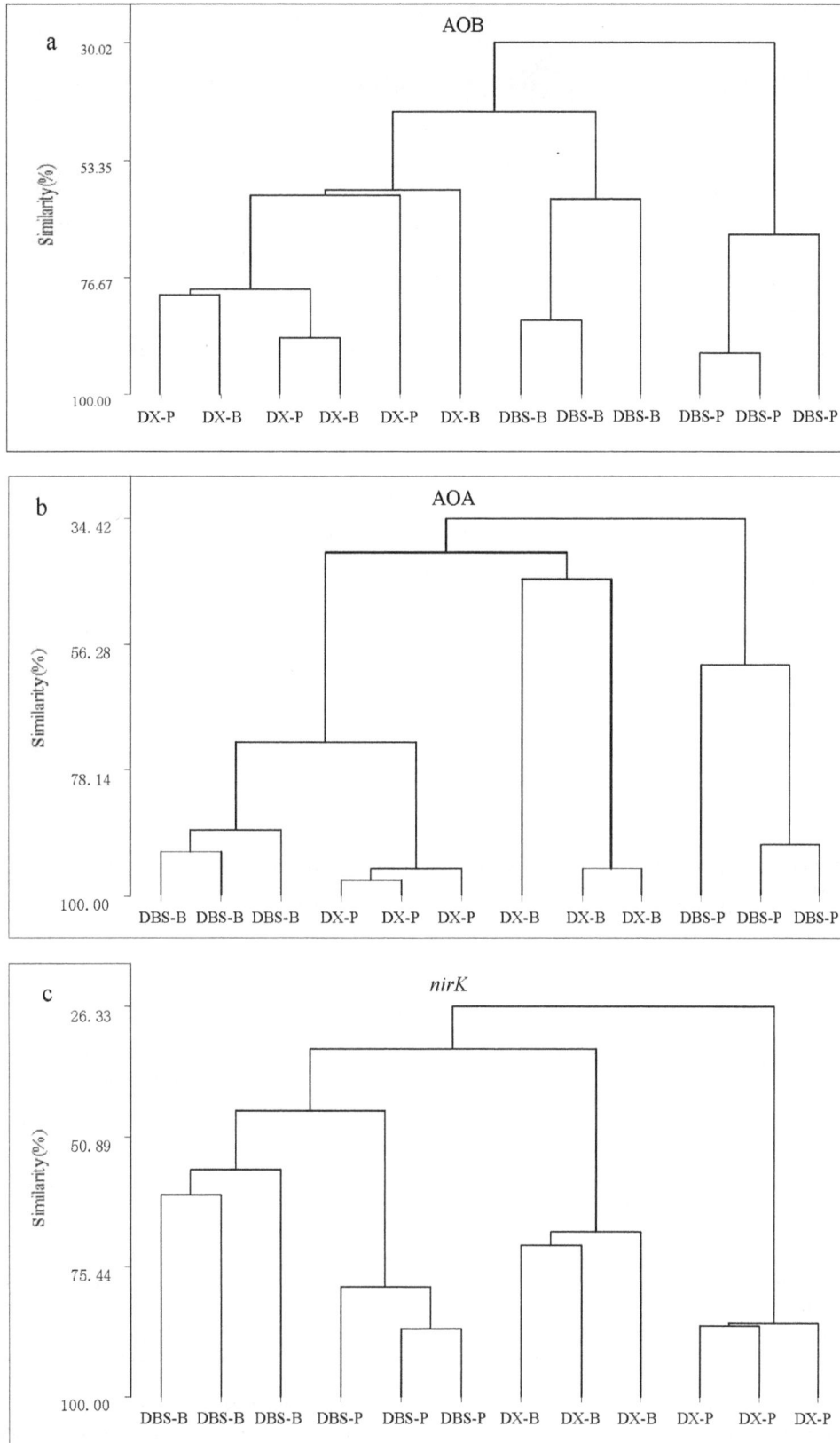

Figure 1. The cluster analysis of DGGE profiles of AOB (a), AOA (b) and *nirK* (c) gene fragments from the soil samples at the two sites. DX-B and DBS-B, background soil from site DX and DBS; DX-P and DBS-P, polluted soil from site DX and DBS. Euclidean distances were calculated from relative positions and intensities of bands, and the samples were clustered using Pearson's product-moment coefficient and a UPGMA (Unweighted Pair Group Methods with Arithmetic mean) algorithm.

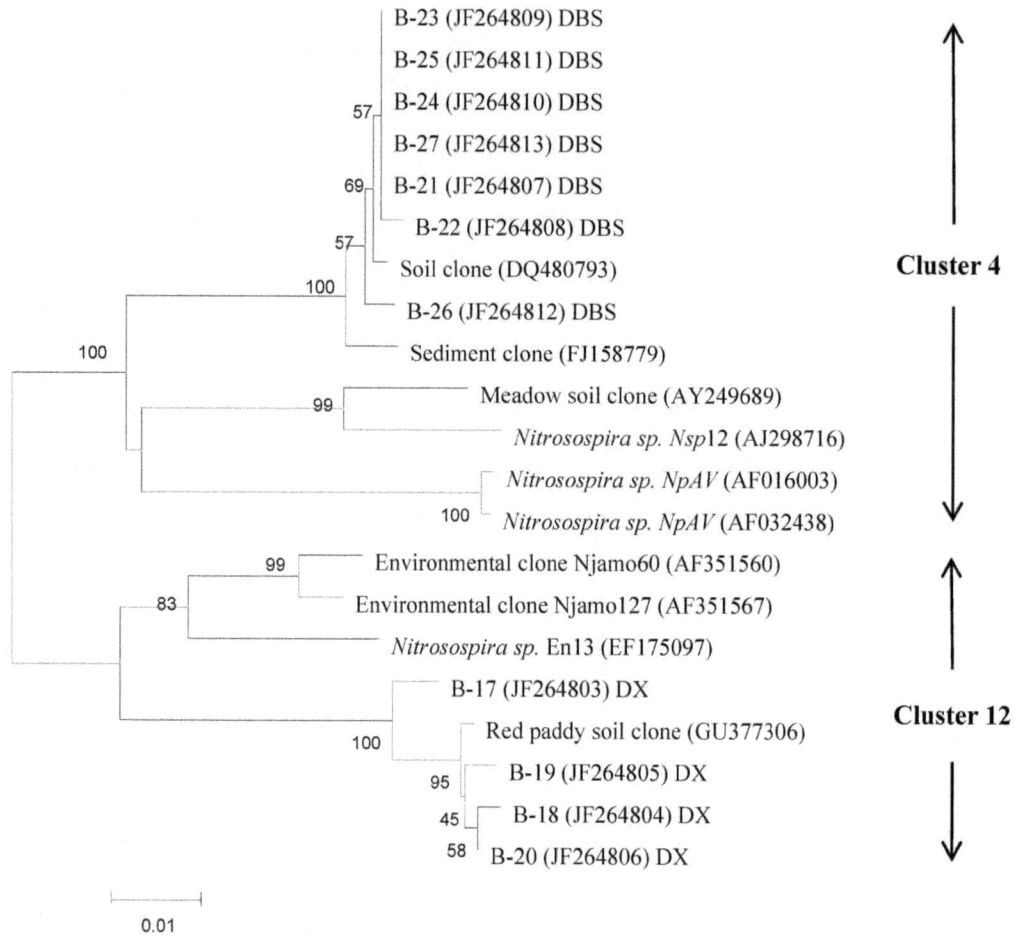

Figure 2. Neighbor-joining phylogenetic tree of AOB *amoA* sequences retrieved from the numbered DGGE bands of Fig. S2(A). Designation of the clones in the dashed line frames includes the following information: excised DGGE band number, accession number in the parentheses, followed by the sampling plot the clone retrieved from. Bootstrap values (>40%) with 1000 replicates are indicated at branch points. Scale bar indicates 1 change per 100 nucleotide positions.

in a sandy loam soil, which recovered within two months after spiking. Our results could be thus in general agreement with the hypothesis that denitrifier populations underwent adaptation to Cd, Pb, Cu and Zn pollution.

Sensitivity of ammonia oxidation to heavy metals had been known dependent on the specific metal or combination of metals. For example, a study by Liu et al. [29] showed no change in AOA abundance and community composition in a mercury-spiked vegetable soil. However, a work by Li et al. [28] showed a significant decrease in AOA abundance in four different soils (grassland and arable soils) with Cu amendments up to 1000 mg kg^{-1}. In a spiked study by Fait et al. [43], the ammonia oxidizer community was shown more vulnerable to Cu pollution than to Ni. The strong effects of copper on AOB populations was also shown in a study by van Beelen et al. [44], who showed that microbial communities more readily adapted to Zn and Ni contamination than to contamination from Cu and Cr in a sludge-amended soil. Therefore, AOA and AOB could be both significantly affected by metal pollution, but that AOA seemed more strongly reduced in Cu-polluted soil.

In this study, microorganisms carrying genes encoding for AOA were more abundant than those expressing AOB, with a ratio of AOA to AOB gene copy numbers ranging from 2.5 to 59.

Moreover, the ratio of AOA/AOB gene copy numbers was 71% lower in PS at DX site but did not significantly changed in those at DBS site (Table 3). Ruyters et al. [25] reported an increase in AOA/AOB ratios in a grassland soil exposed to Zn up to 1300–4000 mg kg^{-1} soil for 12 months under pot experiment. Similarly, Li et al. [28] observed a sharp decrease in AOB populations in a Cu-spiked (1200 mg kg^{-1}) soil during a 21-day long microcosm study with grassland and cropland soils. In line with the potential impact on AOB population, the AOA/AOB ratio was clearly seen decreased in polluted soils while both AOA and AOB abundances were reduced significantly at DX site, where the Cu concentration was as high as 1333 mg kg^{-1}. In a similar fashion, the *nirK/amoA* ratios observed in our study ranged from 0.74 to 7.92 (Table 3), indicating a greater abundance of denitrifiers (*nirK*) relative to ammonia oxidizers (AOA and AOB) in all of the soils except for the BGS at DBS site. Similarly, Hai et al. [45] reported the predominance of *nirK* gene relative to *amoA* genes (AOA and AOB) in the rhizosphere of sorghum in an agricultural soil.

N availability to microbes of soil, usually assessed with C/N ratio, could affect the N removal in soils. Here, N was generally higher in polluted soil than the background soil in both sites for less N consumed by rice with reduced production under metal pollution. However, C/N ratio seemed similar between the two

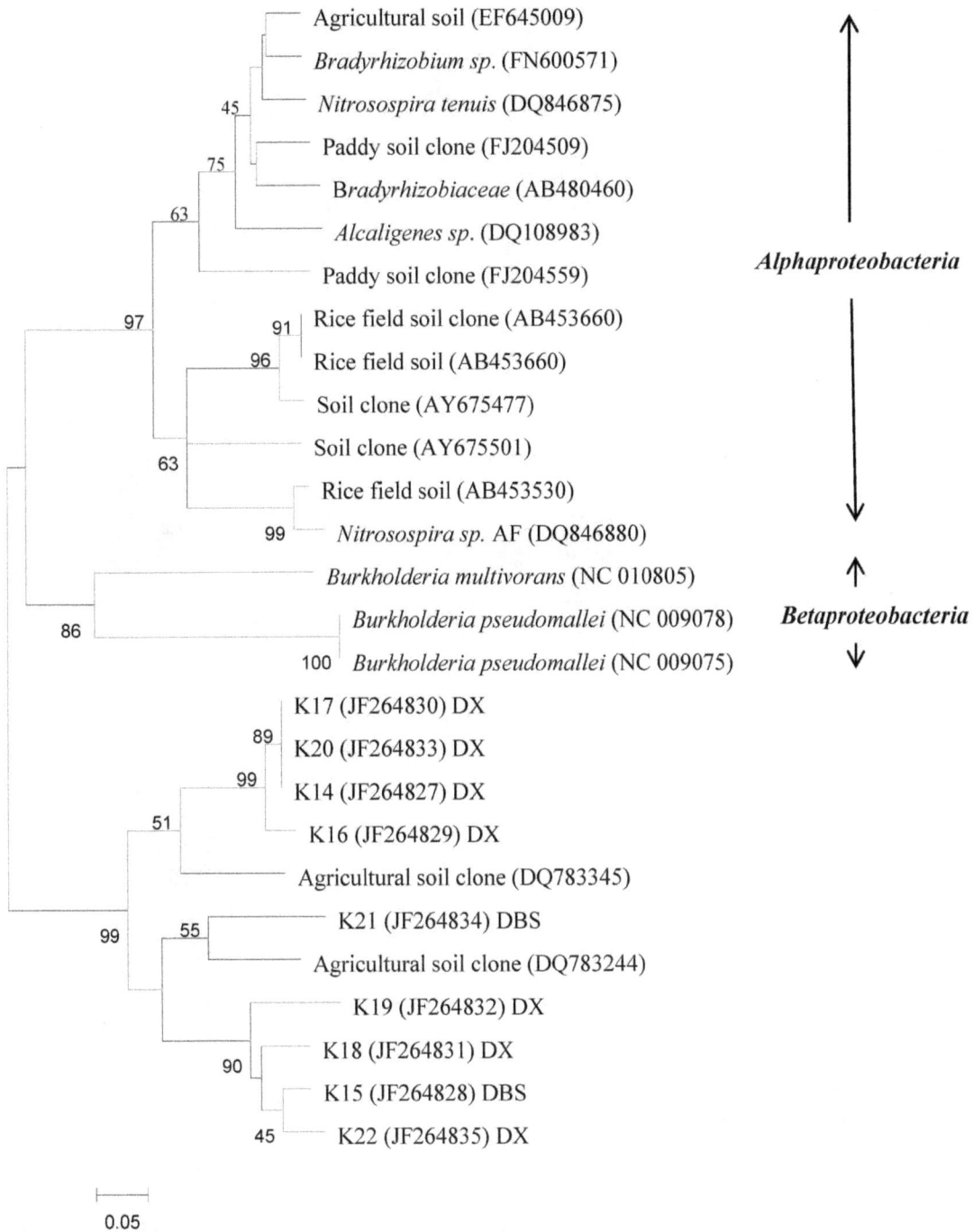

Figure 3. Neighbor-joining phylogenetic tree of *nirK* sequences retrieved from the numbered DGGE bands of Fig. S2(C). Designation of the clones in the dashed line frames includes the following information: excised DGGE band number, accession number in the parentheses, followed by the sampling plot the clone retrieved from. Bootstrap values (>40%) with 1000 replicates are indicated at branch points. Scale bar indicates 1 change per 100 nucleotide positions.

sites both of polluted soils (14.4 and 14.8 respectively for DX and DBS) and of background soils (11.4 and 11.8 respectively for DX and DBS). The finding that, over BGS, a decrease in AOB in PS was higher in DBS than in DX but AOA-to-AOB ratio in PS decreased in DX and increased in DBS suggested a role of metal pollution situation rather than N concentration on the community abundance, This was in line with the finding of a recent study of

the authors group [46] that N_2O production was much higher (by~350%) in metal polluted soil (C/N ratio 11.6) than by 190% in unpolluted soil (C/N ratio 11.8) when amended with polluted straw amendment.

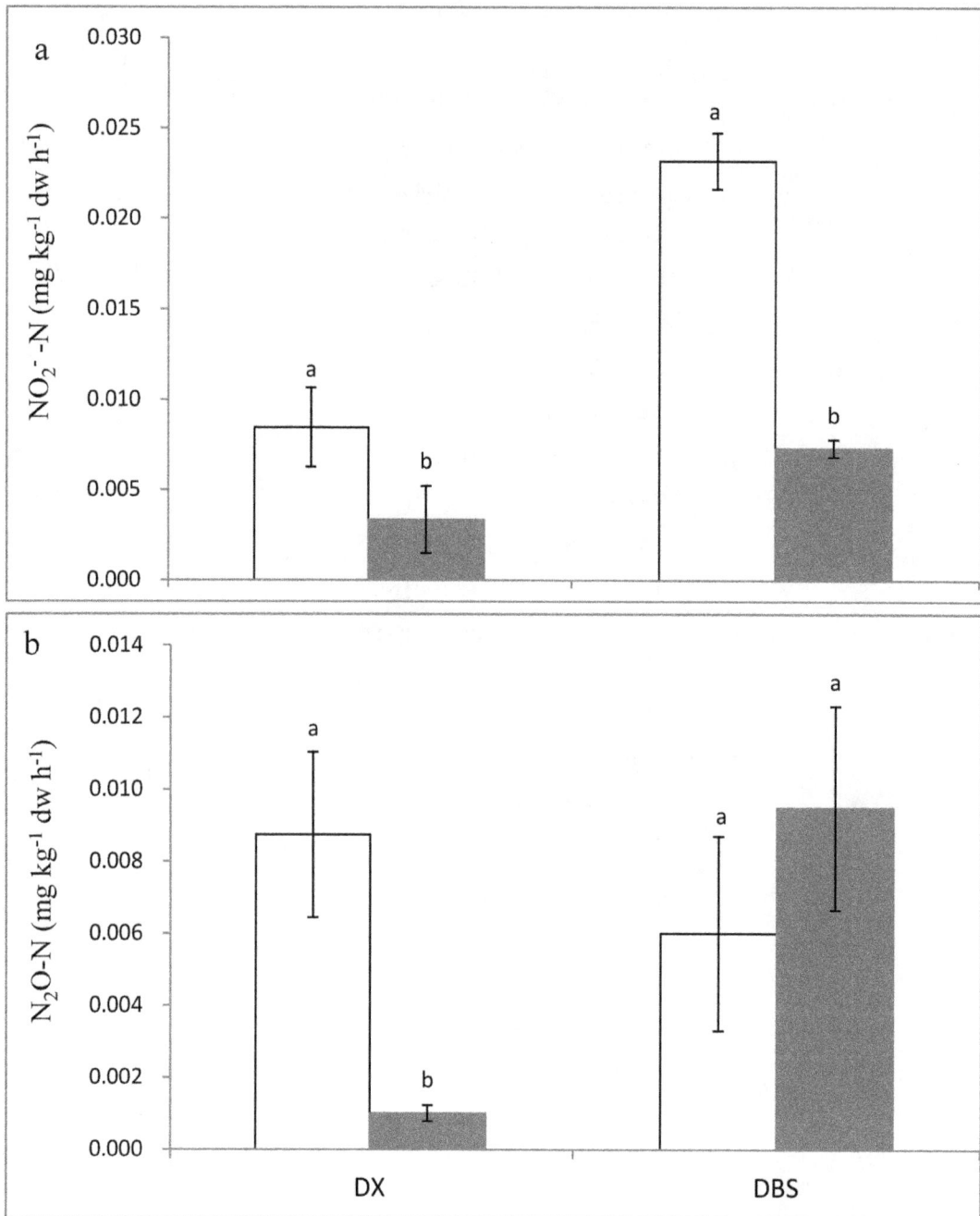

Figure 4. The potential nitrifying activity (a) and denitrifying activity (b) of the background (Blank) and polluted (Shaded) soils in DX and DBS sites. Different lowercase characters indicate significant difference ($p<0.05$) between the background and polluted soils in a single site.

Changes in community structure of ammonia oxidizer and denitrifier in metal polluted soil

Based on DGGE analysis between BGS and PS, both AOB and AOA communities were distinguishable at DBS site with a higher pollution degree by multiple metals but Whereas, only AOA community at DX site with a low pollution degree but predominated by Cu. Phylogenetic analysis revealed that all the AOB sequences in our study belonged to the genus *Nitrosospira*

(Fig. 2). The dominance of *Nitrosospira spp.* both in BGSs and PSs here was already reported in earlier studies [47,48]. The clones from polluted soils in this study were affiliated with several clusters in the genus *Nitrosospira* as identified by Mertens et al. [24]. In the present work, sequences identified as belonging Cluster 4 were obtained only from the unpolluted soil at the DBS site, suggesting that multiple metal pollution of Cd, Pb, Cu and Zn had a strong impact on AOB community composition. However, it was unclear whether the metal tolerant populations in these heavy metal

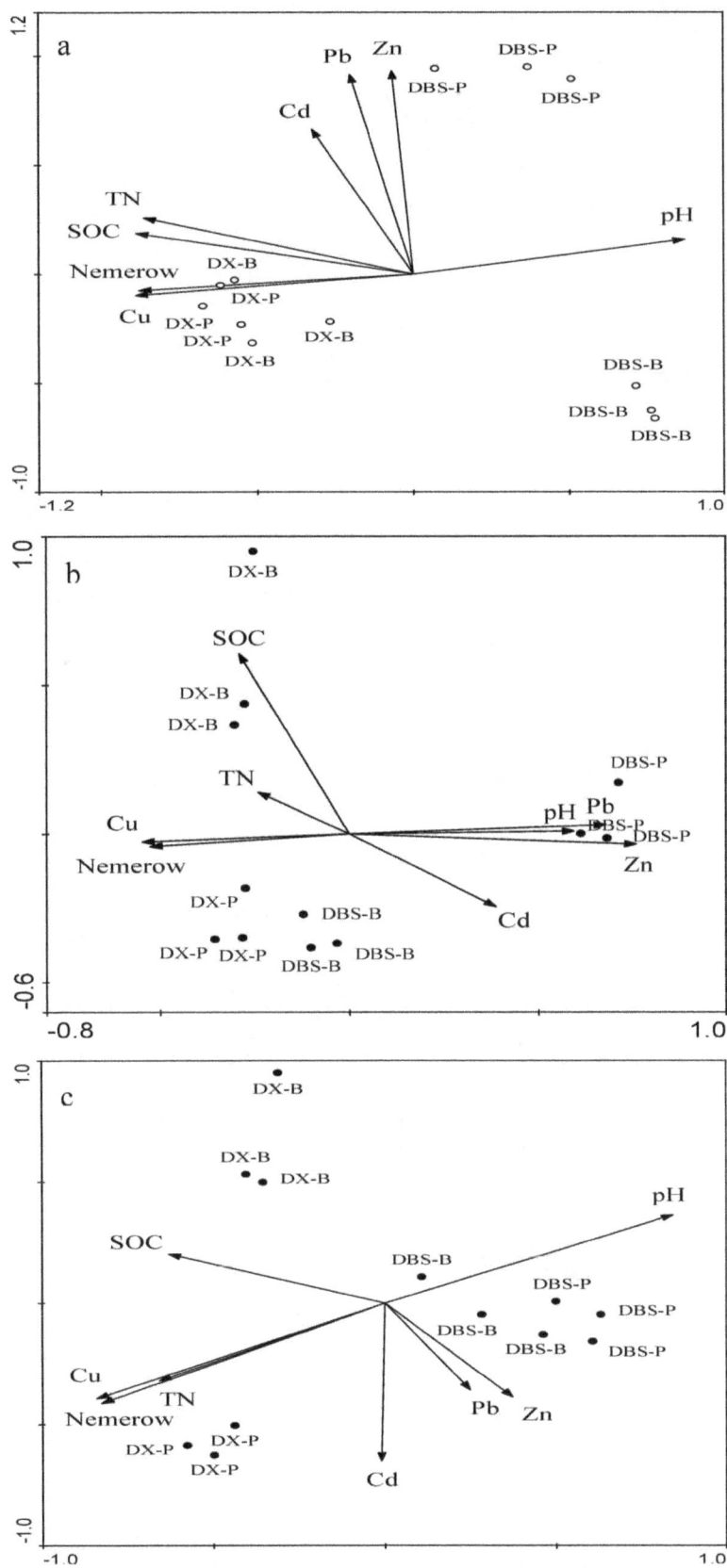

Figure 5. Redundancy analysis ordination plots of relationships between DGGE patterns of AOB (a), AOA (b) and *nirK* (c) gene and soil properties from the soil samples at the two sites. Soil properties included nemerow toxicity index, soil pH, SOC, TN, total concentration of Cd, Pb, Cu and Zn. DX-B and DBS-B, background soil from site DX and DBS; DX-P and DBS-P, polluted soil from site DX and DBS.

polluted soils were intrinsically tolerant to pollution or whether tolerance had been conferred by horizontal gene transfer. Some clones that were common to background soil were phylogenetically close to those that occurred in the polluted soil. As with the AOB community, the mechanistic details of AOA adaptations to heavy metal contamination remains unclear due to the limited ability to cultivate AOA species. Gene encoding AOA were also carried by members of the archaea that had been detected by molecular phylogenetic approaches, but that had not yet been cultured [49,50].

The communities represented by *nirK* were distinguishable between BGS and PS soils at DX site, but not at DBS site. Here, phylogenetic analysis showed that the clones from the polluted and background soils at a single site could be sorted into different groups. Furthermore, the clones in the polluted soils were widespread among samples from both sites. A study by Prasad et al. [51] showed that heavy metal tolerance varied widely among bacterial genera. These include several lineages belonging to the *Proteobacteria* that appeared to be highly tolerant to metal contaminated environments [52,53].

In addition to the influence of metal pollution, ammonia oxidizer and denitrifier community compositions were also affected by soil pH, SOC and TN. Soil organic carbon content is increased in polluted soils probably owing to inhibited decomposition rate [54]. On the other hand, soil pH had been found to affect microbial community structure directly by altering functional microbial groups [55], and indirectly by altering soil factors such as C and N substrate [56] and the availability of metals [57]. In this study, soil pH was lower but N higher in PS than in BGS at both sites, of course, both N and soil pH could be directly or indirectly affected by metal pollution [23]. There should be some interactions of soil chemical condition with metal pollution, for pH and organic matter could alter metal availability in soils to microorganisms.

Relationship between community changes and potential nitrifying and denitrifying activities

Nitrification had been known sensitive to heavy metal pollution [58–60], with heavy metal addition often decreasing soil nitrification rates [26,61]. Here, we observed that nitrifying activity was inhibited significantly in PSs compared to BGSs at both sites, which could be related to an altered soil microbial community composition with more resistant species. This could be partly explained by the remarkable decrease in both AOB and AOA populations with the metal pollution at both sites. AOB community composition was altered but its abundance unchanged in a long term Zn-polluted grassland, leading to an unchanged potential nitrification activity between polluted and unpolluted soils [62]. In contrast, a great decrease (up to 50%) in potential nitrification rates was found in copper spiked soils, with a simultaneous decrease in abundances of AOA and AOB plus a shift in AOB community structure and changes in the composition of the AOA community [63]. Of course, these community compositions could also subject to changes in soil pH and N level in soils under metal pollution [64], which could also be influenced with metal accumulation, as mentioned above. Nevertheless, metal impact on nitrification potential via metal toxicity or via alteration in abiotic conditions associated with C and N availability could be still a question.

Unlike nitrification activity, the potential denitrification activity was unchanged in the metal polluted soils at the DBS site with a soil pH of 5.5 but reduced significantly at DX site with Cu level up to 1333 mg kg^{-1} t and a low pH of 4.1. Apparently, this change was not related to difference in SOC and total N between the two

sites. This seemed contrast to the reports that carbon additions lead to an increase in denitrifying activity but do not change the composition of the *nirK* community [65,66]. An inhibition of denitrification activity had been widely recognized under highly elevated Cu level in soil though soil pH could be an important factor for denitrifying activity. A decrease in denitrification activity in soils spiked or polluted with metals including high Cu was reported for wetland soil [27] and for pasture soil [67]. Attard et al. [68] argued that a change in denitrifying activity could be associated with a change in denitrifier abundance but not with a change in denitrifier community structure under land use changes. Here, a several fold decrease in denitrification activity was seen in DX but not in DBS, which corresponded with a shift in the composition of the *nirK* community in PS from DX but not from DBS (section 4.2). Notably, changes in potential denitrification activity could not be predicted from changes in the abundance of functional gene targeted with the *nirK* primers. As shown by Heylen et al. [69] in a culture-independent study through the *nir* gene sequence analysis of cultivated denitrifier, functional *nir* gene diversity did not match well the denitrifier diversity. There were likely uncertainties regarding the *nirK* gene data. Nitrite reductase was considered the key enzyme in denitrification, containing either cytochrome cd1 encoded gene (*nirS* denitrifier) or copper encoded gene (*nirK* denitrifier), could catalyze the reduction of NO_2^- to nitric oxide (NO). The *nirS* denitrifier appeared more abundant than *nirK* denitrifier, however, the latter could be more sensitive to soil environmental changes [70]. Recently, targeting both *nirK* and *nirS* genes in forest, grassland and agriculture systems had been proposed as an assay to elucidate the abundance and community structure of soil denitrifier [68,70,71]. However, it is still a question if this assay could better track the changes in denitrification activity with changes in denitrifying bacterial abundance in the polluted soils. Meanwhile, gene transcript numbers would be also a potential option to better predict the functional groups responsible for denitrification in these soils since they could reflect the active populations of the community.

Conclusion

Significant changes both in the activities and community structure of ammonia oxidizers and denitrifier existed with metal pollution in an interaction with soil abiotic factors in rice paddies. A consistent decrease in the AOB abundance and nitrifying activity in polluted soil was observed in two sites studied. However, a sharp decrease in AOA abundance and denitrifying activity were seen only in highly Cu-polluted soil though lower pH and higher N was seen in polluted soil compared to the background soil in both sites. By using molecular techniques employing DGGE, we observed a shift in the community structure of AOA, and to a lesser extent, of AOB and denitrifier populations that were associated with different metal composition of the polluted soils. The pollution effects on microbial abundance differed between populations of *amoA* and *nirK* genes in a single site but these changes were not seen correlated to changes in nitrification or denitrification activities. This could suggest either a possible non-specific target of the primers conventionally used in soil study or complex interactions between soil properties and metal contents on the observed community and activity changes. This study suggested that metal pollution could exert impacts on soil microbial communities responsible for N transformation and thus on potential N_2O production in rice paddies though the impacts on different communities could vary with metal composition and the associated changes in soil pH and N availability. However, future works would be required either with new molecular assays and/or on microbial responses to multiple metals under contrasting soil conditions in polluted agricultural soils.

Supporting Information

Figure S1   Principal component analysis (PCA) of DGGE profiles of AOB (a), AOA (b) and *nirK* (c) gene fragment from the soil samples at the two sites. DX-B and DBS-B, background soil from site DX and DBS; DX-P and DBS-P, polluted soil from site DX and DBS. Similar symbols with same color in PCA plot indicate the replicate samples.

Figure S2   DGGE profiles of AOB (A), AOA (B) and *nirK* (C) gene fragment from the soil samples at the two sites. M: 100 bp Marker. DX-B and DBS-B, background soil from site DX and DBS; DX-P and DBS-P, polluted soil from site DX and

DBS. Arrows indicate the excised bands (B17–B27 and K14–K22) for sequencing.

Table S1   Primer sets and thermal profiles used for the absolute quantification of functional target genes.

Acknowledgments

The authors would like to thank the academic editor and the two anonymous reviewers for their helpful comments and suggestions.

Author Contributions

Conceived and designed the experiments: GXP YZL. Performed the experiments: YZL YL TZ. Analyzed the data: YL YZL GXP LQL JWZ XHZ JFZ. Contributed reagents/materials/analysis tools: YJD JFW XYY. Wrote the paper: YL GXP DC.

References

1. Kahrl F, Li Y, Su Y, Tennigkeit T, Wilkes A, et al. (2010) Greenhouse gas emissions from nitrogen fertilizer use in China. Environ Sci Policy 13: 688–694.
2. Luo J, De Klein C, Ledgard S, Saggar S (2010) Management options to reduce nitrous oxide emissions from intensively grazed pastures: a review. Agric Ecosyst Environ 136: 282–291.
3. Ju X, Xing G, Chen X, Zhang S, Zhang L, et al. (2009) Reducing environmental risk by improving N management in intensive Chinese agricultural systems. Proc Natl Acad Sci 106: 3041–3046.
4. Giller KE, Witter E, Mcgrath SP (1998) Toxicity of heavy metals to microorganisms and microbial processes in agricultural soils: a review. Soil Biol Biochem 30: 1389–1414.
5. Faulwetter JL, Gagnon V, Sundberg C, Chazarenc F, Burr MD, et al. (2009) Microbial processes influencing performance of treatment wetlands: a review. Ecol Engineering 35: 987–1004.
6. Wang Y, Shi J, Wang H, Lin Q, Chen X, et al. (2007) The influence of soil heavy metals pollution on soil microbial biomass, enzyme activity, and community composition near a copper smelter. Ecotoxicol Environ Saf 67: 75–81.
7. IPCC (2007) Climate change 2007: the physical science basis: Working group I contribution to the fourth assessment report of the IPCC. UK: Cambridge University Press.
8. Zou J, Huang Y, Jiang J, Zheng X, Sass RL (2005) A 3-year field measurement of methane and nitrous oxide emissions from rice paddies in China: Effects of water regime, crop residue, and fertilizer application. Global Biogeochemical Cycles 19: 10.1029.
9. Zou J, Huang Y, Zheng X, Wang Y (2007) Quantifying direct N2O emissions in paddy fields during rice growing season in mainland China: Dependence on water regime. Atmos Environ 41: 8030–8042.
10. Liu S, Qin Y, Zou J, Liu Q (2010) Effects of water regime during rice-growing season on annual direct N2O emission in a paddy rice-winter wheat rotation system in southeast China. Sci Total Environ 408: 906–913.
11. Jackson LE, Burger M, Cavagnaro TR (2008) Roots, nitrogen transformations, and ecosystem services. Plant Biol 59: 341.
12. Malchair S, De Boeck H, Lemmens C, Ceulemans R, Merckx R, et al. (2010) Diversity-function relationship of ammonia-oxidizing bacteria in soils among functional groups of grassland species under climate warming. Appl Soil Ecol 44: 15–23.
13. Nicol GW, Schleper C (2006) Ammonia-oxidising Crenarchaeota: important players in the nitrogen cycle? TRENDS in Microbiology 14: 207–212.
14. Wuchter C, Abbas B, Coolen MJ, Herfort L, van Bleijswijk J, et al. (2006) Archaeal nitrification in the ocean. Proc Natl Acad Sci 103: 12317–12322.
15. Philippot L, Hallin S, Schloter M (2007) Ecology of denitrifying prokaryotes in agricultural soil. Advan in Agrono 96: 249–305.
16. Frolking S, Qiu J, Boles S, Xiao X, Liu J, et al. (2002) Combining remote sensing and ground census data to develop new maps of the distribution of rice agriculture in China. Global Biogeochemical Cycles 16: 1091.
17. Wu X, Pan G, Li L (2006) Study on soil quality change in the Yangtze River Delta. Geogra Geo-Infor Sci 22: 88–91.
18. Hang X, Wang H, Zhou J, Ma C, Du C, et al. (2009) Risk assessment of potentially toxic element pollution in soils and rice (Oryza sativa) in a typical area of the Yangtze River Delta. Environ Pollut 157: 2542–2549.
19. Ma J, Pa G, Wang H, Xia Y (2004) Investigation on heavy metal pollution in a typical area of the Pearl River Delta. Chin J Soil Science 35: 636–638.
20. Xu C, Xia B, Qin J, He S, Li H, et al. (2007) Analysis and evaluation on heavy metal contamination in paddy soils in the lower stream of Dabaoshan Area, Guangdong Province. J Agro-Environ Sci 26: 549–553.
21. Arao T, Ae N (2003) Genotypic variations in cadmium levels of rice grain. Soil Sci and Plant Nutr 49: 473–479.
22. Liu J, Zhu Q, Zhang Z, Xu J, Yang J, et al. (2005) Variations in cadmium accumulation among rice cultivars and types and the selection of cultivars for reducing cadmium in the diet. J Sci Food Agric 85: 147–153.
23. Liu Y, Zhou T, Crowley D, Li L, Liu D, et al. (2012) Decline in topsoil microbial quotient, fungal abundance and C utilization efficiency of rice paddies under heavy metal pollution across South China. PloS One 7: 10.1371.
24. Mertens J, Broos K, Wakelin SA, Kowalchuk GA, Springael D, et al. (2009) Bacteria, not archaea, restore nitrification in a zinc-contaminated soil. The ISME journal 3: 916–923.
25. Ruyters S, Mertens J, Springael D, Smolders E (2010) Stimulated activity of the soil nitrifying community accelerates community adaptation to Zn stress. Soil Biol Biochem 42: 766–772.
26. Smolders E, Brans K, Coppens F, Merckx R (2001) Potential nitrification rate as a tool for screening toxicity in metal-contaminated soils. Environ Toxicol Chem 20: 2469–2474.
27. Sakadevan K, Zheng H, Bavor H (1999) Impact of heavy metals on denitrification in surface wetland sediments receiving wastewater. Water Sci Technol 40: 349–355.
28. Li X, Zhu Y, Cavagnaro TR, Chen M, Sun J, et al. (2009) Do ammonia-oxidizing archaea respond to soil Cu contamination similarly asammonia-oxidizing bacteria? Plant and soil 324: 209–217.
29. Liu Y, Zheng Y, Shen J, Zhang L, He J (2010) Effects of mercury on the activity and community composition of soil ammonia oxidizers. Environ Sci Pollut Res Int 17: 1237–1244.
30. Ruyters S, Mertens J, T'Seyen I, Springael D, Smolders E (2010) Dynamics of the nitrous oxide reducing community during adaptation to Zn stress in soil. Soil Biol Biochem 42: 1581–1587.
31. Magalhães CM, Machado A, Matos P, Bordalo AA (2011) Impact of copper on the diversity, abundance and transcription of nitrite and nitrous oxide reductase genes in an urban European estuary. FEMS Microbiol Ecol 77: 274–284.
32. Lu R (2000) Methods of Soil and Agro-chemical Aanalysis. Beijing (in Chinese): China Agricultural Science and Technology Press.
33. Nemerow NL (1991) Stream, lake, estuary, and ocean pollution. New York, USA. 0–472 p.
34. McTavish H, Fuchs J, Hooper A (1993) Sequence of the gene coding for ammonia monooxygenase in Nitrosomonas europaea. J Bacteriol 175: 2436–2444.
35. Francis CA, Roberts KJ, Beman JM, Santoro AE, Oakley BB (2005) Ubiquity and diversity of ammonia-oxidizing archaea in water columns and sediments of the ocean. Proc Natl Acad Sci USA 102: 14683–14688.
36. Henry S, Brue D, Stress B, Hallet S, Philippot L (2006) Quantitative detection of the nosZ gene, encoding nitrous oxide reductase, and comparison of the abundances of 16S rRNA, narG, nirK, and nosZ genes in soils. Applied and Environmental Microbiology 72(8): 5181–5189.
37. Muyzer G, Smalla K (1998) Application of denaturing gradient gel electrophoresis (DGGE) and temperature gradient gel electrophoresis (TGGE) in microbial ecology. Antonie van Leeuwenhoek 73: 127–141.
38. Sanguinetti C, Dias Neto E, Simpson A (1994) Rapid silver staining and recovery of PCR products separated on polyacrylamide gels. Biotechniques 17: 914.
39. Schmidt EL, Belser L (1994) Autotrophic nitrifying bacteria. Methods of Soil Analysis: Part 2-Microbiological and Biochemical Properties. 159–177 p.
40. Tiedje JM, Simkins S, Groffman PM (1989) Perspectives on measurement of denitrification in the field including recommended protocols for acetylene based methods: Springer. 217–240 p.
41. Dandie CE, Wertz S, Leclair CL, Goyer C, Burton DL, et al. (2011) Abundance, diversity and functional gene expression of denitrifier communities in adjacent riparian and agricultural zones. FEMS Microbiol Ecol 77: 69–82.

42. Holtan-Hartwig L, Bechmann M, Risnes Høyås T, Linjordet R, Reier Bakken L (2002) Heavy metals tolerance of soil denitrifying communities: N$_2$O dynamics. Soil Biol Biochem 34: 1181–1190.

43. Fait G, Broos K, Zrna S, Lombi E, Hamon R (2006) Tolerance of nitrifying bacteria to copper and nickel. Environ Toxicol Chem 25: 2000–2005.

44. van Beelen P, Wouterse M, Posthuma L, Rutgers M (2004) Location-specific ecotoxicological risk assessment of metal-polluted soils. Environ Toxicol Chem 23: 2769–2779.

45. Hai B, Diallo NH, Sall S, Haesler F, Schauss K, et al. (2009) Quantification of key genes steering the microbial nitrogen cycle in the rhizosphere of sorghum cultivars in tropical agroecosystems. Appl Environ Microbiol 75: 4993–5000.

46. Zhou T, Pan G, Li L Q, Zhang X H, Zheng J W, et al. (2014) Changes in greenhouse gas evolution in heavy metal polluted paddy soils with rice straw return: A laboratory incubation study. Eur J Soil Biol 63:1–6.

47. Stephen JR, Chang Y, Macnaughton SJ, Kowalchuk GA, Leung KT, et al. (1999) Effect of toxic metals on indigenous soil β-subgroup proteobacterium ammonia oxidizer community structure and protection against toxicity by inoculated metal-resistant bacteria. Appl Environ Microbiol 65: 95–101.

48. Koops H, Purkhold U, Pommerening-Röser A, Timmermann G, Wagner M (2006) The lithoautotrophic ammonia-oxidizing bacteria. The prokaryotes 5: 778–811.

49. Könneke M, Bernhard AE, José R, Walker CB, Waterbury JB, et al. (2005) Isolation of an autotrophic ammonia-oxidizing marine archaeon. Nature 437: 543–546.

50. De La Torre JR, Walker CB, Ingalls AE, Könneke M, Stahl DA (2008) Cultivation of a thermophilic ammonia oxidizing archaeon synthesizing crenarchaeol. Environ Microbiol 10: 810–818.

51. Prasad MNV (2001) In Metals in the Environment: Analysis by Biodiversity. CRC Press (New York). 285–321 p.

52. Feris K, Ramsey P, Frazar C, Moore JN, Gannon JE, et al. (2003) Differences in hyporheic-zone microbial community structure along a heavy-metal contamination gradient. Appl Environ Microbiol 69: 5563–5573.

53. Gillan DC, Danis B, Pernet P, Joly G, Dubois P (2005) Structure of sediment-associated microbial communities along a heavy-metal contamination gradient in the marine environment. Appl Environ Microbiol 71: 679–690.

54. Tyler G (1976) Heavy metal pollution, phosphatase activity, and mineralization of organic phosphorus in forest soils. Soil Biol Biochem 8: 327–332.

55. Rousk J, Brookes PC, Bååth E (2009) Contrasting soil pH effects on fungal and bacterial growth suggests functional redundancy in carbon mineralisation. Appl Environ Microbiol 75: 1589–1596.

56. Kemmitt SJ, Wright D, Goulding KWT, Jones DL (2006) pH regulation of carbon and nitrogen dynamics in two agricultural soils. Soil Biol Biochem 38: 898–911.

57. Flis SE, Glenn AR, Dilworth MJ (1993) Interaction between aluminium and root nodule bacteria. Soil Biol Biochem 25: 403–417.

58. Premi P, Cornfield A (1969) Effects of addition of copper, manganese, zinc and chromium compounds on ammonification and nitrification during incubation of soil. Plant and soil 31: 345–352.

59. Doelman P, Haanstra L (1984) Short-term and long-term effects of cadmium, chromium, copper, nickel, lead and zinc on soil microbial respiration in relation to abiotic soil factors. Plant and soil 79: 317–327.

60. Oorts K, Ghesquiere U, Swinnen K, Smolders E (2006) Soil properties affecting the toxicity of CuCl$_2$ and NiCl$_2$ for soil microbial processes in freshly spiked soils. Environ Toxicol Chem 25: 836–844.

61. Nies DH (1999) Microbial heavy-metal resistance. Appl Microbiol Biotechnol 51: 730–750.

62. Mertens J, Springael D, De Troyer I, Cheyns K, Wattiau P, et al. (2006) Long-term exposure to elevated zinc concentrations induced structural changes and zinc tolerance of the nitrifying community in soil. Environ Microbiol 8: 2170–2178.

63. Mertens J, Wakelin SA, Broos K, McLaughlin MJ, Smolders E (2010) Extent of copper tolerance and consequences for functional stability of the ammonia-oxidizing community in long-term copper-contaminated soils. Environ Toxicol Chem 29: 27–37.

64. Balabane M, Faivre D, van Oort F, Dahmani-Muller H (1999) Mutual effects of organic matter dynamics and heavy metals fate in a metallophyte grassland. Environ Pollut 105: 45–54.

65. Henry S, Texier S, Hallet S, Bru D, Dambreville C, et al. (2008) Disentangling the rhizosphere effect on nitrate reducers and denitrifiers: insight into the role of root exudates. Environ Microbiol 10: 3082–3092.

66. Miller M, Zebarth B, Dandie C, Burton D, Goyer C, et al. (2008) Crop residue influence on denitrification, N$_2$O emissions and denitrifier community abundance in soil. Soil Biol Biochem 40: 2553–2562.

67. Bardgett R, Speir T, Ross D, Yeates G, Kettles H (1994) Impact of pasture contamination by copper, chromium, and arsenic timber preservative on soil microbial properties and nematodes. Biol Fertil Soils 18: 71–79.

68. Attard E, Recous S, Chabbi A, De Berranger C, Guillaumaud N, et al. (2011) Soil environmental conditions rather than denitrifier abundance and diversity drive potential denitrification after changes in land uses. Global Change Biol 17: 1975–1989.

69. Heylen K, Gevers D, Vanparys B, Wittebolt L, Geets J, et al. (2006) The incidence of nirS and nirK and their genetic heterogeneity in cultivated denitrifiers. Environ Microbiol 8(11), 2012–2021.

70. Bárta J, Melichová T, Vaněk D, Picek T, Šantrůčková H (2010) Effect of pH and dissolved organic matter on the abundance of nirK and nirS denitrifiers in spruce forest soil. Biogeochemistry 101(1–3): 123–132.

71. Petersen DG, Blazewicz SJ, Firestone M, Herman DJ, Turetsky M, et al. (2012) Abundance of microbial genes associated with nitrogen cycling as indices of biogeochemical process rates across a vegetation gradient in Alaska. Environ Microbiol 14(4): 993–1008.

4

Influence of Soil Type, Cultivar and *Verticillium dahliae* on the Structure of the Root and Rhizosphere Soil Fungal Microbiome of Strawberry

Srivathsa Nallanchakravarthula*, Shahid Mahmood, Sadhna Alström, Roger D. Finlay

Uppsala BioCenter, Department of Forest Mycology and Plant Pathology, Swedish University of Agricultural Sciences, Uppsala, Sweden

Abstract

Sustainable management of crop productivity and health necessitates improved understanding of the ways in which rhizosphere microbial populations interact with each other, with plant roots and their abiotic environment. In this study we examined the effects of different soils and cultivars, and the presence of a soil-borne fungal pathogen, *Verticillium dahliae*, on the fungal microbiome of the rhizosphere soil and roots of strawberry plants, using high-throughput pyrosequencing. Fungal communities of the roots of two cultivars, Honeoye and Florence, were statistically distinct from those in the rhizosphere soil of the same plants, with little overlap. Roots of plants growing in two contrasting field soils had high relative abundance of *Leptodontidium* sp. C2 BESC 319 g whereas rhizosphere soil was characterised by high relative abundance of *Trichosporon dulcitum* or *Cryptococcus terreus*, depending upon the soil type. Differences between different cultivars were not as clear. Inoculation with the pathogen *V. dahliae* had a significant influence on community structure, generally decreasing the number of rhizosphere soil- and root-inhabiting fungi. *Leptodontidium* sp. C2 BESC 319 g was the dominant fungus responding positively to inoculation with *V. dahliae*. The results suggest that 1) plant roots select microorganisms from the wider rhizosphere pool, 2) that both rhizosphere soil and root inhabiting fungal communities are influenced by *V. dahliae* and 3) that soil type has a stronger influence on both of these communities than cultivar.

Editor: Kevin McCluskey, Kansas State University, United States of America

Funding: This work was funded by Grant No. 2007-1689 to RF from The Swedish Research Council for Environment, Agricultural Sciences and Spatial Planning. The funders had no role in study design, data collection and analysis, decision to publish, or preparation of the manuscript.

Competing Interests: The authors have declared that no competing interests exist.

* Email: srivathsa.nallan@gmail.com

Introduction

Plants roots influence the structure and diversity of soil microbial communities through their energy-rich exudates [1,2], creating diverse rhizosphere communities in close proximity to the roots [3]. The roots are colonized by a subset of these microorganisms, creating both a rhizoplane community on the surface and an endophytic compartment. Discrimination between pathogens and beneficial mutualists or commensals is controlled by an immune system [4], but the mechanisms underlying the selection of different core microbiomes [5] by different plant species are still poorly understood. Recent studies of bacteria associated with the roots or rhizosphere of *Arabidopsis thaliana* grown in geochemically distinct soils [6] have shown that they are both strongly influenced by soil type but that the root endophytic compartments from two soils contain overlapping, low complexity communities that are enriched in Actinobacteria and specific families belonging to the Proteobacteria. Differential relative abundance of *Actinocorallia* sp. has also been shown in different ecotypes of *A. thaliana* [7].

Plant host genetics plays a significant role in the selection of microbial communities associated with roots [8–10]. This selection has important implications for plant health and growth since rhizosphere microorganisms are known to have a variety of positive or negative effects. These include plant growth stimulation, antagonism against pathogens and effects on carbon sequestration and phytoremediation. Rhizosphere and root endophyte microorganisms are increasingly being used on a commercial scale for plant protection as biocontrol agents and biofertilizers [11–13]. Management of such microorganisms requires a thorough understanding of their interaction with other microorganisms sharing the same ecological niche. Some attempts have been made to study the effect of biocontrol agents on microbial communities [14–17], however studies focusing on the influence of soil-borne fungal plant pathogens on microbial community structures are less frequent [18–21]. Most of the inoculation studies with pathogens or biocontrol agents have been more focused on bacterial communities than on fungal communities.

The rhizosphere is a dynamic region for many biological processes and interactions [1,11,12,22–25] and empirically-based management strategies for sustainable cultivation need to be based on a more rigorous understanding of the functional basis of the mechanisms underlying the various biotic interactions taking place [26].

Microbial community structure in the rhizosphere and the root endophytic compartment is shaped by soil characteristics as well as by plant genotype [6,7,20,27–31]. Soils in organically managed strawberry fields have been shown to have greater functional activity and higher functional gene diversity than soils in conventionally managed fields [32]. The strawberry rhizosphere has been shown to harbour distinct microbial communities containing *Streptomyces*, *Rhizobium* and *Nocardia* [28,31] and in the presence of *Verticillium dahliae,* a soil-borne fungal pathogen, it is dominated by *Pseudomonas* spp. populations [27]. Studies have also shown that strawberry cultivars respond differently to the inoculation of arbuscular mycorrhizal fungi (AMF) [33,34]. Sporulation of *Phytophthora fragariae* associated with strawberry plants colonized by AMF has also been shown to be reduced in comparison to that in non-AMF colonized plants [35].

Strawberry (*Fragaria ×ananassa*) is grown for its berries in many countries and different cultivars have been developed for commercial production. Rhizosphere bacterial biomass and activity, fruit quality and yield have all been shown to vary under different management systems. Soil microorganisms have been shown to be indirectly involved in determining the fruit quality [32,36]. *Verticillium dahliae* is one of several plant pathogens that affect plant growth and yield in different plant species, including strawberry [37–41]. The pathogen forms conidia and microsclerotia that germinate in the presence of the root exudates, whereupon germ tubes enter the plant roots. The pathogen can persist as microsclerotia on dead plant tissues, or in the soil for more than 10 years [39–41]. In strawberry as few as two microsclerotia per gram of soil can cause 100% wilt [42] and this pathogen is difficult to combat with current management strategies. In a field trial, biocontrol using *Trichoderma harzianum* and *Trichoderma viride* against *V. dahliae* reduced wilting in strawberry by 60%. An indirect effect against *V. dahliae* through induced systemic resistance by *Paenibacillus alvei* K165 has also been observed in *Arabidopsis* sp. [43,44].

Studies have shown that the presence of fungal (*Phytophthora cinnamomi*) and bacterial pathogens, (*Liberibacter asiaticus* and *Erwinia carotovora* (*Pectobacterium carotovorum*)) increases the diversity of Bacteroidetes and Alphaproteobacteria in particular [18,20,21]. Such changes also facilitate the selection, by the plant, of taxa including those with antimicrobial properties [27,45].

Plant resistance is one of the effective strategies in plant protection and attempts have been made to study effects of resistant and susceptible cultivars on microbial communities [46–48]. The results of these studies demonstrate that plant genotype can have a significant impact on soil microbial community structure, and differences in the rhizosphere microbial community have been suggested to contribute to the differences in resistance to disease. Whether this impact is independent of the soil type is not evident from these studies.

Many earlier studies involving soil microbial community analyses have been conducted using either cultivation-dependent approaches or molecular methods such as T-RFLP, DGGE, cloning and Sanger sequencing which are now proven to underestimate the total microbial diversity [49,50]. More recent studies using high throughput techniques have revealed taxa that were not detected with previous methods [51].

In this study we investigated the impact of plant genotype on the fungal community structure in different soil types in the presence and absence of the soil-borne fungal pathogen, *V. dahliae*. We used 454 pyrosequencing to analyze the rhizosphere soil- and root fungal communities in different strawberry cultivars, grown in differently managed soils. We hypothesized that the structure of microbial communities in the rhizosphere soil and roots is determined by both the cultivar and soil type, and may also be re-structured in the presence of the pathogen.

Materials and Methods

Experimental soils and plant material

Four cultivars of strawberry, Florence, Honeoye, Senga Sengana, and Zephyr, and two of the three different soils used in this study were the same as those described in a previous study by Santos-González et al. [30]. The cultivars were selected on the basis of their breeding year, yield level and disease tolerance. Honeoye and Florence were bred after 1978, Zephyr and Senga Sengana before 1966. Honeoye is considered to be a high yielding cultivar and Florence and Senga Sengana are considered to be tolerant to *V. dahliae* [52–54]. In our figures the conventionally managed soil from Kristianstad is referred to as "conventional" and the organically managed soil from Hörby as "organic" [30]. The field in Hörby was managed organically with appropriate crop rotations since 1983 and pre-cropped with potatoes prior to sampling in 2009. The experimental field in Kristianstad is conventionally managed and was planted with strawberry in 2008. The two agricultural fields are located 38 km apart in Southern Sweden. These sites are on private land and permission to collect soil was obtained from the owners Sune Abrahamsson (Kristianstad) and Ingvar Åkesson (Hörby). The soil collection did not involve endangered or protected species. The Kristianstad site at (56° 06′ N, 14° 01′ E) and the Hörby site at (55° 50′ N, 13° 35′ E) have different soil properties. The physico-chemical traits of the two field soils are described by Santos-González et al. [30]. The soils had similar textural properties (8.5–11% clay, 29–35% silt, 51–58% sand) and bulk density (1.05–1.07 g cm^{-3}). However the pH, C, N and P levels of the Hörby soil were significantly (*P*< 0.05) higher than those of the Kristianstad soil (6.36 vs. 5.97; 2.1% vs. 1.61%; 0.17% vs. 0.14%; 181 mg kg^{-1} vs. 113 mg kg^{-1}). The third "soil" was a peat-based growth substrate (Hasselfors Garden Special, Hasselfors Garden AB, Örebro, Sweden) containing 60% light peat, 25% black peat, 15% sand (0.5–4 mm) and 0.050 kg m^{-3} multisport/FTE36, 2.0 kg m^{-3} dolomite and 4.5 kg m^{-3} limestone, amended with 1.3 kg m^{-3} fertilizer (N14:P7:K15, pH-6.0), and is used for commercial greenhouse cultivation in Sweden.

Pathogen inoculum

The pathogen isolate used was *V. dahliae* 12086, originating from strawberry roots and available from the Food and Environment Agency, York, UK (see http://www.q-bank.eu/). It was purified and grown in potato dextrose broth (PDB per L, 12 g Accumedia, Michigan, USA) as stationary cultures for 20 days at 24°C. The mycelial mat thus obtained was macerated carefully in diluted potato dextrose broth (1:10 water v/v). A preparatory study was conducted with the suspension thus obtained to confirm the pathogenicity of the test pathogen. On the basis of the results from the preparatory study (data not shown), 30 ml mycelial suspension was inoculated in *V. dahliae*-treatments. The plant roots were injured by inserting a 0.9 cm diameter cork borer into the root volume in three locations to stimulate the pathogen infection. No dead-cell, or chitin controls were used but the control plants without the pathogen were treated in a similar manner but with suspension medium not containing fungal mycelium.

Experimental design

An outdoor experiment was conducted by growing the four cultivars in the three soils as specified in a previous study [30]. In brief, the soils were prepared before planting with one plantlet per pot (13×13 cm×25 cm deep), placed outdoors and irrigated as needed. The plantlets were taken from storage under refrigeration and were about 5 cm tall with three to four leaves and roots about 15 cm long.

The cumulative fresh berry weight per plant was recorded during the experimental period of 14 weeks. The rhizosphere soil was collected from three replicates 12 weeks after planting and 14 weeks after planting. To sample the rhizosphere soil, the plants were carefully removed from the pots and roots with tightly adhering soil were suspended in phosphate-buffered saline solution (PBS, 0.14 M NaCl, 0.0027 M KCl, 0.010 M phosphate buffer, pH 7.4, Medicago, Sweden) for 30 minutes. The rhizosphere soil thus collected was centrifuged at 8000 RPM for 10 minutes at 4°C (Biofuge PrimoR, Heraeus Sorvall), the roots were once again washed carefully under aerated tap water. The soil pellet and the roots were stored at −20°C for further analyses.

Analysis of fungal communities

DNA extraction and selection of samples for pyrosequencing. DNA extraction methods differ in their efficiency of extraction from different taxa and for the rhizosphere samples, we used two different nucleic acid extraction methods and the resultant DNA extracts were checked individually for quality and reproducibility on a DGGE (denaturing gradient gel electrophoresis) gel following PCR amplifications. The first method was based on 'CTAB' (hexadecyltrimethylammonium bromide) [55] that involved extraction of nucleic acids from 0.5 g of soil. The soil samples were weighed in 2 ml 'Lysing matrix B' tubes (Fisher Scientific, USA) containing 0.5 ml CTAB buffer (120 mM and pH 8.0) and 0.5 ml phenol: chloroform: isoamylalcohol (25:24:1 V/V). The cells were lysed at 5000 RPM for 30 seconds in a bead beater (Precellys 24; Bertin Technologies, France). Proteins and solvent residues were removed using chloroform and isoamyl alcohol (24:1) followed by further precipitation using polyethylene glycol (30%). The precipitated DNA was washed with ice-cold ethanol (70%), air dried and re-suspended in 30 μl of MilliQ water and stored at −20°C for further analysis.

The second nucleic acid extraction method was based on MOBIO 'RNA power soil' total RNA isolation and DNA elution accessory kits (MOBIO laboratories, California, USA) according to manufacturer's recommendations. The concentration and quality of extracted DNA were assessed using NanoDrop (ND-1000 Nanodrop technologies, USA). The banding profiles of these two methods differed so the DNA from these two extraction methods was pooled and used as a mixed template for generating PCR amplicons for pyrosequencing. Prior to pyrosequencing, the control samples were subjected to DGGE-based analysis following PCR amplification. The DGGE banding profiles showed higher numbers of bands at the first sampling, 12 weeks after planting, than at 14 weeks (data not shown). Pyrosequencing was therefore carried out on samples from 12 weeks.

For the analysis of root-associated fungi we chose to analyze two cultivars Honeoye, susceptible to *V. dahliae*, and Florence, tolerant to *V. dahliae*, growing in each of the two field soils. Nucleic acids from roots were extracted using DNeasy plant mini kit (Qiagen, Germany) according to the manufacturer's recommendations. Roots were carefully washed to remove the soil particles attached to the roots and soil removal was microscopically confirmed. They were gently washed three times with 0.1% Triton x100 for two minutes, thereafter washed with sterile milliQ water. All collected roots were subsequently freeze-dried for 48 h (CoolSafe, ScanLaf A/S, Denmark), homogenised by milling in 2 ml tubes containing 2.38 mm diameter stainless steel beads (MOBIO laboratories, California, USA) in a bead-beater for 30 s twice. 50 mg of the homogenised root material was used for nucleic acid extraction. The extracted DNA was quantified and

used as a template for generating PCR amplicons for pyrosequencing.

Pyrosequencing

Pyrosequencing analysis of the fungal communities was carried out as described by Ihrmark et al. [56] using primers fITS7 (5′GTGARTCATCGAATCTTG-3′) and ITS4 (5′TCCTCC-GCTTATTGATATGC-3′) [57]. In brief, each 50 μl reaction mixture consisted of 200 μM of dNTPs, 2.75 mM MgCl₂, primers at 500 nM and 300 nM, 0.025 U polymerase (DreamTaq Green, Thermo Scientific, Waltham, MA) in buffer. The thermocycling conditions were: 5 min at 94°C; 35 cycles (30 s at 94°C; 30 s at 57°C; 30 s at 72°C); 7 min at 72°C. Three technical replicates were run for each sample. To ascertain PCR amplification efficiency and to rule out any possible contamination, all reactions were run with positive and negative controls. The PCR amplification products were analyzed on 1% (w/v) agarose gels prestained with Nancy red-520 (Sigma Aldrich, USA). The triplicate PCR products of each sample were pooled, purified using Agencourt AMPure, PCR purification kit (Beckman Coulter, Massachusetts, USA) and quantified using a Qubit Fluorometer (Invitrogen, USA). Equimolar concentrations of samples were pooled and the resultant mixture was freeze-dried (CoolSafe, ScanLaf A/S, Denmark) overnight. Pyrosequencing was carried out on 2×1/8th of a GS FLX Titanium Pico Titer Plate (LGC Genomics, Germany and MACROGEN, South Korea) according to the manufacturer's instructions (Roche, Branford, CT).

Bioinformatic analysis

The SCATA pipeline (Sequence Clustering and Analysis of Tagged Amplicons, http://www.scata.mykopat.slu.se) was used to analyze the sequence data from 72 rhizosphere samples and 24 root samples. The quality control and data filtering involved removal of singletons, low quality sequences, tags and primer missing sequences, followed by removal of sequences shorter than 200 base pairs (after primer and tag trimming). We used default parameters of the SCATA pipeline except that the proportion of primer match was set at 0.9. Sequences were assembled into clusters by single-linkage clustering with a 98.5% sequence similarity threshold. The most common genotype/sequence of the cluster was used to represent the operational taxonomic units (OTUs). All the representative sequences of the clusters were identified using the NCBI (National Center for Biotechnology Information) database and the nucleotide blast 'BLASTn' program. The most common sequence from each cluster was analyzed manually for blasting. With the exception of rarefaction analysis, only sequences with ≥95% identity were chosen for these analyses.

The pyrosequencing data have been deposited at ENA (European Nucleotide Archive) under accession number PRJEB6260.

Statistical analyses

Statistical analyses were conducted using a significance level of $p \geq 0.05$. The Chao-1 index (PAST (PAleontological STatistics), version 2.17b) was calculated to obtain the number of OTUs present in the four cultivars grown in three different soils with and without *V. dahliae*. The bar diagrams were generated using Microsoft Excel and the significance of the OTU data was estimated using ANOVA with MINITAB. The relative abundance of OTUs was used for Non-metric multi-dimensional scaling (NMDS) ordination performed with 'PAST' using Bray-Curtis dissimilarity indices. Numbers of OTUs were calculated for selected phyla and plotted in Microsoft Excel and their significance was calculated by using ANOVA. SIMPER (similarity percentage) analysis (PAST ver 2.17b) was performed using relative abundance data in order to identify the OTUs that were contributing to the differences among treatments. The rarefaction was estimated using Analytic Rarefaction (Ver 1.3, UGA Stratigraphy Lab, USA).

Results and Discussion

Bioinformatic analysis of root and rhizosphere material associated with the cultivars Honeoye and Florence revealed a total of 577 unique OTUs following quality control filtering and removal of singletons. In total 360 OTUs were associated with rhizosphere soil and 307 OTUs were associated with roots, with an overlap of 90 OTUs between the two compartments. Analysis of the rhizosphere soil associated with roots of all four cultivars revealed 128 OTUs common to both the field soils but only 16 fungal OTUs that were common to all three soils. In conventionally managed soil 63 OTUs were common to all four cultivars, with 28 common OTUs in organically managed soil and 18 common OTUs in peat-based growth substrate.

Cluster analysis of the community structure of the 50 most abundant fungal taxa was performed. These taxa constituted 90% of the reads in roots and 70% of the reads in rhizosphere soil. The analysis (Figure 1) revealed a clear separation between fungal communities from the rhizosphere soil and those from the roots. Within each of these compartments there was a clear separation according to soil type (ANOSIM $p = 0.0001$). The root associated community structure of the organically managed soil showed a statistically significant (ANOSIM $p = 0.015$) separation between Verticillium inoculated and uninoculated treatments. Rhizosphere soil communities in a peat-based substrate formed a distinct cluster with a clear separation between V. dahliae inoculated and equivalent, non-inoculated treatments. In a similar analysis of rhizosphere soil communities associated with four different strawberry cultivars, Honeoye, Florence, Senga-Sengana and Zephyr (Figure S1) clear effects of soil type were visible. Differences due to cultivar were less evident. These results are consistent with the finding of Santos-González et al [30] that soil type, but not cultivar, shapes the assemblages of arbuscular mycorrhizal fungi associated with strawberry roots in these soils.

Additional NMDS analysis of the fungal community structure in the rhizosphere soil of all four cultivars (Figure S2) showed a statistically significant effect of Verticillium inoculation in conventionally managed soil (ANOSIM $p = 0.005$) and peat based substrate (ANOSIM $p = 0.0004$). No such clear separation was evident in organically managed soil.

Rank abundance plots of the 30 most abundant fungal taxa in each of the rhizosphere and root compartments in the absence of Verticillium (Figure 2) revealed that the structure of the fungal microbiome in these two compartments is markedly different, with little overlap. The dominant OTUs from the roots were found at low frequencies in the rhizosphere, while OTUs that were dominant in the rhizosphere were absent or occurred at very low frequencies in the roots, in both field soils (Figure 2a–d). These results are in agreement with other recent studies of bacterial microbiomes of Arabidopsis thaliana and the fungal microbiome of Pisum sativum [6,7,58] that also show distinct communities in the rhizosphere and root endophyte compartments. Within each soil this pattern was the same for both Honeoye and Florence, however there were large differences between the rank abundance distributions of taxa between the two soils. The same lack of overlap between root and rhizosphere soil fungal communities was found in the presence of Verticillium (data not shown).

The top seven most abundant taxa in conventionally managed rhizosphere soil, irrespective of strawberry cultivar were Cryptococcus terreus, Mortierella elongata, Cryptococcus terricola, Panaeolus foenisecii, Phoma eupyrena, Gibellulopsis nigrescens and an unidentified Leptosphaeria species, together accounting for 33% and 37% of the total reads for the Honeoye and Florence treatments respectively. In the organically managed rhizosphere soil the top six fungal taxa accounted for 50% of the reads but Trichosporon dulcitum was the dominant taxon, accounting for 35% and 30% of the total reads in the Honeoye and Florence treatments respectively. Trichosporon dulcitum was also detected in the conventionally managed soil, but at much lower frequencies (3% and 2%). Trichosporon dulcitum is a soil-borne basidiomycetous yeast. Utilization of benzene compounds [59] and biodegradation of phenol [60] have been shown in T. dulcitum, but the ability of this yeast to interact with plants and fungal pathogens is still unclear. Other members of the Trichosporonales have been found in the rhizosphere and ectomycorrhizosphere in soil from a Nothofagus pumilio forest [61]. In the latter study T. dulcitum was found in the bulk soil but not in the rhizosphere and no other information appears to be available concerning the possible associations of this species with plant roots.

The most common fungal taxa found in roots included Entrophospora sp shylm 131, Leptodontidium sp. C2 BESC 319 g, Ilyonectria macrodidyma, Fusarium spp., Phoma sp BEA-2010, Scytalidium lignicola, Acremonium strictum, Exophiala sp. O 1 LILB 26 5 g, and Paraconiothyrium sporulosum. The composition of fungal taxa in roots of plants growing in the organically managed soil was characterised by high abundance of Leptodontidium, whereas both Leptodontidium and Entrophospora were dominant in roots of plants growing in the conventionally managed soil. Differences in the abundance of fungal taxa in relation to soil types and strawberry cultivar are discussed below in a separate SIMPER analysis.

The total numbers of fungal OTUs in the absence of Verticillium inoculation, were generally highest in the rhizosphere of plants grown in conventionally managed soil, intermediate in organically managed soil and lowest in peat-based growth substrate, indicating a strong effect of soil type on fungal diversity. The lowest fungal diversity in the peat based growth substrate is most likely due to its simple composition [62]. In the presence of Verticillium no clear difference between the field soils was evident but numbers of OTUs were lower in the peat-based growth substrate (ANOVA $p \leq 0.05$) (Figure 3a).

In conventionally managed soil, Verticillium inoculation resulted in a consistent reduction (20–45%) in the number of fungal OTUs in the rhizosphere that was statistically significant ($p \leq 0.05$) in Senga Sengana and Zephyr (Figure 3a). In contrast, in the organically managed soil, inoculation with Verticillium resulted in significant ($p \leq 0.05$) increases (114% and 125%) in the total numbers of ascomycete and basidiomycete OTUs respectively in the rhizosphere of Florence (Figure 3a). The reasons for this increase are not clear but may reflect differences in the quality and quantity of root exudates of Florence, which is considered to be a Verticillium tolerant cultivar [63]. Basidiomycota were less abundant in the rhizosphere but also showed a consistent reduction in response to Verticillium in plants grown in conventionally managed soil. This was statistically significant only in Honeoye, but evident as a trend in the other cultivars.

Ascomycota dominated, both in roots and in rhizosphere (Figure 3a & b). Verticillium inoculation had no statistically significant effects on the number of Ascomycotan OTUs in roots of Honeoye and Florence grown in conventional soil but these were significantly reduced by 21% in Honeoye in response to

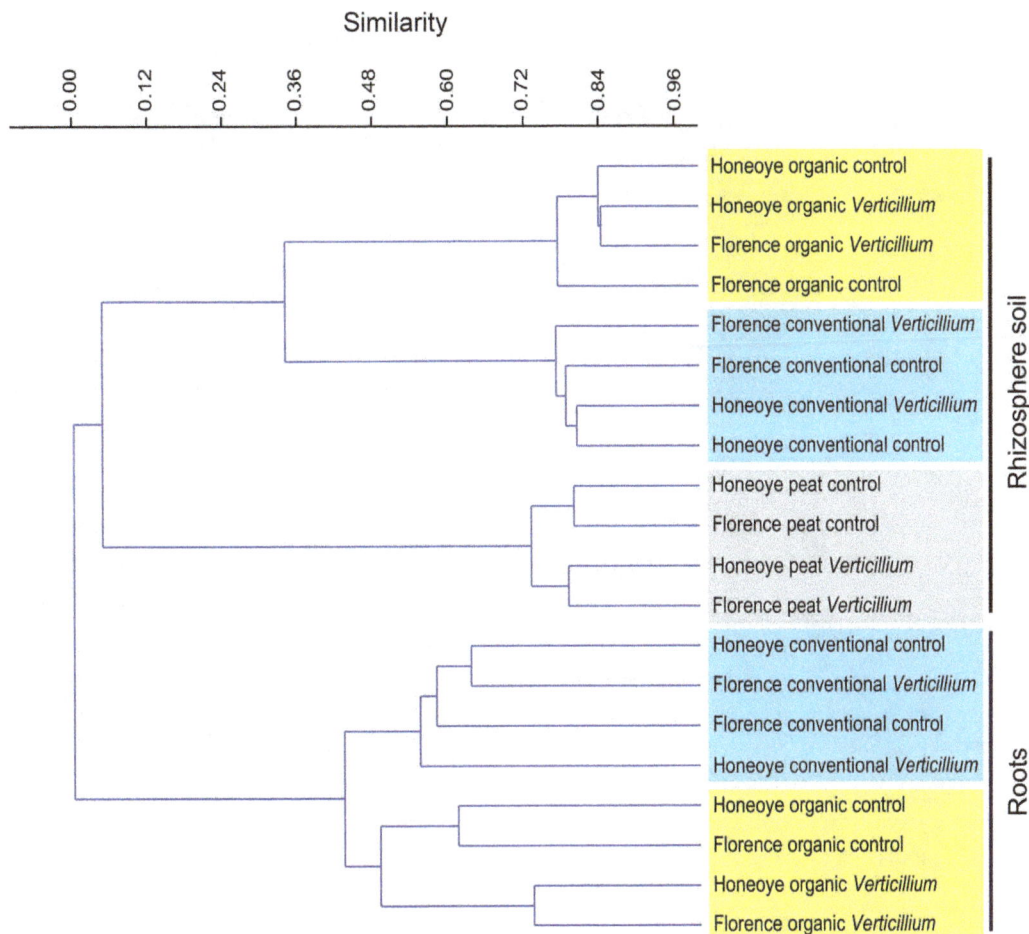

Figure 1. Cluster analysis of the 50 most abundant fungal operational taxonomic units (OTUs) in the rhizosphere and roots of two strawberry cultivars, Honeoye and Florence, grown in conventionally or organically managed soils or a peat based growth substrate, with and without *Verticillium dahliae*. These OTUs constituted 90% of the total reads in roots and 70% of the reads in rhizosphere soil. The clustering is based on paired group linkage using the Bray-Curtis similarity measure. Only OTUs with ≥95% identity are included.

Verticillium inoculation in organically managed soil. The mean number of Ascomycotan taxa in the roots of Florence was 38% lower in the presence of *Verticillium* but this reduction lacked statistical support.

Numbers of Basidiomycotan OTUs in roots were negligible, but 17 Glomeromycotan OTUs were found, including *Entrophospora* sp. shylm 131, *Rhizophagus intraradices* and *Rhizophagus clarus*. Numbers of Glomeromycotan OTUs in Honeoye roots in organically managed soil were significantly increased by 50% (Figure 3b) in response to *Verticillium* inoculation. Xu et al [58] have also shown that the relative abundance of different fungal phyla varies in root and rhizosphere samples of *Pisum sativum* depending upon whether the plant is healthy or infected with the pathogen *Aphanomyces euteiches*.

Rarefraction analysis of the numbers of OTUs associated with different numbers of sequences from root and rhizosphere samples are shown in Figure S3. The root treatment samples approached saturation, with asymptotes between 80 and 120 OTUs, although fewer root treatments were analysed in this study since only the roots of Honeoye and Florence plants were examined. The species accumulation curves for rhizosphere soil samples included three growth substrates and four cultivars and were not saturated, suggesting that not all taxa were detected. However there was a

high degree of consistency between the most abundant taxa in the rhizosphere for both soils (Figure 2), suggesting that the dominant taxa in these communities were adequately described.

SIMPER (Similarity Percentage) analysis of OTUs from rhizosphere and root samples was performed to identify the species primarily contributing to the difference in community structure between the two compartments and the results are summarized in Table S1.

In Florence the effect of *Verticillium* inoculation on fungal community structure was stronger in roots (71% in organically managed soil, 60% in conventionally managed soil) than in the rhizosphere, (46% in organically managed soil, 52% in conventionally managed soil). In Honeoye the contribution to *Verticillium* induced differences was lower but the values for roots (53% in organic soil and 54% in conventionally managed soil) were still higher than in the rhizosphere (41% in organic soil and 48% in conventionally managed soil).

The numbers of reads matching *Verticillium* genera were 564 in roots compared to only 35 in the rhizosphere soil suggesting that *V. dahliae* was more abundant in the roots than the rhizosphere. This is not surprising since *V. dahliae* is a root pathogen but the low number of reads in rhizosphere soil may reflect primer bias against *V. dahliae* or low template concentrations at the time of sampling.

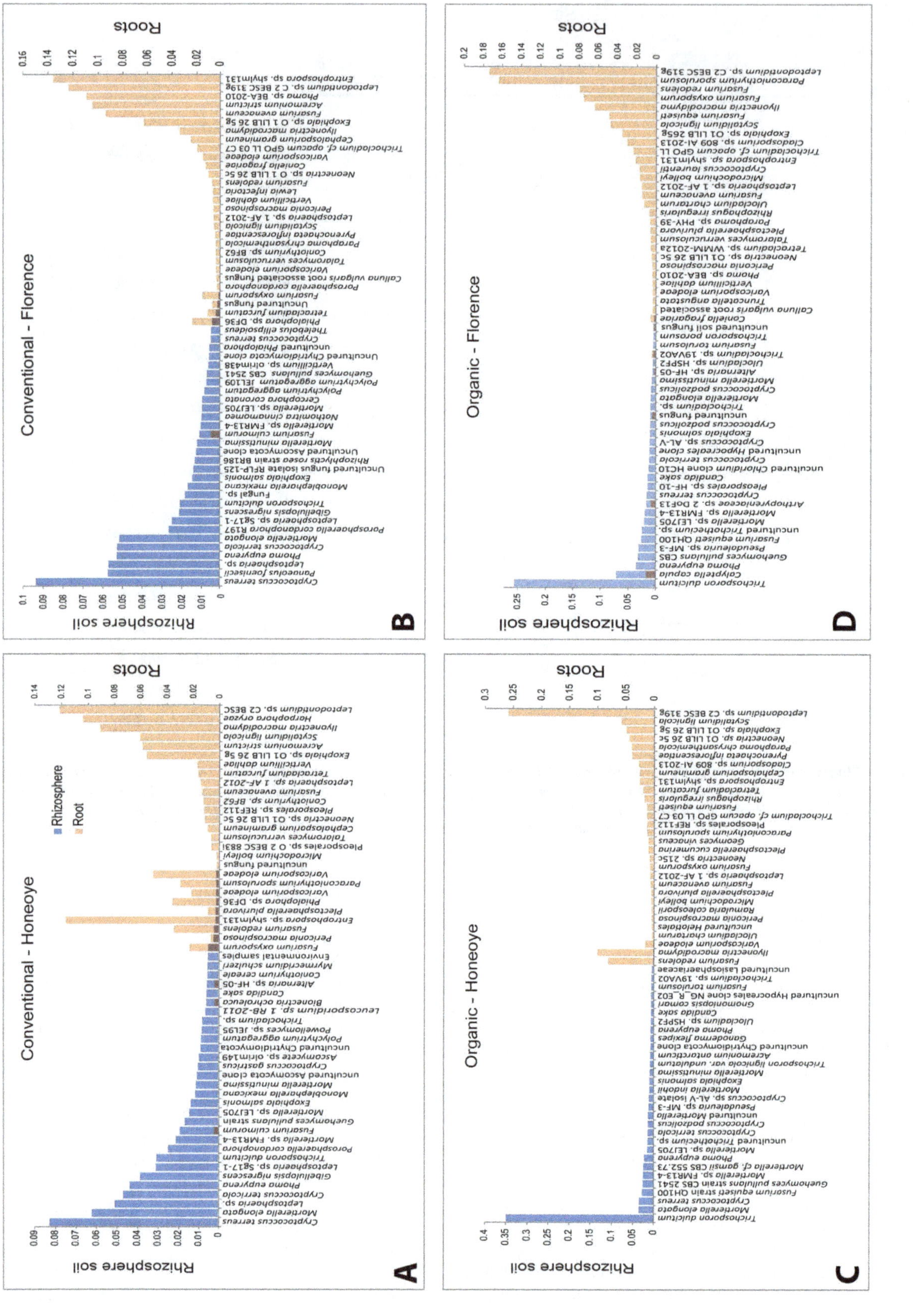

Figure 2. Ranked abundance of the 30 most abundant fungal operational taxonomic units (OTUs) present in the rhizosphere and roots of two strawberry cultivars, Honeoye and Florence, grown in conventionally (A and B) and organically (C and D) managed field soils. Only OTUs with ≥95% identity are included.

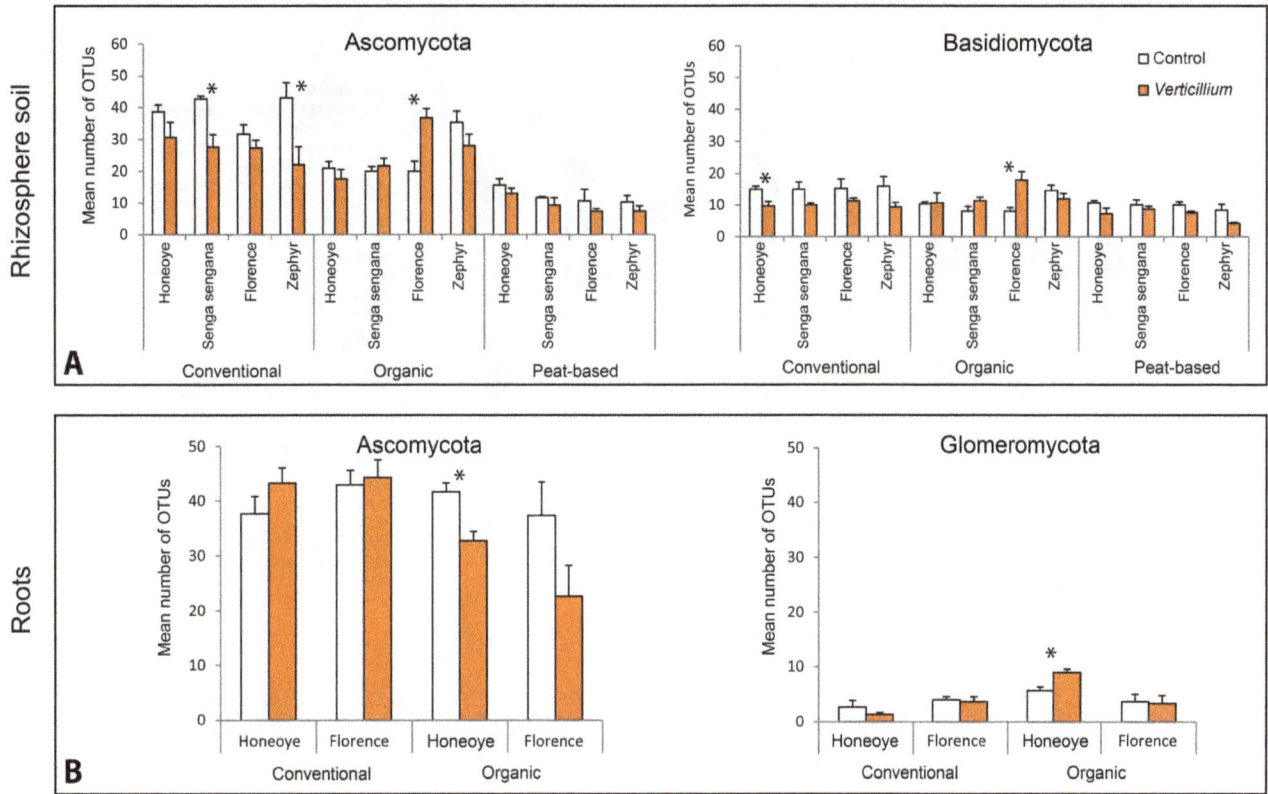

Figure 3. Distribution of fungal operational taxonomic units (OTUs) between different fungal phyla in the rhizosphere (A) and roots (B) of different strawberry cultivars grown in conventionally or organically managed soils or a peat-based growth substrate with and without *Verticillium dahliae*. Only OTUs with ≥95% identity are included. Asterisks indicate statistical significance between control and *Verticillium dahliae* treatments within each soil (*=$p \leq 0.05$, ANOVA). Vertical bars represent mean values and error bars show ±1 SE (n=3).

Van der Mark et al [64] reported detection of significantly more pathogen DNA of *Aphanomyces euteiches* and *Phytophthora medicaginis* in a susceptible variety of alfalfa than in the resistant variety [64]. In our study, in the conventionally managed soil, significantly higher numbers of reads (345) matching *Verticillium* were detected in *Verticillium* inoculated treatments of Honeoye, the susceptible cultivar, than in Florence, a tolerant cultivar (17 reads). In the organically managed soil the numbers of reads were lower than 17 in both cultivars, suggesting that this soil was possibly suppressive in character, while the conventionally managed soil was more conducive to *Verticillium*.

The top ten OTUs contributing to the differences between control and *Verticillium* treatments across all treatments, identified using SIMPER analysis, are shown in Figures 4a–b. The dominant OTUs in roots occur at low frequencies in the rhizosphere soil while OTUs that dominate in the rhizosphere soil are either absent or occur at low frequencies in the roots. This discrepancy between root and rhizosphere soil communities is in agreement with recent observations in other plants [6,7,58]. *Cryptococcus* spp., *Gibellulopsis* sp., *Mortierella* spp. and *Phoma* sp. were abundant in rhizosphere in the conventionally managed soil while *Trichosporon dulcitum* and *Fusarium equiseti* were abundant in rhizosphere of organically managed soil. In contrast, *Leptodontidium* sp. dominated the roots in organically managed soil, while *Entrophospora* sp. dominated the roots of plants in conventionally managed soil (Figure 4a, b).

Verticillium dahliae was among the top ten species contributing to differences in fungal community structure between inoculated and uninoculated roots of Honeoye plants growing in conventionally managed soil and had a higher relative abundance in inoculated treatments than in uninoculated treatments (1.7%, 4.6%). The relative abundance of *Verticillium dahliae* was much lower in roots of Florence plants (0.56%, 0.27%). This result is consistent with the greater tolerance of Florence to *Verticillium* than Honeoye. The fungus was not present amongst the top ten taxa in organically managed soil, irrespective of strawberry cultivar or inoculation treatment.

Leptodontidium sp C2 BESC319g was one of the main contributors to differences in community structure in response to inoculation with *Verticillium*. In conventionally managed soil its relative abundance varied between 12 and 14% and was not affected by *Verticillium* inoculation. However in organically managed soil its relative abundance was much higher, increasing from 28.5% to 45.6% in Honeoye and from 17.3% to 49% in Florence, in the presence of *Verticillium* inoculation. This suggests that *Leptodontidium* sp C2 BESC319g may be one of the biological factors associated with the suppressive character of the organically managed soil. *Leptodontidium orchidicola* has been shown to decrease the negative effect of *V. dahliae* in tomato, but only at low levels of pathogen inoculation [65].

Root endophytes are known to produce secondary metabolites that may have anti-microbial effects on plant pathogens and biological control of *Verticillium dahliae* by root endophytes has been successfully demonstrated [66–70]. In the present study the absence of *V. dahliae* associated with increased abundance of *Leptodontidium* in roots of plants grown in organically managed soil, suggests that this fungus may be of functional significance in

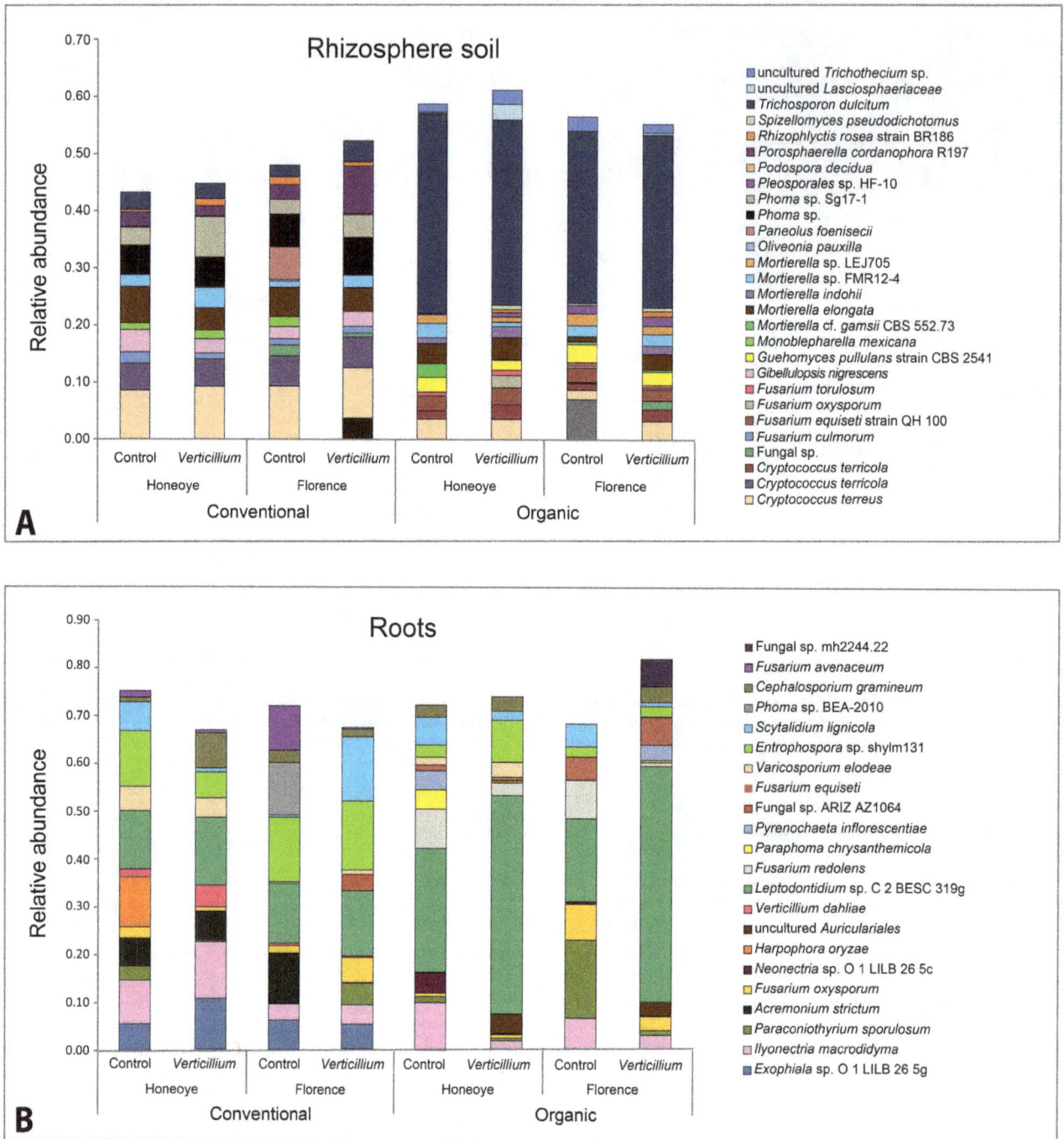

Figure 4. The top ten fungal operational taxonomic units (OTUs) (according to Similarity Percentage analysis) (SIMPER) contributing to the observed differences between control and *Verticillium dahliae* treatments in Honeoye and Florence grown in conventionally and organically managed soils. A) Rhizosphere soil (29–39%), B) Roots (61–74%).

protection against vascular pathogens. Further research on root endophytic communities should provide new biological control strategies for improving plant health [25].

OTUs of other broadly pathogenic taxa were also detected in our study. Examples of these include species belonging to the genera *Scytadilium, Fusarium, Pyrenochaeta, Phoma* and *Cephalosporium* but their pathogenicity to strawberry remains largely unknown. The relative abundance of these taxa was affected by both cultivar and soil type in the presence of *Verticillium* (Figure 4 a–b).

The effect of soil type, cultivar and pathogen inoculation on OTUs was also reflected in the plant performance measured as berry yield in this study (Figure S4). The strawberry yield differed significantly between the four cultivars in all three soils (conventional $p = 0.0014$, organic $p = 0.0087$ and peat $p = 0.0022$) in the absence of *V. dahliae*. Honeoye and Zephyr were selected because of their high yielding potential. In the present study these two cultivars produced the highest yields in the conventionally managed soil and peat based growth substrates but their mean yields decreased in response to inoculation with *V. dahliae*. The response of Florence to inoculation with the pathogen varied according to the soil type demonstrating a significant positive effect in conventionally managed soil and a significantly negative response in organically managed soil. Senga Sengana had the poorest yields in all three soils irrespective of the presence or absence of *V. dahliae* (Figure S4).

Concluding remarks

The results of this study show a marked difference between the fungal community structure in rhizosphere soil and that of fungi growing on or within the root tissues of strawberry plants. This suggests that plant roots select microorganisms from the wider rhizosphere soil pool, but further experiments with sampling at multiple time points are necessary to shed further light on this phenomenon. Strong differences in community structure were also observed between different soils, but differences between fungal communities colonizing roots of different cultivars were less clear. Inoculation with the pathogen *V. dahliae* had a significant influence on community structure and decreased the number of rhizosphere soil- and root-inhabiting fungi in several treatments. Lower numbers of *Verticillium* reads were detected in the roots of Florence, a tolerant cultivar, than in roots of Honeoye. Significant decreases in the number of ascomycete fungal OTUs occurred in roots of both Honeoye and Florence cultivars in response to *V. dahliae* in organically managed soil. The rhizosphere soil of these plants was characterised by high relative abundance of *Trichosporon dulcitum*, while the roots were dominated by *Leptodontidium* sp. C2 BESC 319 g. It is possible that the high abundance of these fungi may cause some suppression of pathogens such as *V. dahliae* but the role of these two fungi in plant protection is still unclear.

Supporting Information

Figure S1   Cluster analysis of the 50 most abundant fungal operational taxonomic units (OTUs) in the rhizosphere of four strawberry cultivars, Honeoye, Florence, Senga Sengana and Zephyr, grown in conventionally and organically managed soils or a peat-based growth substrate, with and without *Verticillium dahliae*. These OTUs constituted 90% of the total reads in roots and 70% of the reads in rhizosphere soil. The clustering was based on paired group linkage using the Bray-Curtis similarity measure. All OTUs with ≥95% identity are included.

Figure   S2   Non-metric   multidimensional   scaling (NMDS) analysis of the effects of inoculation with *Verticillium dahliae* on community structure of fungi colonising the rhizosphere of four different strawberry cultivars, Honeoye, Florence, Senga Sengana and Zephyr, grown in three different soils.

Figure S3   Sample based rarefaction curves of all fungal operational taxonomic units (OTUs) in rhizosphere soil and roots of strawberry using Analytic Rarefaction (Ver 1.3, UGA Stratigraphy Lab, USA).

Figure S4   Fresh berry weight (g/pot) of four different strawberry cultivars, Honeoye, Florence, Senga Sengana and Zephyr, grown in conventionally and organically managed soils or a peat-based growth substrate, with and without *Verticillium dahliae*. Asterisks indicate statistically significant differences ($p < 0.05$) between control and *Verticillium dahliae* inoculated treatments within each soil. Vertical bars represent mean values and error bars indicate ±1SE (n = 6).

Table S1   Similarity percentage (SIMPER) analysis of all fungal operational taxonomic units (OTUs) in rhizosphere and roots of four different strawberry cultivars, Honeoye, Florence, Senga Sengana and Zephyr, grown in conventionally and organically managed soils or a peat-based growth substrate, with and without *Verticillium dahliae*. All OTUs with ≥95% identity are included. (ND = not determined).

Acknowledgments

Prof. Christina Dixelius, Swedish University of Agricultural Sciences, Uppsala is acknowledged for providing the pathogen isolate.

Author Contributions

Conceived and designed the experiments: SN SM SA RDF. Performed the experiments: SN. Analyzed the data: SN SM SA RDF. Contributed reagents/materials/analysis tools: SN SM SA RDF. Contributed to the writing of the manuscript: SN SM SA RDF.

References

1. Morgan JAW, Bending GD, White PJ (2005) Biological costs and benefits to plant-microbe interactions in the rhizosphere. J Exp Bot 56: 1729–1739.
2. Drigo B, Pijl AS, Duyts H, Kielak AM, Gamper HA, et al. (2010) Shifting carbon flow from roots into associated microbial communities in response to elevated atmospheric CO_2. P Natl Acad Sci 107: 10938 −10942.
3. Philippot L, Raaijmakers JM, Lemanceau P, van der Putten WH (2013) Going back to the roots: the microbial ecology of the rhizosphere. Nat Rev Microbiol 11: 789–799.
4. Jones JDG, Dangl JL (2006). The plant immune system. Nature 444: 323–329.
5. Shade A, Handelsman J (2012) Beyond the Venn diagram: the hunt for a core microbiome. Env Microbiol 14: 4–12.
6. Lundberg DS, Lebeis SL, Paredes SH, Yourstone S, Gehring J, et al. (2012) Defining the core *Arabidopsis thaliana* root microbiome. Nature 488: 86–90.
7. Bulgarelli D, Rott M, Schlaeppi K, Ver Loren van Themaat E, Ahmadinejad N, et al. (2012) Revealing structure and assembly cues for *Arabidopsis* root-inhabiting bacterial microbiota. Nature 488: 91–95.
8. Wissuwa M, Mazzola M, Picard C (2008) Novel approaches in plant breeding for rhizosphere-related traits. Plant Soil 321: 409–430.
9. Smith KP, Goodman RM (1999) Host variation for interactions with beneficial plant associated microbes. Ann Rev Phytopathol 37: 473–491.
10. Andreote FD, Rocha UND, Araújo WL, Azevedo JL, Van Overbeek LS (2010) Effect of bacterial inoculation, plant genotype and developmental stage on root-associated and endophytic bacterial communities in potato (*Solanum tuberosum*). Antonie van Leeuwenhoek 97: 389–399.
11. Lodewyckx C, Vangronsveld J, Porteous F, Moore ERB, Taghavi S, et al. (2002) Endophytic bacteria and their potential applications. Crit Rev Plant Sci 21: 583–606.
12. Ryan RP, Germaine K, Franks A, Ryan DJ, Dowling DN (2008) Bacterial endophytes: recent developments and applications. FEMS Microbiol Lett 278: 1–9.
13. Berg G (2009) Plant-microbe interactions promoting plant growth and health: perspectives for controlled use of microorganisms in agriculture. App Microbiol Biotechnol 84: 11–18.
14. Edel-Hermann V, Brenot S, Gautheron N, Aimè S, Alabouvette C, et al. (2009) Ecological fitness of the biocontrol agent *Fusarium oxysporum* Fo47 in soil and its impact on the soil microbial communities. FEMS Microbiol Ecol 68: 37–45.
15. Gao G, Yin D, Chen S, Xia F, Yang J (2012) Effect of biocontrol agent *Pseudomonas fluorescens* 2P24 on soil fungal community in cucumber rhizosphere using T-RFLP and DGGE. PLoS ONE 7(2): e31806.
16. Zhao Y, Li W, Zhou Z, Wang L, Pan Y, et al. (2005) Dynamics of microbial community structure and cellulolytic activity in agricultural soil amended with two biofertilizers. Eur J Soil Bio 41: 21–29.

17. Hoitink H, Boehm M (1999) Biocontrol within the context of soil microbial communities: A substrate-dependent phenomenon. Ann Rev Phytopathol 37: 427–446.

18. Trivedi P, Duan Y, Wang N (2010) Huanglongbing, a systemic disease, restructures the bacterial community associated with citrus roots. Appl Environ Microb 76: 3427–3436.

19. Trivedi P, He Z, Van Nostrand JD, Albrigo G, Zhou J, et al. (2012) Huanglongbing alters the structure and functional diversity of microbial communities associated with citrus rhizosphere. ISME J 6: 363–383.

20. Reiter B, Pfeifer U, Schwab H, Sessitsch A (2002) Response of endophytic bacterial communities in potato plants to infection with *Erwinia carotovora* subsp. atroseptica. Appl Environ Microb 68: 2261–2268.

21. Yang C, Crowley DE, Menge JA (2001) 16S rDNA fingerprinting of rhizosphere bacterial communities associated with healthy and *Phytophthora* infected avocado roots. FEMS Microbiol Ecol 35: 129–136.

22. Kent AD, Triplett EW (2002) Microbial communities and their interactions in soil and rhizosphere ecosystems. Ann Rev Microbiol 56: 211–236.

23. Johansson JF, Paul LR, Finlay RD (2004) Microbial interactions in the mycorrhizosphere and their significance for sustainable agriculture. FEMS Microbiol Ecol 48: 1–13.

24. Hartmann A, Schmid M, Tuinen D van, Berg G (2008) Plant-driven selection of microbes. Plant Soil 321: 235–257.

25. Rodriguez RJ, White JF Jr, Arnold AE, Redman RS (2009) Fungal endophytes: diversity and functional roles. New Phytol 182: 314–330.

26. Mendes R, Kruijt M, de Bruijn I, Dekkers E, van der Voort M, et al. (2011) Deciphering the rhizosphere microbiome for disease-suppressive bacteria. Science 332: 1097–1100.

27. Berg G, Zachow C, Lottmann J, Gotz M, Costa R, et al. (2005) Impact of plant species and site on rhizosphere-associated fungi antagonistic to *Verticillium dahliae* Kleb. Appl Environ Microb 71: 4203–4213.

28. Costa R, Götz M, Mrotzek N, Lottmann J, Berg G, et al. (2006) Effects of site and plant species on rhizosphere community structure as revealed by molecular analysis of microbial guilds. FEMS Microbiol Ecol 56: 236–249.

29. Weller DM, Raaijmakers JM, Gardener BBM, Thomashow LS (2002) Microbial populations responsible for specific soil suppressiveness to plant pathogens. Ann Rev Phytopathol 40: 309–348.

30. Santos-González JC, Nallanchakravarthula S, Alström S, Finlay RD (2011) Soil, but not cultivar, shapes the structure of arbuscular mycorrhizal fungal assemblages associated with strawberry. Microb Ecol 62: 25–35.

31. Smalla K, Wieland G, Buchner A, Zock A, Parzy J, et al. (2001) Bulk and rhizosphere soil bacterial communities studied by denaturing gradient gel electrophoresis: plant-dependent enrichment and seasonal shifts revealed. Appl Environ Microb 67: 4742–4751.

32. Reeve JR, Schadt CW, Carpenter-Boggs L, Kang S, Zhou J, et al. (2010) Effects of soil type and farm management on soil ecological functional genes and microbial activities. ISME J 4: 1099–1107.

33. Fan L, Dalpé Y, Fang C, Dubé C, Khanizadeh S (2011) Influence of arbuscular mycorrhizae on biomass and root morphology of selected strawberry cultivars under salt stress. Botany 89: 397–403.

34. Norman JR, Atkinson D, Hooker JE (1996) Arbuscular mycorrhizal fungal-induced alteration to root architecture in strawberry and induced resistance to the root pathogen *Phytophthora fragariae*. Plant Soil 185: 191–198.

35. Norman JR, Hooker JE (2000) Sporulation of *Phytophthora fragariae* shows greater stimulation by exudates of non-mycorrhizal than by mycorrhizal strawberry roots. Mycol Res 104: 1069–1073.

36. Reganold JP, Andrews PK, Reeve JR, Carpenter-Boggs L, Schadt CW, et al. (2010) Fruit and soil quality of organic and conventional strawberry agroecosystems. PLoS ONE 5: e12346. doi:10.1371/journal.pone.0012346.

37. Ellis MA (2008) *Verticillium* wilt of Strawberry. Ohio State University Extension.

38. Fradin EF, Thomma BPHJ (2006) Physiology and molecular aspects of *Verticillium* wilt diseases caused by *V. dahliae* and *V. albo-atrum*. Mol Plant Pathol 7: 71–86.

39. Lord, Cheryl A S (1994) *Verticillium* wilt of Strawberry. Fact sheet. Durham, New Hampshire, USA: The University of New Hampshire Cooperative extension.

40. Heale J, Karapapa V (1999) The *Verticillium* threat to Canada's major oilseed crop: Canola. Can J Plant Pathol 21: 1–7.

41. Klosterman SJ, Atallah ZK, Vallad GE, Subbarao KV (2009) Diversity, pathogenicity, and management of *Verticillium* Species. Ann Review Phytopathol 47: 39–62.

42. Harris DC, Yang JR (1996) The relationship between the amount of *Verticillium dahliae* in soil and the incidence of strawberry wilt as a basis for disease risk prediction. Plant Pathol 45: 106–114.

43. Meszka B, Bielenin A (2009) Bioproducts in control of strawberry *Verticillium* wilt. Phytopathologia 52: 21–27.

44. Tjamos SE, Flemetakis E, Paplomatas EJ, Katinakis P (2005) Induction of resistance to *Verticillium dahliae* in *Arabidopsis thaliana* by the biocontrol agent K-165 and pathogenesis-related proteins gene expression. Molr Plant-Microb Interact 18: 555–561.

45. Rudrappa T, Czymmek KJ, Paré PW, Bais HP (2008) Root-secreted malic acid recruits beneficial soil bacteria. Plant Physiol 148: 1547–1556.

46. Azad HR, Davis JR, Schnathorst WC, Kado CI (1987) Influence of *Verticillium* wilt resistant and susceptible potato genotypes on populations of antagonistic rhizosphere and rhizoplane bacteria and free nitrogen fixers. Appl Microbiol Biotechnol 26: 99–104.

47. An M, Zhou X, Wu F, Ma Y, Yang P (2011) Rhizosphere soil microorganism populations and community structures of different watermelon cultivars with differing resistance to *Fusarium oxysporum* f. sp. niveum. Can J Microbiol 57: 355–365.

48. Yao H, Wu F (2010) Soil microbial community structure in cucumber rhizosphere of different resistance cultivars to *Fusarium* wilt. FEMS Microbiol Ecol 72: 456–463.

49. Rastogi G, Sani RK (2011) Molecular techniques to assess microbial community structure, function, and dynamics in the environment. In: Ahmad I, Ahmad F, Pichtel J, editors. Microbes and Microbial Technology. New York, NY: Springer New York. 29–57.

50. Torsvik V, Øvreås L (2002) Microbial diversity and function in soil: from genes to ecosystems. Curr Op Microbiol 5: 240–245.

51. Weinert N, Piceno Y, Ding, Meincke R, Heuer H, et al. (2011) PhyloChip hybridization uncovered an enormous bacterial diversity in the rhizosphere of different potato cultivars: many common and few cultivar-dependent taxa. FEMS Microbiol Ecol 75: 497–506.

52. Simpson D, Bell JA, Hammond K, Whitehouse AB (2002) The latest strawberry cultivars from horticultural research international, Tampere, Finland: Acta horticulturae 567: 165–168.

53. Olbricht K, Hanke M (2008) Strawberry breeding for disease resistance in Dresden. Small fruit: breeding and testing of cultivar. Hohenheim, Germany: Eco Fruit. 144–147.

54. Davik J, Daugaard H, Svensson B (2000) Strawberry production in the Nordic countries. Advances in Strawberry Research 19: 13–18.

55. Griffiths RI, Whiteley AS, O'Donnell AG, Bailey MJ (2000) Rapid method for co-extraction of DNA and RNA from natural environments for analysis of ribosomal DNA- and rRNA-based microbial community composition. Appl Environ Microb 66: 5488–5491.

56. Ihrmark K, Bödeker ITM, Cruz-Martinez K, Friberg H, Kubartova A, et al. (2012) New primers to amplify the fungal ITS2 region – evaluation by 454-sequencing of artificial and natural communities. FEMS Microbiol Ecol 82: 666–677.

57. White T, Bruns T, Lee S, Taylor J (1990) Amplification and direct sequencing of fungal ribosomal RNA genes for phylogenetics. PCR protocols: a guide to methods and applications. Innis MA, Gelfland DH, Sninsky JJ and White TJ. San Diego, CA, USA: Academic press. 315–322 p.

58. Xu L, Ravnskov S, Larsen J, Nicolaisen M (2012) Linking fungal communities in roots, rhizosphere, and soil to the health status of *Pisum sativum*. FEMS Microbiol Ecol 82: 736–745.

59. Middelhoven WJ, Koorevaar M, Schuur GW (1992) Degradation of benzene compounds by yeasts in acidic soils. Plant Soil 145: 37–43.

60. Margesin R, Fonteyne PA, Redl B (2005) Low-temperature biodegradation of high amounts of phenol by *Rhodococcus* spp. and basidiomycetous yeasts. Res Microbiol 156: 68–75.

61. Mestre MC, Rosa CA, Safar SVB, Libkind D, Fontenla SB (2011) Yeast communities associated with the bulk-soil, rhizosphere and ectomycorrhizo-sphere of a *Nothofagus pumilio* forest in northwestern Patagonia, Argentina. FEMS Microbiol Ecol 78: 531–541.

62. Girvan MS, Bullimore J, Pretty JN, Osborn AM, Ball AS (2003) Soil type is the primary determinant of the composition of the total and active bacterial communities in arable Soils. Appl Environ Microb 69: 1800–1809.

63. Wu Y, Fang W, Zhu S, Jin K, Ji D (2008) The effects of cotton root exudates on the growth and development of *Verticillium dahliae*. Frontiers of Agriculture in China 2: 435–440.

64. Vandemark GJ, Ariss JJ, Hughes TJ (2010) Real-time PCR Suggests that *Aphanomyces euteiches* is associated with reduced amounts of *Phytophthora medicaginis* in Alfalfa that is co-inoculated with both pathogens. J Phytopathol 158: 117–124.

65. Andrade-Linares DR, Grosch R, Restrepo S, Krumbein A, Franken P (2011) Effects of dark septate endophytes on tomato plant performance. Mycorrhiza 21: 413–422.

66. Raaijmakers JM, Mazzola M (2012) Diversity and natural functions of antibiotics produced by beneficial and plant pathogenic bacteria. Ann Rev Phytopathol 50: 403–424.

67. El-Deeb HM, Arab YA (2013) *Acremonium* as an endophytic bioagent against date palm *Fusarium* wilt. Arch Phytopathol Plant Prot: 1–8.

68. Fávaro LC, Sebastianes FL, Araújo WL (2012) *Epicoccum nigrum* P16, a sugarcane endophyte, produces antifungal compounds and induces root growth. PloS ONE 7(6): e36826. doi:10.1371/journal.pone.0036826.

69. Mousa WK, Raizada MN (2013) The diversity of anti-microbial secondary metabolites produced by fungal endophytes: an interdisciplinary perspective. Frontiers Microbiol 4: 65.

70. Aly AH, Debbab A, Proksch P (2011) Fungal endophytes: unique plant inhabitants with great promises. Appl Microbiol Biotechnol 90: 1829–1845.

5

Estimating Seasonal Nitrogen Removal and Biomass Yield by Annuals with the Extended Logistic Model

Richard V. Scholtz III*, Allen R. Overman

Agricultural & Biological Engineering Department, University of Florida, Gainesville, Florida, United States of America

Abstract

The Extended Logistic Model (ELM) has been previously shown to adequately describe seasonal biomass production and N removal with respect to applied N for several types of annuals and perennials. In this analysis, data from a corn (*Zea mays* L.) study with variable applied N were analyzed to test hypotheses that certain parameters in the ELM are invariant with respect to site specific attributes, like environmental conditions and soil type. Invariance to environmental conditions suggests such parameters may be functions of the crop characteristics and certain other management practices alone (like plant population, planting date, harvest date). The first parameter analyzed was Δb, the difference between the N uptake shifting parameter and the biomass shifting parameter. The second parameter tested was N_{cm}, the maximum N concentration. Both parameters were shown to be statistically invariant, despite soil and site differences. This was determined using analysis of variance with normalized nonlinear regression of the ELM on the data from the study. This analysis lends further evidence that there are common parameters involved in the ELM that do not rely on site-specific or situation-specific factors. More insight into the derivation of, definition of, and logic behind the various parameters involved in the model are also given in this paper.

Editor: Manuel Reigosa, University of Vigo, Spain

Funding: The funding of this analysis was provided by the Florida Experiment Station. The funders had no role in study design, data collection and analysis, decision to publish, or preparation of the manuscript.

Competing Interests: The authors have declared that no competing interests exist.

* E-mail: rscholtz@ufl.edu

Introduction

Effective water and nutrient management plays an essential role in future attempts at sustainable agricultural production. As the world's population continues to grow, the potable water supply is limited and must be guarded from unnecessary withdrawals and contamination from excessive nutrient loads. Strict monitoring and exercises in groundwater modeling of all agricultural operations is cost prohibitive. It therefore becomes necessary to investigate crop nutrient removal from their environment, and to adopt management procedures and rules that are based on a sound scientific foundation.

Overman et al. first proposed the logistic model as a nutrient management tool to describe seasonal biomass yield dependence of forage grasses on applied N [1]. The original application of the logistic model to plant biomass production was based on inductive reasoning, a process where inferences are made from "real world" observations [2]. While inductive reasoning is innately a more empirical method of model development, all models, no matter the complexity, have some element of empiricism [3]. It is because of the application to the "real world," that engineering and the applied sciences are, for practical reasons, inherently more empirical. The logistic model was extended to include seasonal plant N uptake (removal from the environment) dependence on applied N by forage grasses [4] and then for annuals, like corn [5]. The ELM is a five parameter, non-linear, parametric model that is capable of describing the seasonal biomass yields, N uptake, and N concentration with respect to applied N. Work conducted over the years has indicated that the ELM can effectively describe both

annual [6–11] and perennial [8,12–20] crops, for a wide range of nutrient inputs.

The ELM, begins with the simple logistic expression that relates N uptake, N_u, to the N applied, N, which is given by the following relationship:

$$N_u = \frac{A_N}{1+\exp(b_N - cN)} \tag{1}$$

where A_N is the relative maximum N uptake in kg ha^{-1}, b_N is the dimensionless N uptake shifting parameter, and c is the applied N response parameter given in ha kg^{-1}. The phase relationship between biomass production, Y, and N uptake, N_u, is given by

$$Y = \frac{Y_m N_u}{k_N + N_u} \tag{2}$$

where Ym is the maximum potential biomass production in Mg ha^{-1}, and k_N is the N uptake response parameter in kg ha^{-1}. The transformation of Eq. (1) by using Eq. (2) yields the following logistic expression:

$$Y = \frac{A}{1+\exp(b - cN)} \tag{3}$$

that relates biomass production to applied N, where A is the relative maximum biomass production in Mg ha^{-1}, b is the biomass yield shifting parameter, and c is the same applied N response parameter that applies to the N uptake logistic equation. The parameters A_N, A, b_N, b, and c are currently key parameters used with the ELM, and are the easiest to determine from regression analysis.

From the transformation of Eq. (1) into Eq. (3), the relative maximum biomass yield parameter can be written in terms of the maximum potential biomass yield parameter, the relative maximum N uptake parameter, and the N uptake response parameter.

$$A = \frac{Y_m A_N}{k_N + A_N} \tag{4}$$

The biomass shifting parameter can be written in terms of the N uptake shifting parameter, the relative maximum N uptake parameter, and the N uptake response coefficient.

$$b = b_N + \ln\left(\frac{k_N}{k_N + A_N}\right) \tag{5}$$

The difference between the shifting parameter for N uptake and the biomass yield can be written as the following:

$$\Delta b = b_N - b = -\ln\left(\frac{k_N}{k_N + A_N}\right) \tag{6}$$

N concentration is simply defined as the ratio of N uptake to biomass production. This leads to the following relationship:

$$N_c = \frac{N_u}{Y} = N_{cm}\frac{1 + \exp(b - cN)}{1 + \exp(b_N - cN)} \tag{7}$$

The N concentration model suggests that as N applied is increased to exorbitant levels, there is a maximum limit to the N concentration, N_{cm}. This maximum limit is simply the ratio of the relative maximum N uptake with respect to applied N, A_N, to the relative maximum biomass production with respect to applied N, A.

$$\lim_{N \to +\infty} N_c = \frac{A_N}{A} = N_{cm} \tag{8}$$

It seems logical to suggest that as the background amount of N present in the soil is decreased, the N concentration would be reduced to some prescribed lower limit, as there would be lower limit on the percent of proteins present in a given crop to sustain any growth. Mathematically, the model suggests that this lower limit of N concentration, N_{cl}, is a function of the maximum concentration and the difference between the N uptake shifting parameter and the biomass shifting parameter.

$$\lim_{N \to -\infty} N_c = N_{cm}\exp(-\Delta b) = N_{cl} \tag{9}$$

Also, this lower limit of N concentration, N_{cl}, can be found by taking the ratio of the N uptake response coefficient to the maximum potential biomass production parameter.

$$N_{cl} = \frac{k_N}{Y_m} \tag{10}$$

From the phase relationship between biomass production and N uptake, Eq. (2), N concentration can be found from the following equation with respect to N uptake:

$$N_c = \frac{k_N}{Y_m} + \frac{1}{Y_m}N_u \tag{11}$$

This predicts a linear relationship between N concentration, N_c, and N uptake, N_u. The line should have a slope equal to the inverse of the maximum potential biomass production and an intercept that equals the ratio of the N uptake response parameter to maximum potential biomass production. As this is a phase relationship, this is a functional segment that is bounded between N uptake values from 0 to the peak of A_N, and between N concentration values between N_{cl} and N_{cm}.

From the earlier work of Overman et al. [5], it has been shown that for a given site the applied N response parameter, c, and the N uptake intercept parameter, b_N, and the biomass intercept parameter, b, are not unique to the ELM when applied to grain or the whole plant. Meaning that the harvest index is constant for a given site. Their analysis showed that the only differences between the grain and the whole plant appear in the relative maximum N uptake parameter, A_N, and the relative maximum biomass production parameter, A [5]. Because the b_N, b, and c parameters were shown to be constant for both grain fraction and total biomass production, all the differences in both grain N uptake and biomass production, and the differences in the total plant N uptake and biomass production can be estimated with seven model parameters and a value for the seasonal amount of N applied. This is a comparative reduction of three parameters when b_N, b, and c are not held constant between grain and total plant biomass production.

The goal of this work is to continue to elevate the Extended Logistic Model (ELM) beyond the empiricism of it nascent beginnings and achieve a balance between what can be measured and what should be modeled, as called for by Montieth [21]. The intent is to shed new light on the significance of parameters used in the ELM and to contribute to the search for commonality among parameters. Normalized non-linear regression and analysis of variance (ANOVA) were used to show the invariance of two model parameters with respect to environmental differences, namely soil type and water availability.

Methods

Data Set

This analysis uses data collected by Eugene Kamprath from a corn (Pioneer 3320) N-rate field study that was conducted at three regional research stations in North Carolina from 1981 to 1984. A detailed explanation of the field experiment has been previously reported [22]. Supplemental irrigation was provided at the Clayton experiment station for the well-drained Dothan loamy fine sand (*fine-loamy, siliceous, thermic Plinthic Kandiudults*), at a rate of 10 to 12 cm a season, except for 1982 when no additional water was supplied. No irrigation was provided at the Kinston station for the well-drained Goldsboro sandy loam (*fine-loamy, siliceous, thermic Aquic Paleudults*). At the Plymouth experiment station, no irrigation was provided for the poorly-drained Portsmouth very fine sandy

Table 1. Analysis of variance for model parameters for corn grain and total plant biomass production and for corn grain and total plant N uptake, grown on three different soils.

Mode	Parameters Estimated	Degrees of Freedom	Normalized Residual Sum of Squares	Normalized Mean Sum of Squares	F Value
I	21	39	0.0247	0.000633	----
II, Common Δb.	19	41	0.0259	0.000631	----
II–I	--	2	0.00119	0.000595	0.940
III, Common Δb, & N_{cm}.	15	45	0.0267	0.000594	----
III–I	--	6	0.00202	0.000337	0.533

loam (*fine-loamy over sandy or sandy-skeletal mixed, thermic Typic Umbraquults*). The experiments at each station were set up as a RCB design, with four replications. Both total plant and grain fraction biomass were sampled, and every year the experiment was conducted at a new location within the same soil type at each station. This was to limit the impact on the experiment of any residual N in the soil from the previous year. The fertilizer treatments were applied in the form of NH_4NO_3 at rates of 0, 56, 112, 168, and 224 kg ha^{-1} of N. Average values over the four year period were combined for each of the different treatments and the model parameters were evaluated based on those combined averages.

Normalization

Parameters A_N and A for grain, A_N and A for total plant, and b_N; b and c for each site are determined simultaneously, using Newton-Raphson non-linear regression. A detailed description of Newton-Raphson non-linear regression of logistic equation can be found in Overman and Scholtz [8]. The attempt of this methodology is to consistently distribute the standard error amongst all those parameters for further analysis. Because of the unit and an order of magnitude difference between biomass and N uptake parameter values, as well as a subsequent order of magnitude difference between grain and total plant parameter values, a normalization routine is also employed. The error sum of squares for each individual site is initially written as

$$ESS_{NORM} =$$

$$\underbrace{\sum \left(\frac{Y_g}{A_g} - \frac{1}{1+\exp(b-cN)} \right)^2 + \sum \left(\frac{N_{ug}}{A_{Ng}} - \frac{1}{1+\exp(b_N-cN)} \right)^2}_{\text{grain fraction}}$$

$$+ \underbrace{\sum \left(\frac{Y_t}{A_t} - \frac{1}{1+\exp(b-cN)} \right)^2 + \sum \left(\frac{N_{ut}}{A_{Nt}} - \frac{1}{1+\exp(b_N-cN)} \right)^2}_{\text{total plant}} \quad (12)$$

where the total normalized error is resultant from the sum of the normalized error from the three sites. For this study the initial Hessian matrix is 21 by 21 elements and paired with a 21 element Jacobian vector. As a result of the normalization procedure, performing the Newton-Raphson procedure can diverge more readily than a non-normalized procedure. Because of this, it is important to establish a reasonable initial guess for each parameter. First, individual logistic response are evaluated for each data set from all sites, for grain biomass, grain N uptake, for total plant biomass, and total plant N uptake. This results in six A values, six A_N values, six b values, six b_N values, and 12 values for the c parameter. Using the fact that it can be shown for two

straight lines

$$y_1 = m_1 x + b_1 \quad (13)$$

and

$$y_2 = m_2 x + b_2, \quad (14)$$

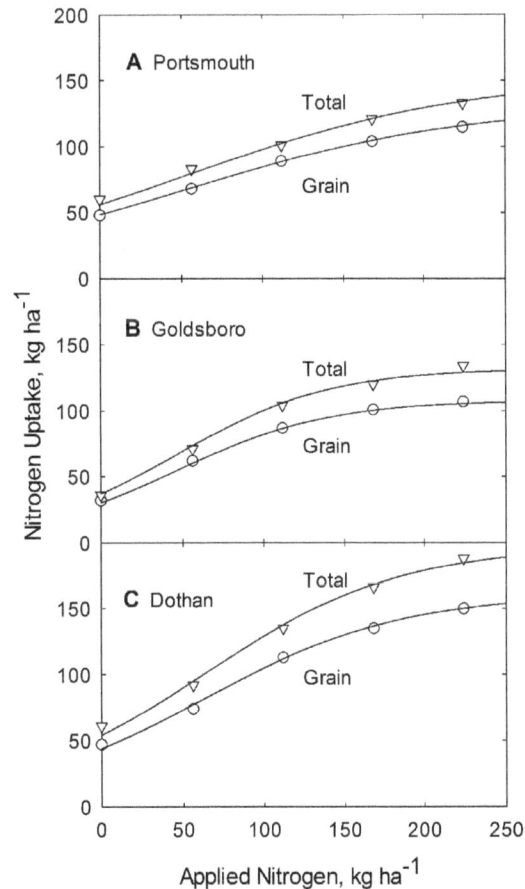

Figure 1. Dependence of grain and total plant N uptake on applied N for corn grown at the Plymouth (A), Kinston (B), and Clayton (C) experiment stations in North Carolina. Curves are constructed from Eq. 1 and from parameters listed in Table 2.

Table 2. Standard logistic model parameters invariant to corn grain and total plant biomass production and for corn grain and total plant N uptake, grown on three different soils.

Soil	Component	Parameter	Estimate
Dothan	Biomass	b	0.319
	N Uptake	b_N	0.978
	Both	c, ha kg^{-1}	0.0161
Goldsboro	Biomass	b	0.282
	N Uptake	b_N	0.941
	Both	c, ha kg^{-1}	0.0209
Portsmouth	Biomass	b	−0.105
	N Uptake	b_N	0.555
	Both	c, ha kg^{-1}	0.0111

who share the same sampling of the independent variable, the best fit single slope shared between them is given by

$$\hat{m} = \frac{m_1 + m_2}{2} \tag{15}$$

and the corresponding intercepts become

$$\hat{b}_1 = b_1 + \frac{1}{2}(m_1 - m_2)\bar{x} \tag{16}$$

and

$$\hat{b}_2 = b_2 + \frac{1}{2}(m_2 - m_1)\bar{x}, \tag{17}$$

the initial guess for the c parameter is the average of all 12 values, and the initial guess for each value of bN can be found from

$$\hat{b}_N = b_N + (c - \bar{c})\bar{N} \tag{18}$$

and for each value of b can be found from

$$\hat{b} = b + (c - \bar{c})\bar{N}. \tag{19}$$

The problem is bounded between the maximum and minimum values of the c parameter and each value of b_N is bounded between

$$\hat{b}_{Nmin} = b_N + (c - c_{max})\bar{N} \tag{20}$$

and

$$\hat{b}_{Nmax} = b_N + (c - c_{min})\bar{N} \tag{21}$$

and each value of b is bounded between

$$\hat{b}_{min} = b + (c - c_{max})\bar{N} \tag{22}$$

and

$$\hat{b}_{max} = b + (c - c_{min})\bar{N}. \tag{23}$$

Analysis

The first hypothesis of this analysis is that the difference between N uptake intercept parameter and the biomass intercept parameter, Δb, is invariant with respect to the differences in soil type and water availability for a given variety of an annual crop. Note that there is no attempt in this work to identify the effects of water availability or site characteristics on the ELM parameters, but to determine which are invariant to those characteristics. For this analysis, the same genetic line of corn is propagated by seeding and harvested at the same relative age. The second hypothesis is that maximum N concentration, N_{cm}, is also invariant with respect to the differences in soil type and water availability. A consequence of both hypotheses being affirmed is that the lower limit to the N concentration, N_{cl}, in the same annual crop is also invariant with respect to soil type and water availability. Parameters were estimated by minimization of the normalized error sum of squares, and analysis of variance (ANOVA) was used to determine the validity of both hypotheses.

For the analysis of variance, three scenarios or modes were used, each with a targeted reduction in the number of parameters used in the ELM to describe the corn data in the Kamprath study. Mode I had 21 separate parameters that were estimated by minimization of the normalized error sum of squares. In Mode I, there are individual values for A, and A_N, for both grain and for total plant, and corresponding values for b, b_N, and c at each of the three sites. For Mode II, the number of parameters estimated dropped to 19, because the Δb parameter was held constant across the three sites. For Mode III, the Δb and the N_{cm} parameters were both held constant across the three sites, reducing the number of estimated parameters to 15.

Nonlinear Coefficients of Determination [23] (Nash-Sutcliffe Model Efficiency Coefficient [24]) will be provided for grain and total plant N uptake and for grain and total plant biomass production just as a relative comparison of fit.

Results

Table 1 contains the summary of the analysis of variance test. The comparison between Modes I and II leads to an increase the *degrees of freedom* to 41 and results in a variance ratio of 0.940. Because the critical $F_{(2,39,95\%)}$ value is 3.24, it is concluded that there is no significant difference between the two modes. Thus in

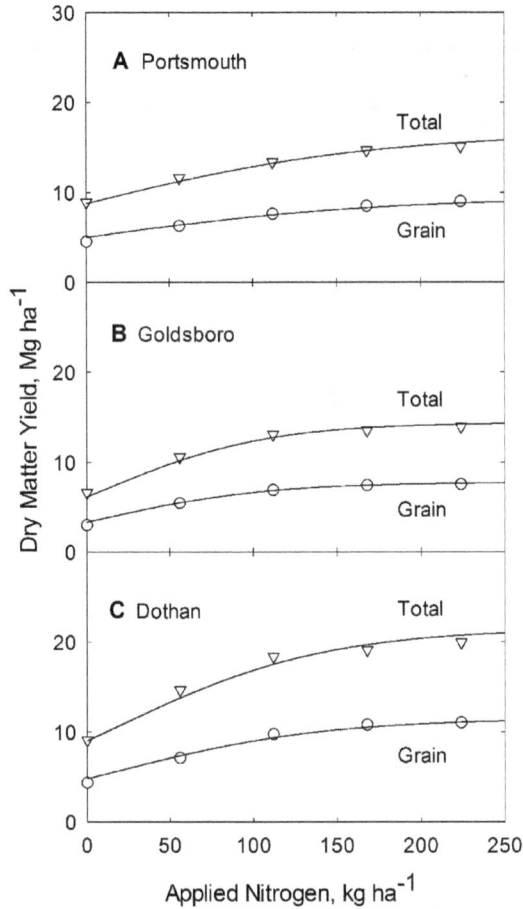

Figure 2. Dependence of grain and total plant biomass production on applied N for corn grown at the Plymouth (A), Kinston (B), and Clayton (C) experiment stations in North Carolina. Curves are constructed from Eq. 3 and from parameters listed in Table 2.

this study the Δb parameter is invariant to all soil and site differences.

Also from Table 1, the comparison between Modes I and III results in an increase the *degrees of freedom* to 45, and in a variance ratio of 0.533. With a critical $F(6,39,95\%)$ value of 2.34, not only is the Δb parameter is invariant, but so are the total plant and grain N_{cm} parameters. Thus, the soil, the field conditions, the environmental constraints, and even water availability play no role in either is the Δb or the two N_{cm} parameters. This leads to an invariance in the total plant and grain N_{cl} parameters, by virtue of Eq. (9).

The dependence of grain and whole plant N uptake on applied N at harvest is represented by Figure 1 for the three soil types. In general there is good agreement between the model line and the data. The resulting N uptake model lines (depicted in Figure 1), are generated from Eq. (1), using parameter values for b_N and c from Table 2 and values for A_{Ng} and A_{Nt} found in Table 3. Equation specific Non-linear Coefficient of Determination values and Error Sum of Squares are provided in Table 4.

Grain and whole plant biomass production versus applied N is shown in Figure 2 for all three soil types. In general there is good agreement between the model line and the data. The resulting biomass model lines (depicted in Figure 2), are generated from Eq. (3), using parameter values for b and c from Table 2 and values for A_g and A_t found in both Table 3. Equation specific Non-linear Coefficient of Determination values and Error Sum of Squares are provided in Table 4.

N concentration dependence on applied N is shown in Figure 3 for all three soil types. The resulting N concentration model lines (depicted in Figure 3), are generated from Eq. (7), using parameter values for b, b_N and c from Table 2 and values for $N_{cm\ g}$ and $N_{cm\ t}$ from Table 6.

The phase relationship between biomass production and N uptake for the corn grain and the whole plant is represented by Figure 4 for each of the three soils. The resulting biomass – N uptake phase model lines (depicted in Figure 4), are generated from Eq. (2), using parameter values for k_{Ng}, k_{Nt}, Y_{mg} and Y_{mt} found in Table 5.

The phase relationship between N Concentration and N uptake for the grain and the whole plant is represented by Figure 4 for each of the three soils. The resulting between N Concentration –

Table 3. Standard logistic model parameters specific to corn grain and total plant biomass production and for corn grain and total plant N uptake, grown on three different soils.

Soil	Component	Parameter	Plant Fraction	Estimate
Dothan	Biomass	A_g, Mg ha^{-1}	Grain	11.5
		A_t, Mg ha^{-1}	Total Plant	21.5
	N Uptake	A_{Ng}, kg ha^{-1}	Grain	161
		A_{Nt}, kg ha^{-1}	Total Plant	198
Goldsboro	Biomass	A_g, Mg ha^{-1}	Grain	7.76
		A_t, Mg ha^{-1}	Total Plant	14.4
	N Uptake	A_{Ng}, kg ha^{-1}	Grain	108
		A_{Nt}, kg ha^{-1}	Total Plant	132
Portsmouth	Biomass	A_g, Mg ha^{-1}	Grain	9.50
		A_t, Mg ha^{-1}	Total Plant	16.7
	N Uptake	A_{Ng}, kg ha^{-1}	Grain	133
		A_{Nt}, kg ha^{-1}	Total Plant	154

Table 4. Standard statistical measures of fit, based on specific component and plant fraction data.

Soil	Component	Plant Fraction	Coefficient of Determination	Specific Error Sum of Squares
Dothan	Biomass	Grain	0.984	0.523
		Total Plant	0.969	2.361
	N Uptake	Grain	0.997	24.5
		Total Plant	0.993	83.9
Goldsboro	Biomass	Grain	0.992	0.145
		Total Plant	0.983	0.468
	N Uptake	Grain	0.998	7.81
		Total Plant	0.993	35.2
Portsmouth	Biomass	Grain	0.977	0.169
		Total Plant	0.985	0.592
	N Uptake	Grain	0.999	5.01
		Total Plant	0.991	19.4

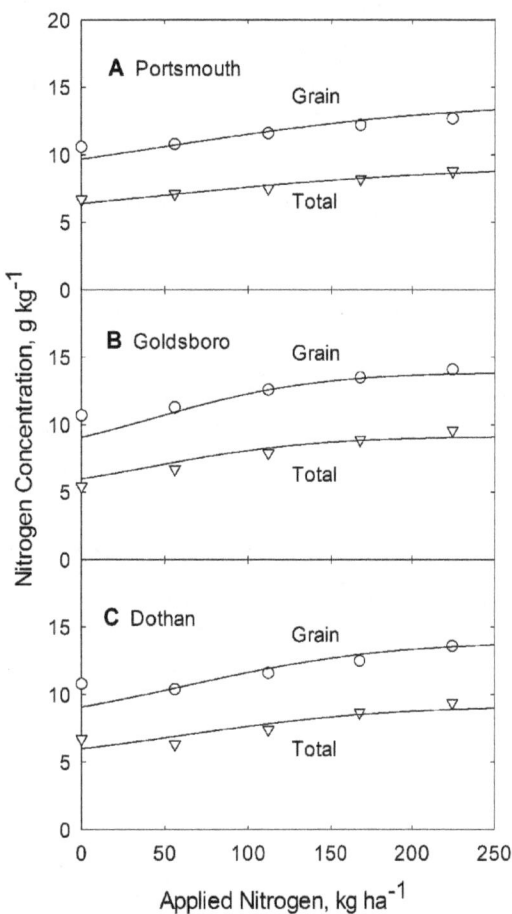

Figure 3. Dependence of grain and total plant N concentration on applied N for corn grown at the Plymouth (A), Kinston (B), and Clayton (C) experiment stations in North Carolina. Curves are constructed from Eq. 7 and from parameters listed in Table 2.

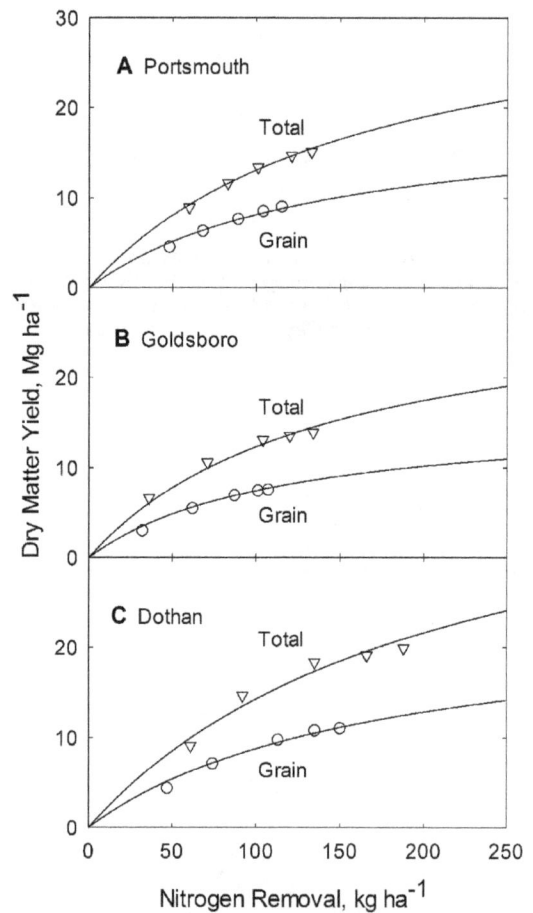

Figure 4. Dependence of grain and total plant biomass production on N uptake for corn grown at the Plymouth (A), Kinston (B), and Clayton (C) experiment stations in North Carolina. Curves are constructed from Eq. 2 and from parameters listed in Table 3.

Table 5. Standard phase model parameters for corn grain and total plant biomass production and for corn grain and total plant N uptake.

Soil	Parameter	Plant Fraction	Estimate
Dothan	k_{Ng}, kg ha^{-1}	Grain	157
	k_{Nt}, kg ha^{-1}	Total Plant	193
	Y_{mg}, Mg ha^{-1}	Grain	22.7
	Y_{mt}, Mg ha^{-1}	Total Plant	42.6
Goldsboro	k_{Ng}, kg ha^{-1}	Grain	104
	k_{Nt}, kg ha^{-1}	Total Plant	125
	Y_{mg}, Mg ha^{-1}	Grain	15.1
	Y_{mt}, Mg ha^{-1}	Total Plant	27.6
Portsmouth	k_{Ng}, kg ha^{-1}	Grain	122
	k_{Nt}, kg ha^{-1}	Total Plant	142
	Y_{mg}, Mg ha^{-1}	Grain	17.7
	Y_{mt}, Mg ha^{-1}	Total Plant	31.4

N uptake phase model lines (depicted in Figure 5), are generated from Eq. (11), using parameter values for k_{Ng}, k_{Nt}, Y_{mg} and Y_{mt} found in Table 5.

Discussion

From this analysis it is concluded that there are aspects of the ELM that are invariant with respect to both soil type and water availability for a given variety of annual crop propagated by seeding and harvested at the same relative age. This analysis has shown, for the Kamprath N-rate study conducted on corn in North Carolina [22] that both the difference between N uptake intercept parameter and the biomass intercept parameter, Δb, and the maximum N concentration, N_{cm}, are in fact invariant with respect to the crop's surrounding environmental conditions. From the model, these facts lead to the conclusion that both the upper limit N concentration, N_{cm}, and the lower limit N concentration, N_{cl}, are both invariant with respect to soil type and water availability in the study analyzed. This further suggests that the N_{cm} and N_{cl} parameters are of more importance to the model. While other parameters, such as

$$A = \frac{A_N}{N_{cm}} \qquad (24)$$

$$\Delta b = \ln\left(\frac{N_{cm}}{N_{cl}}\right) \qquad (25)$$

$$b = b_N - \Delta b = b_N - \ln\left(\frac{N_{cm}}{N_{cl}}\right) \qquad (26)$$

$$Y_m = \frac{A_N}{N_{cm} - N_{cl}} \qquad (27)$$

and

$$k_N = N_{cl} Y_m = \frac{N_{cl} A_N}{N_{cm} - N_{cl}} \qquad (28)$$

can be rewritten, to show the significance that upper and lower limit concentrations have in each parameter and ultimately seasonal plant response to nutrient application. Having upper

Table 6. Parametric factors invariant to site attributes, including soil type.

Plant Fraction	Parameter	Estimate
Both	Δb, kg ha^{-1}	0.660
Grain	$N_{cm\ g}$, g kg^{-1}	14.0
Total Plant	$N_{cm\ t}$, g kg^{-1}	9.18
Grain	$N_{cl\ g}$, g kg^{-1}	7.23
Total Plant	$N_{cl\ t}$, g kg^{-1}	4.75
Grain	$N_{cm\ t} - N_{cl\ t} = A_{Nt}/Y_{mt}$, g kg^{-1}	6.75
Total Plant	$N_{cm\ g} - N_{cl\ g} = A_{Ng}/Y_{mg}$, g kg^{-1}	4.43

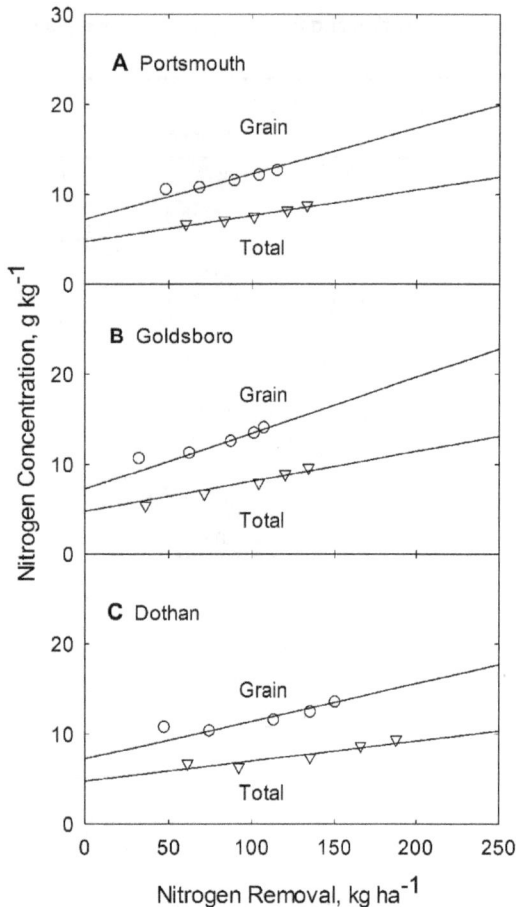

Figure 5. Dependence of grain and total plant N concentration on N uptake for corn grown at the Plymouth (A), Kinston (B), and Clayton (C) experiment stations in North Carolina. Curves are constructed from Eq. 11 and from parameters listed in Table 3.

and season length, the model reduces to three parameters (A_N, b_N, and c) when a crop and season length are chosen. From Overman & Scholtz [26] the logistic response originates within the soil's buffering capacity for P and for K, and the c parameter remains the same from the plant extractable logistic response to nutrient uptake logistic response, and to biomass production. It is here assumed that the c parameter for applied N also originates as the rate response parameter for the soil's buffering capacity of N. The c parameter can be modified by plant population [11]. A Future step should be to analyze various field studies to catalogue soil physical and chemical characteristics and the resulting impact on the c parameter. Mathematically b_N represents shifting parameter which in conjunction with the c parameter as

$$N_{0.5} = \frac{b_N}{c} \qquad (29)$$

$N_{0.5}$ represents the effective level of N necessary to achieve peak N uptake efficiency [10,16,17]. Ultimately, for environmental considerations, setting applied rate of N to the peak uptake level will result in the most N removed per unit N applied. Plus, provided the difference between b and b_N is greater than 0, then the yield will be on the upper portion of the logistic biomass curve to yield

$$Y = \frac{A_N}{N_{cm}[1 + \exp(-\Delta b)]}. \qquad (30)$$

The b_N parameter is affected by changes in plant population [11] and is also influenced by crop type [8,25]. The remaining parameter, A_N, is a linear parameter that is affected by the various environmental conditions, the crop type, the soil type, and various management practices [5–20]. Given Eq. (30), knowing the invariant Δb value for a given crop, and having a reasonable estimate for the background level of N already present in the soil, represented by Eq. (29), exists the beginning of a framework for a more reasonable and more sustainable nutrient management guide. Further analyses are being conducted to verify these findings with other annual propagated by seeding and with perennial crops.

Acknowledgments

The authors would like to acknowledge the hard work of field researchers, like Dr. Kamprath, whose tireless efforts have helped bring about greater understanding of the complexities of agricultural production.

Author Contributions

Conceived and designed the experiments: RVS ARO. Performed the experiments: RVS. Analyzed the data: RVS. Contributed reagents/materials/analysis tools: RVS. Wrote the paper: RVS. Mentorship and review: ARO.

and lower limits to plant nutrient concentration corresponds with plant physiology. Without a minimum level of a given required nutrient, there can be no yield, seasonal or otherwise. There should also be a maximum concentration that can be approached, as there should be diminishing yield increases as higher agronomic rates are applied, or there would be unbounded growth.

If this invariance with respect to soil type and water availability holds for all crops propagated by seeding, the model could be written in terms of parameters that have measurable physiological significance and could give further insight into relationships that govern plant development and nutrient removal. Initial evidence appears promising that perennial crops, such as ryegrass (*Lolium perenne* L.), when held to comparably the same seasonal management practices also exhibit very nearly the same conclusions with regard to both constant values of the N_{cm} and N_{cl} parameters [25].

Given that N uptake and biomass production can be described by five parameters, and if two are invariant to all but crop type

References

1. Overman AR, Martin FG, Wilkinson SR (1990) A logistic equation for yield response of forage grass to nitrogen. Commun Soil Sci Plan 21: 595–609.
2. Ferris T (1989) Coming of age in the milky way. New York: Bantam Doubleday Dell Publishing Group. 495 p.
3. France J, Thornley JHM (1984) Mathematical models in agriculture. London: Butterworth and Company. 352 p.
4. Overman AR, Wilkinson SR, Wilson DM (1994) An extended model of forage grass response to applied nitrogen. Agron J 86: 617–620.
5. Overman AR, Wilson DM, Kamprath EJ (1994) Estimation of yield and nitrogen removal by corn. Agron J 86: 1012–1016.
6. Reck WR, Overman AR (1996) Estimation of corn response to water and applied nitrogen. J Plant Nutr 19: 201-214.

7. Overman AR (1999) Model for accumulation of dry matter and plant nutrient elements by tobacco. J Plant Nutr 22: 1, 81–92.
8. Overman AR, Scholtz RV (2002) Mathematical models of crop growth and yield. New York: Marcel Decker. 342 p.
9. Overman AR, Scholtz RV (2002) Corn response to irrigation and applied nitrogen. Commun Soil Sci Plan 33: 3609–3619.
10. Overman AR, Brock KH (2003) Model analysis of corn response to applied nitrogen and tillage. Commun Soil Sci Plan 34: 2177–2191.
11. Overman AR, Scholtz RV, Brock KH (2006) Model analysis of corn response to applied nitrogen and plant population density. Commun Soil Sci Plan 37: 1157–1172.
12. Overman AR, Howard JC (2012) Model analysis of response to applied nitrogen by cotton. J Plant Nutr 35: 2118–2123.
13. Overman AR, Wilkinson SR (1995) Extended logistic model of forage grass response to applied nitrogen, phosphorus, and potassium. T ASAE 38: 103–108.
14. Overman AR, Stanley RL (1998) Bahiagrass responses to applied nitrogen and harvest interval. Commun Soil Sci Plan 29: 237–244.
15. Overman AR, Wilkinson SR (2003) Extended logistic model of forage grass response to applied nitrogen as affected by soil erosion. T ASAE 46: 1375–1380.
16. Overman AR, Brock KH (2003) Model comparison of coastal bermudagrass and pensacola bahiagrass response to applied nitrogen. Commun Soil Sci Plan 34: 2163–2176.
17. Overman AR, Scholtz RV, Taliaferro CM (2003) Model analysis of response of bermudagrass to applied nitrogen. Commun Soil Sci Plan 34: 1303–1310.
18. Overman AR, Rhoads FM, Brock KH (2005) Model analysis of response of pensacola bahiagrass to applied nitrogen, phosphorus, and potassium. I. Seasonal dry matter. J Plant Nutr 27: 1747–1756.
19. Overman AR, Rhoads FM, Brock KH (2005) Model analysis of response of pensacola bahiagrass to applied nitrogen, phosphorus, and potassium. II. Seasonal P and K uptake. J Plant Nutr 27: 1757–1777.
20. Overman AR, Scholtz RV (2005) Model analysis for response of dwarf elephantgrass to applied nitrogen and rainfall. Commun Soil Sci Plan 35: 2485–2494.
21. Monteith JL (1996) The quest for balance in crop modeling. Agron J 88: 695-697.
22. Kamprath EJ (1986) Nitrogen studies with corn on coastal plain soils, Technical Bulletin 282. Raleigh: North Carolina Agricultural Research Service, North Carolina State University. 15 p.
23. Draper NR, Smith H (1998) Applied regression analysis. 3rd ed. New York: John Wiley and Sons. 736 p.
24. Nash JE, Sutcliffe JV (1970) River flow forecasting through conceptual models part I - A discussion of principles. J Hydrol 10: 282–290.
25. Scholtz RV (2002) Mathematical modeling of agronomic crops: analysis of nutrient removal and dry matter accumulation. Doctor of Philosophy Dissertation. Gainesville: University of Florida. 139 p.
26. Overman AR, Scholtz RV (2012) A memoir on: A model of yield and phosphorus uptake in response to applied phosphorus by potato. Gainesville: University of Florida. 16p. Available: http://ufdc.ufl.edu/IR00001234/00001 Accessed 11 December 2.

6

Climate and Land Use Controls on Soil Organic Carbon in the Loess Plateau Region of China

Yaai Dang[1,2,3◐], Wei Ren[2◐], Bo Tao[2], Guangsheng Chen[2], Chaoqun Lu[2], Jia Yang[2], Shufen Pan[2], Guodong Wang[3], Shiqing Li[1], Hanqin Tian[2]*

1 State Key Laboratory of Soil Erosion and Dryland Farming on the Loess Plateau, Institute of Soil and Water Conservation, Northwest A&F University, Yangling, Shaanxi, China, **2** International Center for Climate and Global Change Research, School of Forestry & Wildlife Sciences, Auburn University, Auburn, Alabama, United States of America, **3** College of Science, Northwest A&F University, Yangling, Shaanxi, China

Abstract

The Loess Plateau of China has the highest soil erosion rate in the world where billion tons of soil is annually washed into Yellow River. In recent decades this region has experienced significant climate change and policy-driven land conversion. However, it has not yet been well investigated how these changes in climate and land use have affected soil organic carbon (SOC) storage on the Loess Plateau. By using the Dynamic Land Ecosystem Model (DLEM), we quantified the effects of climate and land use on SOC storage on the Loess Plateau in the context of multiple environmental factors during the period of 1961–2005. Our results show that SOC storage increased by 0.27 Pg C on the Loess Plateau as a result of multiple environmental factors during the study period. About 55% (0.14 Pg C) of the SOC increase was caused by land conversion from cropland to grassland/forest owing to the government efforts to reduce soil erosion and improve the ecological conditions in the region. Historical climate change reduced SOC by 0.05 Pg C (approximately 19% of the total change) primarily due to a significant climate warming and a slight reduction in precipitation. Our results imply that the implementation of "Grain for Green" policy may effectively enhance regional soil carbon storage and hence starve off further soil erosion on the Loess Plateau.

Editor: Xiujun Wang, University of Maryland, United States of America

Funding: This study was supported by NASA Land Cover and Land Use Change Program (NNX08AL73G), NASA Interdisciplinary Science Program (NNG04GM39C), Chinese Universities Scientific Fund (z10921007), US National Science Foundation Grants (AGS-1243220, CNS-1059376), State Key Laboratory of Soil Erosion and Dryland Farming on the Loess Plateau Foundation (K318009902-1410), and Shaanxi Administration of Foreign Expert Affairs Science and Technology Activities Fundation (201327). Study design was supported by NASA Land Cover and Land Use Change Program (NNX08AL73G), NASA Interdisciplinary Science Program (NNG04GM39C), and Chinese Universities Scientific Fund (z10921007). Data collection and analysis was supported by NASA Land Cover and Land Use Change Program (NNX08AL73G), NASA Interdisciplinary Science Program (NNG04GM39C), and State Key Laboratory of Soil Erosion and Dryland Farming on the Loess Plateau Foundation (K318009902-1410). Preparation of the manuscript was supported by NASA Land Cover and Land Use Change Program (NNX08AL73G), US National Science Foundation Grants (AGS-1243220, CNS-1059376), and State Key Laboratory of Soil Erosion and Dryland Farming on the Loess Plateau Foundation (K318009902-1410).

Competing Interests: The authors have declared that no competing interests exist.

* E-mail: tianhan@auburn.edu

◐ These authors contributed equally to this work.

Introduction

Soil organic carbon (SOC), the major component of soil organic matter, plays a key role in the terrestrial carbon cycle and thus has drawn great attention from scientific community. It is a dynamic component of terrestrial systems, affecting carbon exchange between terrestrial ecosystem and the atmosphere [1,2]. SOC storage is nearly three times as large as carbon storage in vegetation and twice as large as global atmospheric carbon storage [3]. Soil has higher potential to sequester more carbon (such as converting the type of land use) in the future [2,4], therefore, increasing soil carbon storage is one of the most economical and effective ways to alleviate the greenhouse effect, which has become a hot scientific and political issue during the past decades.

Changes in climate and land use, caused by both natural and anthropogenic processes, have greatly influenced the terrestrial carbon balance during the past decades [5,6,7,8]. It was reported that about one fourth of anthropogenic CO_2 emissions were due to land cover and land use change (LCLUC), especially deforestation

[9]. Long-term experimental studies have confirmed that SOC is highly sensitive to land conversion from natural ecosystems, such as forest or grassland, to agricultural land, resulting in substantial SOC loss [6,10]. In addition, LCLUC may also cause carbon depletion by influencing soil respiration [11]. It was estimated that global carbon release from SOC mineralization owing to agricultural activities was approximately 0.80 Pg C/year (1 Pg = 10^{15} g) [12]. Globally, land use change resulted in a carbon release of (1.6 ± 0.8) Pg C per year to the atmosphere during the period of 1990s [13]. However, the effects of conversions from cropland to grassland/forest on the SOC storage have not been fully understood and there still remains large uncertainty.

The Loess Plateau of China (Figure 1), located in the geographic center of China (33°43′N 100°54′E to 41°16′N 114°33′E), covers a total area of 628,000 km^2, which is about 6.5% of China's total land area. The Loess Plateau is characterized by highly erodible soils, steep slopes, being subjected to heavy rain, and low vegetation coverage due to excess exploitation of land resource and improper land use [4,14]. During the past decades, serious soil erosions caused by natural and anthropogenic disturbances (e.g., climate change, natural disasters, LCLUC etc.) occurred in the area of the Loess Plateau. Previous reports also indicated that the warming and drying climate in this region has significantly aggravated soil erosion [15]. As a result, the large amount of fine surface soil eroded from the loess area is transported into the Yellow River and acts as the main source of sediment of this river, which runs through the Loess Plateau and is considered to be the most turbid river in the world. Due to these disturbances, soil carbon storage on the Loess Plateau is much lower compared to other regions in China [16]. In general, adjusting the land use pattern so as to restore the degraded ecosystems and to modify the local rural income structure is regarded as the main measures to control soil and water erosion and to improve farmers' living conditions on the Loess Plateau. Since the 1950s, a series of conservation policies have been implemented in this region, such as extensive tree planting since the 1970s, integrated soil erosion controls on the watershed scale in the 1980s and the 1990s [17,18], and the government-funded project "Grain for Green" in 1999, aiming at transforming the low-yield slope cropland into grassland/forest. The implementation of these policies improved vegetation coverage, altered land use patterns, and changed the SOC storage. Although many field experiments have been performed to explore the impacts of both drying and warming climate and LCLUC on soil carbon storage on the Loess Plateau [19,20], little attention has been paid to the regional impacts of these factors and their interactions.

Over recent decades, many field observations and control experiments have been conducted to explore the effects of climate and land use change on the SOC in this region and make it possible to study the regional effects of climate change and LCLUC on SOC. In addition, many approaches, including eddy covariance flux tower, inventory, remote sensing techniques, forward and inversion models, have been used to examine the regional carbon budget on the Loess Plateau [21,22,23,24,25]. Among them, process-based ecosystem modeling is one of the most effective approaches to estimate regional SOC storage and fluxes in different terrestrial ecosystems driven by multiple global changes factors [7,11,24,26,27]. To address the effects of changes in climate, land use, and other environmental factors on SOC storage in this region, the Dynamic Land Ecosystem Model (DLEM), a highly integrated process-based model [28], was

applied to evaluate the spatial and temporal patterns of SOC storage on the Loess Plateau during 1961–2005. The objectives of this study are: 1) to investigate the temporal and spatial patterns of SOC storage on the Loess Plateau; and 2) to identify the relative contribution of climate and land use changes to the SOC storage changes.

Methods

Model Description

The DLEM is a process-based terrestrial ecosystem model, which aims at simulating the impacts of natural and anthropogenic disturbances on the structure and functions of terrestrial ecosystems over the spatial and temporal contexts. The DLEM has been widely used to simulate the effects of climate variability and change, elevated atmospheric CO_2, tropospheric ozone pollution, land use change, and increasing nitrogen deposition, etc. on terrestrial carbon storage and fluxes in China and other regions across the globe [15,29,30,31,32,33].

In this study, DLEM simulates two kinds of LCLUC: land conversion from natural ecosystem to cropland, and cropland abandonment. In the DLEM model, the balance of soil organic matter depends on the transformation of litter (LIT) to soil organic matter, the fractions of conversion from gross primary production (GPP) to dissolved organic carbon (DOC), the returned organic matter from production decay (PRD) (e.g., manure), the growth of microbe, the methane production from dissolved organic carbon, and the carbon loss from soil organic matter (SOM) decomposition.

$$\frac{dC_{som}}{dt} = k_{tr}LITC_{loss} + k_{gppdoc}GPP + k_{prd}PRD_{docom}$$
$$- k_{rh}SOMC_{docom} - k_{Lucc}C_{som} - DOC_{loss,methane}$$

where k_{tr} is the transfer rate of decomposed LIT to SOM; k_{gppdoc} is the fraction of GPP converted to soil DOC; k_{prd} is the returned rate of decomposed (or consumed) PRD to SOM pools as manure; k_{rh} is the fraction of decomposed SOM that is converted to CO_2 through heterotrophic respiration; k_{Lucc} is coefficient for quick carbon loss from SOM due to land use conversion; $DOC_{loss,methane}$ is DOC consumed for the growth of production of methane. More

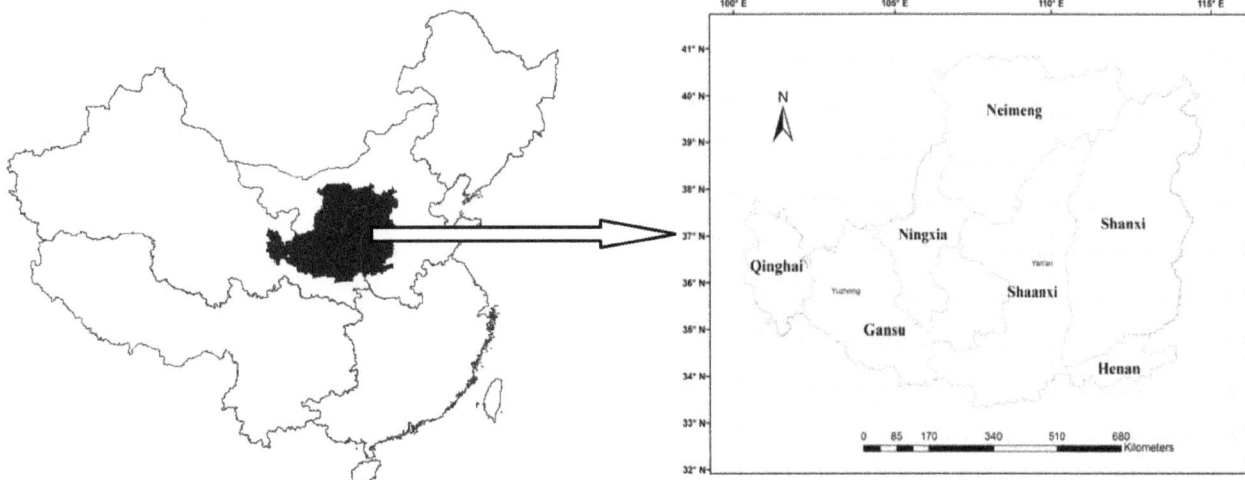

Figure 1. Location of the Loess Plateau, China.

detailed processes were described in our previous papers [15,29,30].

Input Data Description

The major input data in the DLEM include: (1) daily climatic data (i.e. maximum, minimum, average temperature, precipitation, relative humidity, and radiation) and atmospheric chemistry data (i.e. tropospheric O_3, atmospheric CO_2 and nitrogen deposition); (2) soil properties (including soil type, bulk density, depth, pH, soil texture) which are derived from the 1:1 million soil map based on the Second National Soil Survey of China [34,35,36]; (3) contemporary vegetation map for 2000 which was developed from Landsat Enhanced Thematic Mapper (ETM) imagery [37]; (4) long-term land use history which was developed on the basis of three recent (1990, 1995 and 2000) land cover maps and historical census datasets [38,39]. All the input datasets were developed at the spatial resolution of 10 km×0 km. Detailed information about other input data were described in our previous studies [15,29,30,40].

Climate change. Average air temperature on the Loess Plateau increased at a rate of 0.030°C/year from 1961 to 2005 (Figure 2a), higher than those reported for the entire China (about 0.029°C/year) and the global average level (about 0.010°C/year) [41] in the same period. The most rapid increase in temperature occurred during the 1990s. Air temperature increased from the north to the south of the Loess Plateau (Figure 2b), with the greatest increase in the northern Loess Plateau (e.g., the north of Shanxi and the northwest of Inner Mongolia).

The Loess Plateau can be mainly divided into three climate zones according to the precipitation: the northern Loess Plateau

with precipitation below 400 mm, the central Loess Plateau with precipitation between 400 and 500 mm, and the southern Loess Plateau with precipitation above 500 mm [42,43]. The precipitation less than 550 mm/year occurred across most areas of the Loess Plateau. Over the past 45 years, the mean annual precipitation was approximately 423 mm, with the lowest of 288 mm in 1997 and the highest of 661 mm in 1964 (Figure 2c). A slightly decreasing trend at a rate of 1.27 mm/year in precipitation was found on the Loess Plateau from 1961 to 2005. This temporal trend of precipitation was consistent with a previous study based on the meteorological observations according to 99 stations for the period 1956–2005 on the Loess Plateau [44]. Figure 2d further indicated that decreases in precipitation occurred in the most areas of the central and southern Loess Plateau from 1961–1990 to 1991–2005. The regions with the most obvious drying trend were located in the central and southern Loess Plateau, especially in the north of Shaanxi and the center of Shanxi Province with a reduction of more than 80 mm in the recent 15 years (1991–2005) compared to the 1961–1990 average.

LCLUC. Expansion of cropland and pasture was driven by social-economic factors on the Loess Plateau during the past decades, though to some extent, the soil erosion and frequent drought events limited the massive expansion of cropland. Since the 1950s, many measures have been implemented to alleviate and control serious soil erosion on the Loess Plateau. For example, "Grain for Green" project, which was launched in 1999, recommended that cropland with slopes greater than 15° should be converted back to natural vegetation. Driven by such governmental policies, the Loess Plateau experienced a remarkable change in land use, characterizing by a large area of

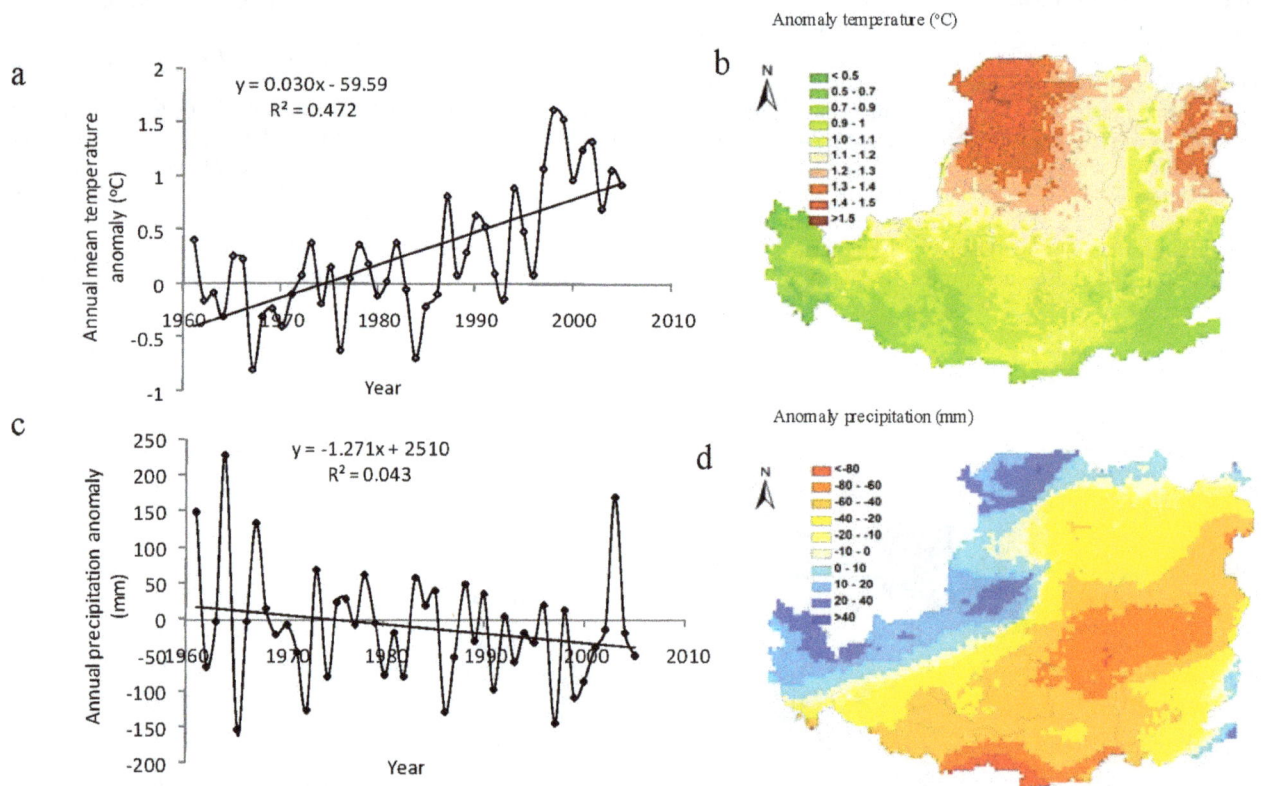

Figure 2. Anomaly of annual mean temperature (a), and precipitation (c) on the Loess Plateau from 1961 to 2005; and spatial distribution of temperature anomaly (b) and precipitation anomaly (d) during 1991–2005 (relative to 1961–1990 average);

conversion from cropland to grassland/forest. In order to well understand the LCLUC history on the Loess Plateau, we analyzed the spatial and temporal variations from 1961 to 2005 (Figure 3–4). Figure 3 showed that cropland area on the Loess Plateau decreased slowly until 1999, and followed by a sharp decrease. However, the grassland area showed an opposite changing trend in the same periods. Since 1961, the area of cropland decreased by 19.61%, while grassland and forest increased by 7.31% and 6.75%, respectively.

Land use patterns exhibited large spatial variations on the Loess Plateau (Figure 4). Grassland was the dominant vegetation type in the northern Loess Plateau. Cropland, grassland and shrubland were the main vegetation type in the middle Loess Plateau. In contrast, cropland and forest occupied large area of the southern Loess Plateau. Compared to 1960, the coverage of cropland decreased mainly due to land conversion from cropland to grassland, especially in the middle and southern part of Loess Plateau. During 1999–2005, cropland largely shrank in some areas of the middle Loess Plateau, especially in the northern Shaanxi and middle of Shanxi Province. These results were consistent with previous studies [18,45].

Model Parameterization and Evaluation

DLEM has been well calibrated and intensively validated against the site-level observed carbon fluxes and pool sizes from Chinese Ecosystem Research Network (CERN) and other previous studies [29,38,46]. In this study, we further compared our simulated results with field observation data and survey data from the Second National Soil Survey (1979–1983) to evaluate the model performance in simulating the SOC storage on the Loess Plateau. The simulated SOC storage as influenced by multiple environmental factors significantly correlated with the observed data ($R^2 = 0.738$, $p<0.01$, Figure 5), indicating that DLEM could capture the spatial and temporal patterns of SOC storage on the Loess Plateau. Several factors may contribute to the difference between model results and field observations. First, the input datasets were developed at a spatial resolution of 10 km×10 km for driving DLEM simulations. Each grid was assumed to have the uniform climate, land use type, and vegetation cover in the model. As a well-known climate-sensitive zone and a fragile ecological belt, some subtle changes in climate, land use or other factors on the Loess Plateau might cause large difference in the SOC storage even within one grid cell. This difference in the same grid will be reflected in the field experiments but neglected in the model

simulations. Second, the shortage of field observation data may weaken the capability of model to realistically capture the magnitude and patterns in SOC changes, which has long been identified as one of the biases in the large-scale model development. In addition, model simplification and neglecting microbial biomass might be another potential reason [13,30].

Simulation Experimental Design and Model Run

In this study, four main simulation experiments were designed to analyze the effects of climate change alone, LCLUC alone, the interaction of climate and LCLUC, and the combined effects of all environmental factors on the SOC storage on the Loess Plateau (Table 1). The experiment I was designed to provide a 'best-estimate' of spatial and temporal patterns of SOC driven by major environmental factors changes including climate variability, LCLUC, elevated CO_2, N fertilizer, N deposition, and O_3 pollution, etc. In experiment II and III, we simulated the contributions of climate variability alone and the LCLUC alone, respectively. In the experiment IV, we tried to understand the interactive effect between climate change and LCLUC on the SOC storage. The model simulations began with an equilibrium run to obtain the baseline carbon pools for each grid. A spin-up of about 100 years was applied if the climate change was included in the simulation experiments. Finally, the model was run in transient mode driven by the daily climate data and other time-variant or invariant input data.

Results and Analysis

Temporal Changes in SOC on the Loess Plateau

Model simulation indicated that the SOC storage over the entire Loess Plateau displayed substantial temporal and spatial variations in the context of multiple environmental factors changes (Figure 6a, b). As a whole, the combination of all these environmental factors considered in this study (e.g. climate, LCLUC and others environment factors) caused a net increase of about 0.27 Pg C in SOC storage from 1961 to 2005. The SOC storage kept relatively stable before the 1970s, and then gradually increased in the following two decades, and rapidly increased since 2000 (Figure 6a). Our further analysis found that the decrease of cropland area was relatively slower from the 1960s to 1999, and became rapid after then. Most of the abandoned cropland were replaced by grassland (Figure 3). Our results implied that temporal pattern in SOC change was partly related to the land use change on the Loess Plateau.

Spatial Variation in SOC on the Loess Plateau

We found that the spatial patterns of SOC storage were primarily controlled by the precipitation distribution. Spatially, the SOC storage increased gradually from the north to the south along an increasing precipitation gradient on the Loess Plateau. Large SOC increases were found in the southern and central Loess Plateau (south of Shaanxi and Shanxi Province in particular); and some other areas show a slight increase of SOC storage in the past decades. Due to changes in multiple environmental factors, the SOC storage increased throughout the majority of the Loess Plateau over the past 45 years (Figure 6b). Some areas showed a significant increase in SOC storage of more than 200 g C/m^2. However, a significant decrease occurred in the middle of central Loess Plateau (especially in the Northern Shaanxi Province), where experienced the most obvious warming and drying tendency in the past decades (Figure 2 b,d), releasing more than 100 g C/m^2 during the study period. We further found that the SOC increase

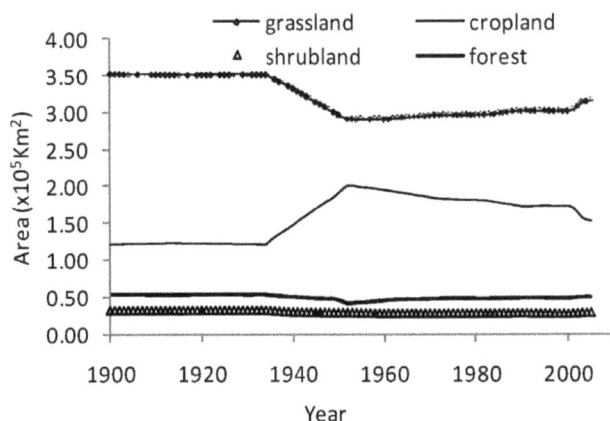

Figure 3. Area of major land use cover on the Loess Plateau during 1900–2005.

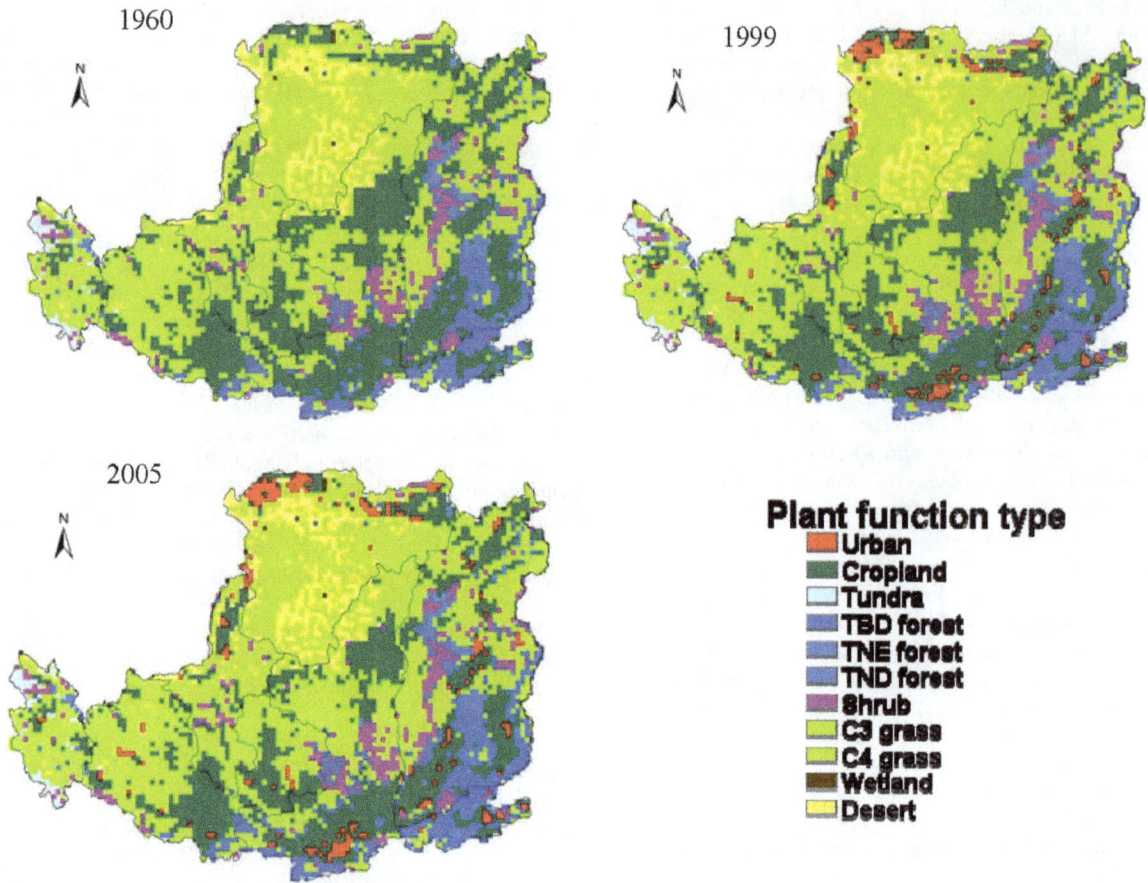

Figure 4. Spatial variations in LCLUC on the Loess Plateau in different year.

in the northern Loess Plateau was lower than that in the southern Loess Plateau during the past decades.

Individual Factorial Contributions to Changes in SOC Storage

To well understand the influence of climate on the SOC storage, we simulated the SOC storage on the Loess Plateau under the influence of climate change alone. DLEM simulated results

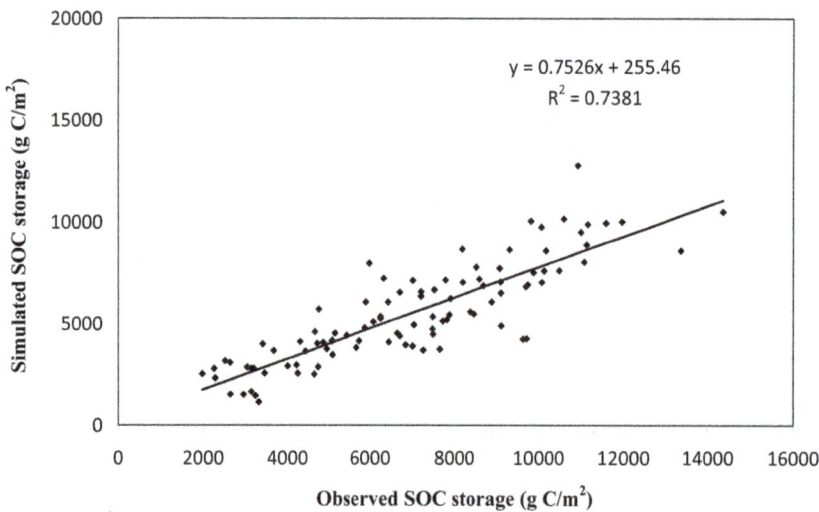

Figure 5. Correlation between simulated and observed SOC storage based on 91 soil samples.

Table 1. Simulation experiments.

Simulation Experiment	Environmental factors		
	Climate	LCLUC	Others
I. All	1960–2005	1960–2005	1960–2005
II. Climate only	1960–2005	1960	1960
III. LCLUC only	1960	1960–2005	1960
IV. Climate-LCLUC	1960–2005	1960–2005	1960

Notes: Climate-LCLUC means the combination effects of climate and LCLUC. Others include the effects of atmospheric CO_2, ozone pollution (AOT 40 index), nitrogen deposition, nitrogen fertilizer application, etc. All means the simulation experiment which includes all above environmental factors.

showed that SOC storage decreased by about 0.05 Pg C with substantial inter-annual fluctuations throughout the Loess Plateau from 1961 to 2005 (Figure 6c). SOC storage showed a slowly decreasing trend before the late 1990s, followed by a sharp decrease until the early 21st century. Since 1961, SOC storage decreased in most areas of the Loess Plateau under the influences of the climate, with a maximum carbon release of 200 g C/m^2 in some areas of the southern and the central Loess Plateau (Figure 6d). Figure 6d also indicated that SOC storage decreased in the southeastern humid monsoon climatic regions, and to a lesser extent in the continental dry climatic regions in the northern Loess Plateau during 1961–2005.

Considering the single effect of LCLUC, our results showed that the SOC storage increased by 0.14 Pg C during 1961–2005 (Figure 6e). We found that the most rapid increase in the SOC storage occurred after the late 1970s (Figure 6e). DLEM simulation results also showed that obvious increases of SOC storage were found in most areas of the central and the southern Loess Plateau (Figure 6f).

Relative Contributions of Climate, LCLUC, and their Interactions

Our simulated results indicated that with multiple environmental changes, SOC continuously increased from 1961 to 2005 on the Loess Plateau, resulting in a net increase in SOC storage by 0.27 Pg C (5.97 Tg C/year) (Figure 7). Among these factors, LCLUC was obviously the major factor affecting the magnitude of SOC change on the Loess Plateau, leading to a significant increase (0.14 Pg C) in SOC storage, accounting for 55% of the net increase in SOC in the past 45 years. However, in the same period, warming and drying climate greatly reduced SOC storage by 0.05 Pg C, approximately 19% of the total SOC storage change. The interaction between climate and LCLUC contributed to the net SOC increases by 3%.

Discussion

SOC Storage Change on the Loess Plateau

In this study, the combination of all environmental factors (climate, LCLUC and others environment factors) caused a net increase of about 0.27 Pg C in SOC storage, indicating that soil acted as a weak carbon sink on the Loess Plateau during the past 45 year. The temporal pattern of SOC storage was largely influenced by land use change and relevant land use policies.

Except the central Loess Plateau, the SOC storage increased throughout the entire region, with a smaller sink in the northern Loess Plateau and a larger sink in the southern Loess Plateau during 1961–2005, which is consistent with previous reports [43,47,48]. The less increase of SOC storage in the northern Loess

Plateau might be due to more sandy soil, lower soil carbon input from plant biomass, and larger soil carbon loss from erosion. This is consistent with previous reports, indicating the soil in the northern Loess Plateau contained more sand which could accumulate less carbon than the soil with more clay [49,50,51,52]. On the other hand, the area of forest and shrubland, characterized by higher aboveground biomass and productivity than cropland and grassland [43,53], decreased from the south to north over the Loess Plateau (Figure 4). This also contributed to realtively less SOC increase in the northern Loess Plateau. In addition, the extensive soil erosions in the northern Loess Plateau was suggested to further explain the lower contents of SOC storage [2,17,54].

Comparisons with Field Observations

Our estimation of SOC storage and its change on the Loess Plateau were comparable to other studies. Based on the 0–100 cm SOC data collected from 382 sampling sites across the entire Loess Plateau in 2008, Liu et al. [55] indicated that mean SOC storage was 7.70 kg C/m^2. Using data gathered by the Second National Soil Survey of China (1979–1983), Xu et al. [56] estimated that SOC storage was 1.07 Pg in 0–20 cm soils layers on the Loess Plateau region (with a land area of 429,800 km^2). A soil survey conducted throughout the Loess Plateau region indicated that the SOC storage amounted to 1.23 Pg C in the 0–20 cm soil layers with land area of 592,900 km^2 during 1985–1988, [55]. Converting SOC storage from different soil depth to the same soil depth according to the summary of the vertical distribution of soil organic in the 0–100 cm soil [31], SOC storage were 6.22 kg C/m^2 and 5.19 kg C/m^2 in Xu et al.'s [57] and Liu et al.'s [55] reports (be mentioned but not published) respectively. In the same soil depth, DLEM-estimated average SOC storage was 6.45 kg C/m^2 in 2005, which falls in the range of previous studies.

Tian et al. [7] estimated an average SOC sink of 94 Tg C/year in the terrestrial ecosystem of China as influenced by multiple global change factors during 1961–2005. Our results showed that soil on the Loess Plateau acted as a weak sink of 5.97 Tg C/year under the combined experiment during the same period. Considering that the Loess Plateau accounts for approximately 6.5% of the country's land surface, implying that the SOC storage on the Loess Plateau was close to the country-level average. At national level, Wang et al. [35] suggested that SOC storage in China decreased by about 5.69 g C/m^2 per year during the 1960s–1980s, while our study showed that the SOC storage on the Loess Plateau significantly increased by 4.30 g C/m^2 per year during the same period. This indicated that successful implementation of land conservation measures was beneficial and effective to reduce soil erosion and improve soil properties, thus enhance the soil carbon sequestration over the Loess Plateau.

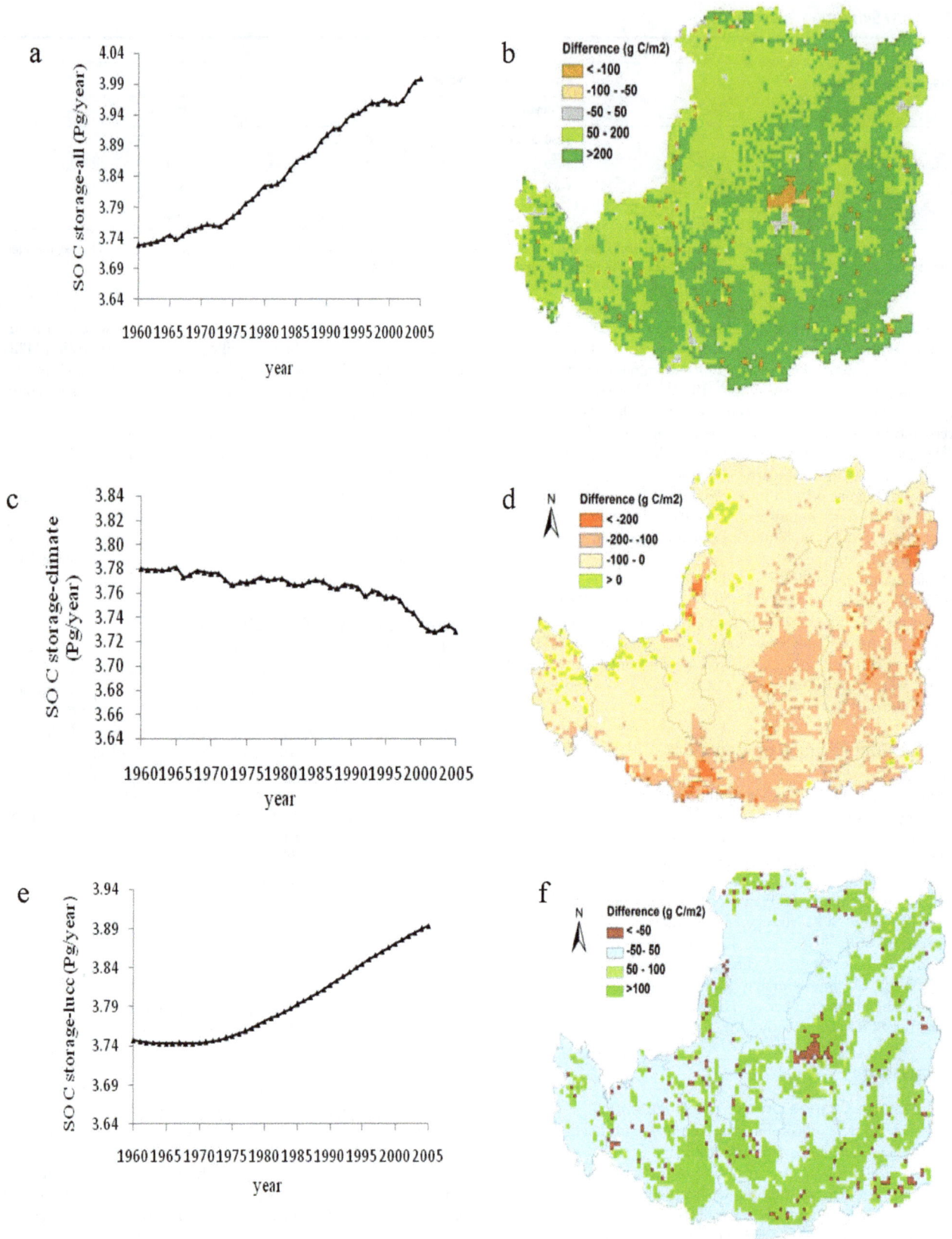

Figure 6. Spatiotemporal variations of SOC storage under experiments I, II, and III on the Loess Plateau from 1960–2005. Note: The "difference" means change in SOC storage between 2005 and 1960.

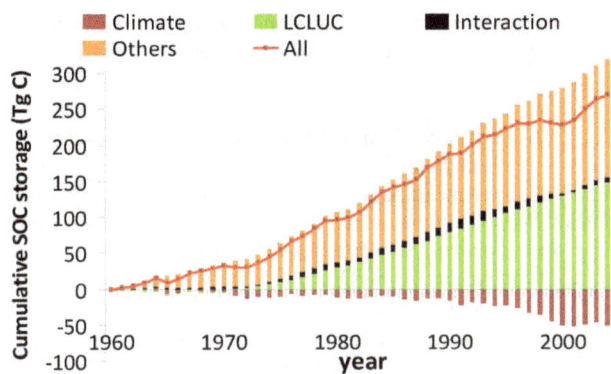

Figure 7. Contributions of multi-factor global changes to accumulative SOC change during 1961–2005. Note: "interaction" referrers to the interactive effects between climate and LCLUC.

Climate Controls on the Spatiaotemporal Patterns of SOC Storage

Climatic factors, especially precipitation and temperature, play an important role in long-term variations of SOC due to their effects on the quantity and quality of organic residue inputs and on the rates of soil organic matter and litter decomposition [1,12,58]. DLEM-simulated results showed that SOC storage decreased by about 0.05 Pg C with a significant inter-annual fluctuations throughout the Loess Plateau from 1961 to 2005 as influenced by climate change alone (Figure 6c). As shown in Figure 2, precipitation decreased at a rate of 1.27 (mm/year) during 1961–2005, which was a notable factor that influenced the change of SOC on the Loess Plateau, especially after 1990. Meanwhile, increased air temperature (0.030°C/year) might accelerate the evapotranspiration and potentially aggravate the water deficiency, thus cause the formation of drying soil layer and suppress the growth of vegetation. Due to the warming and drying climate, dried soil layer was widely distributed in the hilly and gully areas of the Loess Plateau. The development of drying soil layer has been regarded as a key cause for the decrease of SOC storage in some areas on the Loess Plateau in the past decades [59].

Climate also had a substantial effect on the spatial distribution of SOC storage. Our results suggested that SOC storage decreased in most areas of the Loess Plateau under the influences of the climate change since 1961 (Figure 6d). This result could be explained partly by the distribution of temperature and precipitation, which is well-known to have a positive relationship with SOC decomposition [60]. During 1961–2005, both temperature and precipitation were higher in the central and southern Loess Plateau, compared to the northern Loess Plateau. Therefore, the higher loss of soil carbon from decomposition might be an important cause for lower SOC accumulation in the central and southern Loess Plateau. This result was supported by previous studies [43,50,57].

Land Use Controls on the Spatiaotemporal Patterns of SOC Storage

During the past decades, the Loess Plateau has experienced a complex change in land use pattern (Figure 3). Considering the LCLUC alone, DLEM-simulated results showed that the SOC storage increased by 0.14 Pg C during 1961–2005 (Figure 6e), which was significantly larger than the influence of climate change alone. However, in the first few years, the SOC changed slowly, even had a decreasing tendency in the 1960s, and then increased gradually since the 1980s.

Since the 1950s, various soil and water conservation measures including afforestation, cropland abandonment, and terrace construction etc., have been implemented in this region [17,18]. All of these measures directly or indirectly affected the vegetation cover and further influenced the SOC storage on the Loess Plateau. Previous studies demonstrated that soil can lose up to 20–40 percent of organic carbon into the atmosphere when perennial vegetation land was converted into cultivation land [61,62]. On the contrary, conversion from cropland to perennial vegetation land or shrubland was found to accumulate SOC by increasing carbon derived from new vegetation and decreasing carbon loss from decomposition and erosion [11,43,51,63]. Liu et al. [47] reported that SOC in shrubland, which was converted from

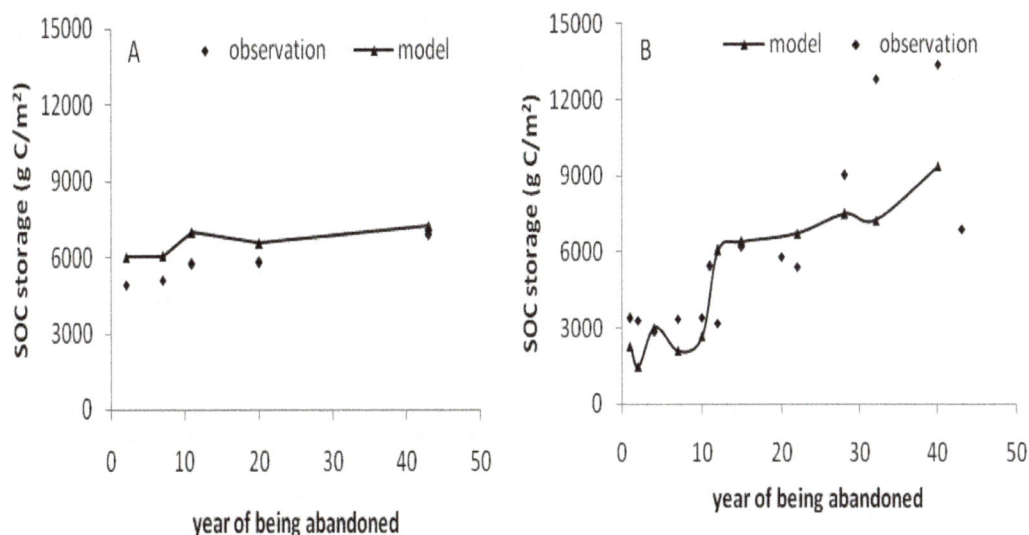

Figure 8. Comparison between simulated (DLEM) and observed SOC storage. Notes: A: cropland abandoned in Yuzhong (Gansu Province), the observation data come from Jinping Jiang et al. [67]; B: cropland abandoned in Yan'an (Shaanxi Province), the observation data come from Junmin Wang et al. [2].

cropland in 1985, were 27.7%–34.8% higher than that of the cropland in 2010 on the Loess Plateau. Fu et al. [64] suggested that cropland abandonment significantly increased the density and stock of SOC in 0–100 cm soil profiles on the Loess Plateau. Feng et al. [48] found a total of 96.1 Tg of additional carbon had been sequestered on the Loess Plateau since China's "Grain for Green" program during 2000–2008, by using remote sensing techniques and ecosystem modelling. Our results are comparable with those previous findings.

LCLUC might induce an immediate change in vegetation coverage but a lagged effect on the change of SOC storage [65]. In this study, the SOC changed slowly, and even had a decreasing trend in the 1960s. We also found that the most rapid increase in the SOC storage occurred after the late 1970s, which might be partly due to extensive and conducted tree planting projects for alleviating soil erosion. The large-scale cropland abandonment in response to the recently implemented "Grain for Green" policy also contributed to the increase of SOC storage on the Loess Plateau. The plantation from "Grain for Green" Project would keep a large proportion of carbon in wood which need long time to return to soil. During the initial period of land conversion from cropland to forests, leaf biomass of trees were very low, so the litters on the ground decreased and resulted in a slight change (even decrease) in SOC. During the period of implementing "Grain for Green" project, litters on the ground accumulated gradually with the growth of planted trees, resulting in increasing SOC storage after a certain time period. Compared with simulated results in the context of multiple environmental factors, a similar temporal pattern for the SOC change was found when considering LCLUC alone from 1961 to 2005 (Figure 6a,e). This further implied that LCLUC was the dominant factor controlling temporal variations of SOC storage.

DLEM simulated results also showed obvious increases of SOC storage in most areas of the central and southern Loess Plateau (Figure 6f). Further analysis found that changes in SOC storage were smaller in the LCLUC alone simulation experiment than that with the combination of all environmental change (Figure 6b). However, SOC storage changed slightly in most areas of the northern Loess Plateau from 1961 to 2005, which were also suggested by other research results (e.g., [18,43]).

Interactive Effects of LCLUC and Climate on SOC Storage

Based on DLEM simulations, Tian et al. [7] found that LCLUC accounted for 17% of the net carbon increase, but climate change reduced it by 4% in China in the past decades. However, we found that LCLUC increased SOC storage by 0.14 Pg C, accounting for 55% of the net increase in SOC in the past 45 years over the Loess Plateau. In the same period, warming and drying climate greatly reduced SOC storage by 0.05 Pg C, approximately 19% of the total SOC storage change. These results implied that the SOC was more sensitive to LCLUC and climate change on the Loess Plateau comparing to other regions in China. In the past decades, LCLUC, particularly in the southern Loess Plateau, enhanced the vegetation coverage and reduced the anthropogenic disturbance, which further enhanced the SOC storage.

The interaction between climate and LCLUC contributed to the net SOC increases by 3%. Our results also showed that the interactive effects among environmental factors can't be neglected in attributing the changes of SOC storage in response to environmental factors on the Loess Plateau. Although the interaction among environmental factors has been recognized long before [66], most of the field experiments still overlook it. This study further demonstrated that the modeling approach may

serve as one complementary tool for the field experiments in addressing interactive effects among multiple environmental factors.

Effects of Cropland Abandonment on SOC Storage

Cropland abandonment and natural vegetation recovery were important implementations to mitigate soil loss on the Loess Plateau. Previous studies demonstrated that former land use types, soil property, climate change, and soil management practices were crucial to changes of SOC storage in the establishment of perennial vegetation type [49,50]. Other factors, such as the abandonment age, have also been found to play a significant role in SOC accumulation and should not be ignored [4,50].

In this study, we chose two sites (Yuzhong City and Yan'an City) (Marked in Figure 1) located on the Loess Plateau to explore SOC storage pattern at different abandonment stages (or restoration age) and to further evaluate the DLEM simulated results against the observations (Figure 8). It showed that DLEM results could well capture the distribution characteristics of the SOC storage in different abandonment stages. Both model results and field observations demonstrated that the restoration age played a key role in SOC accumulation. Jiang et al. [67] indicated that SOC storage changed little in early abandonment stage in a semiarid hilly area of Yuzhong City, followed by an obvious increase after 9–12 years, and then increased stably (Figure 8A). Wang et al. [2] found the similar temporal pattern through studying the change of SOC storage during different successional stages of rehabilitated grassland in Yan'an City. Compared to cropland, the rehabilitated grassland had a lower SOC storage at the early stage (1–12 years). However, SOC storage was higher than that in cropland after 15 years and then increased steadily (Figure 8B). Generally, long-term abandonment (>10 years) and the following colonization of natural vegetation could lead to substantial increase of SOC storage. The difference of the SOC storage between two sites over the early abandonment stage might be attributed to difference in environmental factors such as climate and soil property. Yan'an City is located in a semi-arid and warm temperate zone with less precipitation and more obvious drying and warming tendency, compared with Yuzhong City during the past years. This also partly explained the lower SOC storage in cropland in this climate zone than others over the Loess Plateau.

Our results indicated that cropland abandonment could increase SOC storage, improve soil quality and promote ecosystem restoration, especially in the warm and dry climate zone. However, these effects will emerge after a relative longer time period (e.g., >10 years) under local environmental conditions.

Uncertainties

This study examined temporal and spatial patterns of SOC storage and attributed these patterns to multiple environmental factors on the Loess Plateau during 1961–2005. There were several uncertainties which need to be addressed in our future work. First, impacts of some ecological processes, such as soil erosion, were not separated from other environmental factors in this study. The soil erosion area could be as high as 45.4×10^4 km^2 (72.3% of the land area) over the Loess Plateau. Soil erosion has long been identified as one of key factors controlling the change of SOC storage [47]. Further efforts should be put on interactions among soil erosion and other environmental factors. Second, the input datasets were developed at the spatial resolution of 10 km × 10 km which was the finest dataset for the Loess Plateau. However, it is still difficult to capture subtle change of SOC storage due to high heterogeneity in land surface processes. In the long run, finer gidded datasets would be greatly helpful for further quantifying temporal and spatial changes in SOC storage on the Loess Plateau. In addition, the uncertainties from other input data, model structure, and parameterization need to be further specified in the furture efforts.

Conclusions

This study examined effects of climate and land use changes on SOC storage on the Loess Plateau of China in the context of multiple global changes by using an integrated ecosystem model. The results showed that temperature on the Loess Plateau has significantly increased, while precipitation slightly decreased during 1961–2005. Meanwhile, this region experienced a remarkable change in land cover and land use, characterized by conversions from cropland to grassland/forest owing to the government policies to alleviate soil and water losses during the past decades. The overall change in SOC storage due to multiple environmental factors was estimated to be a net increase of SOC storage by 0.27 Pg C during 1961–2005, indicating that soil on the Loess Plateau acted as a carbon sink in this period. Among multiple factors, LCLUC led to a significant increase in SOC storage of 0.14 Pg C, accounting for 55% of the net increase in SOC. In contrast, climate change reduced SOC storage by 0.05 Pg C (approximately 19% of the total SOC change). The interaction of climate and LCLUC accounted for 3% of the net increase in SOC. Our results were consistent with field observation data and both of them suggested that SOC storage could be enhanced significantly by the conversion of cropland to grassland along with the increasing abandonment age on the Loess Plateau.

However, the magnitude could be influenced by local environmental conditions.

This study provides the first attempt to quantify relative effects of multiple environmental factors (climate and LCLUC in particular) on regional SOC storage on the Loess Plateau over the past decades. The results drawn from this study provide insight for land management as well as policy-making to enhance carbon sequestration and alleviate the serious soil erosion conditions on the Loess Plateau. To reduce uncertainties in estimating effects of climate and land use changes on SOC storage, it is needed to put further efforts in developing more reliable and fine-resolution input data and improve model representation of some other processes relevant to SOC, such as soil erosion.

Acknowledgments

This study was supported by NASA Land Cover and Land Use Change Program (NNX08AL73G), NASA Interdisciplinary Science Program (NNG04GM39C), Chinese Universities Scientific Fund (z10921007), US National Science Foundation Grants (AGS-1243220, CNS-1059376), State Key Laboratory of Soil Erosion and Dryland Farming on the Loess Plateau Foundation (K318009902-1410), and Shaanxi Administration of Foreign Expert Affairs Science and Technology Activities Fundation (201327). Thank Mingliang Liu, Xiaofeng Xu, and Jiyuan Liu for their contributions in the development of spatial data and the DLEM model.

Author Contributions

Conceived and designed the experiments: HT YD WR. Performed the experiments: YD WR. Analyzed the data: YD WR BT GC. Contributed reagents/materials/analysis tools: GC BT JY SP. Wrote the paper: YD WR HT BT GC CL JY SP GW SL.

References

1. Lal R (2004) Soil carbon sequestration impacts on global climate change and food security. Science 304: 1623–1627.
2. Wang Y, Fu B, Lü Y, Chen L (2011) Effects of vegetation restoration on soil organic carbon sequestration at multiple scales in semi-arid Loess Plateau, China. Catena 85: 58–66.
3. Post WM, Emanuel WR, Zinke PJ, Stangenberger AG (1982) Soil carbon pools and world life zones.
4. Zhang K, Dang H, Tan S, Cheng X, Zhang Q (2010) Change in soil organic carbon following the 'Grain-for-Green' programme in China. Land Degradation & Development 21: 13–23.
5. Eglin T, Ciais P, Piao S, Barre P, Bellassen V, et al. (2010) Historical and future perspectives of global soil carbon response to climate and land-use changes. Tellus B 62: 700–718.
6. Martin D, Lal T, Sachdev C, Sharma J (2010) Soil organic carbon storage changes with climate change, landform and land use conditions in Garhwal hills of the Indian Himalayan mountains. Agriculture, Ecosystems & Environment 138: 64–73.
7. Tian H, Melillo J, Lu C, Kicklighter D, Liu M, et al. (2011) China's terrestrial carbon balance: Contributions from multiple global change factors. Global Biogeochemical Cycles 25.
8. Wang X, Feng Z, Ouyang Z (2001) The impact of human disturbance on vegetative carbon storage in forest ecosystems in China. Forest ecology and management 148: 117–123.
9. Barnett TP, Adam JC, Lettenmaier DP (2005) Potential impacts of a warming climate on water availability in snow-dominated regions. Nature 438: 303–309.
10. Paul EA (1997) Soil organic matter in temperate agroecosystems: long term experiments in North America: CRC PressI Llc.
11. Post WM, Kwon KC (2000) Soil carbon sequestration and land-use change: processes and potential. Global change biology 6: 317–327.
12. Lal R, Bruce J (1999) The potential of world cropland soils to sequester C and mitigate the greenhouse effect. Environmental Science & Policy 2: 177–185.
13. Schimel J (2001) 1.13-Biogeochemical Models: Implicit versus Explicit Microbiology. Global Biogeochemical Cycles in the Climate System: 177–183.
14. Liu S, Bliss N, Sundquist E, Huntington TG (2003) Modeling carbon dynamics in vegetation and soil under the impact of soil erosion and deposition. Global Biogeochemical Cycles 17.
15. Zhang C, Tian H, Pan S, Liu M, Lockaby G, et al. (2008) Effects of forest regrowth and urbanization on ecosystem carbon storage in a rural–urban gradient in the southeastern United States. Ecosystems 11: 1211–1222.
16. Li Z, Liu W-z, Zhang X-c, Zheng F-l (2009) Impacts of land use change and climate variability on hydrology in an agricultural catchment on the Loess Plateau of China. Journal of hydrology 377: 35–42.
17. Chen L, Gong J, Fu B, Huang Z, Huang Y, et al. (2007) Effect of land use conversion on soil organic carbon sequestration in the loess hilly area, loess plateau of China. Ecological Research 22: 641–648.
18. Xin Z, Xu J, Zheng W (2008) Spatiotemporal variations of vegetation cover on the Chinese Loess Plateau (1981–2006): impacts of climate changes and human activities. Science in China Series D: Earth Sciences 51: 67–78.
19. Fu B-J, Wang Y-F, Lu Y-H, He C-S, Chen L-D, et al. (2009) The effects of land-use combinations on soil erosion: a case study in the Loess Plateau of China. Progress in Physical Geography 33: 793–804.
20. Wang G, Huang J, Guo W, Zuo J, Wang J, et al. (2010) Observation analysis of land-atmosphere interactions over the Loess Plateau of northwest China. Journal of Geophysical Research: Atmospheres (1984–2012) 115.
21. Zhao M, Heinsch FA, Nemani RR, Running SW (2005) Improvements of the MODIS terrestrial gross and net primary production global data set. Remote Sensing of Environment 95: 164–176.
22. Fang J, Guo Z, Piao S, Chen A (2007) Terrestrial vegetation carbon sinks in China, 1981–2000. Science in China Series D: Earth Sciences 50: 1341–1350.
23. Peters W, Jacobson AR, Sweeney C, Andrews AE, Conway TJ, et al. (2007) An atmospheric perspective on North American carbon dioxide exchange: CarbonTracker. Proceedings of the National Academy of Sciences 104: 18925–18930.
24. Piao S, Ciais P, Friedlingstein P, de Noblet-Ducoudré N, Cadule P, et al. (2009) Spatiotemporal patterns of terrestrial carbon cycle during the 20th century. Global Biogeochemical Cycles 23.
25. Li K, Wang S, Cao M (2004) Vegetation and soil carbon storage in China. SCIENCE IN CHINA SERIES D EARTH SCIENCES-ENGLISH EDITION- 47: 49–57.
26. King J, Bradley R, Harrison R, Carter A (2004) Carbon sequestration and saving potential associated with changes to the management of agricultural soils in England. Soil Use and Management 20: 394–402.
27. Yan H, Cao M, Liu J, Tao B (2007) Potential and sustainability for carbon sequestration with improved soil management in agricultural soils of China. Agriculture, Ecosystems & Environment 121: 325–335.
28. Tian H, Chen G, Liu M, Zhang C, Sun G, et al. (2010) Model estimates of net primary productivity, evapotranspiration, and water use efficiency in the terrestrial ecosystems of the southern United States during 1895–2007. Forest ecology and management 259: 1311–1327.
29. Ren W, Tian H, Liu M, Zhang C, Chen G, et al. (2007) Effects of tropospheric ozone pollution on net primary productivity and carbon storage in terrestrial ecosystems of China. Journal of Geophysical Research: Atmospheres (1984–2012) 112.

30. Xu X, Tian H, Zhang C, Liu M, Ren W, et al. (2010) Attribution of spatial and temporal variations in terrestrial methane flux over North America. Biogeosciences Discussions 7: 5383–5428.

31. Tian H, Xu X, Lu C, Liu M, Ren W, et al. (2011) Net exchanges of CO2, CH4, and N2O between China's terrestrial ecosystems and the atmosphere and their contributions to global climate warming. Journal of Geophysical Research: Biogeosciences (2005–2012) 116.

32. Ren W, Tian H, Tao B, Huang Y, Pan S (2012) China's crop productivity and soil carbon storage as influenced by multifactor global change. Global Change Biology 18: 2945–2957.

33. Tian H, Chen G, Zhang C, Liu M, Sun G, et al. (2012) Century-scale responses of ecosystem carbon storage and flux to multiple environmental changes in the southern United States. Ecosystems 15: 674–694.

34. Shi H, Shao M (2000) Soil and water loss from the Loess Plateau in China. Journal of Arid Environments 45: 9–20.

35. Wang S, Tian H, Liu J, Pan S (2003) Pattern and change of soil organic carbon storage in China: 1960s–1980s. Tellus B 55: 416–427.

36. Zhang XP, Zhang L, McVicar TR, Van Niel TG, Li LT, et al. (2008) Modelling the impact of afforestation on average annual streamflow in the Loess Plateau, China. Hydrological Processes 22: 1996–2004.

37. Liu J, Tian H, Liu M, Zhuang D, Melillo JM, et al. (2005) China's changing landscape during the 1990s: Large-scale land transformations estimated with satellite data. Geophysical Research Letters 32.

38. Liu M, Tian H (2010) China's land cover and land use change from 1700 to 2005: Estimations from high-resolution satellite data and historical archives. Global Biogeochemical Cycles 24.

39. Tian H, Chen G, Zhang C, Melillo JM, Hall CA (2010) Pattern and variation of C: N: P ratios in China's soils: a synthesis of observational data. Biogeochemistry 98: 139–151.

40. Chen H, Tian H, Liu M, Melillo J, Pan S, et al. (2006) Effect of land-cover change on terrestrial carbon dynamics in the southern United States. Journal of environmental quality 35: 1533–1547.

41. Trenberth K, Jones P, Ambenje P, Bojariu R, Easterling D, et al. (2007) Observations: Surface and Atmospheric Climate Change, chap. 3 of Climate Change 2007: The Physical Science Basis. Contribution of Working Group I to the Fourth Assessment Report of the Intergovernmental Panel on Climate Change [Solomon, S., Qin, D., Manning, M., Marquis, M., Averyt, KB, Tignor, M., Miller, HL and Chen, Z.(eds.)]., 235–336. Cambridge University Press, Cambridge, UK and New York, NY, USA.

42. Li R, Yang W, Li B (2008) Research and future prospects for the Loess Plateau of China. SciencePress, Beijing (in Chinese).

43. Chang R, Fu B, Liu G, Liu S (2011) Soil carbon sequestration potential for "Grain for Green" project in Loess Plateau, China. Environmental management 48: 1158–1172.

44. Xin Z, Xu J, Ma Y (2009) Spatio-temporal variation of erosive precipitation in Loess Plateau during past 50 years. Scientia Geographica Sinica 29: 89–104.

45. Song Y, Ma M-g (2007) Study on vegetation cover change in Northwest China based on SPOT VEGETATION data. Journal of Desert Research 27: 89–93.

46. Ren W, Tian H, Xu X, Liu M, Lu C, et al. (2011) Spatial and temporal patterns of CO2 and CH4 fluxes in China's croplands in response to multifactor environmental changes. Tellus B 63: 222–240.

47. Liu X, Li F-M, Liu D-Q, Sun G-J (2010) Soil organic carbon, carbon fractions and nutrients as affected by land use in semi-arid region of Loess Plateau of China. Pedosphere 20: 146–152.

48. Feng X, Fu B, Lu N, Zeng Y, Wu B (2013) How ecological restoration alters ecosystem services: an analysis of carbon sequestration in China's Loess Plateau. Scientific reports 3.

49. Guo Z, Ruddiman WF, Hao Q, Wu H, Qiao Y, et al. (2002) Onset of Asian desertification by 22 Myr ago inferred from loess deposits in China. Nature 416: 159–163.

50. Paul K, Polglase P, Nyakuengama J, Khanna P (2002) Change in soil carbon following afforestation. Forest ecology and management 168: 241–257.

51. LAGANIÈRE J, ANGERS DA, PARÉ D (2010) Carbon accumulation in agricultural soils after afforestation: a meta-analysis. Global Change Biology 16: 439–453.

52. Cheng L, Zhao X (2011) Soil mineralized nutrients changes and soil conservation benefit evaluation on 'green project grain' in ecologically fragile areas in the south of Yulin city, Loess Plateau. African Journal of Biotechnology 10: 2230–2237.

53. Xiao Y (1990) Comparative studies on biomass and productivity of Pinus tabulaeformis plantations in different climatic zones in Shaanxi Province [China]. Acta Phytoecologica et Geobotanica Sinica 14.

54. Fu X, Shao M, Wei X, Horton R (2010) Soil organic carbon and total nitrogen as affected by vegetation types in Northern Loess Plateau of China. Geoderma 155: 31–35.

55. Liu Z, Shao Ma, Wang Y (2011) Effect of environmental factors on regional soil organic carbon stocks across the Loess Plateau region, China. Agriculture, Ecosystems & Environment 142: 184–194.

56. Xu X, Zhang K, Peng W (2003) Spatial distribution and estimating of soil organic carbon on Loess Plateau. J Soil Water Conserv 17: 13–15.

57. Xu X, Liu W, Kiely G (2011) Modeling the change in soil organic carbon of grassland in response to climate change: effects of measured versus modelled carbon pools for initializing the Rothamsted Carbon model. Agriculture, Ecosystems & Environment 140: 372–381.

58. Schlesinger WH (1990) Evidence from chronosequence studies for a low carbon-storage potential of soils. Nature 348: 232–234.

59. Han F, Hu W, Zheng J, Du F, Zhang X (2010) Estimating soil organic carbon storage and distribution in a catchment of Loess Plateau, China. Geoderma 154: 261–266.

60. Lehmann J, Skjemstad J, Sohi S, Carter J, Barson M, et al. (2008) Australian climate–carbon cycle feedback reduced by soil black carbon. Nature Geoscience 1: 832–835.

61. Houghton R, Hackler J, Lawrence K (1999) The US carbon budget: contributions from land-use change. Science 285: 574–578.

62. van der Werf GR, Morton DC, DeFries RS, Olivier JG, Kasibhatla PS, et al. (2009) CO2 emissions from forest loss. Nature Geoscience 2: 737–738.

63. Richter DD, Markewitz D, Trumbore SE, Wells CG (1999) Rapid accumulation and turnover of soil carbon in a re-establishing forest. Nature 400: 56–58.

64. Fu B-J, Zhang Q-J, Chen L-D, Zhao W-W, Gulinck H, et al. (2006) Temporal change in land use and its relationship to slope degree and soil type in a small catchment on the Loess Plateau of China. Catena 65: 41–48.

65. Kuzyakov Y, Gavrichkova O (2010) Review: Time lag between photosynthesis and carbon dioxide efflux from soil: a review of mechanisms and controls. Global Change Biology 16: 3386–3406.

66. Dermody O (2006) Mucking through multifactor experiments; design and analysis of multifactor studies in global change research. New Phytologist 172: 598–600.

67. Jiang J-P, Xiong Y-C, Jiang H-M, Ye D-Y, Song Y-J, et al. (2009) Soil microbial activity during secondary vegetation succession in semiarid abandoned lands of Loess Plateau. Pedosphere 19: 735–747.

7

Effect of Abandonment on Diversity and Abundance of Free-Living Nitrogen-Fixing Bacteria and Total Bacteria in the Cropland Soils of Hulun Buir, Inner Mongolia

Huhe[1¤], Shinchilelt Borjigin[1], Yunxiang Cheng[2], Nobukiko Nomura[1], Toshiaki Nakajima[1], Toru Nakamura[1], Hiroo Uchiyama[1*]

1 Graduate School of Life and Environmental Sciences, University of Tsukuba, Tsukuba, Ibaraki, Japan, 2 State Key Laboratory of Grassland Agro-Ecosystems, College of Pastoral Agriculture Science and Technology, Lanzhou University, Lanzhou, China

Abstract

In Inner Mongolia, steppe grasslands face desertification or degradation because of human over activity. One of the reasons for this condition is that croplands have been abandoned after inappropriate agricultural management. The soils in these croplands present heterogeneous environments in which conditions affecting microbial growth and diversity fluctuate widely in space and time. In this study, we assessed the molecular ecology of total and free-living nitrogen-fixing bacterial communities in soils from steppe grasslands and croplands that were abandoned for different periods (1, 5, and 25 years) and compared the degree of recovery. The abandoned croplands included in the study were natural restoration areas without human activity. Denaturing gradient gel electrophoresis and quantitative PCR (qPCR) were used to analyze the *nifH* and 16S rRNA genes to study free-living diazotrophs and the total bacterial community, respectively. The diversities of free-living nitrogen fixers and total bacteria were significantly different between each site ($P<0.001$). Neither the total bacteria nor *nifH* gene community structure of a cropland abandoned for 25 years was significantly different from those of steppe grasslands. In contrast, results of qPCR analysis of free-living nitrogen fixers and total bacteria showed significantly high abundance levels in steppe grassland ($P<0.01$ and $P<0.03$, respectively). In this study, the microbial communities and their gene abundances were assessed in croplands that had been abandoned for different periods. An understanding of how environmental factors and changes in microbial communities affect abandoned croplands could aid in appropriate soil management to optimize the structures of soil microorganisms.

Editor: Bas E. Dutilh, Radboud University Medical Centre, NCMLS, Netherlands

Funding: This work was supported by a Grant-in-Aid for Scientific Research (B) (No. 21310050 to HU) and CREST program of Japan Science and Technology Corporation (JST; http://www.jst.go.jp/EN/index.html). The funders had no role in study design, data collection and analysis, decision to publish, or preparation of the manuscript.

Competing Interests: The authors have declared that no competing interests exist.

* Email: uchiyama.hiroo.fw@u.tsukuba.ac.jp

¤ Current address: Institute of Soil and Fertilizer and Save Water Agricultural, Gansu Academy of Agricultural Sciences, Anning District, Lanzhou, Gansu, China

Introduction

Steppe grasslands are distributed over vast areas in the arid and semiarid regions of the Eurasian continent [1]. The Inner Mongolia steppe is an important part of Eurasia and has been used by pastoral nomads for long periods of time. However, vast areas were converted to cropland, and farmers have increased the size of these croplands during the past 40 years because of a rapid increase in the human population. Subsequently, many of the croplands were abandoned because of soil degradation and desertification caused by inappropriate agricultural management [2,3]. Further, some croplands were abandoned to restore natural vegetation, as in the case where the Chinese government proposed to restore farmlands to grasslands or forests [4]. Some plant species, including grasses, shrubs, and trees, are introduced into the abandoned croplands to restore the vegetation in some locations [5–8]. In most areas, the plant community of abandoned cropland is likely to be restored naturally by controlling human activities [9]. Many researchers have evaluated the abandoned

cropland ecosystem changes by studying the vegetation changes [9–12]. There are interactions between plants and microorganisms. Plants exude diverse compounds, such as organic acids, enzymes, and polysaccharides, from the roots. Further, plants can recognize microbe-derived compounds and adjust their defense and growth responses according to the type of microorganisms encountered. Conversely, microorganisms can detect suitable plant hosts and initiate their colonization strategies in the rhizosphere by producing canonical plant growth-regulating substances such as auxins or cytokinins [13]. However, most studies have not evaluated changes in soil microbial community and its gene abundance in croplands that have been abandoned for different periods of time.

Cultivated soil or grassland soil contains an estimated 2×10^9 prokaryotic cells per gram [14]. Soil microbial communities are important factors of agriculturally managed systems because they are responsible for most nutrient transformations in soil and influence the aboveground plant diversity and productivity [15]. Next to water, nitrogen is the second-most limiting factor for plant growth [16]. Nitrogen cycling in natural ecosystems and during traditional agricultural production rely on nitrogen fixation of diazotrophic bacteria [17]. Diazotrophs are highly diverse and widely distributed across bacterial and archaeal taxa [18]. Approximately 80% of biological nitrogen fixation is performed by diazotrophs in symbiosis with legumes [19]. However, under specific conditions, the free-living bacteria in soil (e.g., cyanobacteria, *Pseudomonas*, *Azospirillum*, and *Azotobacter*), may fix significant amounts of nitrogen $(0–60 \text{ kg·N·ha}^{-1} \text{·year}^{-1})$ [20,21]. This may be particularly important in abandoned field soils, where legume plants have not been cultivated and there are very few symbiotic plants (e.g., Leguminosae, *Azolla*, *Myrica*, and *Alnus.*).

In this study, the diazotrophic population was monitored by PCR-denaturing gradient gel electrophoresis (DGGE) exploiting the *nifH* gene. The *nifH* gene is the most conserved gene in the *nif* operon and encodes the Fe subunit of the nitrogenase enzyme [22]. Because of the conserved nature of the *nifH* gene, it has been possible to identify primer sets that can be used to analyze nitrogen fixers so that this community can be analyzed by a PCR-DGGE–based technique [20,23–25]. We have tested the diversity and abundances of free-living nitrogen fixers and the total bacterial population changes that occurred over time in soils belonging to artificially disrupted environments (abandoned cropland soils).

Materials and Methods

Ethics statement

No specific permissions are required for our conducting field survey in this area, since land in China belongs to the public and our field studies did not involve any endangered or protected plant species within.

Site description and sample collection

The study area is located in the Hulun Buir grassland (115°31′–126°04′ E, 47°05′–53°20′ N) in northeastern Inner Mongolia, China (Figure 1). The Hulun Buir grassland area is about $2.6 \times 10^5 \text{ km}^2$, with a west to east distribution of arid steppe, semi-arid steppe, and meadow steppe. The study area is located in the semi-arid areas.

We established 4 sites, 3 abandoned croplands and a light-grazing steppe grassland (LGSG) (Figure 1) that had an intensity of about 1.4 sheep ha^{-1}. The 3 croplands were abandoned for 1, 5, and 25 years (Y1, Y5, and Y25, respectively), and the control area was a LGSG. In the Y1, Y5, and Y25 sites, *Zea mays*, *Helianthus annuus*, and *Elymus cylindricus* were rotated for approximately 40 years. The sites were subsequently abandoned because of land degradation; soil fertility including both organic C and total N had decreased by approximately 70%.

Plant surveying and soil sampling were conducted in August 2010. All of the sites were selected for their similar topography (flat). Each site contained 5 replicates in a randomized plot (1 × 1 m) design: Y1 (site 1), plots 1–5; Y5 (site 2), plots 6–10; Y25 (site 3), plots 11–15; and LGSG (site 4), plots 16–20. The coordinates and elevations of the sampled sites are as follows: Y1, 48° 38′ 43″ N, 116° 57′ 56″ E, 545 m; Y5, 48° 38′ 50″ N, 117° 00′ 48″ E, 550 m; Y25, 48° 38′ 45″ N, 117° 01′ 56″ E, 545 m and LGSG, 48° 32′ 00″ N, 116° 40′ 18″ E, 568 m. The mean temperature and precipitation from 2000 to 2009 for each site [26] were as follows: LGSG, Y1, Y5, and Y25, 1.6°C and 213 mm. In each plot, the species composition was recorded. Plant communities were classified on the basis of their differential species [27,28]; all species were identified and measured for cover, height, and density, and Shannon-Wiener diversity index was calculated. Soil moisture was measured with a TRIME-FM (Ettlingen, Germany). Above-ground plant biomass was also determined by clipping the plants at ground level, sorting by species, drying at 60°C for 48 h, and weighing the samples (Table 1 and Table S1).

In each plot, the soil samples were collected from 5 randomly selected points (0 to 10 cm deep) and mixed into 1 sample. After carefully removing the surface organic materials and fine roots, each mixed sample was divided into 2 parts. One part was air-dried for the analysis of soil physicochemical properties. The other was sifted through a 2-mm sieve, sealed in sample vials, kept on ice for transport to the laboratory, and stored at −20°C for microbial assays.

The soil texture was determined by mechanical analysis using the pipette method [29], and the soil texture was used to estimate the saturated hydraulic conductivity [30] for each site (Table S2). The soil texture was classified according to the International

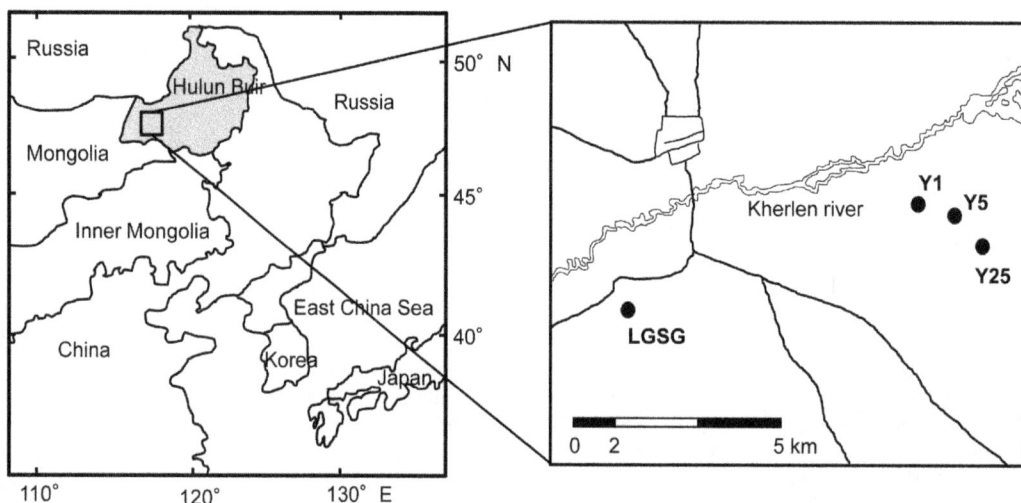

Figure 1. Map of the study area and the 4 study sites (·) in Hulun Buir.

Table 1. Changes to pH, available NO$_3$-N, available NH$_4$-N, available P, soluble Fe, organic C, total N, plant diversity (P-H'), plant biomass (P-B), soil moisture, and hydraulic conductivity (HC) across the field trial and Pearson's product-moment correlation analysis comparing data to $nifH$ and 16S rRNA diversity and gene copies.

Abandoned Cropland or significance parameter	pH	NO$_3$-N (mg kg^{-1})	NH$_4$-N (mg kg^{-1})	P (mg kg^{-1})	Fe (g kg^{-1})	Organic C (g kg^{-1})	Total N (g kg^{-1})	H$_2$O (%)	P-H'	P-B	HC (×10^{-3} cm s^{-1})
Abandoned Cropland*											
Y1	7.72±0.13 a	3.35±0.42 a	1.59±0.64 a	12.38±4.09 a	0.43±0.06 a	6.55±0.96 a	0.7±0.09 a	8.0±0 a	1.63±0.34 a	48.5±9.77 a	2.93±0.07 a
Y5	8.57±0.37 b	3.99±0.67 b	0.76±0.04 b	12.92±2.68 a	0.56±0.04 a	9.16±1.51 a	0.94±0.12 b	9.06±0.13 ab	1.71±0.43 a	17.16±4.44 b	2.95±0.03 a
Y25	7.82±0.17 a	3.54±0.53 a	1.09±0.21 a	12.5±1.8 a	0.46±0.04 b	7.23±0.64 a	0.75±0.06 ab	10.7±1.09 b	1.82±0.21 a	79.65±13.5 c	2.79±0.02 b
LGSG	6.21±0.14 c	4.21±0.06 c	4.84±0.37 b	24.48±5.67 b	1.12±0.06 c	24.84±3.09 b	2.3±1.71 c	12.12±1.4 bc	0.94±0.02 b	141.9±5.54 d	2.75±0.02 b
Correlation (P)**											
with $nifH$ DGGE H'	+++	++	NS	++	NS	NS	NS	NS	NS	NS	NS
with $nifH$ copy number	NS	NS	++	+++	+++	+++	++	++	—	++	NS
with 16S rRNA DGGE H'	NS	NS	NS	NS	+++	++	++	+++	NS	++	—
with 16S rRNA copy number	NS	NS	+++	NS	NS	++	++	NS	NS	++	NS

*Y1, Y5 and Y25 mean field abandoned for 1 year, 5 years and 25 years, respectively. LGSG is light grazing steppe grassland. The values shown for management factors are means ± standard errors.
**P, Pearson's product-moment correlation coefficient; NS, not significant; ++/—, significant positive or negative correlation at $P<0.05$; +++/—, significant positive or negative correlation at $P<0.01$. Significant differences are indicated by different letters.

Society of Soil Science (ISSS) classification system. The soils of all 4 sites were sandy loam.

Concentrations of NO_3-N and NH_4-N in the KCl extracts were determined with the zinc reduction-naphthylethylenediamine method for NO_3-N [31] and the indophenol blue colorimetric method for NH_4-N [32]. The soil phosphorus content was determined by the Truog method [33]. Ferrous iron was measured by the o-phenanthroline method [34,35]. The organic C and total N were determined by the dry combustion method using an NC analyzer (Sumigraph NC-900; Sumika Chemical Analysis Service, Tokyo, Japan). The soil samples were previously treated with acid to eliminate water and inorganic carbonates [36]. The soil pH was obtained by measuring the equilibrium pH of soil pastes containing 1 g of soil homogenized in 1 mL of H_2O (Table 1).

DNA extraction and PCR

DNA from each of the 20 samples (4 sites×5 replicates) was extracted in 3 subsamples from 0.5 g of soil with the FastDNA Spin Kit for soil (MP Biomedicals, Illkirch, France) according to the manufacturer's protocol. The quality and quantity of the DNA extracts were checked with a SmartSpec Plus spectrophotometer (Bio-Rad Laboratories, United States). The samples were pooled and stored at -20°C until use.

A fragment of the *nifH* gene (approximately 360 bp) was amplified with a nested PCR strategy. First-round reactions were performed with the primers nifH32F (TGAGACAGATAGCTA-TYTAYGGHAA) and nifH623R (GATGTTCGCGCGGCAC-GAADTRNATSA) as described previously [37]. The genomic DNA extract (15–40 ng) was added to PCR mixtures containing 5 μL of 10×ExTaq buffer (Takara, Madison, WI, United States), 4 μL of a mix of deoxynucleoside triphosphates (2.5 mM each), 37.5 μL of water, 0.5 μL of 100 μM nifH32F, 0.5 μL of 100 μM nifH623R, and 0.5 μL of ExTaq DNA polymerase (5 U/μL; Takara, Madison, WI, United States). The reaction mixtures were amplified by 1 denaturation step (5 min at 94°C), followed by 30 cycles of 94°C for 1 min, 50°C for 1 min, and 72°C for 1 min, and 1 final 7 min extension cycle at 72°C. For the second round of the nested amplification, 1 μL of this reaction mixture was used as the template in a 50 μL reaction mixture containing the same reagent mixture described above, but with 39 μL of water, 0.25 μL of 100 μM nifH1-GC (in order to clamp the products for DGGE, the primer nifH1 [CTGYGAYCCNAARGCNGA] was added to the GC-clamp [CGCCCGCCGCGCGCGGCGGGCGGGGCGGG-GGGCACGGGGGG] on the 5′ side), and 0.25 μL of 100 μM nifH2 (ADNGCCATCATYTCNCC) [38]. The thermal cycling protocol for the nested reactions was the same as above except that the annealing temperature was raised to 57°C.

Fragments of approximately 200 bp, corresponding to the V3 region of the 16S rRNA gene [39], were amplified using a reaction mixture that contained the same reagent mixture described above, except with 0.25 μL of 100 μM 357F-GC (CCTACGGGAGG-CAGCAG-GC-clamp) and 0.25 μL of 100 μM 518R (GTAT-TACCGCGGCTGG); products were amplified using a touch-down thermocycling program [39]. All of the PCR amplicons were electrophoresed on an agarose gel to ascertain the sizes and purified using the UltraClean PCR Clean-Up Kit (MO BIO Laboratories, Carlsbad, CA, United States).

DGGE

DGGE was performed using the D-Code system (Bio-Rad Laboratories, Hercules, CA, United States) as described by Baxter and Cummings [40]. The polyacrylamide concentration in the gel was 8%, and the linear denaturing gradient was 30% to 60% (100% denaturant corresponds to 7 M urea and 40% deionized formamide). The gel was run at 36 V for 18 h at 60°C in 0.5×TAE buffer. The gel was then stained for 30 min with 1:10,000 (v/v) SYBR Gold, rinsed with 0.5×TAE, and scanned on a transilluminator. Bands were identified, and relative intensities were calculated based on the percentage of intensity of each band in a lane. This was done with an image-analyzing system (Image Master; Amersham Pharmacia Biotech, Uppsala, Sweden). Shannon-Wiener diversity index (H') was calculated by the formula $H' = -\Sigma\, p_i \ln(p_i)$, where p_i is the ratio of relative intensity of band i compared with the relative intensity of the lane. The used of different gradient gel (30% to 70%) to assessed the reproducibility of the DGGE results, similar results were obtained (data not shown).

Real-time PCR assay

Reactions were set up using SYBR green (Bio-Rad Laborato-ries, the Netherlands) according to Baxter and Cummings [41] with the LightCycler 1.5 system (Roche Applied Sciences, Indianapolis, IN, United States). Reaction mixtures were heated to 95°C for 15 min to denature the DNA before completing 40 cycles of denaturation (95°C for 45 s/15 s [*nifH*/16S rRNA]), annealing (55°C for 45 s/65°C for 15 s [*nifH*/16S rRNA]), and extension (72°C for 45 s/15 s [*nifH*/16S rRNA]). Soil DNA extracts were diluted 1:100 to prevent inhibition of PCR by soil contaminants (e.g., by co-extracted humic substances), and each run included triplicate reactions for each DNA sample, the standard curve, and the no template control. The average copy number was converted into copies of the gene per gram of soil. nifHF (AAAGGYGGWATCGGYAARTCCACCAC) and nifHR (TTGTTSGCSGCR TACATSGCCATCAT) primers [42] were used for *nifH* quantitative PCR (qPCR), and 357F and 518R primers were used for total bacteria qPCR. Dilution series of pGEM-T Easy vector (Promega, Madison, WI, United States) DNA with cloned bacterial *nifH* (*Azospirillum brasilense* ATCC 29729) and 16S rRNA gene (*Pseudomonas aeruginosa* PAO1) fragments were used to generate standard curves ranging from 10^1 to 10^7 gene copies·μL^{-1} for DNA quantification. The specificity of the amplified products was checked by the observation of a single melting peak and the presence of a unique band of the expected size in a 2% agarose gel stained with ethidium bromide. The standard curve produced was linear ($r^2 > 0.98$), and the PCR efficiency (Eff $= 10^{(-1/slope)}$-1) was>0.90.

Statistical analysis

In all tests, significant effects/interactions were those with a P value that was <0.05. Statistical analysis was performed using SPSS statistical software package (version 19.0; SPSS, Inc., Chicago, IL, United States). Variables of each group must be normally distributed to perform Pearson's product-moment correlations and analysis of variance (ANOVA). The normal distribution of the residuals was evaluated using the Kolmogorov-Smirnov test. If the requirement was not met, data were log-transformed prior to analysis. The homogeneity of the variances was checked by Levene's test. For pairwise comparison of means, Tukey's test was applied. Differences between main effects were tested by ANOVA. Correlations between *nifH* and 16S rRNA diversity and gene copies and environmental factors were tested by Pearson's product-moment correlations.

The choice between a linear or unimodal species response model depends on the underlying gradient length, which is measured in standard deviation units along the first ordination axis and can be estimated by detrended correspondence analysis (DCA). It is recommended to use linear methods when the gradient length is <3, unimodal methods when it is>4, and any method for intermediate gradient lengths [43]. The DCA gradient length for *nifH* gene patterns was 1.77, and that for 16S rRNA patterns was 1.11. Therefore, linear species response models such as partial least-squares regression (PLSR) and redundancy analysis (RDA) were used for multivariate statistical analysis. PLSR is an extension of multiple regression analysis in which the effects of linear combinations of several predictors on a response variable (or multiple response variables) are analyzed. PLSR is especially useful when the number of predictor variables is similar to or higher than the number of observations and/or predictors are highly correlated [44,45]. For PLSR model generation, the software assigns a value known as the variable influence on projection (VIP) to each environmental variable. The VIP indicates the relative importance to the model. Significance was assessed using the VIP parameter (a VIP>1 indicated a significant contribution of the environmental factor to the statistical model) [46,47]. RDA can be considered an extension of principal component analysis (PCA) in which the main components are constrained to be linear combinations of the environmental variables. RDA not only represents the main patterns of species variation as much as they can be explained by the measured environmental variables but also displays correlations between each species and each environmental variable in the data [48]. DCA, RDA, and PCA were performed using the Canoco program for Windows 4.5 (Biometris, Wageningen, the Netherlands). PLSR was performed using the Simca-P 11.0 (Umetrics AB, Umea, Sweden).

Results

Diversity of *nifH*

DGGE gels are shown in Figure S1A. Analysis of the *nifH* DGGE Shannon-Wiener diversity index values for the whole data set indicated that the abandoned period significantly affected the *nifH* diversity ($P<0.001$) by a separate ANOVA. For these reasons, multiple comparisons of the *nifH* DGGE Shannon-Wiener diversity index values were conducted.

Soils of Y1 (2.8 ± 0.02) and Y5 (2.83 ± 0.06) showed significantly higher *nifH* diversity than soils of Y25 (2.72 ± 0.08) and LGSG (2.72 ± 0.1) (Y1×Y25, $P=0.04$; Y1×LGSG, $P=0.02$; Y5×Y25, $P=0.009$; Y5×LGSG, $P=0.025$) (Figure 2A). The Y25 and

LGSG soils did not show significant differences in *nifH* diversity. Pearson's product-moment correlation indicated that pH, NO_3-N, and P were positively correlated with the *nifH* Shannon-Wiener diversity index (Table 1).

qPCR of the *nifH* gene

We used qPCR to compare copy numbers of the functional gene (*nifH*) at the 4 sites (Figure 3A). The *nifH* gene copy number in LGSG soil (5.5×10^5 copies·g^{-1} of soil) was the higher compared to the soils of the other 3 abandoned croplands (average number of copies·g^{-1} of soil, 1.9×10^5 in Y1, 3.4×10^5 in Y5, and 2.3×10^5 in Y25). In the abandoned cropland soils, the *nifH* gene copy number did not significantly change over time ($P>0.05$) (Figure 3A).Using Pearson's product-moment correlation, we found that NH_4-N, P, Fe, organic C, total N, soil moisture, and plant biomass were positively correlated, and the Shannon-Wiener diversity index was negatively correlated with the *nifH* copy number (Table 1).

Total bacterial diversity

There are clearly differences in the nitrogen-fixing communities of soil from long and short abandoned periods. In order to ensure that these factors are affecting the nitrogen-fixing community specifically and not the bacterial community as a whole, the 16S rRNA gene diversity and abundance were also analyzed. DGGE gels showing the diversity of total bacteria are shown in Figure S1B. As with the free-living nitrogen-fixing community, ANOVA results for the Shannon-Wiener diversity index of the 16S rRNA gene indicated that the abandoned period ($P<0.001$) was a significant factor. Therefore, each sample was analyzed separately by multiple comparisons. Soils of Y25 and LGSG (both almost equal, 3.17 ± 0.07) showed higher 16S rRNA diversity than soils of Y1 (2.8 ± 0.04) and Y5 (3.06 ± 0.01) (Figure 2B). Pearson's product-moment correlation indicated that organic C, total N, Fe, soil moisture, and plant biomass were positively correlated, and hydraulic conductivity was negatively correlated with the 16S rRNA gene Shannon-Wiener diversity index (Table 1).

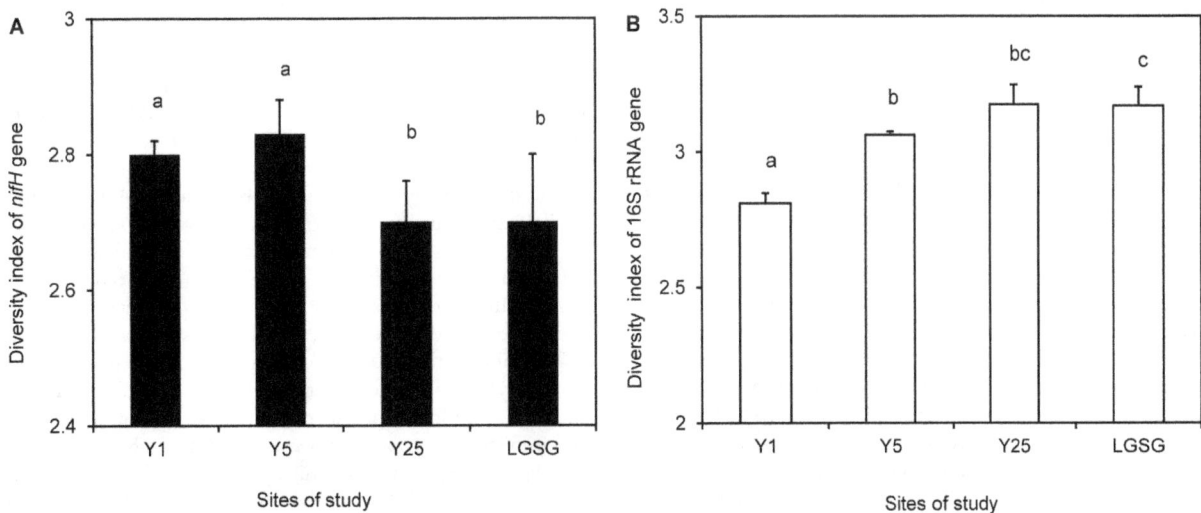

Figure 2. Shannon-Wiener diversity index values for *nifH* (A) and 16S rRNA genes (B). Data sets and results of analysis of variance (ANOVA) in the abandoned cropland (Y1, Y5, and Y25) and light-grazing steppe grassland (LGSG) soils (n = 5; error bars represent standard deviations). Significant differences are indicated by different letters.

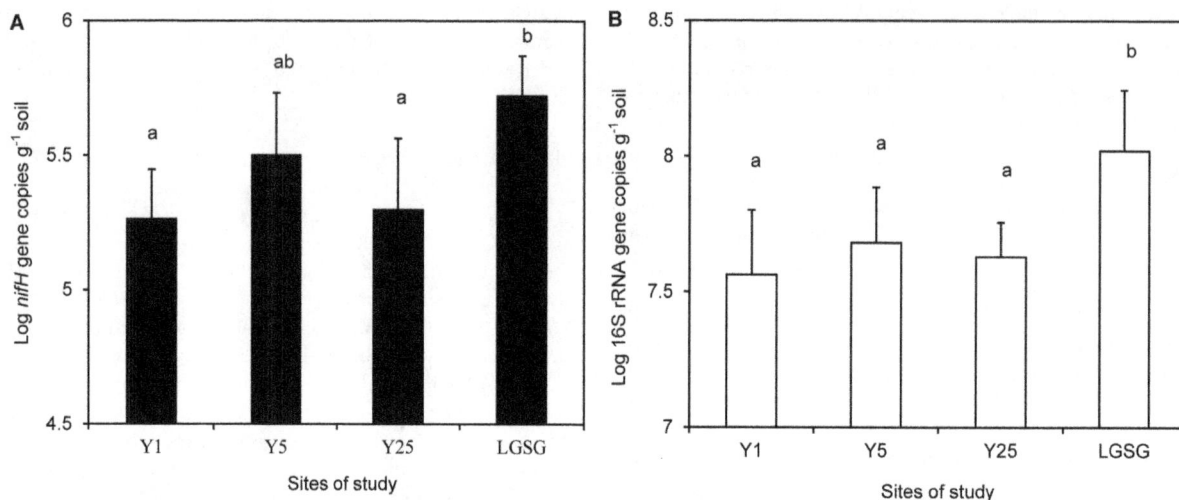

Figure 3. Copy numbers of the *nifH* **(A) and 16S rRNA (B) genes.** Results of analysis of variance (ANOVA) in the abandoned cropland (Y1, Y5, and Y25) and light-grazing steppe grassland (LGSG) soils (n = 5; error bars represent standard deviations). Significant differences are indicated by different letters.

qPCR of the 16S rRNA gene

By performing the qPCR analysis, we aimed to compare the differences between each site rather than attain an absolute quantification. Similar to *nifH* results presented above, the gene copy numbers of 16S rRNA in LGSG soil (1.2×10^8 copies·g^{-1} of soil) was higher than that of 3 other abandoned cropland soils (average numbers of copies·g^{-1} of soil, 8.2×10^7 in Y1, 5.2×10^7 in Y5, and 4.4×10^7 in Y25) (Figure 3A and B). The 16S rRNA gene copy numbers did not change significantly over time in the abandoned cropland soils (P>0.05) (Figure 3B). Pearson's product-moment correlation indicated that NH$_4$-N, organic C, total N, and plant biomass were positively correlated with the 16S rRNA gene copy number (Table 1).

Effects of environmental variables on the microbial community structure

Because of multicollinearity among environmental variables (Table S3), potential effects of environmental variables on free-living nitrogen-fixing bacteria and the total bacteria community composition were assessed by PLSR (Table 2). Plant biomass, hydraulic conductivity, pH, Fe, and soil moisture, and NH4-N predominantly affected on both free-living nitrogen-fixing bacteria and the total bacteria community composition in this study area. Moreover, C and N also exerted effects on the total bacteria community composition (Table 2).

These correlations of microbial diversity with environmental parameters were also supported by RDA (Figure S2). The ordination plots for the *nifH* and 16S rRNA genes are given in Figure S2A and S2B, respectively. In these figures, the first 2 canonical axes explained 43.6% and 59.1% of the variance of the species data (functional genes) and 51.0% and 63.4% of the variance of the species–environment relationship, respectively. The projection of environmental variables with respect to the free-living nitrogen-fixing community composition revealed that the first canonical axis is positively correlated with soil moisture, Fe, NH4-N, and plant biomass, and it is negatively correlated with pH and hydraulic conductivity (Figure S2A). With respect to the total bacteria community composition, the first canonical axis is positively correlated with Fe, NH4-N, plant biomass, C, and N,

and it is negatively correlated with pH and hydraulic conductivity (Figure S2B).

Discussion

This study allowed a detailed analysis of the effects of key components of abandoned cropland with the same land use history (permanent grassland was turned into arable land by cultivation about 40 years ago and had rotated crops of *Zea mayscorn*, *Helianthus annuus* and *E. cylindricus*) and light-grazing steppe grassland systems on soil bacterial and free-living nitrogen-fixing bacterial population structure and gene copy number. Cultivation imparted a strong effect on both total and free-living nitrogen-fixing bacterial population structures (measured by DGGE profiles) and abundances (measured by gene copy numbers).

The copy numbers of *nifH* per gram of LGSG soil was higher than that in Y1, Y5, and Y25 soil. In this study, markedly lower levels of nutrient (e.g., organic C, P, and Fe) in the soil of abandoned cropland (Table 1) may suppress the growth of free-living nitrogen-fixing bacteria. Organic C, P, and Fe and *nifH* copy number have a significant positive Pearson's product-moment correlation (Table 1). The source of carbohydrate is important to allow N$_2$ fixation activity, which requires large amounts of energy and reducing equivalents [49]. Phosphorus availability can increase nitrogen fixation because nitrogen fixation process requires large amounts of adenosine triphosphate (ATP) [50]. The [4Fe-4S] cluster of the *nifH* gene requires the Fe [22,51]; therefore, a low copy number of *nifH* might be related to the content of organic C, P, and Fe. In addition, Coelho et al. [52,53] found that 30% more free-living diazotrophs could be isolated from soil in the presence of low levels of nitrogen fertilizer than from soil in the presence of high levels of nitrogen fertilizer. In contrast, low copy number of *nifH* was obtained from the abandoned cropland of low nitrogen levels (NO$_3$-N and NH$_4$-N) (Table 1). The result may be caused by nutrients content of abandoned cropland soils.

Shannon-Wiener diversity index of *nifH* of LGSG was lower than that of Y1 and Y5 (Figures 2A and 3A). Previous studies indicated that pH is an important factor for the structure of the bacterial community [54,55]. Belnap [56] thought that most

Table 2. Results of partial least squares regression (PLSR) analysis with the explanatory capacity of the first component and the variable influence on projection (VIP) of each predictor within each component to estimate significant predictors.

Gene	nifH	16S rRNA
Explained variance in fingerprinting pattern (%)	15.2	22.5
Explained variance of component predictors (%)	7.80	15.40
Environment variables*	Variable Influence on Projection (VIP)	
P-B	1.29	1.23
HC	1.17	1.07
pH	1.15	1.07
Fe	1.13	1.17
H$_2$O	1.07	1.00
NH$_4$-N	1.04	1.02
ELE	0.97	0.95
N	0.94	1.01
C	0.93	1.01
P	0.77	0.73
P-H'	0.66	0.75
NO$_3$-N	0.61	0.70

*All environmental variables are shown and include pH, NO$_3$-N, NH$_4$-N, organic carbon (C), total nitrogen (N), available phosphorus (P), soluble iron (Fe), elevation (ELE), hydraulic conductivity (HC), soil moisture (H$_2$O), plant biomass (P-B), and plant diversity (P-H').

nitrogen-fixing microorganisms have an optimum soil pH of 7 or above. In this study, the samples from abandoned croplands with pH>7 had a higher diversity of free-living nitrogen-fixing microorganisms than the samples from the LGSG with pH<7.

The nifH gene profile showed there was a higher variability between replicate samples than that of the 16S rRNA gene profile. Therefore, this result suggested that the free-living nitrogen-fixing bacteria community was susceptible in this study area. An explanation for this may be the local changes at the microsite/aggregate scale because of spatial and temporal variation in exudation along plant roots [57–59]. Similar approaches have demonstrated that there are differences in ammonia oxidizer populations in sediment and soil [60].

The RDA revealed that the distribution of the factors that influence both the free-living nitrogen-fixing bacteria and the broader bacterial community distribution are very similar. This suggests that soil factors that affect the free-living nitrogen fixers are likely to affect the community as a whole in steppe grassland soil. Larkin et al. [61] reported that plant effects are the most important drivers of soil microbial community characteristics within a given site and soil type. In our study, PLSR and RDA showed that plant biomass strongly influenced (Table 2) and was positively correlated with the first axis (Figure S2) in both the free-living nitrogen-fixing bacteria and the broader bacterial community distribution. This is caused by interactions between plants and microorganisms. Above-ground net primary productivity was expected to increase soil carbon input by enhancing the turnover of plant biomass and enhancing root exudation and may therefore influence carbon-limited microbial communities in the soil [62,63]. Concurrently, microorganisms also affect plants by producing canonical plant growth-regulating substances such as auxins or cytokinins [13]. However, the Shannon-Wiener diversity index was not significantly influenced in either the free-living nitrogen-fixing bacteria or the broader bacterial community distribution in this study area. Pearson's product-moment corre-

lations showed that plant diversity was not significantly correlated with nifH gene diversity, or 16S rRNA gene diversity and copy number, but did negatively correlate with the copy number of the nifH gene (Table 1). Carney and Matson [64] found that plant diversity had a significant effect on the microbial community composition through alterations in microbial abundance rather than community composition. However, several studies report that plant diversity has little direct effect on bacterial community composition [65,66]. Therefore, this effect may depend on the type of plant community examined.

A number of additional factors such as hydraulic conductivity, pH, Fe, soil moisture, and NH4-N significantly influenced both the free-living nitrogen-fixing bacteria and the broader bacterial community distribution (Table 2). In our study, the soil from different sites was the same type, but the water holding capacity function of each research soil was different. The soil moisture content is in the order of LGSG>Y25>Y5>Y1 (Table 1). This suggests that a balance of macropores and micropores has not been fully recovered in the abandoned crop soils. A balance of macropores and micropores in soil influences the permeability and water holding capacity of soil [67]. In addition, the biological decomposition of organic materials produces natural glues, which bind and strengthen soil aggregates [67], and helps soils hold water and nutrients, which may change the balance of macropores and micropores. Organic matter also is a long-term, slow-release storehouse of nitrogen, phosphorus, and sulfur [67]. Accordingly, organic matter significantly influenced the total bacterial community distribution (Table 2), which may have resulted because of the significant difference in soil organic matter content in the abandoned cropland and LGSG ($P<0.05$) (Table 1). Additionally, the soil texture and hydraulic conductivity is closely correlated, and soil texture influences the balance between macropores and micropores [67]. The hydraulic conductivities of sites Y1 and Y5 were significantly different from those of sites Y25 and LGSG (Table 1), and the RDA also indicated that soil moisture was negatively correlated with hydraulic conductivity (Figure S2). Therefore, the hydraulic conductivity of soil also showed that the water holding capacity function of soils from the research sites is different.

The significance of NH4-N indicates that the fixed nitrogen of the free-living nitrogen-fixing bacteria will be the main nitrogen source in the soil of steppe grasslands, where there are few plants such as legumes. Nitrogen cycling in natural ecosystems and traditional agricultural production relies on biological nitrogen fixation primarily by diazotrophic bacteria [16].

In our study, higher copy number/diversity of the *nifH* gene/ 16S rRNA gene was observed in the LGSG soil, where the Fe content was significantly higher, than in the abandoned cropland soil. According to Pearson's product-moment correlation and PLSR, the Fe content affects the nitrogen-fixing community. This may result because Fe is a required material for the [4Fe-4S] cluster of the *nifH* gene [22,51], and it is possible that the Fe content affects the 16S rRNA gene diversity in other microorganisms with the iron-containing enzyme [68-70] as it does the nitrogen-fixing bacteria.

Gaby and Buckley [71] comprehensively evaluated primers of *nifH* gene and found that *nifH* gene different primer sets amplified different groups in the *nifH* phylogeny. In this study, we used primer sets for DGGE and qPCR analyses of the *nifH* gene that targeted different sequence positions, thus the results obtained were not comparable. However, environmental factors differed greatly between the abandoned cropland and LGSG soils, thus having differential effects on the soil microbial communities. Furthermore, the trends in the changing copy number and diversity of the *nifH* gene were similar to that of the 16S rRNA gene. Therefore, the results we obtained regarding the copy number and diversity of *nifH* gene are able to explain the dynamic change in the trend of free-living nitrogen-fixing bacteria communities in abandoned cropland during different periods. Further research on molecular ecology could provide more detail analysis of microbial diversity, such as by sequencing of DGGE band and high-throughput sequencing technology etc., which would contribute to more accurate insights into the black box of the dynamic change of microbial communities.

We discovered that microbial communities of the abandoned cropland in this study area are strongly influenced by plant biomass, soil moisture, Fe, and NH4-N. Robust information about the mechanisms that regulate the diversity, structure, and composition of natural communities is urgently needed to help conserve ecosystem function and mitigate biodiversity loss from current and future environmental changes. Conversely, the results of this study suggest that advances in desertification may be prevented by adjusting environmental factors of the abandoned cropland, such as the soil moisture content, Fe, and NH4-N, which will enhance the function of a microbial community and possibly increasing plant biomass production. In addition, *nifH* gene copy number had significant positive Pearson's product-moment correlation with soil moisture, organic C, P, and Fe etc., therefore, by adjusting these environmental factors may also increase the

abundance of free-living nitrogen-fixing bacteria in abandoned cropland.

Supporting Information

Figure S1 Denaturing gradient gel electrophoresis (DGGE) profiles of the *nifH* (A) and 16S rRNA genes (B). For all images, the numbers refer to the plot numbers in the sample areas.

Figure S2 Redundancy analysis (RDA) of *nifH* (A) and 16S rRNA (B) genes data. Ordination plots of *nifH* (A) and 16S rRNA (B) genes associated with abandoned croplands for different times: Y1 (•), Y5 (▲), and Y25 (■), and *light-grazing* steppe grassland (LGSG, ○). The plots were generated by redundancy analysis (RDA) of the denaturing gradient gel electrophoresis (DGGE) profiles. All environmental variables are shown, including pH, NO_3-N, NH_4-N, organic carbon (C), total nitrogen (N), available phosphorus (P), soluble iron (Fe), elevation (ELE), hydraulic conductivity (HC), soil moisture (H_2O), plant biomass (P-B), and plant species richness (P-H'). Values on the axes indicate the percentages of total variation explained by each axis.

Table S1 Total plant cover, plant Shannon's diversity index, plant biomass, plant types, and their coverage in each research plot.

Table S2 Soil texture (clay, silt, and sand) content and saturated hydraulic conductivity of soil samples at each site.

Table S3 Collinearity among environmental parameters as determined by Spearman's rank correlation coefficient rho. Significantly correlated parameters show Spearman's rank correlation coefficient rho >0.6 and <-0.6 in bold. All environmental variables are shown, such as pH, NO_3-N, NH_4-N, organic carbon (C), total nitrogen (N), available phosphorus (P), soluble iron (Fe), elevation (ELE), hydraulic conductivity (HC), soil moisture (H_2O), plant biomass (P-B), and plant diversity (P-H').

Acknowledgments

We thank Dr. Kenji Tamura (University of Tsukuba) for his valuable suggestions on the research content and soil sampling, and Takashi Kanda (University of Tsukuba) for excellent technical support in measuring organic C and total N concentrations.

Author Contributions

Conceived and designed the experiments: H SB HU. Performed the experiments: H SB. Analyzed the data: H YXC. Contributed reagents/materials/analysis tools: HU NN T. Nakajima T. Nakamura. Wrote the paper: H.

References

1. Archibold OW (1995) Ecology of World Vegetation. London: Chapman & Hall. pp. 1–522.
2. He C, Zhang Q, Li Y, Li X, Shi P (2005) Zoning grassland protection areas using remote sensing and cellular automata modeling – A case study in Xilingol steppe grassland in northern China. J Arid Environ 63: 814–826.
3. Tong C, Wu J, Yong S, Yang J, Yong W (2004) A landscapescale assessment of steppe degradation in the Xilin River basin, Inner Mongolia, China. J Arid Environ 59: 133–149.
4. Li JH, Fang XW, Jia JJ, Wang G (2007) Effect of legume species introduction to early abandoned field on vegetation development. Plant Ecol 191: 1–9.

5. Van Der Putten WH, Mortimer SR, Hedlund K, Van Dijk C, Brown VK, et al. (2000) Plant species diversity as a driver of early succession in abandoned fields: a multi-site approach. Oecologia 124: 91–99.
6. Xue ZD, Hou QC, Han RL, Wang SQ (2002) Trails and research on ecological restoration by *Sophora viciifolia* in Gullied Rolling Loess Region. J Northwest Forestry Univ 17: 26–29.
7. Zhang JD, Qiu Y, Chai BF, Zheng FY (2000) Succession analysis of plant communities in Yancun low middle hills of Luliang Mountains. J Plant Resour. Environ 9: 34–39.

8. Zou HY, Chen JM, Zhou L, Hongo A (1998) Natural recoverage succession and regulation of the Prairie vegetation on the Loess Plateau. Res. Soil. Water Conserv 5: 126–138.

9. Cheng YX, Nakamura T (2007) Phytosociological study of steppe vegetation in east Kazakhstan. Grassland Sci 53: 172–180.

10. EI-Sheikh MA (2005) Plant succession on abandoned fields after 25 years of shifting cultivation in Assuit, Egypt. J Arid Environ 61: 461–481.

11. Prévosto B, Kuiters L, Bernhardt-Römermann M, Dölle M, Schmidt W, et al. (2011) Impacts of land abandonment on vegetation: successional pathways in European habitats. Folia Geobot 46: 303–325.

12. Štolcová J (2002) Secondary succession on an early abandoned field: vegetation composition and production of biomass. Plant Protection Sci 38: 149–154.

13. Ortiz-Castro R, Contreras-Cornejo HA, Macias-Rodriguez L, Lopez-Bucio J (2009) The role of microbial signals in plant growth and development. Plant Signal. Behav 4: 701–712.

14. Daniel R (2005) The metagenomics of soil. Nat Rev Microbiol 3: 470–478.

15. Van Der Heijden M, Bardgett R, van Straalen N (2008) The unseen majority: soil microbes as drivers of plant diversity and productivity in terrestrial ecosystems. Ecol Lett 11: 296–310.

16. Vitousek PM, Aber J, Howarth RW, Likens GE, Matson PA, et al. (1997) Human alteration of the global nitrogen cycle: causes and consequences. Ecol Appl 7: 737–750.

17. Orr CH, James A, Leifert C, Cooper JM, Cummings SP (2011) Diversity and activity of free-living nitrogen-fixing bacteria and total bacteria in organic and conventionally managed soils. Appl Environ Microbiol 77: 911–919.

18. Dixon R, Kahn D (2004) Genetic regulation of biological nitrogen fixation. Nat Rev Microbiol 2: 621–631.

19. Peoples MB, Herridge DF, Ladha JK (1995) Biological nitrogen fixation: an efficient source of nitrogen for sustainable agricultural production? Plant Soil 174: 3–28.

20. Burgmann H, Widmer F, Von Sigler W, Zeyer J (2004) New molecular screening tools for analysis of free-living diazotrophs in soil. Appl Environ Microbiol 70: 240–247.

21. Kahindi JHP, Woomer P, George T, de Souza Moreira FM, Karanja NK, et al. (1997) Agricultural intensification, soil biodiversity and ecosystem function in the tropics: the role of nitrogen-fixing bacteria. Appl Soil Ecol 6: 55–76.

22. Roeselers G, Stal LJ, van Loosdrecht MCM, Muyzer G (2007) Development of a PCR for the detection and identification of cyanobacterial nifD genes. J Microbiol Methods 70: 550–556.

23. Poly F, Monrozier LJ, Bally R (2001) Improvement in the RFLP procedure for studying the diversity of nifH genes in communities of nitrogen fixers in soil. Res Microbiol 152: 95–103.

24. Rosado AS, Duarte GF, Seldin L, Van Elsas JD (1998) Genetic diversity of nifH gene sequences in Paenibacillus azotofixans strains and soil samples analyzed by denaturing gradient gel electrophoresis of PCR-amplified gene fragments. Appl Environ Microbiol 64: 2770–2779.

25. Widmer F, Shaffer BT, Porteous LA, Seidler RJ (1999) Analysis of nifH gene pool complexity in soil and litter at a Douglas fir forest site in the Oregon Cascade Mountain range. Appl Environ Microbiol 65: 374–380.

26. Matsuura K, Willmott CJ (2009) Terrestrial precipitation: 1900-2010 gridded monthly time series. Available: http://climate.geog.udel.edu/~climate/. Accessed 2014 Aug 27.

27. Braun-Blanquet J (1964) Pflanzensoziologie, 3rd revised edn. New York: Springer-Verlag. pp. 1–865.

28. Mueller-Dombois D, Ellenberg H (1974) Aims and Methods of Vegetation Ecology. New York: John Wiley and Sons. pp. 1–547.

29. Day PR (1965) Particle fraction and particle-size analysis. In: Black CA, editor. Methods of soil analysis. Madison: American Society of Agronomy. pp. 545–566.

30. Siosemarde M, Byzedi M (2011) Studding of number of dataset on precision of estimated saturated hydraulic conductivity. World Academy Sci, Eng Technol 74: 521–524.

31. Leonardo M (2009) Development and validation of a method for determination of residual nitrite/nitrate in foodstuffs and water after zinc reduction. Food Anal Methods 2: 212–220.

32. Motsara MR, Roy RN (2008) Guide to laboratory establishment for plant nutrient analysis. FAO Fertilizer and Plant Nutrition Bulletin No. 19, FAO, Rome. Pp.17–76.

33. Truog E, Meyer AH (1929) Improvements in the deniges colorimetric method for phosphorus and arsenic. Indus, and Engin Chem, Analyt Ed 1: 136–139.

34. Saywell LG, Cunningham BB (1937) Determination of iron: colorimetric o-phenanthroline method. Ind Eng Chem Anal 9: 67–69.

35. Sugio T, Taha T, Kanao T, Takeuchi F (2007) Increase in Fe^{2+}-producing activity during growth of Acidithiobacillus ferrooxidans ATCC 23270 on sulfur. Biosci Biotechnol Biochem 71: 2663–2669.

36. Schumacher BA (2002) Methods for the determination of total organic carbon (TOC) in soils and sediments. U.S. Environmental Protection Agency, Washington, Ecological Risk Assesment Support Center.

37. Steward GF, Zehr JP, Jellison RP, Montoya JP, Hollibaugh JT (2004) Vertical distribution of nitrogen-fixing phylotypes in a meromictic, hypersaline lake. Microb Ecol 47: 30–40.

38. Zehr JP, McReynolds LA (1989) Use of degenerate oligonucleotides for amplification of the nifH gene from the marine cyanobacterium Trichodesmium thiebautii. Appl Environ Microbiol 55: 2522–2526.

39. Muyzer G, De Waal EC, Uitterlinden AG (1993) Profiling of complex microbial populations by denaturing gradient gel electrophoresis analysis of polymerase chain reaction-amplified genes coding for 16S rRNA. Appl Environ Microbiol 59: 695–700.

40. Baxter J, Cummings SP (2006) The impact of bioaugmentation on metal cyanide degradation and soil bacteria community structure. Biodegradation 17: 207–217.

41. Baxter J, Cummings SP (2008) The degradation of the herbicide bromoxynil and its impact on bacterial diversity in a top soil. J Appl Microbiol 104: 1605–1616.

42. Rösch C, Mergel A, Bothe H (2002) Biodiversity of denitrifying and dinitrogen-fixing bacteria in an acid forest soil. Appl Environ Microbiol 68: 3818–3829.

43. ter Braak CJF, Smilauer P (2002) CANOCO reference manual and CanoDraw for Windows user's guide: software for canonical community ordination (version 4.5). Ithaca, USA (url: www.canoco.com): Microcomputer Power. 500 p.

44. Carrascal LM, Galvan I, Gordo O (2009) Partial least squares regression as an alternative to current regression methods used in ecology. Oikos 118: 681–690.

45. Naether A, Foesel BU, Naegele V, Wüst PK, Weinert J, et al. (2012) Environmental factors affect acidobacterial communities below the subgroup level in grassland and forest soils. Appl Environ Microbiol 78: 7398–7406.

46. Umetrics AB (2005) SIMCA-P v. 11 Analysis Advisor. Umeå, Sweden: Umetrics AB.

47. Tremaroli V, Workentine ML, Weljie AM, Vogel HJ, Ceri H, et al. (2009) Metabolomic investigation of the bacterial response to a metal challenge. Appl Environ Microbiol 75: 719–728.

48. Ramette A (2007) Multivariate analyses in microbial ecology. FEMS Microbiol Ecol 62: 142–160.

49. Chan YK, Barraquio WL, Knowles R (1994) N_2-fixing pseudomonas and related soil bacteria. FEMS Microb Rev 13: 95–118.

50. Reed SC, Seastedt TR, Mann CM, Suding KN, Townsend AR, et al. (2007) Phosphorus fertilization stimulates nitrogen fixation and increases inorganic nitrogen concentrations in a restored prairie. Appl Soil Ecol 36: 238–242.

51. Gavini N, Burgess BK (1992) FeMo cofactor synthesis by a nifH mutant with altered MgATP reactivity. J Biol Chem 267: 21179–21186.

52. Coelho MRR, de Vos M, Carneiro NP, Marriel IE, Paiva E, et al. (2008) Diversity of nifH gene pools in the rhizosphere of two cultivars of sorghum (Sorghum bicolor) treated with contrasting levels of nitrogen fertilizer. FEMS Microbiol Lett 279: 15–22.

53. Coelho MRR, Marriel IE, Jenkins SN, Lanyon CV, Seldin L, et al. (2009) Molecular detection and quantification of nifH gene sequences in the rhizosphere of sorghum (Sorghum bicolor) sown with two levels of nitrogen fertilizer. Appl Soil Ecol 42: 48–53.

54. Noll M, Wellinger M, (2008) Changes of the soil ecosystem along a receding glacier: Testing the correlation between environmental factors and bacterial community structure. Soil Biol Biochem 40: 2611–2619.

55. Wakelin SA, Macdonald LM, Rogers SL, Gregg AL, Bolger TP, et al. (2008) Habitat selective factors influencing the structural composition and functional capacity of microbial communities in agricultural soils. Soil Biol Biochem 40: 803–813.

56. Belnap J (2001) Factors influencing nitrogen fixation and nitrogen release in biological soil crusts. In: Belnap J, Lange OL, editors. Biological soil crusts: structure, function, and management. Berlin: Springer-Verlag. pp. 241–261.

57. Clayton SJ, Clegg CD, Murray PJ, Gregory PJ (2005) Determination of the impact of continuous defoliation of Lolium perenne and Trifolium repens on bacterial and fungal community structure in rhizosphere soil. Biol Fertil Soils 41: 109–115.

58. Marilley L, Vogt G, Blanc M, Aragno M (1998) Bacterial diversity in the bulk soil and rhizosphere fractions of Lolium perenne and Trifolium repens as revealed by PCR restriction analysis of 16S rDNA. Plant Soil 198: 219–224.

59. Marschner P, Yang CH, Liebere R, Crowley DE (2001) Soil and plant specific effects on bacterial community composition in the rhizosphere. Soil Biol Biochem 33: 1437–1445.

60. Stephen JR, McCaig AE, Smith Z, Prosser JI, Embley TM (1996) Molecular diversity of soil and marine 16S rRNA gene sequences related to beta-subgroup ammonia-oxidizing bacteria. Appl Environ Microbiol 62: 4147–4154.

61. Larkin RP, Honeycutt CW (2006) Effects of different 3-year cropping systems on soil microbial communities and Rhizoctonia diseases of potato. Phytopathology 96: 68–79.

62. Niklaus PA, Alphei J, Ebersberger D, Kampichler C, Kandeler E, et al. (2003) Six years of in situ CO_2 enrichment evokes changes in soil structure and soil biota of nutrient-poor grassland. Glob Chang Biol 9: 585–600.

63. Zak DR, Holmes W, White DC, Peacock A, Tilman D (2003) Plant diversity, soil microbial communities, and ecosystem function: are there any link? Ecology 84: 2042–2050.

64. Carney KM, Matson PA (2005) Plant communities, soil microorganisms, and soil carbon cycling: does altering the world belowground matter to ecosystem functioning? Ecosystems 8: 928–940.

65. Kennedy N, Brodie E, Connolly J, Clipson N (2004) Impact of lime, nitrogen and plant species on bacterial community structure in grassland microcosms. Environ Microbiol 6: 1070–1080.

66. Nunan N, Daniell TJ, Singh BK, Papert A, McNicol JW, et al. (2005) Links between plant and rhizoplane bacterial communities in grassland soils, characterized using molecular techniques. Appl Environ Microbiol 71: 6784–6792.

67. Cogger C (2000) Soil Management for Small Farms. EB 1895. Pullman, WA: Washington State University Cooperative Extension. 24 p.

68. Sze IS, Dagley S (1984) Properties of salicylate hydroxylase and hydroxyquinol 1,2-dioxygenase purified from *Trichosporon cutaneum*. J Bacteriol 159: 353–359.

69. Conway T, Ingram LO (1989) Similarity of *Escherichia coli* propanediol oxidoreductase (fucO product) and an unusual alcohol dehydrogenase from *Zymomonas mobilis* and *Saccharomyces cerevisiae*. J Bacteriol 171: 3754–3759.

70. Drennan CL, Heo J, Sintchak MD, Schreiter E, Ludden PW (2001) Life on carbon monoxide: X-ray structure of Rhodospirillum rubrum Ni-Fe-S carbon monoxide dehydrogenase. Proc Natl Acad Sci USA 98: 11973–11978.

71. Gaby JC, Buckley DH (2012) A comprehensive evaluation of PCR primers to amplify the nifH gene of nitrogenase. PLoS ONE 7: e42149.

8

Phthalic Acid Esters in Soils from Vegetable Greenhouses in Shandong Peninsula, East China

Chao Chai[1], Hongzhen Cheng[1], Wei Ge[2], Dong Ma[1], Yanxi Shi[1]*

1 College of Resources and Environment, Qingdao Agricultural University, Qingdao, China, 2 College of Life Sciences, Qingdao Agricultural University, Qingdao, China

Abstract

Soils at depths of 0 cm to 10 cm, 10 cm to 20 cm, and 20 cm to 40 cm from 37 vegetable greenhouses in Shandong Peninsula, East China, were collected, and 16 phthalic acid esters (PAEs) were detected using gas chromatography-mass spectrometry (GC-MS). All 16 PAEs could be detected in soils from vegetable greenhouses. The total of 16 PAEs (Σ_{16}PAEs) ranged from 1.939 mg/kg to 35.442 mg/kg, with an average of 6.748 mg/kg. Among four areas, including Qingdao, Weihai, Weifang, and Yantai, the average and maximum concentrations of Σ_{16}PAEs in soils at depths of 0 cm to 10 cm appeared in Weifang, which has a long history of vegetable production and is famous for extensive greenhouse cultivation. Despite the different concentrations of Σ_{16}PAEs, the PAE compositions were comparable. Among the 16 PAEs, di(2-ethylhexyl) phthalate (DEHP), di-n-octyl phthalate (DnOP), di-n-butyl phthalate (DnBP), and diisobutyl phthalate (DiBP) were the most abundant. Compared with the results on agricultural soils in China, soils that are being used or were used for vegetable greenhouses had higher PAE concentrations. Among PAEs, dimethyl phthalate (DMP), diethyl phthalate (DEP) and DnBP exceeded soil allowable concentrations (in US) in more than 90% of the samples, and DnOP in more than 20%. Shandong Peninsula has the highest PAE contents, which suggests that this area is severely contaminated by PAEs.

Editor: Raffaella Balestrini, Institute for Plant Protection (IPP), CNR, Italy

Funding: This work was supported by the "Science and Technology Plan Projects of Qingdao (No. 12-1-3-64-nsh), the Two Districts" Foundation of Shandong Province, China (No. 2011-Yellow-19) and the Talent Foundation of Qingdao Agricultural University (No. 630642). The funders had no role in study design, data collection and analysis, decision to publish, or preparation of the manuscript.

Competing Interests: The authors have declared that no competing interests exist.

* E-mail: yanxiyy@126.com

Introduction

Phthalic acid esters (PAEs) are used extensively as plasticizers of plastic products, such as polyvinyl chloride, and as nonplasticizers in consumer products, including medical devices, building materials, paints, pesticides, fertilizes, food packaging, and so on [1]. The large-scale production and application of 6.0 million tons/yr [2] of PAEs have made these materials ubiquitous environment pollutants [3–8]. Some PAEs have endocrine disruptive effects [9], and six PAEs are categorized as priority environmental pollutants by the United States Environmental Protection Agency [10].

Greenhouse cultivation has expanded dramatically in China since the 1980s, reaching up to 3.5 million ha by 2011 [11]. Greenhouse cultivation is mainly for vegetable production in China, and plastic greenhouses account for more than 99% of greenhouse cultivation relative to glass greenhouses [12–13]. Several studies detected PAEs in soils of vegetable greenhouses in Nanjing and Hangzhou [14–15], as well as in other agricultural soils, such as vegetable soils in Guangzhou and paddy soils in Leizhou Peninsula in China [16–17]. The buildup of PAEs in agricultural soils may contaminate agricultural products, and further raise the human health risk [18].

Shandong Peninsula is the largest Peninsula in China with rapid urbanization and high population density of 550 people/km². The Peninsula includes the cities of Qingdao, Yantai, Weifang, and Weihai. Shandong Peninsula has a long history of vegetable greenhouse cultivation and is a main vegetable-producing region,

with its greenhouse coverage accounting for approximately 50% of that of China. The vegetable greenhouses in this peninsula are close to the highly populated urban areas, and plastic film is widely used. Plastic film of 30000 tons/yr is estimated to be used only in one county, i.e., Shouguang in Weifang of Shandong Peninsula [15]. PAEs account for 10 wt% to 60 wt% of plastic products [9,19], thus giving rise to concerns about the potential risk of PAEs in recent years. However, few studies focused on the characteristics of PAEs in soils of vegetable greenhouses in Shandong Peninsula.

This study provides information on the concentrations, compositions, and distributions of 16 PAEs in soils from vegetable greenhouses in Shandong Peninsula and discusses possible sources, influence factors, and potential environment risk.

Materials and Methods

Chemicals and materials

Mixed standard solutions of 16 PAEs containing dimethyl phthalate (DMP), diethyl phthalate (DEP), diisobutyl phthalate (DiBP), di-n-butyl phthalate (DnBP), dimethylglycol phthalate (DMGP), di(4-methyl-2-pentyl) phthalate (DMPP), di(2-ethylhexyl) phthalate (DEHP), di(2-ethoxyethyl) phthalate (DEEP), dipentyl phthalate (DPP), di-n-hexyl phthalate (DHXP), butylbenzyl phthalate (BBP), di(2-n-butoxyethyl) phthalate (DBEP), dicyclohexyl phthalate (DCHP), di-n-octyl phthalate (DnOP), diphenyl phthalate (DPhP), and di-n-nonyl phthalate (DNP) were supplied by O2SI, Inc. (USA). The concentration of each PAE in this mixture solution was 1000 mg/L. Glassware was steeped with

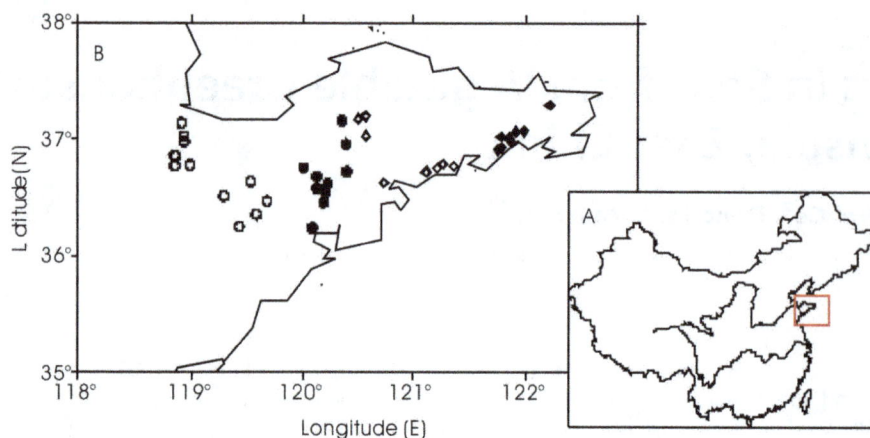

Figure 1. Schematic map showing the geographical location of (A) Shandong Peninsula and (B) the vegetable soil sampling sites in 4 regions in the Shandong Peninsula (solid round: Qingdao; solid diamond: Weihai; circle: Weifang; diamond: Yantai).

K_2CrO_7/H_2SO_4 solution for 12 h, washed with redistilled water, and then baked at 300°C for 4 h. Acetone, petroleum ether, and diethyl ether were of analytical grade and re-distilled before use to avoid PAEs contamination. Hexane was of HPLC grade and purchased from Anpel Company Inc. Florisil (60 mesh to 80 mesh) was activated at 650°C, and anhydrous sodium sulfate was baked at 420°C for 4 h.

Sampling

No specific permissions were required for sampling locations/activities. The field studies did not involve endangered or protected species. A total of 111 soil samples were collected from 37 vegetable greenhouses in Qingdao (number of samples: 30), Weihai (number of samples: 24), Weifang (number of samples: 33), and Yantai (number of samples: 24) in Shandong Peninsula in from 28 to 30 May 2012. The sampling locations are shown in Fig. 1.

Each sampling site consisted of five sub-samples (0.2 kg each) in the middle and four corners at depths of 0 cm to 10 cm, 10 cm to 20 cm, and 20 cm to 40 cm. The five sub-samples were mixed immediately after sampling, and then the soils were collected using aluminum foil envelopes through a pre-cleaned stainless steel auger and transported to laboratory in an ice box. Soils were stored in glass bottles at −20°C until analysis after being freeze-dried, ground, and homogenized with a stainless steel sieve (60 mesh). PAE contamination was avoided during sampling and further processing.

Soil physical and chemical analyses

Soil pH was measured using a pH meter with a soil/water ratio of 1:2.5. Soil cation exchange capacity (CEC) was analyzed using the Ba^{2+} compulsive exchange method [20]. Particle-size fraction was determined using the pipette method, and the soil texture was classified according to the Soil Survey Division Staff [21]. Total organic carbon (TOC) was determined using the wet oxidation method with chromate [22] and total nitrogen (TN) using micro-Kjeldahl digestion method [23].

Table 1. The main characteristics of the soils from vegetable greenhouses in Shandong Peninsula.

Area	Soil depth (cm)	pH	TOC (g/kg)	TN (g/kg)	C/N	CEC (mol/kg)	Sand (%)	Silt (%)	Clay (%)
Qingdao	0~10	6.62±0.64	31.7±9.8	1.3±0.4	26.44±0.90	0.14±0.05	54.1±3.8	27.6±6.6	17.8±3.7
	10~20	6.52±0.56	29.6±13.3	1.1±0.3	26.02±0.36	0.15±0.06	53.2±5.4	25.9±4.6	15.5±4.6
	20~40	6.64±0.42	25.1±7.1	0.8±0.3	33.49±9.12	0.11±0.05	46.7±4.6	24.6±7.8	15.9±4.2
Weihai	0~10	6.31±0.56	30.0±3.8	1.4±0.6	23.89±6.93	0.07±0.03	50.7±3.4	22.1±0.7	17.1±2.7
	10~20	6.10±0.43	27.7±4.1	1.0±0.1	27.33±2.72	0.09±0.01	55.3±6.2	29.6±9.9	15.6±1.4
	20~40	5.90±0.55	19.3±7.2	0.7±0.2	27.32±9.13	0.08±0.04	52.6±4.5	33.2±1.6	16.0±2.5
Weifang	0~10	6.88±0.46	32.2±6.0	1.7±3.1	40.49±5.39	0.13±0.09	61.3±0.2	37.8±7.7	19.9±6.0
	10~20	6.96±0.36	27.9±12.1	0.9±0.9	41.81±8.63	0.13±0.06	57.0±9.8	33.6±2.6	17.9±5.6
	20~40	6.99±0.49	23.4±4.3	1.2±2.3	42.55±5.78	0.14±0.08	59.2±2.1	28.1±5.9	21.1±7.4
Yantai	0~10	6.46±0.66	24.0±6.4	1.3±0.4	18.95±5.92	0.10±0.02	57.8±3.4	38.7±9.5	18.9±4.6
	10~20	6.56±0.96	23.7±6.3	1.1±0.4	22.99±5.90	0.10±0.03	57.4±4.2	38.6±4.4	18.0±3.0
	20~40	7.10±0.99	20.5±5.0	0.7±0.1	29.22±9.46	0.11±0.03	55.1±6.1	33.6±6.6	16.3±1.7

Table 2. The detection rate and concentration of PAEs in all soil samples from vegetable greenhouses in Shandong Peninsula (n = 111).

PAEs	Detection rate (%)	Mean (mg/kg)	SD (mg/kg)	Minimum (mg/kg)	Maximum (mg/kg)
DMP	99.1	0.364	0.276	ND	1.245
DEP	100	0.108	0.169	0.002	1.051
DiBP	96.4	1.118	1.928	ND	11.434
DnBP	100	1.471	2.715	0.016	15.722
DMGP	23.4	0.015	0.031	ND	0.170
DMPP	48.6	0.246	0.405	ND	1.971
DEHP	100	1.465	1.207	0.073	5.323
DEEP	23.4	0.041	0.243	ND	2.556
DPP	58.6	0.088	0.098	ND	0.516
DHXP	64.9	0.084	0.157	ND	1.448
BBP	86.5	0.194	0.557	ND	5.691
DBEP	18.9	0.015	0.038	ND	0.267
DCHP	44.1	0.035	0.048	ND	0.204
DnOP	97.3	1.239	1.796	ND	14.397
DPhP	82.0	0.240	0.290	ND	2.371
DNP	19.8	0.026	0.060	ND	0.251
Σ_{16}PAEs		6.748	5.716	1.939	35.442

ND: not detected. The data labeled as "ND" were treated as zero in further statistical treatment.

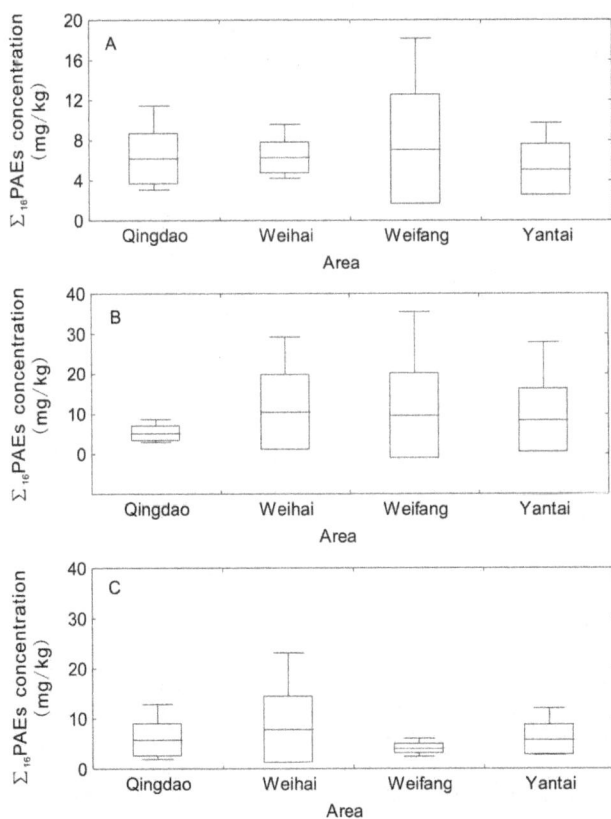

Figure 2. The concentrations of Σ_{16}PAEs in (A) soils of 0–10 cm, (B) soils of 10–20 cm and (C) soils of 20–40 cm from vegetable greenhouses in Shandong Peninsula.

Sample extraction of PAEs and instrumental analysis

PAEs extraction was conducted according to Wang's methods [24]. 5.0 g soil was spiked with surrogate standard (benzyl benzoate) and extracted through ultrasonication for 15 min thrice with 90 mL of acetone/petroleum ether (1:3, v:v). The extracts were combined, filtered, and concentrated to approximately 1 mL. The extracts were cleaned with anhydrous sodium sulfate (3 g), florisil (6 g), and anhydrous sodium sulfate (3 g) on a glass column (1 cm i.d.). The column was washed with 10 mL of petroleum ether/diethyl ether (10:0.4, v:v), and then PAEs were eluted with 90 mL of petroleum ether/diethyl ether (10:3, v:v). The extracts were reduced to 1.0 mL in hexane, and internal standard (diisophenyl phthalate) was added before instrumental analysis.

Instrumental analysis was performed on an Agilent 6890 GC-5973 MSD gas chromatography-mass spectrometry system (GC-MS) in electron impact and selective ion monitoring modes according to Zeng et al. [25]. The GC column used was DB-5MS capillary column (30 m×0.25 mm i.d. ×0.25 mm film thickness, J&W Scientific). The column temperature program was 80°C (1 min), to 180°C (10°C/min, 1 min), to 300°C (2°C/min, 10 min). The carrier gas was helium with flow rate of 0.8 mL/min. Then, 1 μL of the extracts was injected into GC-MS in splitless injection mode, and the injector temperature was 250°C. The GC-MS transfer line was 280°C, and the post run temperature was 285°C for 2 min.

Quality control and quality assurance

Quality assurance was performed by analyzing a procedural and solvent blank, a spiked blank every 10 samples, surrogate standards for each sample, and sample duplicate. DiBP, DnBP, and DEHP were subtracted from those in the soil samples because of the small amount in procedural blanks. The surrogate recoveries were 84.1%±8.5%, and no surrogate corrections were

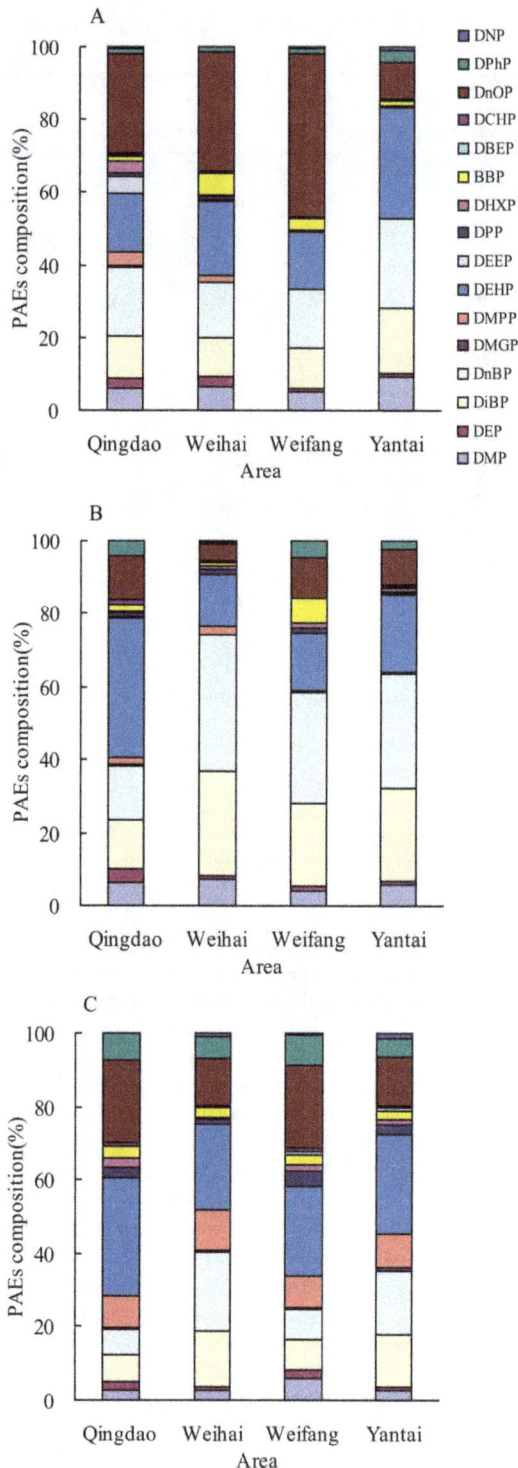

Figure 3. The compositions of PAEs in (A) soils of 0–10 cm, (B) soils of 10–20 cm and (C) soils of 20–40 cm from vegetable greenhouses in Shandong Peninsula.

made to the final PAE concentrations reported. The calibration curves were used with six concentration levels of a standard mixture for PAEs quantitation. The standard mixture was analyzed every 10 samples to determine instrument stability and to confirm the calibration curve. The instrumental detection limits

ranged from 1~9 pg, calculated by signal to noise ratio of 10. The method detection limits for PAEs were determined as mean field blanks plus three times the standard deviation of the field blanks [25], ranging from 0.002 mg/kg for DEP to 0.024 mg/kg for DEHP.

Results

Properties of the soils from vegetable greenhouses

The major characteristics of soils from vegetable greenhouses in Shandong Peninsula are presented in Table 1. The pH (H_2O) of soils was neutral in three sample areas, but was less than 6.5 in Weihai, which was moderately acidic. The TOC ranged from 19.3 g/kg to 32.2 g/kg and presented a decreasing trend with soil depth. The C/N ratio was approximately 20 to 30, except in Weifang, which had a value of more than 40 on average. This value indicated that the organic matter content was more C-rich. The C/N ratio presented an increasing trend with soil depth. The CEC followed a similar pattern as pH, with lower values in soil samples from Weihai. Most of the soils were sandy loam, and some were loam.

PAE concentrations in soils from vegetable greenhouses in Shandong Peninsula

All 16 PAEs were detected in soils from vegetable greenhouses (Table 2). Among them, three PAEs, namely, DEP, DnBP, and DEHP, were detected in all samples. The detection rates of another three PAEs (DMP, DiBP, and DnOP) were more than 90%. By contrast, the detection rates of DBEP and DNP were lower than 20%. The mean concentrations of DiBP, DnBP, DEHP, and DnOP were more than 1 mg/kg, higher than other PAEs. On the whole, the mean was almost systematically inferior to standard deviation, suggesting a very high heterogeneity between soils. The total of 16 PAEs (Σ_{16}PAEs) in Shandong Peninsula ranged from 1.939 mg/kg to 35.442 mg/kg, with an average of 6.748 mg/kg.

The concentrations of Σ_{16}PAEs in soils from vegetable greenhouses in different areas of Shandong Peninsula are presented in Fig. 2. The variability of Σ_{16}PAEs was high in Weifang in the two upper layers, but low in Qingdao (10 cm to 20 cm) and Weifang (20 cm to 40 cm). The maximum value of Σ_{16}PAEs in soils at 0 cm to 10 cm and 10 cm to 20 cm both appeared in Weifang, with values of 18.179 and 35.442 mg/kg, respectively.

PAE composition in soils of vegetable greenhouses from Shandong Peninsula

Despite the different concentration of Σ_{16}PAEs, the PAE compositions in soils from vegetable greenhouses in Shandong Peninsula were comparable (Fig. 3). DnOP had the highest proportion (27.1% to 45%) in soils at 0 cm to 10 cm in Qindao, Weihai, and Weifang, whereas DEHP had the highest proportion (30.4%) in Yantai. The proportion of DnBP and DiBP ranged from 10.6% to 24.5%, suggesting significant proportion. In soils at 10 cm to 20 cm, DEHP had the largest proportion of 38.4% in Qingdao, but DnBP had the largest and DiBP had the second largest proportion in the other three areas. In soils at 20 cm to 40 cm, DEHP was a dominant congener, ranging from 23.4% to 31.8% in four areas. DnOP in Qingdao and Weifang had the second highest proportion, whereas DnBP in Weihai and Yantai had the second highest proportion. Therefore, DEHP, DnOP, DnBP, and DiBP had the largest proportion in soils in all areas. In addition, DMP accounted for about 5~10% in the upper two

Table 3. Comparisons of PAEs contents in agricultural soils in China (mg/kg).

Area	DMP	DEP	DnBP	DEHP	BBP	DnOP	Σ6PAEs	Σ16PAEs	Type of soils	References
Shandong Peninsula	0.36 (ND~1.24)	0.11 (ND~1.21)	1.47 (0.02~15.72)	1.47 (0.07~5.32)	0.19 (ND~5.69)	1.24 (ND~14.40)	4.48 (1.18~23.35)	6.73 (1.94~35.44)	Soils of vegetable greenhouses	In this study
Nanjing	0.006 (ND~0.016)	0.005 (ND~0.012)	0.19 (ND~1.41)	1.72 (0.034~9.031)	0.003 (ND~0.038)	0.158 (ND~1.739)	1.89 (0.15~9.68)		Soils of vegetable greenhouses	[14]
Hangzhou	ND	0.59 (0.06~1.49)	0.21 (0.14~0.35)	1.48 (0.81~2.20)	0.05 (0.03~0.16)	0.14 (0.10~0.25)	2.47	2.75 (1.90~4.36) (Σ11PAEs)	Soils of vegetable greenhouses	[15]
Guangzhou, Shenzhen,	(ND~0.68)	(ND~1.77)	(3.75~18.45)	(2.82~25.11)	(ND~1.48)	(ND~0.92)	(3.00~45.67)		vegetable Soils	[16]
Zhangjiang,	0.02 (ND~0.45)	0.09 (ND~1.06)	0.23 (ND~7.65)	0.15 (ND~6.38)	0.05 (ND~2.83)	0.03 (ND~0.32)	0.56 (0.01~9.30)		Vegetable soil	[50]
Dongguan,	0.02 (ND~0.86)	0.05 (ND~1.60)	0.49 (ND~17.51)	0.17 (ND~4.22)	0.13 (ND~5.89)	0.06 (ND~1.12)	0.92 (0.01~25.99)		Paddy soil	
Zhongshan,	0.03 (ND~0.12)	0.19 (ND~2.50)	0.41 (ND~4.13)	0.12 (ND~2.69)	0.02 (ND~0.26)	0.02 (ND~0.08)	0.81 (0.05~5.92)		Banana soil	
Zhuhai,	0.03 (ND~0.18)	0.06 (ND~0.44)	0.30 (ND~1.77)	0.07 (ND~0.26)	0.01 (ND~0.03)	0.02 (ND~0.07)	0.49 (ND~2.10)		Sugarcane soil	
Shunde	0.02 (ND~0.06)	0.06 (ND~0.40)	0.20 (ND~0.56)	0.11 (ND~0.99)	0.01 (ND~0.06)	0.02 (ND~0.15)	0.43 (0.04~1.38)		Orchard soil	
Leizhou Peninsula	0.02	0.02	0.45	0.01	0.01	0.02	0.53	1.11 (0.02~5.45)	Sugarcane soil	[17]
	0.01	0.01	0.24	0.24	0.01	0.02	0.53	0.86 (ND~2.78)	Paddy soil	
	0.02	0.01	0.27	0.12	0.01	0.02	0.45	0.61 (0.02~1.87)	Vegetable soil	
	0.03	0.01	0.28	0.10	0.01	0.01	0.44	0.60 (0.28~1.05)	Orchard soil	
Huizhou	0.004 (ND~0.03)	0.01 (ND~0.22)	0.15 (ND~0.39)	0.09 (ND~0.44)	0.002 (ND~0.04)	0.01 (ND~0.06)	0.31 (0.09~0.75)	0.60 (0.18~2.04)	Agricultural soil	[51]
							0.28 (0.06~0.64)	0.59 (0.18~2.04)	Vegetable soil	
							0.24 (0.08~0.64)	0.65 (0.43~1.21)	Paddy soil	

Table 3. Cont.

Area	DMP	DEP	DnBP	DEHP	BBP	DnOP	Σ₆PAEs	Σ₁₆PAEs	Type of soils	References
							0.22	0.51	Orchard soil	
							(0.08~0.64)	(0.38~0.63)		

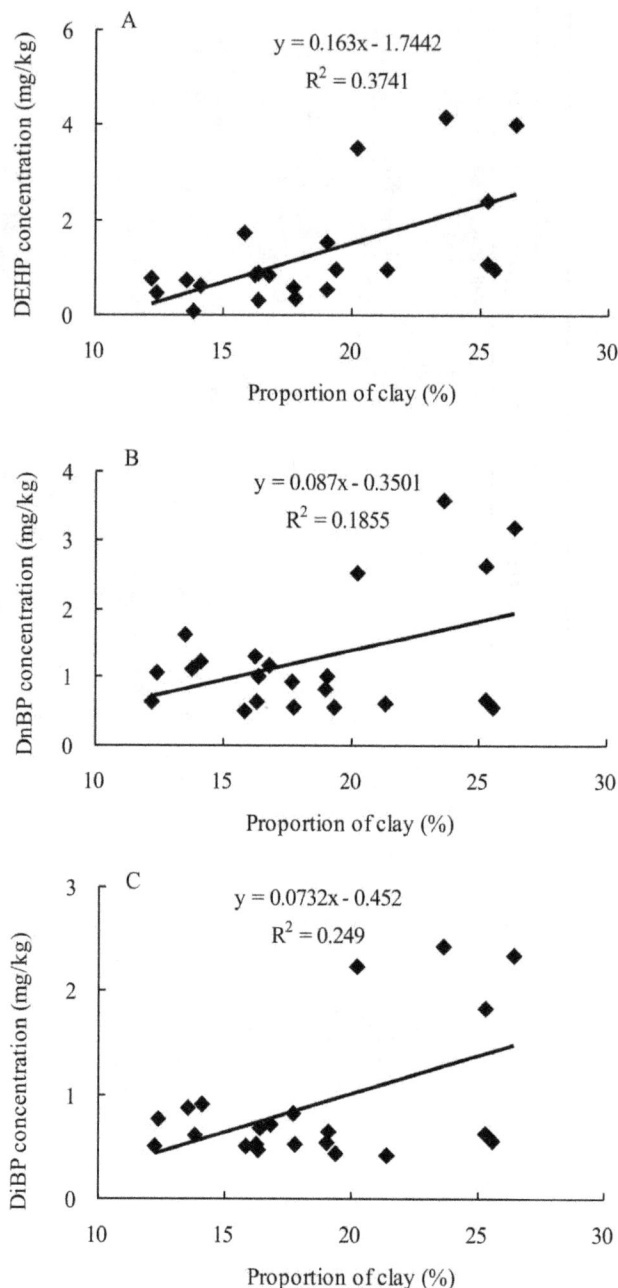

Figure 4. Correlations of the concentrations of (A) DEHP, (B) DnBP and (C) DiBP with the proportions of clay in 0–10 cm soils of vegetable greenhouses from Shandong Peninsula.

layers; by contrast, DEP, DMPP, BBP, and DPhP accounted for only approximately 1% to 5%, suggesting a small proportion.

Discussion

Potential sources of PAEs in soils from vegetable greenhouses

Soils that are being used or were used for vegetable greenhouses had higher PAE concentrations (Table 3), which suggests that PAEs are widespread in soils from vegetable greenhouses. Various PAE sources exist in soils from vegetable greenhouses. The plastic film used in vegetable greenhouses is a major source of PAEs. The

Table 4. Ratio of PAEs in samples exceeding allowable and cleanup concentrations in soils from vegetable greenhouses in Shandong Peninsula.

Soil depth (cm)	Ratio of samples exceeding allowable concentrations (%)						Ratio of samples exceeding cleanup concentrations (%)					
	DMP	DEP	DnBP	BBP	DEHP	DnOP	DMP	DEP	DnBP	BBP	DEHP	DnOP
0~10	94.6	89.2	97.3	0	2.7	45.9	0	0	0	0	0	0
10~20	100	94.6	94.6	2.7	5.4	27	0	0	13.5	0	0	0
20~30	89.2	97.3	89.2	0	0	21.6	0	0	2.7	0	0	0

maximum value of Σ_{16}PAEs in soils at 0 cm to 10 cm and 10 cm to 20 cm appeared in Weifang (Fig. 2), which has a long history of vegetable production and is famous for extensive greenhouse cultivation. Even more remarkably, the plastic film used in greenhouse cultivation in Shandong Peninsula is replaced annually, which may result in a higher concentration of PAEs in soils from vegetable greenhouses than in other soils. In addition, PAEs are found in organic fertilizers in China [26] and in compost of sewage sludge with rice straw [27]. The amount of fertilizers used for vegetable planting in greenhouses is more than that used for field crops, and the proportion of organic fertilizers has increased since 2007 [28]. Moreover, PAEs are found in the groundwater in China [29–31], and groundwater is used for irrigation in vegetable greenhouse, which may result in the buildup of PAEs in vegetable greenhouse soils. More importantly, Zeng et al. [32] found a declining trend of PAEs in agricultural soils that were far from urban centers. The highest PAE contents are found in soils close to architectural markets, where plastic materials are sold, and those close to large chemical manufacturing factories. Most vegetable greenhouses in this study are near industrialized cities with large populations. Over 300 plastic manufacturers that produced 0.3 million tons/yr of plastic existed in Shandong Peninsula by 2003. All these factors may have resulted in the high concentration of PAEs in vegetable greenhouses in Shandong Peninsula.

Among 16 PAEs, DEHP and DnBP are found to be the two most abundant PAEs in agricultural soils in Guangzhou, Shenzhen, Leizhou Peninsula, and Huizhou (Table 3). Moreover, DiBP is found to be abundant in Guangzhou agricultural soils [32], whereas DnOP is abundant in the soils of vegetable greenhouses in Nanjing [14]. Similarly, DEHP and DnOP are the two most abundant PAEs in soils of vegetable greenhouses in Shandong Peninsula, followed by DnBP and DiBP (Fig. 3). The relative contribution of PAEs in agricultural soils is consistent with that in sediment [33], air [5,34], and waters [35]. The global consumption of PAEs is about 6.0 million tones/yr, mainly as plasticizers in the plastic industry. Among plasticizers of PAEs, DEHP, DnOP, and DiBP/DnBP are widely used. It is found that DEHP and DnBP are two dominant PAE components in white and black mulch film used in vegetable production systems, ranging from 48.0~115.6 mg/kg and 2.3~3.2 mg/kg, respectively [14]. We also found that besides DEHP, DnOP and DiBP were two dominant PAEs in polyvinyl chloride (PVC) plastics mainly used in vegetable greenhouses in Shandong Peninsula, accounting for 20% and 10% of total of 16 PAEs, respectively. DMP and DEP are also detected in the plastics film, though their contents are low. Therefore, the plastics film may be a major potential source of some PAEs. Furthermore, PAEs are found in fertilizer and manure. DEHP, DnBP, DMP and DEP are the major organic pollutants in fertilizers, with contents more than

2.5 mg/kg [27]. Similarly, six PAEs (DEHP, DnBP, DnOP, DMP, DEP and BBP) are found in chicken, pig, cow and duck manure, in the range of 2.24~6.84 mg/kg [14]. These potential sources may lead to the high detection rates of DMP, DEP DnBP, DiBP, DEHP and DnOP (Table 2).

Relationship between PAEs and soil properties or age of vegetable greenhouses

Soil properties, such as pH, organic matter, texture, and redox potential, have a certain effect on the migration of hydrophobic organic compounds (HOCs) in soil [36–37]. A positive correlation between HOCs and TOC is found in several research [32,38]; however, it is not in this study. Katsoyiannis [39] reported that no correlation can be found between HOCs and TOC if continuous sources of HOCs exist in soils. In this study, several continuous inputs of PAEs in vegetable greenhouse soils, including plastic film, fertilizers, and irrigation, may hinder the achievement of equilibrium between PAEs and TOCs.

The relationship among the major PAEs, including DEHP, DnBP and DiBP, with the proportions of clay in 0 cm to 10 cm soils was analyzed (Fig. 4). DiBP, DnBP, and DEHP have a significantly positive correlation with the proportion of clay ($r = 0.431 \sim 0.611$, $p < 0.05$). A similar relationship of HOCs, such as PAEs, organic chlorinated pesticides, and PAHs, with clay is also found in soils [32,40–41]. The clay of sediment or soil shows stronger capability to adsorb HOCs than sand and silt, due to small granulometry but high specific surface area [42–43]. Besides, the aging of organic matter, such as humic material, distributes around clay complexes, resulting in the formation of films of organic material [44]. These films of organic material are very difficult to remove, and so organic matter builds up and becomes a permanent part of the clay complexes. The clay-organic complexes supply rich reactive sites for the adsorption of organic pollutants [45].

A positive correlation between Σ_{16}PAE concentration and age of vegetable greenhouses was found in this study ($r = 0.294$, $p < 0.01$), suggesting that PAEs in soils may be related with the cumulative use of potential PAE sources over years in greenhouse vegetable cultivation. However, the correlation coefficient is not high. Studies demonstrate the biodegradability of some PAEs in aerobic condition [46–47], though PAEs are resistant to degradation through hydrolysis and photolysis. The biodegradability of PAE congeners is different. Shanker et al found the degradation rates of DMP and DBP were greater than that of DEHP under aerobic conditions [48]. Additionally, PAEs migrates deeper in soils profiles, and the TOC of soil and volume of leaching water can affect the migration of PAEs [49]. These factors may result in the low correlation coefficient between PAE contents and age of vegetable greenhouses.

Comparison of PAE concentrations in the different soils in China

Comparisons of PAE contents in agricultural soils from China are presented in Table 3. The average Σ_6PAE (DMP, DEP, DnBP, DEHP, BBP, and DnOP) contents in the soils from vegetable greenhouses in Shandong Peninsula, Nanjing, and Hangzhou are approximately 2 mg/kg to 6 mg/kg, higher than other types of soils. High Σ_6PAE concentrations are also found in vegetable soils from Guangzhou and Shenzhen, where soils are previously used to plant greenhouse vegetables. In comparison, the Σ_6PAE concentrations in vegetables, paddy, banana, sugarcane, or orchard soil are low, ranging from 0.2 mg/kg to 1 mg/kg.

Potential risk assessment of soils of vegetable greenhouses from Shandong Peninsula

PAEs have a variety of toxic effects. Long term exposure to PAEs results in decreased fertility in females, fetal defect, altered hormone levels, uterine damage and male reproduction abnormalities such as reduced sperm production and motility, cell damage, cell tumors, etc [52–55]. According to human health based levels that correspond to excess lifetime cancer risks and human health based levels for systemic toxicant calculated from reference doses, allowable soil concentrations and cleanup levels of PAEs have been recommended in New York, USA [56]. The allowable soil concentrations are 0.02, 0.071, 0.081, 1.125, 4.35, and 1.20 mg/kg for DMP, DEP, DnBP, BBP, DEHP, and DnOP, respectively; and soil cleanup levels are 2, 7.1, 8.1, 50, 50, and 50 mg/kg, respectively [56]. PAEs exceeding allowable and cleanup concentrations may be a menace to human health. According to these criteria, the ratios of PAE concentration in this study exceeding allowable and cleanup concentrations are

presented in Table 4. The ratios of DMP, DEP, and DnBP exceeding allowable concentrations are 90% to 100% at different soil depths, suggesting high PAE pollution. Moreover, the ratios of DnOP exceeding allowable concentrations are also high, particularly in soils at 0 cm to 10 cm; however, the ratios of BBP and DEHP are low. Similarly, in agricultural soils around Guangzhou, DMP, DEP, and DnBP also exceed allowable concentrations [32]. Notably, DnBP in some samples is approximately twice to thrice higher than the recommended cleanup concentration. These soil samples are mostly from the vegetable greenhouses with ages of approximately 10 years, suggesting that long-term application of plastic film or manure in vegetable greenhouses may increase environmental and health risks.

The cultivated vegetables can uptake and accumulate PAEs, but the difference is found in accumulated amount of PAE congeners by vegetables. Compared with DEHP, more DBP in soils is accumulated in stalk and leaf of carrot, cucumber and tomato [24]. The physical and chemical properties, such as molecular weight and octanol/water partition coefficient (K_{ow}), have effects on the accumulation of PAEs by plants. Due to the smaller molecular weight and lower K_{ow}, DBP is more easily absorbed and transported by vegetables than DEHP. Furthermore, several studies report a positive correlation between accumulated PAE amount by vegetables and contents in soils [24,57]. Thus, mitigation of PAEs in soils is important to lower the risks of PAEs to human health.

Author Contributions

Conceived and designed the experiments: CC HC YS. Performed the experiments: CC HC. Analyzed the data: HC WG. Contributed reagents/materials/analysis tools: WG DM. Wrote the paper: CC HC YS.

References

1. Staples CA, Peterson DR, Parkerton TF, Adams WJ (1997) The environmental fate of phthalate esters, a literature review. Chemosphere 4: 667–749.
2. Xie Z, Ebinghaus R, Temme C, Lohmann R, Caba A, et al. (2007) Occurrence and air-sea exchange of phthalates in the Arctic. Environmental Science & Technology 41: 4555–4560.
3. Lin C, Lee C, Mao W, Nadim F (2009) Identifying the potential sources of di-(2-ethylhexyl) phthalate contamination in the sediment of the Houjing River in southern Taiwan. Journal of Hazardous Materials 161: 270–275.
4. Buszka PM, Yeskis DJ, Kolpin DW, Furlong ET, Zaugg SD, et al. (2009) Waste-indicator and pharmaceutical compounds in landfill-leachate-affected ground water near Elkhart, Indiana, 2000–2002. Bulletin of Environmental Contamination and Toxicology 82: 653–659.
5. Wang P, Wang SL, Fan CQ (2008) Atmospheric distribution of particulate- and gas-phase phthalic esters (PAEs) in a Metropolitan City, Nanjing, East China. Chemosphere 72(10):1567–1572.
6. Zhang LF, Dong L, Ren IJ, Shi SX, Zhou L, et al. (2012) Concentration and source identification of polycyclic aromatic hydrocarbons and phthalic acid esters in the surface water of the Yangtze River Delta, China. Journal of Environmental Sciences 24(2): 335–342.
7. Bauer MJ, Herrmann R (1997) Estimation of the environmental contamination by phthalic acid esters leaching from household wastes. Science of the Total Environment 208(1–2): 49–57.
8. Amir S, Hafidi M, Merlina G, Hamdi H, Jouraiphy A, et al. (2005) Fate of phthalic acid esters during composting of both lagooning and activated sludges. Process Biochemistry 40(6): 2183–2190.
9. Hens GA, Caballos AMP (2003) Social and economic interest in the control of phthalic acid esters. Trends in Analytical Chemistry 22, 847–857.
10. United States Environmental Protection Agency (USEPA) (2013) Electronic Code of Federal Regulations, Title 40-Protection of Environment, Part 423-Steam Electric Power Generating Point Source Category. Appendix A to Part 423–126, Priority Pollutants. http://www.ecfr.gov/cgi-bin/text-idx?c¼ ecfr&SID¼b960051a53c9015d817718d71f1617b7&rgn¼div5&view¼text& node¼40.30.0.1.1.23&idno¼40#40:30.0.1.1.23.0.5.9.9
11. Li ZH, Wang GZ, Qi F (2012) Current Situation and Thinking of Development of Protected Agriculture in China. Chinese Agricultural Mechanization 239(1): 7–10(in Chinese with English abstract).
12. Costa JM, Heuvelink E (2004) Protected cultivation rising in China. Fruit & Vegetable Technology 4: 8–11.
13. Zou ZR (2002) Facility Horticulture Science. China agriculture press, Beijing (in Chinese).
14. Wang J, Luo YM, Teng Y, Ma WT, Christie P, et al. (2013) Soil contamination by phthalate esters in Chinese intensive vegetable production systems with different modes of use of plastic film. Environmental Pollution 180: 265–273.
15. Chen YS, Luo YM, Zhang HB, Song J (2011) Preliminary study on PAEs pollution of greenhouse soils. ACTA Pedologica Sinica 48(3): 516–523(in Chinese with English abstract).
16. Cai QY, Mo CH, Li YH, Zeng QY, Wang BG, et al. (2005) The study of PAEs in soils from typical vegetable fields in areas of Guangzhou and Shenzhen, South China. ACTA Ecologica Sinica 25(2): 283–288(in Chinese with English abstract).
17. Guan H, Wang JS, Wan HF, Li PX, Yang GY (2007) PAEs Pollution in soils from typical agriculture area of Leizhou Peninsula. Journal of Agro-Environment Science 26(2): 622–628(in Chinese with English abstract).
18. Mariko M, Mutsuko HK, Makoto E (2008) Potential adverse effects of phthalic acid esters on human health: A review of recent studies on reproduction. Regulatory Toxicology and Pharmacology 50(1): 37–49.
19. Chou K, Robert OW (2006) Phthalates in food and medical devices. Journal of Medical Toxicology 2: 126–135.
20. Bascomb CL (1964) Rapid method for the determination of cation-exchange capacity of calcareous and non-calcareous soils. Journal of the Science of Food and Agriculture 15(12): 821–823.
21. Soil Survey Division Staff. Soil Survey Manual (1993) In: Agriculture Handbook, Revised Edition, vol. 18. United States Department of Agriculture, Washington DC.
22. Schwartz V (1995) Fractionated combustion analysis of carbon in forest soils - new possibilities for the analysis and characterization of different soils. Fresenius' journal of analytical chemistry 351(7): 629–631.
23. Flowers TH, Bremner JM (1991) A rapid dichromate procedure for routine estimation of total nitrogen in soils. Communications in Soil Science and Plant Analysis 22(13–14): 1409–1416.
24. Wang ML (2007) Research on analytical method and environmental behavior of PAEs in vegetable greenhouse. Shandong Agricultural University. Doctor thesis (in Chinese with English abstract).
25. Zeng F, Cui KY, Xie ZY, Wu L, Luo DL, et al. (2009) Distribution of phthalate esters in urban soils of subtropical city, Guangzhou, China. Journal of Hazardous Materials 164: 1171–1178.

26. Cai QY, Mo CH, Wu QT, Zeng QY, Katsoyiannis A (2007) Quantitative determination of organic priority pollutants in the composts of sewage sludge with rice straw by gas chromatography coupled with mass spectrometry. Journal of Chromatography A 1143: 207–214.

27. Mo CH, Cai QY, Li YH, Zeng QY (2008) Occurrence of priority organic pollutants in the fertilizers, China. Journal of Hazardous Materials 152: 1208–1213.

28. Liu ZH, Jiang LH, Zhang WJ, Zheng FL, Wang M, et al. (2008) Evolution of fertilization rate and variation of soil nutrient contents in greenhouse vegetable cultivation in shandong. ACTA Pedologica Sinica 45 (2): 296–303 (in Chinese with English abstract).

29. Zhang D, Liu H, Liang Y, Wang C, Liang HC, et al. (2009) Distribution of phthalate esters in the groundwater of Jianghan plain, Hubei, China. Frontiers of Earth Science in China 3(1): 73–79.

30. Xiong PX, Gong X, Deng L (2008) Analysis of PAE Pollutants in Farm Soil and Water Samples in Nanchang City. Chemistry 8: 636–640 (in Chinese with English abstract).

31. Wang C, Liu H, Cai HS, Liang Y, Liang HC, et al. (2009) Source Analysis and Detection of Trace Phthalate Esters in Groundwater in Wuhan. Environmental Science & Technology 32(10): 118–123 (in Chinese with English abstract).

32. Zeng F, Cui KY, Xie ZY, Wu LN, Liu M, et al. (2008) Phthalate esters (PAEs): Emerging organic contaminants in agricultural soils in peri-urban areas around Guangzhou, China. Environmental Pollution 156: 425–434.

33. Liu H, Liang HC, Liang Y, Zhang D, Wang C, et al. (2010) Distribution of phthalate esters in alluvial sediment: A case study at JiangHan Plain, Central China. Chemosphere 78(4): 382–388.

34. Zeng F, Lin YJ, Cui KY, Wen JX, Ma YQ, et al. (2010) Atmospheric deposition of phthalate esters in a subtropical city. Atmospheric Environment 44(6): 834–840.

35. He W, Qin N, Kong XZ, Liu WX, He QS, et al. (2013) Spatio-temporal distributions and the ecological and health risks of phthalate esters (PAEs) in the surface water of a large, shallow Chinese lake. Science of The Total Environment 461–462: 672–680.

36. Hitch RK, Day HR (1992) Unusual persistence of DDT in some western USA soils. Bulletin of Environmental Contamination and Toxicology 48: 259–264. http://www.dec.ny.gov/docs/remediation_hudson_pdf/cpsoil.pdf.

37. Cousins IT, Bondi G, Jones KC (1999) Measuring and modelling the vertical distribution of semivolatile organic compounds in soils. I: PCB and PAH soil core data. Chemosphere 39: 2507–2518.

38. Jiang YF (2009) Preliminary study on composition, distribution and source indentification of persistent organix pollutants in soil of Shanghai. Shandong University. Doctor thesis (in Chinese with English abstract).

39. Katsoyiannis A (2006) Occurrence of polychlorinated biphenyls (PCBs) in the Soulou stream in the power generation area of Eordea, northwestern Greece. Chemosphere 65: 1551–1561.

40. Wang L (2013) Pollution characteristics of organochlorine pesticides in Daling River estuary. Dalian Maritime University. Master thesis (in Chinese with English abstract).

41. Chen J, Wang XJ, Tao S (2005) The Influences of soil total organic carbon and clay contents on PAHs vertical distributions in soils in Tianjin area. Research of Environmental Sciences 18(4): 79–83 (in Chinese with English abstract).

42. Amellal N, Portal JM, Berthelin J (2001) Effect of soil structure on the bioavailability of polycyclic aromatic hydrocarbons within aggregates of a contaminated soil. Applied Geochemistry 16: 1611–1619.

43. Benlahcen KT, Chaoui A, Budzinski H, Bellocq J, Garrigues Ph (1997) Distribution and sources of polycyclic aromatic hydrocarbons in some Mediterranean coastal sediments. Marine Pollution Bulletin 34(5): 298–305.

44. Gjessing ET (1976) Physical & Chemical Characteristics of Aquatic Humus. Ann Arbor Science Publishers Inc. (Ann Arbor), Mich.

45. Evans KM, Gill RA, Robotham PWJ (1990) The PAH and organic content of sediment particle size fractions. Water, Air, and Soil Pollution 51: 13–31.

46. Shelton DR, Boyd SA, Tiedje JM (1984) Anaerobic biodegradation of phthalic acid esters in sludge. Environmental Science & Technology 18: 93–97.

47. Ejlertsson J, Meyerson U, Svensson BH (1996) Anaerobic degradation of phthalic acid esters during digestion of municipal solid waste under landfilling conditions. Biodegradation 7: 345–352.

48. Shanker R, Ramakrishna C, Seth PK (1985) Degradation of some phthalic acid esters in soil. Environmental Pollution Series A, Ecological and Biological 39(1):1–7.

49. Wan TT, He GX, Zhang ZH, Zhu L (2013) Simulation on soil column leaching of oxygen nonhydrocarbon migration in soil profiles. Acta Scientiae Circumstantiae 33(10): 2795–2806(in Chinese with English abstract).

50. Yang GY, Zhang TB, Gao ST, Huo ZX, Wan HF, et al. (2007) Distribution of phthalic acid esters in agricultural soils in typical regions of Guangdong Province. Chinese Journal of Applied Ecology 18(10): 2308–2312(in Chinese with English abstract).

51. Tan Z, Li CH, Mo CH (2012) Distribution of Phthalic Acid Esters in Agricultural Soils of Huizhou City. Environmental Science and Management 37(5): 120–123(in Chinese with English abstract).

52. Biscardi D, Monarca S, De Fusco R, Senatore F, Poli P, et al. (2003) Evaluation of the migration of mutagens/carcinogens from PET bottles into mineral water by Tradescantia/micronuclei test, Comet assay on leukocytes and GC/MS. Science of the Total Environment 302:101–108.

53. Sharpe RM, Fisher JS, Millar MM, Jobling S, Sumpter JP (1995) Gestational and lactational exposure of rats to xenoestrogens results in reduced testicular size and sperm production. Environmental Health Perspectives 103:1136–1143.

54. Jones HB, Garside DA, Liu R, Roberts JC (1993) The influence of phthalate-esters on leydig-cell structure and function in-vitro and in-vivo. Experimental and Molecular Pathology 58:179.

55. Giuseppe L, Alberto V, Claudio DF (2004) Di-2-ethylhexyl phthalate and endocrine disruption: A review. Current Drug Targets-Immune, Endocrine & Metabolic Disorders 4: 37–40.

56. Department of Environmental Conservation, New York, USA (1994) Determination of soil cleanup objectives and cleanup levels (TAGM 4046). http://www.dec.ny.gov/regulations/2612.html.

57. Chiou CT, Sheng GY, Manes M (2001) A partition-limited model for the plant uptake of organic contaminants from soil and water. Environmental Science Technology 35: 1437–1444.

9

Unravelling the Carbon and Sulphur Metabolism in Coastal Soil Ecosystems Using Comparative Cultivation-Independent Genome-Level Characterisation of Microbial Communities

Basit Yousuf[1,2], Raghawendra Kumar[1,2], Avinash Mishra[1,2]*, Bhavanath Jha[1,2]*

1 Discipline of Marine Biotechnology and Ecology, CSIR-Central Salt and Marine Chemicals Research Institute, Bhavnagar, Gujarat, India, 2 Academy of Scientific and Innovative Research, CSIR, New Delhi, India

Abstract

Bacterial autotrophy contributes significantly to the overall carbon balance, which stabilises atmospheric CO_2 concentration and decelerates global warming. Little attention has been paid to different modes of carbon/sulphur metabolism mediated by autotrophic bacterial communities in terrestrial soil ecosystems. We studied these pathways by analysing the distribution and abundance of the diagnostic metabolic marker genes *cbbM*, *apsA* and *soxB*, which encode for ribulose-1,5-bisphosphate carboxylase/oxygenase, adenosine phosphosulphate reductase and sulphate thiohydrolase, respectively, among different contrasting soil types. Additionally, the abundance of community members was assessed by quantifying the gene copy numbers for 16S rRNA, *cbbL*, *cbbM*, *apsA* and *soxB*. Distinct compositional differences were observed among the clone libraries, which revealed a dominance of phylotypes associated with carbon and sulphur cycling, such as *Gammaproteobacteria* (*Thiohalomonas*, *Allochromatium*, *Chromatium*, *Thiomicrospira*) and *Alphaproteobacteria* (*Rhodopseudomonas*, *Rhodovulum*, *Paracoccus*). The rhizosphere soil was devoid of sulphur metabolism, as the *soxB* and *apsA* genes were not observed in the rhizosphere metagenome, which suggests the absence or inadequate representation of sulphur-oxidising bacteria. We hypothesise that the novel *Gammaproteobacteria* sulphur oxidisers might be actively involved in sulphur oxidation and inorganic carbon fixation, particularly in barren saline soil ecosystems, suggesting their significant putative ecological role and contribution to the soil carbon pool.

Editor: Gabriel Moreno-Hagelsieb, Wilfrid Laurier University, Canada

Funding: This study was supported by the Council of Scientific and Industrial Research (CSIR; www.csir.res.in), Government of India, New Delhi [CSC0102–TapCoal; Senior Research Fellowship to BY]. The funders had no role in study design, data collection and analysis, decision to publish, or preparation of the manuscript.

Competing Interests: The authors have declared that no competing interests exist.

* Email: avinash@csmcri.org (AM); bjha@csmcri.org (BJ)

Introduction

Soil microbial communities are indispensable for the health of the Earth as they drive major biogeochemical cycles, play a critical role in agriculture and have a significant impact on climatic change [1–2]. Autotrophic soil microorganisms are an integral component of the ecosystem and facilitate the availability of otherwise unavailable CO_2 to other organisms. This assimilation process occurs through various complex biochemical pathways [3]. The distribution of carbon fixation strategies are widespread among prokaryotes and depends on the individual autotrophic organism and is also determined by different habitat characteristics, such as the energy demand, the availability of inorganic compounds (sulphide, elemental sulphur, thiosulphate and sometimes ferrous iron and hydrogen), usage of coenzymes and the oxygen sensitivity of enzymes [4].

The predominant route of this fixation process is the Calvin–Benson–Bassham (CBB) reductive pentose phosphate pathway, which is ubiquitously prevalent among the aerobic members of the *Alpha-*, *Beta-* and *Gammaproteobacteria* [5]. The key enzyme of this pathway, ribulose 1,5-bisphosphate carboxylase/oxygenase (RuBisCO), occurs in forms I and II, whose large subunits are encoded by the *cbbL* and *cbbM* genes, respectively. The *cbbL* gene is found in plants, green algae, *Cyanobacteria* and many chemolithoautotrophs, whereas *cbbM* is reported to occur in several photosynthetic bacteria, aerobic and facultative anaerobic chemoautotrophic bacteria and dinoflagellates [6]. The occurrence of the *cbbM* gene has been exclusively investigated for chemolithoautotrophy from aquatic habitats such as hydrothermal vents [5], hypersaline habitats [7], soda lake sediments [8], thermal Springs [9], Movile Cave in Romania [10], with only one study from a terrestrial ecosystem reported so far [11].

Microbial sulphur oxidation is one of the vital processes for the biogeochemical sulphur cycle and is closely linked to carbon cycling. The sulphur-based assimilation of inorganic carbon occurs via complex sulphur oxidation mechanisms. One pathway involves the complete oxidation of reduced sulphur compounds to sulphate, which occurs in the SOX pathway, whereas the APS pathway implicates adenosine-5-phosphosulphate as an intermediate [12]. The key genes include those encoding adenosine-5-phosphosulphate, reductase alpha-subunit (*apsA*) and thiosulphate-oxidizing complex (*sox*). The Sox pathway operates in a wide range of photo- and chemoautotrophic bacteria [13]. A fully functional Sox complex involves SoxB, SoxXA, SoxYZ and SoxCD components, among which SoxB is considered to be the key constituent [12]. The *soxB* gene is regarded as a useful phylogenetic marker and its presence has been demonstrated in various environments, particularly from aquatic habitats such as marine sediments, hydrothermal vents and soda lakes [13–15]. The *apsA* gene can reveal the occurrence of both sulphate-reducing prokaryotes as well as sulphur-oxidising prokaryotes [16–17], which have been investigated in different environments, such as hydrothermal vents, saline alkaline soil and Qinghai-Tibetan Lakes [15,18–19]. Sulphur-oxidising bacterial communities encompass physiologically and phylogenetically diverse members of *Alpha-*, *Beta-*, *Gamma-* and *Epsilonproteobacteria*, *Chlorobia* and *Chloroflexi* [12]. The sulphur-reducing bacteria mostly belong to *Deltaproteobacteria* [20–21].

This study represents the first comparative molecular analysis of metabolic marker genes (*cbbM*, *apsA* and *soxB*) performed by targeted metagenomics of key enzymes of different complex biochemical pathways involved in autotrophy in coastal saline, agricultural and rhizosphere soil niches. This study is complementary to a previously performed determination of community structure at these sites based on 16S rRNA and *cbbL* genes [22–23]. The aim of the present work was to broaden our view on the diversity and abundance of alternative modes of autotrophic metabolism.

Experimental Procedures

Ethics Statement

Sampling locations are not the part of any national parks or protected areas and do not require any specific permits. It is further to confirm that the field studies did not involve endangered or protected species and the specific location of sampling sites was given with their respective description.

Soil samples and physicochemical characteristics

The study was conducted on three bulk soil types (high saline, low saline and agriculture) and one rhizosphere soil type, situated along the Arabian Sea coast, Gujarat, India. There was no crop in the agricultural field at the time of sample collection. However, farmers grew cotton and groundnut regularly in the field. The bulk soil types (0–10 cm of topsoil) were collected from nine transects of each site and sieved through 2 mm mesh to make three composite soil samples per site. The rhizosphere sample was taken from nine randomly selected plantlets, by separating soil tightly adhered to the roots to make three composite rhizosphere samples. The four sampling sites were designated as (i) SS1- saline soil samples collected from the barren land away from the sea coast (N 21°35.711′, E 72°16.875′); (ii) SS2- saline soil samples collected from barren land near the sea coast (N 21°45.402′, E 72°14.156′); (iii) AS- soil samples collected from the agricultural field (N 20°53.884′, E 70°29.730′); (iv) RS- soil samples collected from the rhizosphere (N 20°53.884′, E 70°29.730′). These soil samples were transported to the laboratory immediately and frozen at −20°C for further processing. The composite soil samples were imperilled to physical and chemical analyses for determining the major soil characteristics. The soil pH and salinity were measured on air-dried soil in deionised water by using a 1:2 (w/v) soil:liquid ratio using the Seven Easy pH and Conductivity meter (Mettler-Toledo

AG, Switzerland). The total soil organic carbon was analysed by Liqui TOC (Elementar, Germany) while total carbon, nitrogen and sulphur contents were determined by CHNS analyser (Perkin Elmer series ii, 2400, USA).

Metagenome extraction and PCR amplification

Total soil DNA was extracted in triplicate from 500 mg of each soil samples [22–23] and was used as template to amplify targeted genes (*cbbM*, *apsA*, *aclB* and *soxB*) using gene specific primers following their respective PCR conditions (**Table S1**).

Generation of clone libraries

The functional genes (*cbbM*, *apsA* and *soxB*) were amplified individually from each site (in triplicate). The products were processed by excising expected amplicon size from the gel, purified by using QIAquick gel extraction kit (Qiagen, Hilden, Germany) and cloned into the pGEM-T/pGEM-T Easy vector (Promega, Madison, WI, USA). Clones were selected randomly, screened for the presence of correct inserts sizes (520, 380 and 753 of *cbbM*, *apsA* and *soxB* genes, respectively) and the positive clones were sequenced (at M/s Macrogen Inc., S. Korea).

Alignment and phylogenetic reconstruction

The nucleotide sequences showing anomalous short or longer lengths and poor quality, and chimeric sequences were removed from the data to eliminate the inaccuracy in assessment of community structure. The taxonomic affiliation of these functional genes was assessed by using a BLASTn identity and BLASTx, similarity but the affiliations were given based on BLASTn identity search tool.

The multiple nucleotide sequence alignment was performed using Clustal Omega to estimate the number of representative operational taxonomic units (OTUs), generated using Mothur program [24]. The evolutionary history of all the genes was inferred by the Maximum Likelihood method of by *MEGA* v.5.2 using bootstrap resampling method with 500 bootstrap replications [25]. Model selection analysis was conducted to calculate the best-fit model of nucleotide substitution by *MEGA* v.5.2 based on lowest Bayesian Information Criterion [25]. The codon positions included were 1st+2nd+3rd+Noncoding and the positions containing gaps and missing data were eliminated from the datasets.

Community structure determination based on functional genes

The threshold for OTUs generally varies amongst different genes. Total four thresholds 92, 95, 97 and 99% nucleotide sequence identity were tested (data not shown) to cluster nucleotide sequences into operational taxonomic units (OTUs). Among these, sequence identity cut-off of 95% was used further in the study to define an OTU. It uses the furthest neighbour method to assort similar sequences into groups at arbitrary levels of taxonomic identity. Multiple sequence alignment was performed with Clustal Omega. The Jukes-Cantor evolutionary distance matrices were calculated by the DNADIST program within the PHYLIP [26]. Rarefaction curves, coverage richness estimators (Chao and ACE) and diversity indices (Shannon and Simpson index) were determined using Mothur [24].

Assessment of environmental clustering and statistical analysis

The rooted phylogenetic tree was generated and imported into UniFrac along with the environmental labels [27]. Phylogenetic tree based analysis of community diversity was performed using

Table 1. Biodiversity and predicted richness of the *cbbM*, *apsA* and *soxB* gene sequences.

Genes	No of clones	Sobs[1] (OTU)	Shannon Weiner (H)	Simpson (1-D)	Chao	ACE	Jackknife	Coverage (%)	No of Singletons
cbbM									
SS1	102	44	3.0	0.90	89.1	167.0	94.8	71	28
SS2	113	35	2.6	0.86	50.5	93.5	54	84	18
AS	101	16	2.02	0.82	21.2	42.1	23	93	29
RS	129	16	1.5	0.62	23	50.3	24	93	8
apsA									
SS1	117	46	3.1	0.90	113.7	129.5	159.8	75	29
SS2	140	69	3.8	0.97	151.5	269.8	160.4	67	79
AS	131	53	3.4	0.96	97	215.2	92.4	73	34
soxB									
SS1	107	35	3.0	0.91	43.3	49.5	49	86	14
SS2	117	47	3.4	0.97	101	147.6	102.6	75	28
AS	93	46	3.6	0.98	59.1	67.7	67	77	21

the UniFrac significance test and P test. The P tests were also corrected for multiple comparisons (Bonferonni correction) which indicate phylogenetic distribution between samples by pairwise comparisons and also determine whether environments are significantly different.

The relationships between the major taxonomic groups and environmental factors (e.g. pH, Electrical Conductivity- EC, Total Carbon- TC, Total Nitrogen- TN and Total Sulphur- TS) were analysed by stepwise canonical correspondence analysis (CCA) using PAST [28]. Hierarchical clustering of physicochemical data was built with complete linkage method using euclidean distances between data points. Permutation tests were carried out using two-way ANOVA for all analyses and P value of 0.05 ($P \leq 0.05$) was considered significant.

Quantitative real time-PCR (qRT-PCR)

Gene copy number (of 16S rRNA, *cbbL*, *cbbM*, *apsA*, and *soxB* per g of soil) was enumerated by qRT PCR using gene specific primers and standardised annealing temperatures [22] (**Table S1**). The experiments were repeated three times independently.

GenBank submission and accession numbers

All the validated nucleotide sequence data reported in this study were deposited in the GenBank database with accession numbers as KF788311-KF788755- *cbbM* gene sequences; KF788756-KF789143- *apsA* gene sequences; KF789144-KF789459- *soxB* gene sequences from all clone libraries.

Results

Soils collected from four different sites: three bulk soil types (SS1-low saline, SS2-high saline, AS-agricultural) and a rhizosphere soil (RS), showed variations in water content, pH, salinity, organic carbon, nitrogen and sulphur contents (**Table S2**). Sites were selected to understand the occurrence of different biochemical pathways in these ecologically distinct niches and to compare bulk and rhizosphere soil types.

Functional communities based on the *cbbM* gene

The *cbbM* clone libraries resulted in 102, 113, 101 and 129 clone sequences from SS1, SS2, AS and RS respectively. These *cbbM* clone sequences were grouped into 44, 35, 16 and 16 unique OTUs (phylotypes) within their respective clone libraries (**Table 1**). The low richness of *cbbM* genes was detected as 0.12 and 0.15 OTU per clone for RS and AS clone libraries, however, it was higher for SS1 and SS2 (0.43, 0.30 OTU per clone). The most dominant phylotypes of the SS1 clone library showed affiliation to *Gammaproteobacteria* (46 clones), *Rhodopseudomonas palustris* (18 clones) and *Thiohalorhabdus denitrificans* (nine clones); fourteen clones were related to uncultured bacteria. Other phylotypes were allied to numerous genera, which were represented by fewer clones. The majority of clones from the SS2 library were attributed to RuBisCO genes from *Thiohalomonas denitrificans* (22 clones), *Rhodopseudomonas palustris* (17 clones), *Thiohalomonas nitratireducens* (15 clones) and *Rhodovulum sulfidophilum* (five clones). The second-largest group comprised uncultured bacterium consisting of twenty six clones. *Magnettospirillium magnetotacticum*, *Gammaproteobacteria* and *Halothiobacillus* were the other genera represented in this library. The agricultural soil was dominated by clones associated with *Rhodopseudomonas palustris* (22 clones), *Rhodovulum sulfidophilum* (17 clones) and *Thiobacillus denitrificans* (nine clones). The maximum number of clones was assigned to *Gammaproteobacteria* (53 clones). The soil rhizosphere was exclusively dominated by

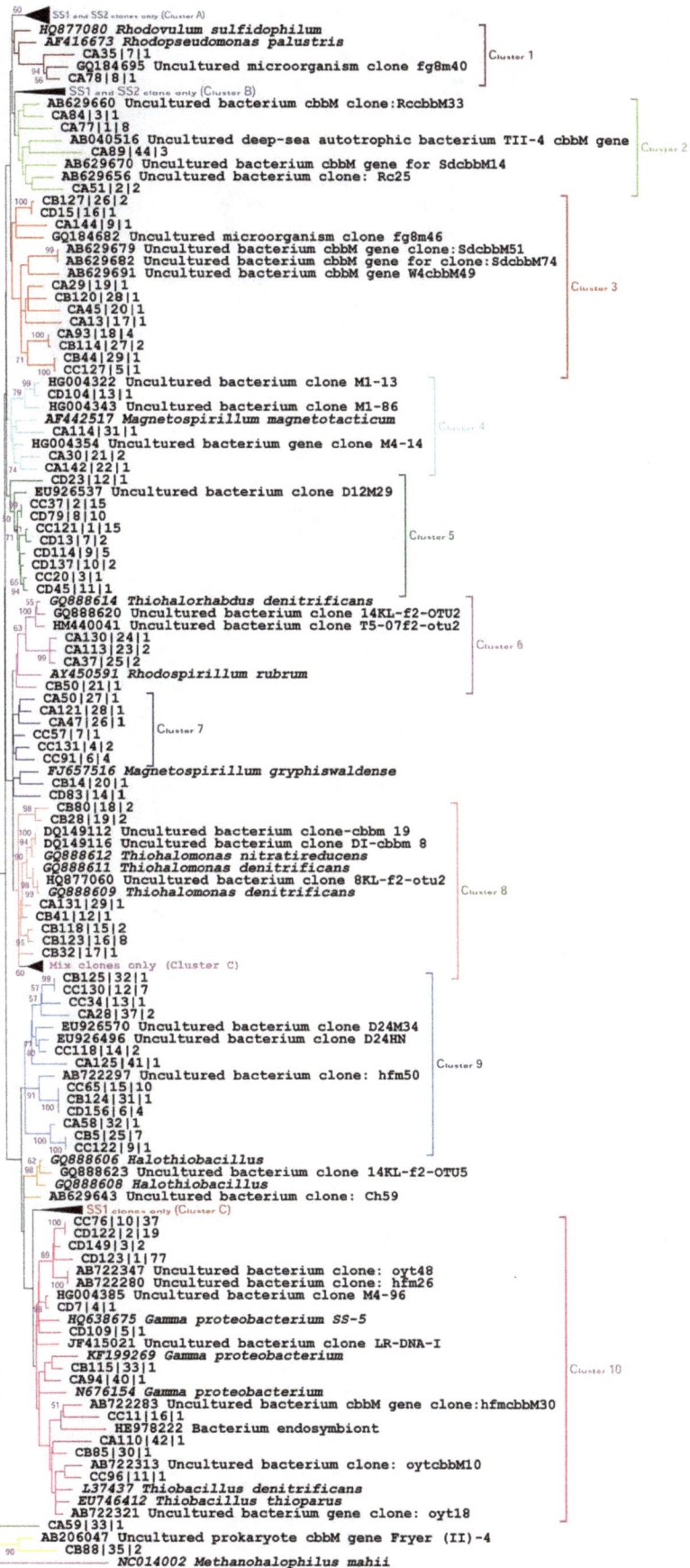

Figure 1. Phylogenetic analysis of the *cbbM* gene clones. The final dataset of 168 nucleotide sequences include sequences from low saline (SS1), high saline (SS2), agricultural (AS) and rhizosphere (RS) soil clone libraries, coded as 'CA', 'CB', 'CC', and 'CD' respectively, and closely related *cbbM* gene sequences from known cultured representatives and environmental clones. The scale bar indicates 0.5 substitutions per site. The *cbbM* gene sequence of *Methanohalophilus mahii* was used as outgroup for tree calculations.

Gammaproteobacteria (103 clones) followed by *Thiohalomonas nitratireducens* (17 clones). Although all the *cbbM* clone libraries consisted primarily of *Gammaproteobacteria* and *Alphaproteobacteria* phylogenetic groups, the relative abundance of the functional microbial groups varied considerably (**Figure S1**). Most nucleotide sequences displayed 75 to 88% sequence identity to their closest relatives according to a BLASTn identity search, but some *cbbM* genes showed greater sequence identity (91–96%) and were related to *Thiohalomonas denitrificans*, *Thiobacillus denitrificans* and uncultured clones. We also tested the occurrence of the *aclB* gene using different PCR conditions, but were unable to amplify it, suggesting it is rare or absent (**Figure S2**).

Functional community based on the *apsA* gene

To investigate the distribution of sulphur-oxidising photo- and chemoautotrophs, we established the *apsA* clone libraries from SS1, SS2 and AS, which comprised 117, 140, 131 valid sequences. These sequences could be grouped into 46, 69 and 53 phylotypes, respectively (**Table 1**). Despite of repeated attempts to amplify the *apsA* gene from the metagenome of rhizospheric soil using previously described and various modified PCR conditions, the gene could not be amplified, which revealed the absence or inadequate representation of sulphur-oxidising bacteria (**Figure S2**). These three habitats were characterised by phylotypes allied to few cultured signature genera, which totally dominated the community, for example: *Chromatium okenii* (SS1, SS2, AS; 3, 3, 39 clones), *Allochromatium minutissimum* (5, 2, 33 clones) exclusively dominated at the agricultural site; similarly, *Deltaproteobacteria- Desulfarculus baarsii*, *Desulfococcus oleovorans*, *Desulfovibrio giganteus*, *Desulfofustis glycolicus*, *Desulfomicrobium baculatum* (6, 3, 0 clones), *Gammaproteobacteria* (1, 17, 1 clones), *Robbea* (3, 18, 1 clones), *Olavius algarvensis* (9, 5, 0 clones), *Thiobacillus plumbophilus*, *T. thioparus*, *T. denitrificans* (6, 5, 6 clones), *Thiochromatium tepidum* (1, 2, 2 clones), *Halochromatium glycolicum* (0, 1, 1 clones) and *Thiococcus pfennigii* (0, 7, 2 clones) dominated in saline soil ecosystems. Uncultured bacteria (76, 53, 43 clones) were the most dominant group and were represented all habitats. The other saline soil clone libraries were affiliated to genera, represented by fewer clones. The levels of nucleotide sequence identity ranged from 71 to 90% for most clone sequences. A few clones showed a nucleotide identity up to 93% and one clone from the SS1 library displayed 100% nucleotide identity with an uncultured *Gammaproteobacterium* clone isolated from the gut microflora, reported from Vanuatu.

Functional community based on the *soxB* gene

To further analyse the potential of sulphur oxidation at these four sites, the *soxB* gene was amplified from the SS1, SS2 and AS metagenomes. The lack of amplification in the RS metagenome further confirmed the absence or inadequate representation of sulphur-oxidising bacteria (**Figure S2**). The *soxB* gene clone libraries resulted in 107, 117 and 93 valid sequences which were grouped into 35, 47 and 46 phylotypes, respectively (**Table 1**). The SS1 clone library was represented by phylotypes allied to *Rhodothalassium salexigens* (15 clones), *Thiomicrospira crunogena* (18 clones), *Paracoccus* (12) and a further 12 clones were related to uncultured bacteria. A large number of clones (30) were related to

Endosymbionts of *Ifremeria* (74% sequence identity). The SS2 clone library consisted of dominant phylotypes that were ascribed to genera such as *Spirochaeta* (25 clones), *Rhodovulum adriaticum* (17 clones) and *Thiomicrospira crunogena* (11 clones). A total of thirty three clones showed no affiliation to any recognised cultured genera. The other bacterial groups were similar to those in SS1, except for *Hydrogenophaga* (2 clones), *Azospirillum* (1 clone), sulphur-oxidising bacteria (1 clone), *Thiohalomonas denitrificans* (1 clone) and *Pandoraea* (3 clones). The agricultural soil clone library was not dominated by any particular group, but had an equal distribution of numerous bacterial genera which differed from those in saline soil clone libraries and include: *Marichromatium purpuratum* (8 clones), *Thiocystis violacea* (5 clones), *Thiocapsa roseopersicina*, *Thiobacillus denitrificans* and *Thiobacillus aquaesulis*. These bacterial genera belonged to *Gammaproteobacteria*, *Betaproteobacteria* and a few clones were related to *Alphaproteobacteria* phylogenetic groups (**Figure S1**). Most phylotypes were related to uncultured clones (28) and showed 70 to 85% sequence identity to their closest relatives according to BLASTn identity searches. Two clones from the AS library exhibited 100% nucleotide sequence identity with uncultured clones reported from coastal sediments from Janssand, Germany.

Comparative molecular phylogeny

The composite phylogenetic trees were generated from representative phylotypes of *cbbM*, *apsA* and *soxB* functional genes, together with closely related reference nucleotide sequences retrieved from NCBI. The Maximum Likelihood phylogeny of *cbbM* gene sequences allocated the phylotypes into site-specific clusters that did not show any affiliation to cultured representatives and can be explained by Cluster A, B and C (**Figure 1**). In addition to these clusters, clades 1, 2 and 6 showed a SS1 site-specific distribution and were related to cultured representatives such as *Rhodovulum sulfidophilum* and *Rhodopseudomonas palustris* (Figure 1a). Cluster 8 mostly showed a SS2 site-specific distribution, but also contained one phylotype at the SS1 site and showed a close affiliation to *Thiohalomonas nitratireducens* and *Thiohalomonas denitrificans*.

The *apsA* nucleotide sequences of all three clone libraries revealed site-specific partitioning and could be distributed into six major clusters (**Figure 2**). The majority of phylotypes (60%) grouped together with uncultured bacteria and showed no affiliation to any cultured representatives, and therefore, could be considered as novel lineages. Cluster 6 consisted of sequences that grouped together with sulphur-reducing bacterial groups such as *Desulfofustis glycolicus*, *Desulfomicrobium baculatum* and *Desulfococcus oleovorans*.

The *soxB* nucleotide sequences grouped into ten major clusters in the phylogenetic tree and showed affiliation to *soxB* genes from different cultured references (**Figure 3**). The majority of sequences (40% of clones) also formed site-specific clusters, such as one group of five clusters that only possessed sequences from saline soils and showed no affiliation to cultured representatives. Similarly, three clusters were specific for the AS site (67% clones), and were also not allied to any representative cultured bacteria. Cluster numbers 1, 4 and 9 contained sequences specific to saline soils SS1 and SS2, and showed affiliation to *soxB* genes of different cultured representatives, including *Thiomicrospira crunogena*,

PA100|37|3
PB147|12|10
PB61|13|2
PB81|38|1
PB140|36|2
PB138|37|1
PB63|35|1
PA48|33|3
JQ823129 Uncultured bacterium clone A364
PB74|20|1
PA69|35|3
PB109|34|1
EU864041 Robbea sp. 3 SB-2008
PB127|19|1
JX908345 Uncultured bacterium clone
PB50|18|1
PA84|34|1
PB105|17|1
AM234055 Olavius ilvae Gamma 3 endosymbiont
AM234051 Olavius algarvensis Gamma 3 endosymbiont
PA113|29|1
PB65|16|1
PA74|31|1
PB40|14|3
PA119|30|3
PA96|26|1
PA66|27|1
PA141|21|2
PA109|22|1
PA53|25|1
PA75|24|1
PA103|28|1
PA105|23|7
PB93|15|4
FJ468532 Uncultured bacterium clone C6507
PA43|32|2
SS1 and SS2 Clones only

Cluster 1

PB62|31|4
PB158|32|1
DQ995799 Uncultured prokaryote clone S8CL3
PC105|7|1
PC143|9|1
PC147|33|1
FMB78990 Uncultured gamma proteobacterium
FMB78989 Uncultured gamma proteobacterium
PC156|8|2
PB18|29|3
PB120|30|1
PC90|11|2
PC111|10|7
PC75|12|2

Cluster 2

O AS Clones only
O AS Clones only
SS1 and SS2 Clones
SS2 and AS Clones
SS1 and SS2 Clone
PA95|10|1
PC128|2|1
PC145|3|1
PC49|1|9
PC7|4|1
JN934449 Uncultured alpha proteobacterium
PB59|61|1
HQ675103 Gamma proteobacterium
HQ675091 Gamma proteobacterium
PA13|8|1
PB29|51|2
PA39|9|1
PA18|14|1
PB38|62|3
PC95|47|1
PA108|16|2
PA143|15|4
PB149|63|1
PB126|58|1
PB77|60|1
PB100|59|2
PB110|49|1
PB75|52|1
PB139|46|20
PB26|47|1
PB108|48|1
PA88|4|6
PB80|50|2

Cluster 3

Cluster 4

PC53|51|1
PC84|52|1
PC124|53|1
SS1 and SS2 Clones
PB70|11|2
AF196341 Uncultured sulphate-reducing
PB57|4|2
PA138|41|5
PB48|5|8
PB12|6|1
PB79|7|1
JN860259 Uncultured gamma proteobacterium
PA60|42|1
PA11|39|1
EU864035 Gamma proteobacterium Robbea sp.
JX908334 Uncultured bacterium clone P_A10
PA87|40|1
PB46|8|1

Cluster 5

PA64|46|1
AF418149 Desulfarculus baarsii
AF418126 Desulfococcus oleovorans
PA62|43|1
AF418141 Desulfovibrio giganteus
PA68|44|1
AF418120 Desulfomicrobium baculatum
AF418130 Desulfofustis glycolicus
PA130|45|1
AF418146 Desulfobulbus elongates
PB118|10|1
JX908348 Uncultured bacterium clone Ds_B1
PB142|9|2

Cluster 6

EF641954 Thiobacillus thioparus
EF641952 Thiobacillus denitrificans
EF641953 Sulfur-oxidizing bacterium AII2
EF641951 Sulfur-oxidizing bacterium DIII5
EF641957 Endosymbiont of Inanidrilus exumae clone 37
EF641933 Halochromatium salexigens
EF641935 Chromatium okenii strain 6210
EF641934 Halochromatium glycolicum
EF641963 Allochromatium minutissimum
EF641949 Thiocystis violacea
EF641936 Thermochromatium tepidum
EF641947 Rhabdochromatium marinum
EF641944 Thiococcus pfennigii
EF641932 Thiohalocapsa halophile
EF641943 Thiococcus pfennigii
EF641942 Thiococcus pfennigii
Thermodesulfobacterium hveragerdense

Figure 2. Phylogenetic analysis of the *apsA* gene clones. The final dataset of 215 nucleotide sequences include sequences from low saline soil (SS1), high saline soil (SS2), agricultural soil (AS) clone libraries, coded as 'PA', 'PB' 'PC' respectively, and closely related *apsA*-gene sequences from known cultured representatives and environmental clones. 500 bootstrap analyses were performed and percentages are shown at nodes. The scale bar indicates 1.0 substitutions per site. The *apsA* gene sequence of *Thermodesulfobacterium hveragerdense* was used as outgroup for tree calculations.

Spirochaeta, Rhodobacter, Paracoccus and sulphur-oxidising bacteria. Clusters 5, 6, 7 and 8 comprised sequences from all three libraries and were allied to different cultured representatives, including *Rhodothalassium salexigens, Marinobacter* sp. and *Marichromatium purpuratum*. Cluster 10 was specific for the AS site and tightly clustered with *Thiovirga sulfuroxydans, Azospirillum* sp. and *Thiobacillus aquaesulis*.

Diversity indices and community structure based on *cbbM*, *apsA* and *soxB*

The α-diversity indices such as ACE, Chao, the number of observed OTUs and the Shannon and Simpson index were evaluated for all clone libraries. The analysis of these diversity indices indicated the predominance of bacteria harbouring the *cbbM* gene at saline soils (SS1 & SS2), whereas the agricultural soil systems (AS & RS) represented less diverse niches (**Table 1**). The diversity index assessment of the *apsA* and *soxB* gene-clone libraries revealed the dominance of sulphur oxidising bacteria in SS2 and AS habitats and less diversity at SS1. This α-diversity is often represented by a rarefaction curve, which is a plot of the number of observed OTUs as a function of the total number of clones captured. This plot of *cbbM* gene sequences (distance = 0.05) reached an asymptote in the AS and RS clone libraries, but did not reach saturation in the SS1 & SS2 clone libraries (**Figure S3**). In the *apsA* and *soxB* gene libraries, the rarefaction curves inclined towards an asymptote for SS1, but non-asymptotic curve were produced by the SS2 and AS gene libraries (**Figure S3**).

We employed phylogenetic tree-based comparisons, the UniFrac metric and phylogenetic *P*-test to *cbbM*, *apsA* and *soxB* clone libraries to investigate the β-diversity, which is a measure of the community structure comparison. Weighted UniFrac environmental clustering analysis indicated that the assemblages of bacterial sequences at all four habitats are highly differentiated (UniFrac $P \leq 0.03$). To determine whether the samples clustered in two dimensional space, PCA was applied to the UniFrac metric. The ordination diagram (**Figure 4a**) of *cbbM* clone libraries revealed that the strongest variation in the data set was between agricultural and saline soils, as these were separated on the first axis of the ordination diagram, which explains the high percentage of total variation (55.51%). For the *apsA* and *soxB* gene clone libraries, the first axis separated agricultural and saline soils, which explains the total community variability (57.78%) among three sample sites (**Figure 4b, c**). The Unifrac analysis revealed the differentiation in community structure and diversity in quite different soil ecosystems which was supported by the *p* significance ($P \leq 0.03$). The uniqueness at these habitats was further supported by a Venn diagram, which indicated a very low overlap of phylotypes (**Figure S4**). The relative abundance of different phylotypes within the respective clone libraries was depicted by Heatmap (**Figure S5**).

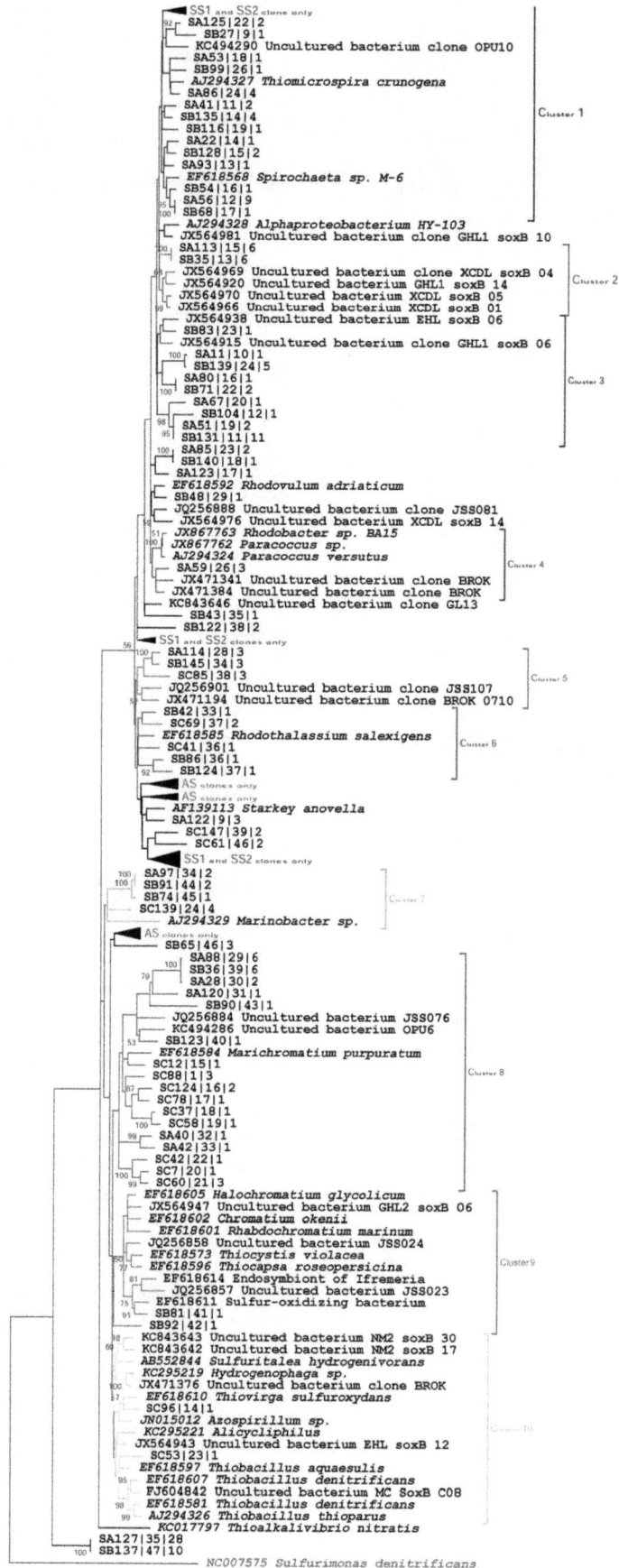

SS1 and SS2 clone only
SA125|22|2
SB27|9|1
KC494290 Uncultured bacterium clone OPU10
SA53|18|1
SB99|26|1
AJ294327 Thiomicrospira crunogena
SA86|24|4
SA41|11|2
SB135|14|4
SB116|19|1
SA22|14|1
SB128|15|2
SA93|13|1
EF618568 Spirochaeta sp. M-6
SB54|16|1
SA56|12|9
SB68|17|1
AJ294328 Alphaproteobacterium HY-103
JX564981 Uncultured bacterium clone GHL1 soxB 10
SA113|15|6
SB35|13|6
JX564969 Uncultured bacterium clone XCDL soxB 04
JX564920 Uncultured bacterium GHL1 soxB 14
JX564970 Uncultured bacterium XCDL soxB 05
JX564966 Uncultured bacterium XCDL soxB 01
JX564938 Uncultured bacterium EHL soxB 06
SB83|23|1
JX564915 Uncultured bacterium clone GHL1 soxB 06
SA11|10|1
SB139|24|5
SA80|16|1
SB71|22|2
SA67|20|1
SB104|12|1
SA51|19|2
SB131|11|11
SA85|23|2
SB140|18|1
SA123|17|1
EF618592 Rhodovulum adriaticum
SB48|29|1
JQ256888 Uncultured bacterium clone JSS081
JX564976 Uncultured bacterium XCDL soxB 14
JX867763 Rhodobacter sp. BA15
JX867762 Paracoccus sp.
AJ294324 Paracoccus versutus
SA59|26|3
JX471341 Uncultured bacterium clone BROK
JX471384 Uncultured bacterium clone BROK
KC843646 Uncultured bacterium clone GL13
SB43|35|1
SB122|38|2
SS1 and SS2 clones only
SA114|28|3
SB145|34|3
SC85|38|3
JQ256901 Uncultured bacterium clone JSS107
JX471194 Uncultured bacterium clone BROK 0710
SB42|33|1
SC69|37|2
EF618585 Rhodothalassium salexigens
SC41|36|1
SB86|36|1
SB124|37|1
AS clones only
AS clones only
AF139113 Starkey anovella
SA122|9|3
SC147|39|2
SC61|46|2
SS1 and SS2 clones only
SA97|34|2
SB91|44|2
SB74|45|1
SC139|24|4
AJ294329 Marinobacter sp.
AS clones only
SB65|46|3
SA88|29|6
SB36|39|6
SA28|30|2
SA120|31|1
SB90|43|1
JQ256884 Uncultured bacterium JSS076
KC494286 Uncultured bacterium OPU6
SB123|40|1
EF618584 Marichromatium purpuratum
SC12|15|1
SC88|1|3
SC124|16|2
SC78|17|1
SC37|18|1
SC58|19|1
SA40|32|1
SA42|33|1
SC42|22|1
SC7|20|1
SC60|21|3
EF618605 Halochromatium glycolicum
JX564947 Uncultured bacterium GHL2 soxB 06
EF618602 Chromatium okenii
EF618601 Rhabdochromatium marinum
JQ256858 Uncultured bacterium JSS024
EF618573 Thiocystis violacea
EF618596 Thiocapsa roseopersicina
EF618614 Endosymbiont of Ifremeria
JQ256857 Uncultured bacterium JSS023
EF618611 Sulfur-oxidizing bacterium
SB81|41|1
SB92|42|1
KC843643 Uncultured bacterium NM2 soxB 30
KC843642 Uncultured bacterium NM2 soxB 17
AB552844 Sulfuritalea hydrogenivorans
KC295219 Hydrogenophaga sp.
JX471376 Uncultured bacterium clone BROK
EF618610 Thiovirga sulfuroxydans
SC96|14|1
JN015012 Azospirillum sp.
KC295221 Alicycliphilus
JX564943 Uncultured bacterium EHL soxB 12
SC53|23|1
EF618597 Thiobacillus aquaesulis
EF618607 Thiobacillus denitrificans
FJ604842 Uncultured bacterium MC SoxB C08
EF618581 Thiobacillus denitrificans
AJ294326 Thiobacillus thioparus
KC017797 Thioalkalivibrio nitratis
SA127|35|28
SB137|47|10
NC007575 Sulfurimonas denitrificans

Cluster 1
Cluster 2
Cluster 3
Cluster 4
Cluster 5
Cluster 6
Cluster 7
Cluster 8
Cluster 9
Cluster 10

0.2

Figure 3. Phylogenetic analysis of the *soxB* gene clones. The final dataset of 186 nucleotide sequences include sequences from low saline (SS1), high saline (SS2), agricultural (AS) soil clone libraries coded as 'SA', 'SB' 'SC' respectively, and closely related *soxB* gene sequences from known cultured representatives and environmental clones. 500 bootstrap analyses were performed and percentages are shown at nodes. The scale bar indicates 0.2 substitutions per site. The *apsA* gene sequence of *Sulfurimonas denitrificans* was used as outgroup for tree calculations.

Abundance of 16S rRNA and functional gene(s) using Real-Time PCR

The 16S rRNA and functional genes involved in carbon (C) cycling and sulphur (S) are important for sustainable ecosystems. The abundance of these genes (copy number per g soil) was determined in all four soil ecosystems using extracted metagenomes and Real-time PCR. The qPCR results showed a heterogeneous distribution of 16S rRNA gene densities among the four sample sites. The number of gene copies was highest in the rhizosphere and agricultural soil samples, followed by in saline soil SS1 and SS2, with mean values ranging from 7.82×10^9 to 2.8×10^9 16S rRNA copies per g soil (**Table S3**). The abundance of 16S rRNA genes was three-fold higher than that of functional genes. The abundance of *cbbL* genes (per g soil) was significantly higher ($P \leq 0.05$) in AS than in saline soils (SS1 and SS2). A trend towards an increase in the abundance of *cbbM* gene copies per g soil was observed from rhizospheric soil to saline soil (RS>AS>SS2>SS1) (**Figure 5**). Regarding sulphur cycles, the gene copies of *apsA* and *soxB* genes was significantly higher for the SS1 soil type, followed by in the AS and SS2 types (SS1>AS>SS2) (**Figure 5**).

Distribution of functional gene OTUs in relation to habitat physicochemical properties

The distribution of the abundant microbial communities among different sites was analysed using Canonical Correspondence Analysis (CCA) plots (**Figure 6**), which revealed that all the functional genes (*cbbM*, *apsA* and *soxB*) of microbial communities were marginally significantly ($P \leq 0.05$) correlated with all the selected environmental variables. The best P value that was obtained was 0.06, with an eigen-value of 0.5 following 999 Monte Carlo permutations for any stepwise iteration of the environmental factor and gene sequence data sets. In the *cbbM* clone library, axes

1 and 2 showed variation of 56% and 28.3% respectively (**Figure 6a**). Similarly, in the *apsA* clone library, SS2 strongly positively correlated with EC and TS with a variation for axis 1 of 75.3% (**Figure 6b**). An axis 2 (24.7% of the variance) was correlated with TC and TN, which in turn, was correlated with the ordination of *Allochromatium minutissimum* and *Chromatium okenii*. In the *soxB* clone library, both CCA axes 1 and 2 (58.4%, 41.56% of the variance respectively), were strongly positively correlated with TS, EC and pH and were strongly but negatively correlated with TC and TN (**Figure 6c**). The ordination of *soxB* OTUs designated to taxonomic groups such as *Allochromatium minutissimum*, *Spirochaeta* sp. and *Rhodobacter*, highly correlated with the SS2 site and with TS, EC and pH environmental variables.

Discussion

This study provides a useful insight into the comparative microbial community structure and the occurrence of potential autotrophic bacterial groups/pathways in coastal saline soils and agricultural/rhizosphere ecosystems. To the best of our knowledge, this study represents the first comprehensive information on functional diversity and the quantification of genes (*cbbM*, *apsA* and *soxB*) associated with autotrophic bacterial biota at these sites using gene-targeted metagenomics.

Gammaproteobacteria as predominant group involved in *cbbM* gene based CO_2 fixation

The RuBisCO form I, which is based on autotrophic metabolism at these four selected sites [22,23], revealed that *cbbL*-harbouring autotrophs were exclusively dominated by *Alpha-* and *Betaproteobacteria*. In this study, *cbbM*-based metabolism was envisioned, which is prevalent in microaerobic/anaerobic autotrophs in diverse

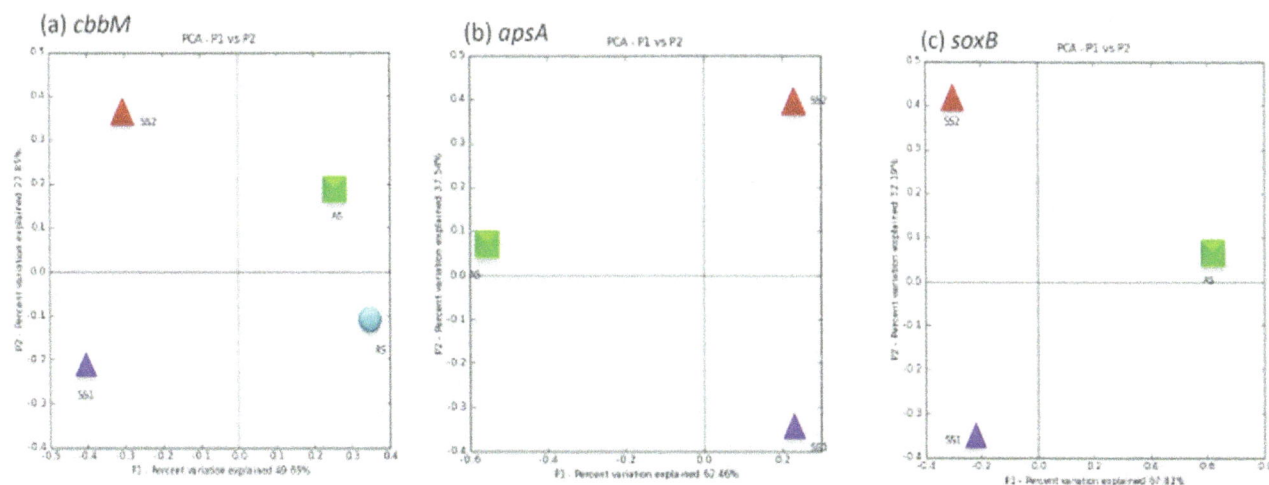

Figure 4. UniFrac PCA of the *cbbM*, *apsA* and *soxB* gene clone libraries. The ordination plots for the first two dimension to show the relationship between saline soils, agricultural and rhizosphere soil types for (a) *cbbM* (b) *apsA* and (c) *soxB* gene assemblages. The saline soils are represented by purple diamond (SS1) and green circle (SS2), agricultural soil (AS) is represented by circle and rhizosphere soil by blue square. Each axis indicates the fraction of the variance in the data that the axis accounts for.

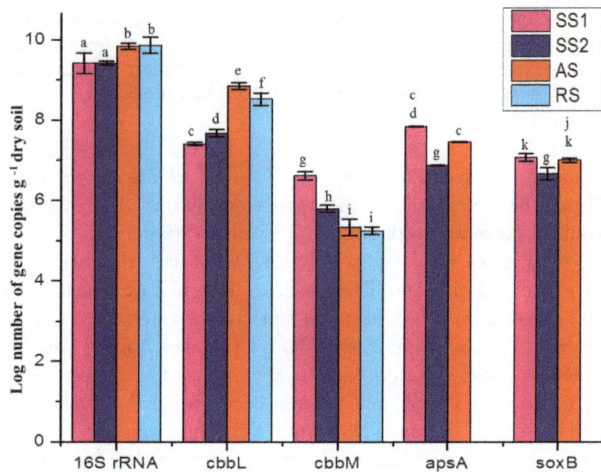

Figure 5. Abundance of the 16S rRNA, *cbbL*, *cbbM*, *apsA* and *soxB* genes in metagenomes of four different soil types determined by qPCR.

environments [29]. Similar studies have been performed in variable environments, particularly in aquatic sites such as groundwater, extreme hypersaline habitats and hydrothermal sites [6–8,11,15,30–31], but have been overlooked in terrestrial habitats. Notably, this gene mostly occurs in organisms that possess the *cbbL* gene [30], which might be highly advantageous to autotrophic bacteria, because the dissimilar kinetic properties of the enzyme allows them to assimilate CO_2 efficiently in both aerobic and anaerobic conditions [30].

In all four *cbbM* libraries, we recovered a number of gene clones that were related to *Gammaproteobacteria*, which are hypothesised to oxidise sulphur in marine sediments, based on their 16S rRNA phylogenetic relationship to other uncultured sulphur-oxidising bacteria [14]. The large number of these most consistently occurring phylotypes were ascribed to different carbon fixing genera that are reported to be associated with biogeochemical transformations in extreme environments [29]. Clones related to *Rhodopseudomonas palustris*, which is a metabolically versatile bacterium capable of anoxygenic photosynthesis under anaerobic conditions using a variety of carbon sources [32], were observed in SS1, SS2 and AS clone libraries. *Thiohalorhabdus denitrificans* is an extremely halophilic, sulphur-oxidising, deep-lineage *Gammaproteobacteria* from hypersaline habitats [33]. The nine clones from the AS library showed affiliation to *Thiobacillus denitrificans*, which shows obligate autotrophic metabolism and obtains energy for CO_2 fixation by combining the oxidation of inorganic sulphur compounds with denitrification [34]. This bacterium has a wide distribution and plays a major role in biogeochemical cycles on a global scale, especially linking sulphur and nitrogen cycles. *Rhodovulum sulfidophilum*-related sequences were exclusively dominant in the AS library, but these microbes are commonly reported from marine and high salt environments and can utilise both reduced sulphur compounds; sulphide and thiosulphate [35]. The presence of sequences belonging to *Gammaproteobacteria* in the 16S rRNA gene library in our previous study [23] as well as in the *cbbM* gene libraries, demonstrates its importance in saline soil ecosystems, which suggests that the *Gammaproteobacteria*-related RuBisCO might contribute to primary production in such environments [14]. The results of 16S rRNA, *cbbL* and *cbbM* clone libraries collectively indicate that sulphur might be the

governing element in supporting chemolithoautotrophic communities in saline ecosystems. The majority of the clone sequences from saline (75%) and agricultural soil (72%) clone libraries did not cluster with cultured representatives in the molecular phylogenetic tree (**Figure 1**) and thus could be considered to be novel genes. The identification of these novel genes will contribute to knowledge about the genetic pool of these genes at these habitats, as well as to the databases.

Gammaproteobacteria and Alphaproteobacteria as the dominant sulphur oxidisers

Because sulphur metabolism occurs via multiple oxidation pathways, this study targeted the key genes (*apsA* & *soxB*) of two important sulphur oxidation pathways [12]. The study revealed the lack of *soxB* gene amplification in the RS metagenome (**Table S1** and **Figure S2**), although this gene has been reported in various cultured rhizobacteria isolated from crop plants [36]. This might be due to the absence or inadequate representation of sulphur-oxidising bacteria in rhizosphere soil and primer bias cannot be completely ruled out. The saline soils were dominated by phylotypes affiliated to *Alpha*- and *Gammaproteobacteria*, such as *Rhodothalassium salexigens*, *Thiomicrosporra crunogena*, *Paracoccus*, *Spirochaeta* and *Rhodovillum adriaticum*; however, *Betaproteobacteria*-related phylotypes dominated the AS site and were putative sulphide oxidisers, which agrees with our previous report on the AS 16S rRNA clone library [24]. The *soxB* gene has been detected in anaerobic anoxygenic phototrophic members of the *Alphaproteobacteria*, such as *Rhodothalassium*, *Rhodospirillum* and *Rhodovulum* [16]. Notably, *soxB* phylotypes related to those of *Rhodothalassium salexigens* (15 clones) and *Rhodovulum adriaticum* (17 clones) were most dominant putative sulphide oxidisers in SS1 and SS2 clone libraries, respectively. The *Thiomicrospira crunogena*- and *Paracoccus*- like phylotypes were the other dominant group in the SS1 clone library, which are obligately chemolithoautotrophic sulphur-oxidising *Gammaproteobacteria* and *Alphaproteobacteria*, respectively, and were restricted to saline soil ecosystems only. The *Paracoccus*-like phylotypes were also retrieved from saline soil *cbbL* clone libraries, as was observed in our previous study [23], reflecting their active role in saline habitats. *Thiomicrospira* sp. has been demonstrated to oxidise thiosulphate to sulphate using the *Paracoccus pantotrophus* homologous *sox* cluster, despite the absence of *aps* genes [37]. Representatives of the *Thiomicrospira* group behave as obligate autotrophs, which use form I RuBisCO (green-like) and form II RuBisCO as the key enzymes for inorganic carbon assimilation [38]. It is notable that the green-like *cbbL* gene could only be amplified from the SS2 site in our previous study [23]. This, together with the results here, suggests that the *Thiomicrospira* group might play an active role at this site. These dominant, recognised genera, can oxidise sulphur, and can thus play a prominent role in biogeochemical sulphur and carbon cycling. The AS clone library was dominated by *Gammaproteobacteria*, *Betaproteobacteria*-related phylotypes, including *Marichromatium purpuratum* and *Thiobacillus*. *Thiobacilli* spp. are aerobic and anaerobic sulphide-oxidising bacteria that possess *sox* genes [34] and were here restricted to agriculture soil.

Previously, the *Chloroflexi* group was observed in saline-soil sites [23], but in the present study, *soxB* gene libraries showed no affiliation to members of the *Chloroflexi* group. This agrees with the report that the primers used for *soxB* gene amplification did not amplify the *soxB* genes of *Epsilonproteobacteria* and *Chloroflexi* [16]. Moreover, *soxB* genes are not reported in members of the *Chloroflexi* group, and *Epsilonproteobacteria* were not represented in our 16S rRNA library. The *soxB* genes identified

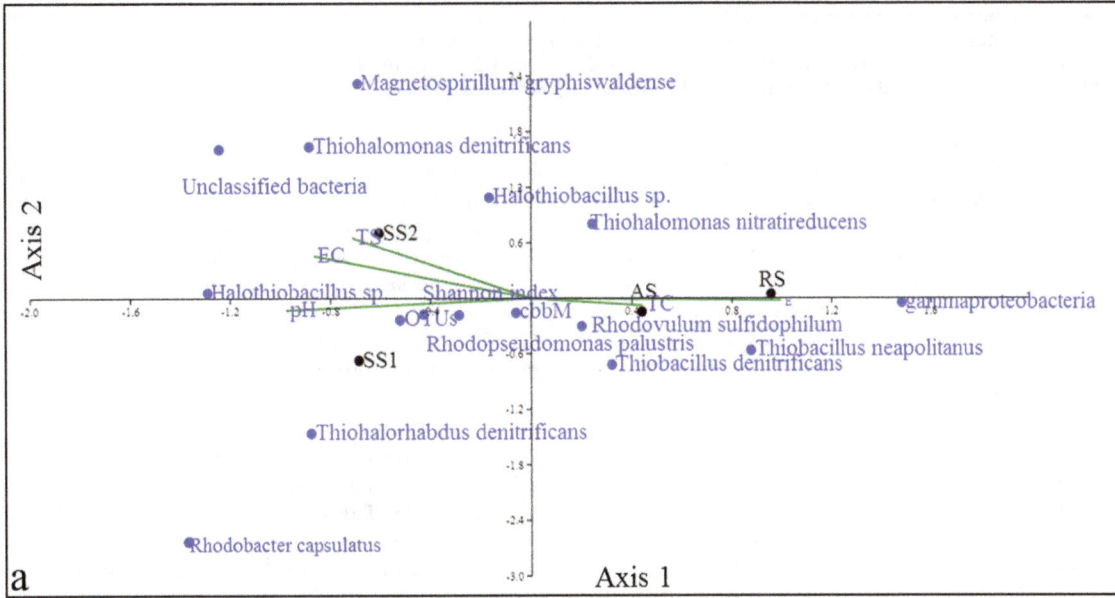

a

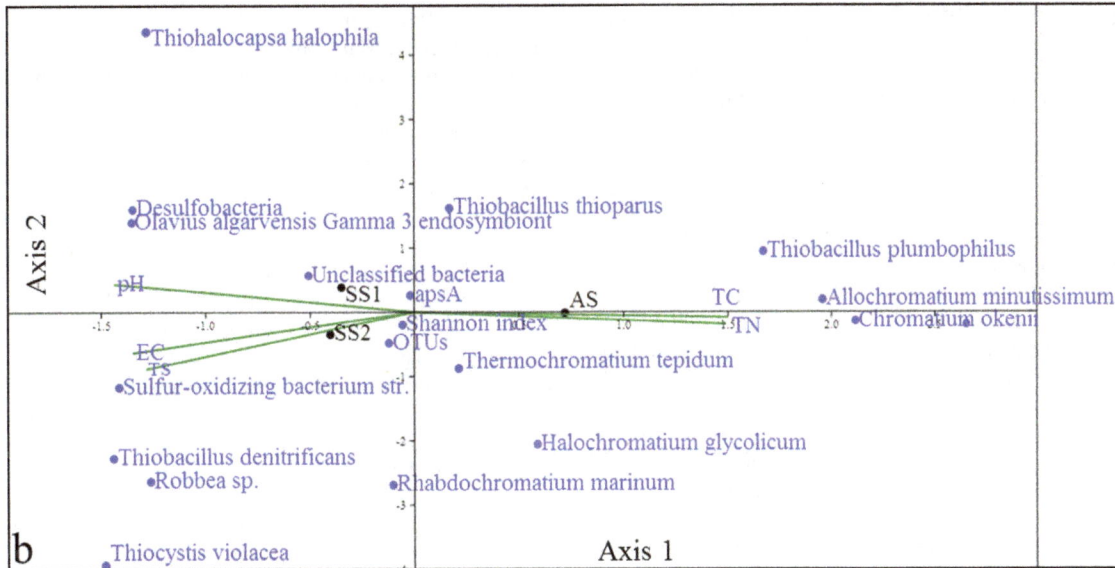

b

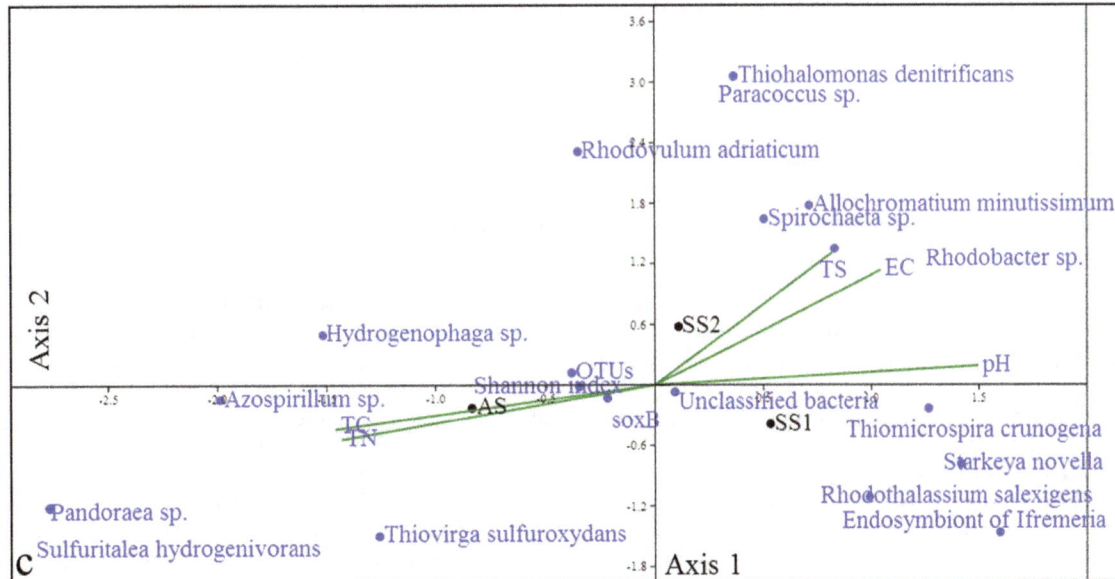

c

Figure 6. Canonical correspondence analysis. Canonical correspondence analysis (CCA) of nucleotide sequence data sets retrieved from saline soils (SS1 & SS2), agricultural soil (AS) and rhizosphere soil (RS) systems along with selected environmental variables of these four sites. The selected environmental variables include electrical conductivity (EC), pH, total C content (TC), total nitrogen content (TN) and total sulphur content (TS) for four soil types. The CCA ordination diagrams for (a) *cbbM* (b) *apsA* and (c) *soxB* clone libraries.

from all the analysed libraries led us to conjecture about the global distribution of these genes. This widespread occurrence could be attributed to horizontal gene transfer, as revealed by *soxB*-based phylogenies [16,39]. In the *soxB* clone library, *Betaproteobacteria* predominantly represented the AS site, which agrees with data from the previous study, where *Betaproteobacteria* was predominant in freshwater/low-salinity environments [40] and also reinforced previous results with the 16S rRNA clone library, in which the *Betaproteobacteria* group was highly abundant at this site.

The composite molecular phylogeny envisages that 40% of saline and 60% of agricultural soil clones were not affiliated to any cultured representatives harbouring *soxB*, which indicated a high unprecedented novel diversity (**Figure 3**). Numerous phylotypes showed a very clear connection to potential sulphur oxidisers such as *Thiomicrospira crunogena*, *Rhodothalassium* and *Paracoccus*, suggesting the potential for sulphur-based metabolism for energy generation at these sites, in addition to Calvin cycle.

The key enzyme aps reductase, performs sulphur metabolism in both reductive and oxidative pathways [41] and the *apsA* gene is considered to be a remarkable tool for investigating sulphur-oxidising prokaryotes. The majority of the recovered sequences in all three *apsA* clone libraries grouped with *Gammaproteobacteria* (SS1, SS2, AS; 29, 70, 81 clones), many of which were affiliated with potential sulphur oxidisers. Most of these AS *Gammaproteobacteria* SOPs were anaerobic anoxygenic phototrophs (*Chromatiaceae*), indicating that aerobic or microaerobic conditions prevailed at this site. Only few saline soil *apsA* gene clone sequences were related to those from these anaerobic anoxygenic phototrophs. The *apsA* saline-soil clone libraries most frequently contained phylotypes (SS1 and SS2-76 and 53 clones), which were related to those of uncultured clones and could be considered novel genes from yet un-described sulphur oxidisers. The results illustrate the under-exploration of the sulphur-oxidising microflora at these sites, which could be a pool for potential novel strains involved in sulphur metabolism. Many of the phylotypes were related to the symbiotic SOP-like *Olavius ilvae*, *O. algarvensis* and *Robbea* Gamma 3 endosymbionts, which are gutless oligochaete worms from Mediterranean seagrass sediments [42]. These co-occurring symbionts are sulphur-oxidising and sulphate-reducing bacteria and can assimilate CO_2 autotrophically, which enables them to provide multiple sources of nutrition to the particular host [43–44]. The absence of *apsA* gene amplification in the RS metagenome corresponded to *soxB* gene, further confirmed the lack of sulphur-oxidising representatives in this habitat. The numerically largest group of *apsA* phylotypes from the AS library were related to *Chromatium okenii* (39 clones) and *Allochromatium minutissimum* (33 clones). The ecological versatility of *Allochromatium* strains causes them to inhabit numerous habitats, such as stagnant freshwater ditches, ponds, lakes, sewage lagoons, estuaries and salt marshes. Five phylotypes from the SS2 library were related to *Allochromatium vinosum*, which can switch from anaerobic to aerobic sulphur oxidation. This genus was formerly referred to as *Chromatium vinosum*, a representative of the *Chromatiaceae* and possesses two sets of divergent genes that encode the green-like RuBisCO enzyme [45]. The saline-soil libraries (SS1 & SS2) possessed limited phylotypes that represented *Deltaproteobacteria* members such as *Desulfarculus baarsii* and a

few members of *Alpha-* and *Betaproteobacteria*. The *Beta-* and *Gammaproteobacteria* were the most frequent sulphur oxidisers, whereas sulphur-reducing bacteria most frequently belonged to *Deltaproteobacteria*, as reported in previous studies [46]. The composite phylogeny of *apsA* genes revealed that the majority of clone sequences (60%) did not show affiliation to well-recognized sulphur-oxidising bacteria, prompting us to presume their novelty and uniqueness (**Figure 2**).

Gene abundance

Bacterial 16S rRNA gene copy numbers ranged from 2.6×10^9 to 7.8×10^9 per g of dry soil; SS2 had the fewest copies, followed by SS1, AS and RS (**Table S3**), which is comparable to the numbers measured in other soil ecosystems using real-time PCR [18,47–48]. Previously, about 1×10^{11} 16S rRNA copies were reported from paddy soils [11]. The 16S rRNA gene copy number was two to four orders of magnitude higher than that of all studied functional genes. The precise estimation of copy number was difficult, as the number of 16S rRNA gene copies per bacterial cell ranged from 1–15 [49]. The results showed that the *cbbL* gene copy numbers significantly ($P \leq 0.05$) outnumbered *cbbM*. This result agrees with results from the present clone library analysis, as well as with research performed in paddy soils [11]. The copy number of the *cbbL* gene was comparable to that in the study on paddy soil [50–51] and somewhat higher than that reported in other bulk soil niches [52–53]. Moreover, because the number of copies of the *cbb* operon in bacteria varies [38], the number of copies calculated could not be precisely defined.

The abundance of *soxB* has not been established in terrestrial ecosystems, but the gene has been reported from extreme water systems such as Qinghai-Tibetan Lakes; hypersaline lakes [19] and the copy number in sediment was comparable with that in the present data (**Table S3**). To the best of our knowledge, there has only been one report on *apsA* gene abundance in soil ecosystems to date, [18] and the copy number of the *apsA* gene observed in saline–alkaline soil contrasted with that detected in this study. The differential abundance of these metabolic marker genes can be attributed to phylogenetically diverse copies of functional genes such as *apsA*, which can be present in a single species [17]. The abundance of *apsA* and *soxB* genes was comparable at each site, which corresponded with the results observed for *apsA* and *dsrB* [54], where the copy number for both genes was almost equal at all sites.

Conclusions

In this study, we have reported the comparative occurrence of functional genes involved in different modes of autotrophy among four contrasting terrestrial habitats and envisaged their composite phylogeny for the first time. The collective data obtained revealed that *Gammaproteobacteria* and *Alphaproteobacteria* were the abundant groups and represented major fractions of the sequences of the *cbbM*, *apsA* and *soxB* clone libraries from saline-soil ecosystems. Thus, we hypothesise from the results, that novel *Gammaproteobacteria* sulphur-oxidisers might play a prominent role in primary production and might be heavily involved in the cycling of carbon and sulphur compounds, principally at saline-soil sites. This study requires other collaborative approaches to confirm the hypothesis concerning metabolic functioning in these ecosystems, such as novel bacterial isolation and stable isotope probing. Nevertheless, this study detected a pattern in the distribution of functional microbial guilds across these sites, which is a prerequisite for the ecological understanding and ascertaining of autotrophic physiology at these sites.

Supporting Information

Figure S1 Distribution of functional microbial groups across four different soil habitats.

Figure S2 PCR amplification of targeted functional genes. PCR amplification of *cbbM*, *apsA*, *aclB* and *soxB*, using different soil DNA as template following their respective gene specific primers and PCR conditions (Table S1). M: Marker (100 bp DNA ladder); SS1, SS2, AS, and RS: Soil Samples; PC: Positive control.

Figure S3 Rarefaction curves for targeted functional genes clone library. Rarefaction curves for (a) *cbbM* (b) *apsA* and (c) *soxB* gene clone libraries at 0.05 cut-off. Bacterial richness in SS1, SS2, AS and RS soils is indicated by slopes of the rarefaction curves.

Figure S4 Venn diagrams for targeted functional genes. Venn diagrams representing the observed overlap of OTUs for (a) *cbbM*, (b) *apsA* and (c) *soxB* gene libraries (distance = 0.05). Venn diagrams show overall overlap of representative genera between

soils. The values in the diagram represent the number of genera that were taxonomically classified.

Figure S5 Heat map analysis. Heat map showing abundance of OTUs in (a) *cbbM*, (b) *apsA* and (c) *soxB* gene clone libraries (distance = 0.05). Each row in the heatmap represents a different OTU and the colour of the OTU in each group scaled between black and red according to the relative abundance of that OTU within the group.

Table S1 Primer sets used for the study of microbial community structure and gene abundance.

Table S2 Physicochemical characteristics of soil samples.

Table S3 Copy no. of gene(s) determined by qPCR in soil metagenomes.

Acknowledgments

CSIR-CSMCRI Communication No. PRIS-029/2014.

Author Contributions

Conceived and designed the experiments: BY AM BJ. Performed the experiments: BY RK. Analyzed the data: BY AM. Contributed to the writing of the manuscript: BY AM.

References

1. Falkowski P, Scholes RJ, Boyle E, Canadell J, Canfield D, et al. (2000) The global carbon cycle: a test of our knowledge of earth as a system. Science 290: 291–296.
2. Falkowski PG, Fenchel T, DeLong EF (2008) The microbial engines that drive Earth's biogeochemical cycles. Science 320: 1034–1039.
3. Thauer RK (2007) A fifth pathway of carbon fixation. Science 318: 1732–1733.
4. Berg IA, Kockelkorn D, Buckel W, Fuchs G (2007) A 3-hydroxypropionate/4-hydroxybutyrate autotrophic carbon dioxide assimilation pathway in archaea. Science 318: 1782–1786.
5. Tourova TP, Kovaleva OL, Sorokin DY, Muyzer G (2010) Ribulose-1,5-bisphosphate carboxylase/oxygenase genes as a functional marker for chemo-lithoautotrophic halophilic sulfur-oxidizing bacteria in hypersaline habitats. Microbiol 156: 2016–2025.
6. Morse D, Salois P, Markovic P, Hastings JW (1995) A nuclear encoded form II RuBisCO in dinoflagellates. Science 268: 1622–1624.
7. Kato S, Nakawake M, Ohkuma M, Yamagishi A (2012) Distribution and phylogenetic diversity of *cbbM* genes encoding RuBisCO form II in a deep-sea hydrothermal field revealed by newly designed PCR primers. Extremophiles 16: 277–283.
8. Kovaleva OL, Tourova TP, Muyzer G, Kolganova TV, Sorokin DY (2011) Diversity of RuBisCO and ATP citrate lyase genes in soda lake sediments. FEMS Microbiol Ecol 75: 37–47.
9. Hall JR, Mitchell KR, Jackson-Weaver O, Kooser AS, Cron BR, et al. (2008) Molecular characterization of the diversity and distribution of a thermal spring microbial community by using rRNA and metabolic genes. Appl Environ Microbiol 74: 4910–4922.
10. Chen Y, Wu L, Boden R, Hillebrand A, Kumaresan D, et al. (2009) Life without light: microbial diversity and evidence of sulfur- and ammonium-based chemolithotrophy in Movile Cave. ISME J 3: 1093–1104.
11. Xiao KQ, Bao P, Bao QL, Jia Y, Huang FY, et al. (2013) Quantitative analyses of ribulose-1,5-bisphosphate carboxylase/oxygenase (RuBisCO) large-subunit genes (*cbbL*) in typical paddy soils. FEMS Microbiol Ecol 87: 89–101.
12. Ghosh W, Dam B (2009) Biochemistry and molecular biology of lithotrophic sulfur oxidation by taxonomically and ecologically diverse bacteria and archaea. FEMS Microbiol Rev 33: 999–1043.
13. Tourova TP, Slobodova NV, Bumazhkin BK, Kolganova TV, Muyzer G, et al. (2012) Analysis of community composition of sulfur- oxidizing bacteria in hypersaline and soda lakes using *soxB* as a functional molecular marker. FEMS Microbiol Ecol 84: 280–289.
14. Lenk S, Arnds J, Zerjatke K, Musat N, Amann R, et al. (2011) Novel groups of *Gammaproteobacteria* catalyse sulfur oxidation and carbon fixation in coastal, intertidal sediment. Environ Microbiol 13: 758–774.
15. Hügler M, Gärtner A, Imhoff JF (2010) Functional genes as markers for sulfur cycling and CO2 fixation in microbial communities of hydrothermal vents of the Logatchev field. FEMS Microbiol Ecol 73: 526–537.
16. Meyer B, Imhoff JF, Kuever J (2007) Molecular analysis of the distribution and phylogeny of the *soxB* gene among sulfur-oxidizing bacteria- evolution of the Sox sulfur oxidation enzyme system. Environ Microbiol 9: 2957–2977.
17. Meyer B, Kuever J (2007) Molecular analysis of the distribution and phylogeny of dissimilatory adenosine-5-phosphosulfate reductase-encoding genes (*aprBA*) among sulfur- oxidizing prokaryotes. Microbiol 153: 3478–3498.
18. Keshri J, Mishra A, Jha B (2013) Microbial population index and community structure in saline-alkaline soil using gene targeted metagenomics. Microbiol Res 168: 165–173.
19. Yang J, Jiang H, Dong H, Wu G, Hou W, et al. (2013) Abundance and diversity of sulfur-oxidizing bacteria along a salinity gradient in four Qinghai-Tibetan lakes, China. Geomicrobiol J 30: 851–860.
20. Muyzer G, Stams AJ (2008) The ecology and biotechnology of sulphate-reducing bacteria. Nat Rev Microbiol 6: 441–454.
21. Headd B, Engel AS (2012) Evidence for niche partitioning revealed by the distribution of sulfur oxidation genes collected from areas of a terrestrial sulfidic spring with differing geochemical conditions. Appl Environ Microbiol 79: 1171–1182.
22. Yousuf B, Keshri J, Mishra A, Jha B (2012) Application of targeted metagenomics to explore abundance and diversity of CO2-fixing bacterial community using *cbbL* gene from the rhizosphere of *Arachis hypogaea*. Gene 1: 18–24.
23. Yousuf B, Sanadhya P, Keshri J, Jha B (2012) Comparative molecular analysis of chemolithoautotrophic bacterial diversity and community structure from coastal saline soils, Gujarat, India. BMC Microbiol 12: 150.
24. Schloss PD, Handelsman J (2008) A statistical toolbox for metagenomics: assessing functional diversity in microbial communities. BMC Bioinfo 9: 34–48.
25. Tamura K, Peterson D, Peterson N, Stecher G, Nei M, et al. (2011) MEGA5: Molecular evolutionary genetics analysis using maximum likelihood, evolutionary distance, and maximum parsimony methods. Mol Biol Evol 28: 2731–2739.
26. Felsenstein J (1989) Phylip: Phylogeny inference package, version 3.2. Cladistics 5: 164–166.
27. Lozupone C, Hamady M, Knight R (2006) UniFrac- An online tool for comparing microbial community diversity in a phylogenetic context. BMC Bioinfo 7: 371–384.

28. Hammer O, Harper DAT, Ryan PD (2001) PAST: Paleontological statistics software package for education and data analysis. Palaeontol Electron 4: 1–9.

29. Badger MR, Bek EJ (2008) Multiple RuBisCO forms in proteobacteria: their functional significance in relation to CO_2 acquisition by the CBB cycle. J Exp Bot 59: 1525–1541.

30. Alfreider A, Schirmer M, Vogt C (2011) Diversity and expression of different forms of RuBisCO genes in polluted groundwater under different redox conditions. FEMS Microbiol Ecol 79: 649–660.

31. Kong W, Dolhi JM, Chiuchiolo A, Priscu J, Morgan-Kiss RM (2012) Evidence of form II RubisCO (*cbbM*) in a perennially ice-covered Antarctic lake. FEMS Microbiol Ecol 82: 491–500.

32. Larimer FW, Chain P, Hauser L, Lamerdin J, Malfatti S, et al. (2004) Complete genome sequence of the metabolically versatile photosynthetic bacterium *Rhodopseudomonas palustris*. Nat Biotechnol 22: 55–61.

33. Sorokin DY (2008) Diversity of halophilic sulfur-oxidizing bacteria in hypersaline habitats. In Microbial sulfur metabolism (Dahl C & Friedrich CG eds). Proceedings of the international symposium on microbial sulfur metabolism, pp. 225–237. Springer-Berlin, Munster, Germany.

34. Beller HR, Letain TE, Chakicherla A, Kane SR, Legler TC, et al. (2006) Whole-genome transcriptional analysis of chemolithoautotrophic thiosulfate oxidation by *Thiobacillus denitrificans* under aerobic versus denitrifying conditions. J Bacteriol 188: 7005–7015.

35. McDevitt CA, Hanson GR, Noble CJ, Cheesman MR, McEwan AG (2002) Characterization of the redox centers in dimethyl sulfide dehydrogenase from *Rhodovulum sulfidophilum*. Biochemistry 41: 15234–15244.

36. Anandham R, Indiragandhi P, Madhaiyan M, Ryu KY, Jee HJ, et al. (2008) Chemolithoautotrophic oxidation of thiosulfate and phylogenetic distribution of sulfur oxidation gene (*soxB*) in rhizobacteria isolated from crop plants. Res Microbiol 159: 579–589.

37. Brinkhoff T, Muyzer G, Wirsen CO, Kuever J (1999) *Thiomicrospira kuenenii* sp. nov. & *Thiomicrospira frisia* sp. nov., two mesophilic obligately chemo-lithoautotrophic sulfur-oxidizing bacteria isolated from an intertidal mud flat. Int J Syst Bacteriol 49: 385–392.

38. Tourova TP, Spiridonova EM, Berg IA, Kuznetsov BB, Sorokin DY (2006) Occurrence, phylogeny and evolution of ribulose-1,5-bisphosphate carboxylase/oxygenase genes in obligately chemolithoautotrophic sulfur-oxidizing bacteria of the genera *Thiomicrospira* and *Thioalkalimicrobium*. Microbiology 152: 2159–2169.

39. Petri R, Podgorsek L, Imhoff JF (2001) Phylogeny and distribution of the *soxB* gene among thiosulfate-oxidizing bacteria. FEMS Microbiol Lett 197: 171–178.

40. Wu QL, Zwart G, Schauer M, Agterveld MPK, Bahn MW (2006) Bacterioplankton community composition along a salinity gradient of sixteen high-mountain lakes located on the Tibetan Plateau, China. Appl Environ Microbiol 72: 5478–5485.

41. Geets J, Borremans B, Diels L, Springael D, Vangronsveld J, et al. (2006) *DsrB* gene-based DGGE for community and diversity surveys of sulfate-reducing bacteria. J Microbiol Methods 66: 194–205.

42. Ruehland C, Blazejak A, Lott C, Loy A, Erséus C, et al. (2008) Multiple bacterial symbionts in two species of co-occurring gutless oligochaete worms from Mediterranean sea grass sediments. Environ Microbiol 10: 3404–3416.

43. Woyke T, Teeling H, Ivanova NN, Huntemann M, Richter M, et al. (2006) Symbiosis insights through metagenomic analysis of a microbial consortium. Nature 443: 950–955.

44. Bayer C, Heindl NR, Rinke C, Lücker S, Ott JA, et al. (2009) Molecular characterization of the symbionts associated with marine nematodes of the genus Robbea. Environ Microbiol Rep 1: 136–144.

45. Viale AM, Kobayashi H, Akazawa T (1989) Expressed genes for plant-type ribulose 1,5-bisphosphate carboxylase/oxygenase in the photosynthetic bacterium *Chromatium vinosum*, which possesses two complete sets of the genes. J Bacteriol 171: 2391–2400.

46. Andreote FD, Jiménez DJ, Chaves D, Dias ACF, Luvizotto DM, et al. (2012) The microbiome of Brazilian mangrove sediments as revealed by metagenomics. PLoS One 7: e38600.

47. Henry S, Bru D, Stres B, Hallet S, Philippot L (2006) Quantitative detection of the *nosZ* gene, encoding nitrous oxide reductase and comparison of the abundances of 16S rRNA, *narG*, *nirK* and *nosZ* genes in soils. Appl Environ Microbiol 72: 5181–5189.

48. Smith CJ, Nedwell DB, Dong LF, Osborn AM (2006) Evaluation of quantitative polymerase chain reaction-based approaches for determining gene copy and gene transcript numbers in environmental samples. Environ Microbiol 8: 804–815.

49. Acinas SG, Marcelino LA, Klepac CV, Polz MF (2004) Divergence and redundancy of 16S rRNA sequences in genomes with multiple rrn operons. J Bacteriol 186: 2629–2635.

50. Yuan H, Ge T, Chen C, O'Donnell AG, Wu J (2012) Significant role for microbial autotrophy in the sequestration of soil carbon. Appl Environ Microbiol 78: 2328–2336.

51. Yuan H, Ge T, Wu X, Liu S, Tong C, et al. (2012) Long-term field fertilization alters the diversity of autotrophic bacteria based on the ribulose-1,5-biphosphate carboxylase/oxygenase (RuBisCO) large-subunit genes in paddy soil. Appl Microbiol Biol 95: 1061–1071.

52. Videmšek U, Hagn A, Suhadolc M, Radl V, Knicker H, et al. (2009) Abundance and diversity of CO_2-fixing bacteria in grassland soils close to natural carbon dioxide springs. Microbiol Ecol 58: 1–9.

53. Selesi D, Pattis I, Schmid M, Kandeler E, Hartmann A (2007) Quantification of bacterial RuBisCO genes in soils by *cbbL* targeted real-time PCR. J Microbiol Methods 69: 497–503.

54. Blazejak A, Schippers A (2011) Real-time PCR quantification and diversity analysis of the functional genes *aprA* and *dsrA* of sulfate-reducing prokaryotes in marine sediments of the Peru continental margin and the Black Sea. Front Microbiol 2: 253.

10

Microbial Diversity of a Mediterranean Soil and Its Changes after Biotransformed Dry Olive Residue Amendment

José A. Siles[1]*, Caio T. C. C. Rachid[2,3], Inmaculada Sampedro[1,4], Inmaculada García-Romera[1], James M. Tiedje[3]

1 Department of Soil Microbiology and Symbiotic Systems, Estación Experimental del Zaidín, Consejo Superior de Investigaciones Científicas (CSIC), Granada, Spain, **2** Institute of Microbiology Paulo de Góes, Federal University of Rio de Janeiro, Rio de Janeiro, Brazil, **3** Center for Microbial Ecology, Michigan State University, East Lansing, Michigan, United States of America, **4** Thayer School of Engineering, Dartmouth College, Hanover, New Hampshire, United States of America

Abstract

The Mediterranean basin has been identified as a biodiversity hotspot, about whose soil microbial diversity little is known. Intensive land use and aggressive management practices are degrading the soil, with a consequent loss of fertility. The use of organic amendments such as dry olive residue (DOR), a waste produced by a two-phase olive-oil extraction system, has been proposed as an effective way to improve soil properties. However, before its application to soil, DOR needs a pre-treatment, such as by a ligninolytic fungal transformation, e.g. *Coriolopsis floccosa*. The present study aimed to describe the bacterial and fungal diversity in a Mediterranean soil and to assess the impact of raw DOR (DOR) and *C. floccosa*-transformed DOR (CORDOR) on function and phylogeny of soil microbial communities after 0, 30 and 60 days. Pyrosequencing of the 16S rRNA gene demonstrated that bacterial diversity was dominated by the phyla Proteobacteria, Acidobacteria, and Actinobacteria, while 28S-rRNA gene data revealed that Ascomycota and Basidiomycota accounted for the majority of phyla in the fungal community. A Biolog EcoPlate experiment showed that DOR and CORDOR amendments decreased functional diversity and altered microbial functional structures. These changes in soil functionality occurred in parallel with those in phylogenetic bacterial and fungal community structures. Some bacterial and fungal groups increased while others decreased depending on the relative abundance of beneficial and toxic substances incorporated with each amendment. In general, DOR was observed to be more disruptive than CORDOR.

Editor: Boris Alexander Vinatzer, Virginia Tech, United States of America

Funding: This study has been funded by the Spanish Ministry of Science and Innovation (Project AGL2008–572). The sequencing work was done while J.A. Siles was a visiting scientist at the Center for Microbial Ecology, Michigan State University. J.A. Siles and I. Sampedro were supported by the JAE program, which is co-financed by the Consejo Superior de Investigaciones Científicas (CSIC) and the European Social Fund. The funders had no role in study design, data collection and analysis, decision to publish, or preparation of the manuscript.

Competing Interests: The authors have declared that no competing interests exist.

* Email: josesimartos@gmail.com

Introduction

The Mediterranean basin is one of the 25 most important biodiversity hotspots on Earth due to its particular climatic and geological characteristics [1]. This region has thus been identified as one of the priority regions for conservation in Europe, as human activity is causing a dramatic crisis in biodiversity [2]. However, as knowledge of soil microbial diversity in this part of the world is limited, it is essential to broaden our understanding of this diversity in order to achieve a balance between conservation and human development [3].

Olives are among the most important and widespread crops in the Mediterranean region, where they occupy a highly stable area of cultivation [4]. The olive oil industry generally produces vast amounts of wastes [5], and in Spain, the residues produced by the two-phase centrifuging olive oil extraction system has been highlighted. This technology produces a liquid phase (olive oil) and an organic waste sludge. Through a heating process and the

use of organic solvents, this primary waste is then revalorized into low quality olive oil and a final waste product known as "alpeorujo" or dry olive residue (DOR) [6]. In Spain alone, 5 million tons of DOR are produced annually in a short period of time [7]. This waste, until now, has been used for energy and co-generation purposes [8]. However, the international regulations on limiting CO_2 emissions and the presence of polyaromatic hydrocarbons in DOR combustion gases are restricting these practices [6,8]. An alternative for DOR revalorization is its exploitation as an organic amendment as it contains high concentrations of organic matter and minerals of agricultural importance [9]. Its use as an organic amendment could be especially beneficial in the Mediterranean region, where many soils are experiencing degradation and fertility loss due to agro-chemical treatments, excessive and deep tillage, continuous cropping, overgrazing and luxury irrigation [10]. Organic amendments, which improve the physical, chemical and biological properties of soil, have thus been proposed as an effective way of maintaining and improving soil fertility [11]. However, DOR contains polyphenols and other organic components which are capable of inhibiting microbial growth, plant germination and morphogenesis [12], and therefore needs to be treated before being applied to soil in high doses. The transformation of DOR by ligninolytic fungi has been demonstrated to be a rapid and effective technique to stabilize organic matter, enhance C/N ratio, reduce phenolic concentration and to eliminate the phytotoxic effects of the waste in order to facilitate its use as an organic amendment [13–15].

Soil bacteria and fungi play a pivotal role in biogeochemical cycles and are responsible for nutrient cycling by mineralizing and decomposing of organic matter [3,16,17]. These communities may also influence nutrient availability for crops through solubilization, chelation and oxidation/reduction processes [18]. Furthermore, soil microorganisms establish symbiotic and antagonist relationships with plants that affect their status and perform other functions such as soil structure maintenance [19] and the degradation of pollutants [20]. Thus, microbial communities govern soil quality and constitute an important component of this ecosystem. In this sense, the implementation of sustainable soil strategies such as the use of biotransformed DOR as an organic amendment requires knowledge of microbial community behaviour under these conditions. To date, only Sampedro et al. [21] have made a preliminary study under "in vitro" conditions of the impact of *Phlebia* sp.-transformed DOR on soil microbiology using denaturing gradient gel electrophoresis (DGGE) and phospholipid fatty acid (PLFA) analysis. Other works have also assessed the effect of low doses of raw DOR on the physico-chemical properties of soil [9,22]. Thus, to the best of our knowledge, no studies have been made about the effect of raw and fungi-transformed DOR on soil microbiology, using more accurate and informative tools such as high-throughput sequencing techniques. In this survey, pyrosequencing was used to analyze the diversity of bacterial and fungal communities in a Mediterranean soil and their responses to raw and fungi-transformed DOR amendments. This work complements two other reports using the same experimental treatments but evaluated by culture-dependent approaches [23,24]. In the present study, we aimed to: i) describe the bacterial and fungal diversity in an agricultural Mediterranean soil by means of 16S and 28S rRNA gene pyrosequencing, respectively; ii) obtain an insight into the functional changes produced by untransformed and *Coriolopsis floccosa*-transformed DOR on microbial communities over a short time period (0, 30 and 60 days) using a Biolog EcoPlate system; iii) and to investigate the effects of amendments containing these two types of DOR on soil fungal and bacterial communities.

Materials and Methods

Soil sampling

The soil studied, which was obtained from the "Cortijo Peinado" field (Granada, Spain, 37°13'N, 3° 45'W), was classified as loam according to the USDA system [25] and presented a low organic matter content (10 g kg^{-1} total organic carbon), which is typical of Mediterranean agricultural soils [4]. The main soil properties have been summarized in Table S1.

The climate in the region is typically Mediterranean with annual rainfall average of 357 mm, the wettest month is December with 53 mm and the driest one is August with 3 mm. The mean annual temperature is 15.1°C; the coldest month is January (mean 6.7°C) and warmest one is July (mean 24.8°C) (http://www.aemet.es).

The plot from which soil samples were collected has been used for agricultural purposes and, in recent years, fruit trees have been cultivated on this land. The area is not part of a conservation zone, does not contain any protected species and does not belong to a private land. Permission to sample the soil was obtained directly from the farm owners and technical experts. At sample collection time, the soil in the field had recently been ploughed and plants were not present. To collect the soil samples, the plot (10,000 m^2) was divided into 10 sub-plots of equal size. Five 1 kg subsamples were collected randomly from the Ap horizon (at a depth of 0–20 cm) of each sub-plot and the subsamples were combined into a single pooled sample. Subsequently, the different composited samples were sieved (5 mm sterilized mesh) and mixed. The soil was stored for three days at room temperature until the experiment was performed.

DOR

DOR was supplied by an olive oil manufacturer (Sierra Sur S.A., Granada, Spain) and was stored at −20°C until use.

DOR biotransformation

DOR was transformed using the fungus *Coriolopsis floccosa* (Spanish Type Culture Collection, CECT 20449), formerly known as *Coriolopsis rigida*. The transformation was carried out according to Siles et al. [26]. Briefly, sterilized polyurethane sponge (PS) cubes were placed in Erlenmeyer flasks and 25 ml culture medium was added. *C. floccosa* inoculum was then added to the PS cubes and incubated at 28°C for 7 days. After this period of time, 25 g of DOR was placed above colonized PS. Solid-state cultures on DOR were carried out at 28°C for 30 days. Non-inoculated DOR samples were prepared as controls. Then, untransformed DOR (DOR) and *C. floccosa*-transformed DOR (CORDOR) were autoclaved three times for complete sterilization; sterility was confirmed by no observed growth on potato dextrose agar after 2 weeks. Finally, untransformed DOR (DOR) and *C. floccosa*-transformed DOR (CORDOR) were sieved, homogenized and stored at 4°C until the soil amendment experiment began. The main chemical properties of DOR and CORDOR have previously been reported by Siles et al. [27].

Soil amendment

The experiment was carried out in 0.5 L pots. DOR and CORDOR were added to soil pots at concentrations of 50 g kg^{-1} (equivalent to 150 Mg ha^{-1}). Soil samples without the residue were also prepared (control soil). One sorghum plant (*Sorghum bicolor*), of homogeneous size, was planted in each pot. The experiment was performed in a greenhouse with supplementary light at 25/19°C and 50% relative humidity. Manual regular watering was provided during the experiment, with soil humidity being maintained at 15–20%.

The untreated soil and soil amended with sterilized DOR and CORDOR were analysed at day 0, 30 and 60 of treatment. The experiment consisted of five pots for each treatment at each time. At each sampling, the soil from the five pots was mixed, homogenized and sieved (2 mm sterilized mesh). Three 100 g soil subsamples for each treatment were then placed in sterile Falcon tubes and stored at −80°C until the samples were analyzed.

Community-level physiological profile

Community level physiological profiles (CLPP) were assessed using a Biolog EcoPlate system (BIOLOG. Inc., CA, USA). Each Biolog EcoPlate contains 31 different kinds of carbon sources in triplicate (seven types of carbohydrates, nine carboxylic acids, four polymers, six amino acids, two amines/amides and three miscellaneous types). To determine the CLPP for each sample, 1 g of soil was shaken in 10 ml of sterile saline solution (0.85% w/v NaCl) at 150 rpm for 1 h. Soil suspensions were then serially diluted on the basis of the viable cell counts obtained for each sample in Siles et al. [23], in order to avoid interference from the number of cells in the oxidation of substrates. 130 μl soil solutions were used for each well and Ecoplates were incubated at 25°C for 9 days. All analyses were performed in triplicate. The rate of use of C sources was indicated by the reduction in tetrazolium salts, which changed from colourless to purple [28]. Colour development for each well was obtained in terms of optical density (OD) at 590 nm every 24 h using an automated plate reader (Eon Microplate Spectrophotometer, BioTek Instruments, Inc., Germany). Microbial activity was then calculated as average well colour development (AWCD) as described by Insam et al. [29]. The 168 h OD data for each sample in triplicate, divided by their AWCD to normalize the values, were selected in order to determine substrate richness (S_f), Shannon's functional diversity index (H_f), substrate evenness (J_f) and principal coordinate analysis (PCoA) using the PAST program ver. 2.17 [30]. PCoA of 9 samples according to their CLPP was carried out using normalized AWCD data for each substrate using a Euclidean distance matrix. Statistical differences between the treatments were analysed by ANOVA, and Tukey's honest significance difference (HSD) test was used for multiple comparison of means at a 95% confidence interval.

DNA extraction, PCR amplification and pyrosequencing

Soil DNA was extracted using the MoBio Ultra Clean Soil DNA Isolation Kit (MoBio Laboratories Inc., Solana Beach, CA, USA) following the manufacturer's instructions. For each sample, three different DNA extractions were carried out, each of which was taken from a subsample. Afterwards, a pooled DNA sample for each treatment was prepared. All DNA templates were quantified with the aid of a Qubit 2.0 Fluorometer (Life Technologies, Grand Island, NY), and sample DNA concentrations were homogenised. Fungal and bacterial PCR amplifications were then carried out.

For bacteria, a 16S rRNA gene fragment was amplified capturing the hypervariable V4 region using the primers 577F and 926R designed with eight-base barcodes and pyrosequencing adapters [31]. Triplicate amplification reactions were performed in 20-μl volumes containing: 2 μl Roche 10 × Fast Start High Fidelity buffer with 18 mM $MgCl_2$ (Roche Applied Sciences), 0.5 μl Roche Fast Start High Fidelity Taq (5 U/μl), 0.75 μl Invitrogen 10 mM deoxynucleoside triphosphate (dNTP) mix, 1 μl of each primer (10 pmol μl^{-1}), 0.2 μl New England BioLabs 10 mg ml^{-1} bovine serum albumin (BSA), 3 μl DNA template (8 ng μl^{-1}) and 11.55 μl H_2O. Negative controls using sterilized water instead of soil DNA extract were included to check for primer and sample DNA contamination. The cycling conditions were as follows: initial denaturation at 94°C for 3 min, followed by 30 cycles of denaturation at 94 °C for 45 s, primer annealing at 56 °C for 45 s, extension at 72 °C for 1 min and final extension for 7 min. Reactions were then combined and purified using gel electrophoresis followed by the QIAquick gel extraction kit (QIAGEN Inc., Valencia, CA) and the QIAquick PCR Purification kit (QIAGEN Inc., Valencia, CA) according to the manufacturer's recommendations.

For fungi, a 625 bp fragment of the 28S rRNA gene was PCR-amplified in three replicate 20-μl reactions for each sample using primers LR3 and LR0 which included barcodes for sample discrimination [32]. PCR amplifications included: 4 μl Promega GoTaq buffer, 0.5 μl GoTaq DNA polymerase, 1.5 μl Roche 25 mM $MgCl_2$, 1 μl Invitrogen 10 mM dNTP mix, 1 μl of each primer (10 pmol μl^{-1}), 0.2 μl New England BioLabs 10 mg ml^{-1} BSA, 3 μl DNA template (8 ng μl^{-1}) and 7.8 μl H_2O. The thermal cycling program was: an initial denaturation at 94 °C for 3 min followed by 30 cycles at 94 °C for 1 min, at 51 °C for 40 s, and at 72 °C for 1 min, followed by an extension at 72 °C for 10 min. The reactions of each sample were then pooled and purifications were performed as for bacteria.

Amplicons from all samples for bacteria and fungi were composited together in equimolar concentrations and sequenced using a Roche Sequencer GS FLX Titanium series (454 Life Sciences, Branford, CT) at Utah State University.

Pyrosequencing data analysis

Raw fungal and bacterial sequences were processed using Mothur version 1.31.0 [33]. Briefly, sequencing errors were reduced using the AmpliconNoise algorithm and low-quality sequences were removed [minimum length of 150 base pairs (bp), allowing 1 mismatch in the barcode, 2 mismatches in the primer, and homopolymers no longer than 8 bp]. Sequences were then aligned using the package's internal alignment feature and the SILVA database as template [34]. The chimera.uchime function was then used to identify potentially chimeric sequences which were subsequently removed [35]. Finally, the high-quality fungal and bacterial sequences were separately clustered into operational taxonomic units (OTUs) at a 3% dissimilarity distance. The number of sequences per sample was normalized before OTU definition based on the number of sequences obtained from the smallest library. OTU (phylogenetic richness – S_p) distribution among samples was used to calculate rarefaction curves, the phylogenetic Shannon diversity index (H_p), phylogenetic evenness (J_p), Chao 1 and ACE (abundance-based coverage estimation) diversity estimator indices as well as Good's coverage by Mothur. Significant differences in Shannon diversity indices between control and amended samples at a given sampling time were assessed using the diversity t test, with $p < 0.05$ being regarded as statistically significant [36]. On the other hand, differences in the fungal and bacterial community composition of each pair of samples were determined using the unweighted UniFrac metric (1,000 permutations). The unweighted UniFrac distances between samples were then used to model PCoA for each community. Finally, representative sequences from the 14 most abundant bacterial and fungal OTUs were obtained using Mothur. These sequences were identified by manually blasting in the EzTazon-e database (http://eztaxon-e.ezbiocloud.net/) [37] for bacteria and in the CBS-KNAW Fungal Biodiversity Center (http://www.cbs.knaw.nl/) for fungi.

To examine changes in the relative abundance of the different microbial groups mediated by amendments, the non-normalized bacterial and fungal sequences were classified using the Ribosomal Database Project (RDP) bacterial 16S rRNA gene and the fungal 28S rRNA gene classifier (http://rdp.cme.msu.edu/classifier/classifier.jsp) at a 50% bootstrap confidence level for both communities [32,38,39]. To describe the bacterial and fungal diversity of the soil studied, all 16S rRNA gene and 28S rRNA gene sequences were merged into a single file for each community, which was then subjected to the RDP classifier at the same bootstrap confidence level.

The raw pyrosequencing data were deposited in the MG-RAST public database (http://metagenomics.anl.gov/) under accession number 4552035.3 for bacteria sequences and 4552036.3 for fungi sequences.

Pearson's method was used to examine trends between functional and phylogenetic properties with respect to the chemical characteristics of the different soil samples reported in a previous article [27]. Normality of data was assessed by Kolmogorov–Smirnov test.

Results

Soil microbial diversity

Bacterial diversity. After pyrosequencing analysis, a total of 17,322 sequences across the 9 samples passed through the high quality filters with an average read length of 311 bp. The number of sequences per sample ranged from 2,248 (C-T0) to 1,674 (CORDOR-T1) (Table 1). The average number of bacterial sequences was $1,924 \pm 160$ (mean$\pm$SD) per sample. These sequences were grouped into 2,267 different OTUs at 97% sequence similarity. This total number of OTUs consisted of 1,143 nonsingleton OTUs and 1,124 singletons. The rarefaction curves of the different treatments did not reach a plateau for any sample (Fig. S1). Good's coverage values (ranging from 0.76 to 0.81) (Table 1) also indicated that the sequences obtained were insufficient to fully capture bacterial diversity.

Phylogenetic assignment analysis enabled the classification of ~86% sequences at phylum level. The soil's bacterial diversity was distributed among 17 different phyla. The most common phyla were Proteobacteria, Acidobacteria, Actinobacteria, Gemmatimonadetes, Firmicutes and Verrucomicrobia (Fig. 1A). Approximately 83% of reads could be classified into 42 different classes, in particular, *Alphaproteobacteria* (with *Skermanella*, *Microvirga*, *Phenylobacterium* being the most common genera in this class), *Gp6* (*Gp6*), *Actinobacteria* (*Nocardioides* and *Solirubrobacter*), *Gemmatimonadetes* (*Gemmatimonas*), *Gammaproteobacteria* (*Steroidobacter*) and *Deltaproteobacteria* (*Geobacter*) (Fig. 1B).

Fungal diversity. A total of 38,410 valid 28S rRNA gene sequences, with an average read length of 338 bp, were obtained from the 9 samples. The average number of reads per sample was $4,267 \pm 2,258$ (mean$\pm$SD), with sample C-T0 having the highest number of sequences (9,405) and DOR-T1 having the lowest number of reads (1,230) (Table 2). The total number of sequences represented a total of 1,160 different OTUs at 97% confidence threshold; 720 of these were nonsingleton OTUs and the rest (440) were singletons. The rarefaction curves (Fig. S2) and Good's coverage values (Table 2) indicated that sampling was not fully exhaustive for any sample. However, according to the coverage data, the fungal community was more thoroughly characterized than the bacterial community.

Fungal RDP sequence classification (50% confidence threshold) yielded ~80% classified sequences among 5 different phyla, particularly Ascomycota, Basidiomycota and Chytridiomycota (Fig. 2A). On the other hand, the fungal diversity of the fungal community consisted of 18 different classes (71% of total sequences). The most abundant classes were *Sordariomycetes* (with the most numerous genera in this class being *Chaetomium*, *Fusarium* and *Stachybotrys*), *Pezizomycetes* (*Ascobolus* and *Peziza*), *Dothideomycetes* (*Alternaria*, *Lophiostoma* and *Cladosporium*), *Chytridiomycetes* (*Nowakowskiella*), *Eurotiomycetes* (*Aspergillus* and *Eupenicillium*) and *Agaricomycetes* (*Coprinellus*) (Fig. 2B).

Effects of DOR and CORDOR on soil microbial communities

Community level physiological profile (CLPP). The functional indices S_f and H_f based on CLPPs significantly decreased ($p<0.05$) in the samples amended with DOR and CORDOR at 30 and 60 days (Table 3). On the other hand, S_f and H_f did not vary between samples at initial sampling time. The

Table 1. Phylogenetic bacterial diversity characteristics obtained from unamended soil (C) and soil amended with untransformed DOR (DOR) or *C. floccosa*–transformed DOR (CORDOR) at 0 (T0), 30 (T1) and 60 (T2) days.

Soil sample	Sequence number	S_p	H_p	J_p	Chao 1	ACE	Good's Coverage
C-T0	2248	649	5.82(5.76;5.89)	0.899(0.892;0.907)	1381(1202;1617)	2132(1946;2344)	0.76
DOR-T0	1957	580	5.59(5.52;5.66)	0.879(0.870;0.888)	1240(1070;1471)	1913(1739;2113)	0.79
CORDOR-T0	1981	592	5.59(5.52;5.66)	0.876(0.867;0.884)	1351(1156;1612)	2262(2059;2494)	0.78
C-T1	1858	613	5.70(5.63;5.77)	0.888(0.880;0.896)	1379(1185;1638)	2030(1845;2243)	0.77
DOR-T1	1904	522	5.34*(5.26;5.41)	0.853(0.843;0.863)	1103(945;1320)	1569(1416;1747)	0.81
CORDOR-T1	1674	646	5.81(5.74;5.88)	0.898(0.890;0.906)	1428(1235;1685)	2137(1949;2351)	0.76
C-T2	1787	560	5.84(5.77;5.90)	0.901(0.894;0.909)	1382(1203;1618)	2106(1922;2317)	0.76
DOR-T2	1888	638	5.81(5.74;5.87)	0.899(0.892;0.907)	1330(1160;1555)	2198(2006;2417)	0.77
CORDOR-T2	2025	651	5.86(5.80;5.93)	0.905(0.897;0.912)	1236(1094;1424)	1756(1602;1934)	0.77

Values in parenthesis are 95% confidence intervals as calculated using MOTHUR.
S_p –Phylogenetic richness.
H_p–Phylogenetic Shannon index.
J_p–Phylogenetic evenness.
Diversity t test was performed for each amended sample with its control (* significant differences, $p \leq 0.05$).

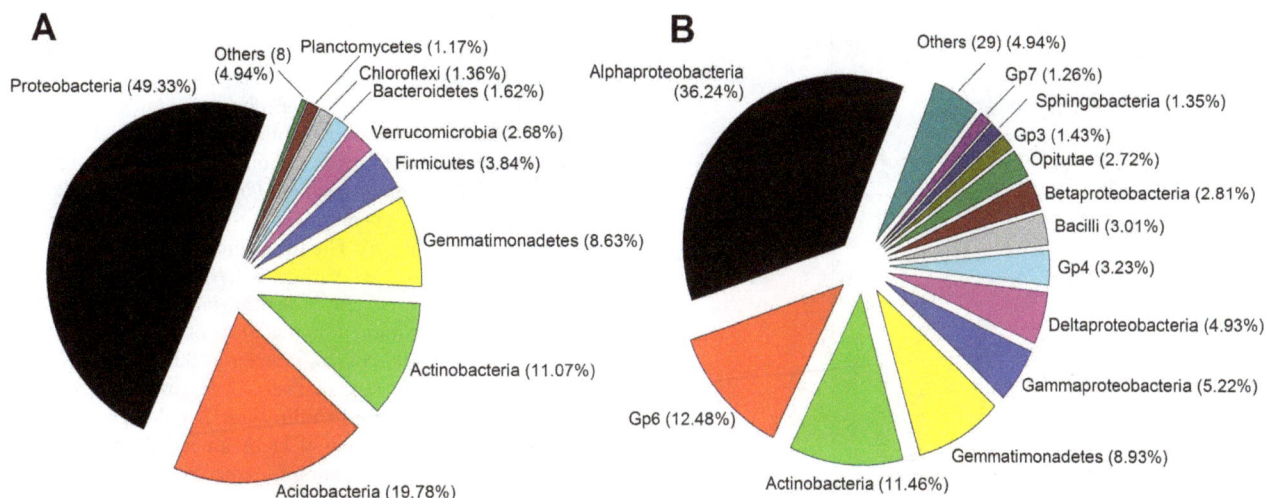

Figure 1. Composition of the bacterial community in the soil studied based on 16S rRNA gene pyrosequencing at phylum (A) and class (B) level.

lowest microbial physiological diversity was found in the soil treated with CORDOR at 30 days (Table 3).

PCoA of the CLPP dataset showed that around 53% of the variability was due to two principal coordinates, the first (PC1) accounting for 33.39% and the second (PC2) accounting for 19.09% (Fig. 3). These two coordinates grouped the samples in two clusters and one sample was situated alone. The values for correlating each C source with PC1 and PC2 are shown in the Table S2. One of the clusters, situated in the lower-left quadrant, contained all the samples at initial sampling time and control samples at 30 and 60 days, with this group being closely associated with carbohydrates and polymers (D-cellobiose, cyclodextrin and glycogen). Another cluster, located in the upper quadrants, consisted of the samples amended with DOR for 30 and 60 days and soil amended with CORDOR for 60 days. This group was highly weighted by carboxylic acids (malic, itaconic and D-galacturonic acid) and carbohydrates (Beta-methyl-D-glucoside). Finally, the soil treated with CORDOR for 30 days was situated in the lower-right quadrant; some carbohydrates (D-xylose and i-erythritol), amines/amides (putrescine and phenylethylamine) and amino acids (L-arginine) were the most oxidized substrates in this sample.

Phylogenetic bacterial community. The DOR amendment significantly reduced H_p (diversity t test, p<0.001) at 30 days with respect to the unamended soil (Table 1). This amendment also caused a diminution in the S_p, J_p, Chao 1 and ACE indices at this sampling time. However, DOR did not alter bacterial diversity at the other sampling times. In addition, no changes were observed in the diversity characteristics of samples amended with COR-DOR at any sampling time (Table 1).

PCoA of bacterial pyrosequencing data based on the unweighted UniFrac metric revealed that amendments caused variations in community structure (Fig. 4). The analysis showed that 66.78% of variance can be explained by two principal coordinates, one accounting for 53.18% and the other for 13.60% of the variation. The nine samples grouped in two clusters and one sample did not cluster. One of the groups was situated to the left of PC1 and consisted of all samples at initial sampling time and control samples at 30 and 60 days. The pairwise unweighted UniFrac test did not find any significant differences between the samples of this group (p>0.05). Another group was situated in the lower right

quadrant and was made up of the samples amended with CORDOR for 30 and 60 days as well as the soil treated with DOR for 60 days. The bacterial community structure in these samples was significantly different with respect to their control samples (pairwise unweighted UniFrac test, p<0.001). Finally, soil amended with DOR for 30 days was located in the upper right quadrant. This sample showed significant differences with respect to unamended soil at 30 days (p<0.001).

Changes in bacterial community structures mediated by amendments were probably caused by alterations in the relative abundance of the Proteobacteria, Acidobacteria, Actinobacteria and Gemmatimonadetes phyla (Fig. 5A). In the Proteobacteria phylum, the most significant changes occurred in *Alphaproteo-bacteria* due to its predominance in this phylum (Fig. S3A). It is interesting to note that the orders belonging to *Alphaproteobacteria* responded differently to amendments (Fig. 5B), since *Rhodospirillales* [represented in the top 14 most abundant bacteria with OTU 1 (*Skermanella stibiiresistens*) and OTU 9 (*Skermanella aerolata*) (Table S3)] decreased their relative abundances after DOR and CORDOR application at 30 and 60 days while *Rhizobiales* [OTU 2 (*Microvirga aerophila*) and OTU 7 (*Rhizobium rosettiformans*) (Table S3)], *Caulobacterales* [OTU 6 (*Phenylobacterium* sp.) (Table S3)] and *Sphingomonadales* showed considerably greater abundance after these treatments at the same sampling times. With regard to the Acidobacteria phylum, an overall decrease in its abundance in amended samples was observed, with the reduction being more evident in samples amended with DOR (Fig. 5A). *Gp6* and *Gp7* were the classes most affected by inputs although clear evidences of the specific effect of each amendment on these groups were not found (Fig. S3B). Four of the most abundant OTUs (3, 8, 10 and 12) were identified as uncultured *Acidobacteria* although a more detailed identification of these OTUs was not possible (Table S3). In the *Actinobacteria* phylum, the suborder *Propionibacterinae* [OTU 5 (*Nocardoides mesophilus*) (Table S3)] responded negatively to both types of amendment at 30 and 60 days (Fig. S3C). Finally, Gemmatimonadetes were also affected by the application of DOR. OTU 13, associated with *Gemmatimonadaceae*, the only family present in this phylum, experienced a drastic decrease in its relative abundance after treatment with DOR at 30 and 60 days (Table S3).

Table 2. Phylogenetic fungal diversity characteristics obtained from unamended soil (C) and soil amended with untransformed DOR (DOR) or *C. floccosa*–transformed DOR (CORDOR) at 0 (T0), 30 (T1) and 60 (T2) days.

Soil sample	Sequence number	S_p	H_p	J_p	Chao 1	ACE	Good's Coverage
C-T0	9405	245	4.02(3.92;4.12)	0.731(0.726;0.757)	630(485;862)	994(854;1167)	0.87
DOR-T0	3658	240	3.91(3.80;4.01)	0.742(0.737;0.758)	625(479;890)	982(852;1092)	0.88
CORDOR-T0	2321	239	3.92(3.79;4.02)	0.700(0.698;0.728)	664(474;989)	918(784;1083)	0.89
C-T1	4093	247	4.16(4.07;4.26)	0.755(0.741;0.769)	557(443;737)	1055(917;1220)	0.88
DOR-T1	1230	235	4.01(3.91;4.11)	0.645(0.625;0.666)	586(455;792)	1272(1090;1494)	0.87
CORDOR-T1	4007	292	4.52*(4.61;4.77)	0.826(0.814;0.837)	610(499;782)	890(778;1027)	0.86
C-T2	4238	249	4.10(4.00;4.21)	0.744(0.728;0.760)	551(438;731)	838(724;979)	0.88
DOR-T2	4234	241	4.05(3.94;4.15)	0.738(0.722;0.755)	616(471;850)	1020(875;1197)	0.86
CORDOR-T2	5224	271	4.43*(4.34;4.52)	0.791(0.779;0.803)	735(565;1004)	1296(1128;1497)	0.86

Values in parenthesis are 95% confidence intervals as calculated using MOTHUR.
S_p – Phylogenetic richness.
H_p – Phylogenetic Shannon index.
J_p – Phylogenetic evenness.
Diversity t test was performed for each amended sample with its control (* significant differences, $p \leq 0.05$).

Phylogenetic fungal community. With regard to fungal community, CORDOR caused a significant increment in H_p as compared to unamended samples (diversity t test, $p < 0.05$) after 30 and 60 days (Table 2). The other diversity indices were also affected by this amendment. By contrast, DOR did not alter the diversity characteristics of fungal community at any sampling time.

Fungal PCoA based on unweighted UniFrac distances indicated that this community structure was altered depending on the amendment applied (Fig. 6). The two principal PCoA coordinates explained 61.34% of the variations (40.77% and 20.57%, respectively) and separated the 9 samples into three groups; one group, in the left quadrant, was made up of all the samples at initial sampling time and control samples at 30 and 60 days. No significant differences ($p > 0.05$) between these samples were found using the pairwise unweighted UniFrac test. Another group, in the upper right quadrant, consisted of DOR amended samples at 30 and 60 days; the fungal community structure of these samples differed significantly from their respective control samples (pairwise unweighted UniFrac test, $p < 0.001$). The last group, in the lower right quadrant, included the samples amended with CORDOR for 30 and 60 days, which also presented a significantly different fungal community structure from that of the respective unamended samples (pairwise unweighted UniFrac test, $p < 0.001$).

The most striking changes in this community caused by soil amendments occurred in the Ascomycota phylum which dominated fungal diversity (Fig. S4). In this sense, the *Sordariomycetes* class responded positively to both types of amendment after 30 and 60 days, with a more important increase being observed following treatment with DOR (Fig. 7A). This increment in the samples amended with DOR for 30 and 60 days was due to an increase in the relative abundance of *Hypocreales* (Fig. 7B), which was probably caused by OTU 1 (*Fusarium* sp.) (Table S4). Curiously, this group decreased following CORDOR amended treatment for 30 and 60 days (Fig. 7B). However, in these treatments, *Sordariales* increased, probably due to a proliferation of genera such as *Podospora* sp. (OTU 11) and *Cercophora* sp. (OTU12), although, in this group, CORDOR caused a reduction in *Chaetomium* sp. (OTU 2) (Table S4). DOR also caused a reduction in this fungal group after 30 and 60 days. On the other hand, the *Eurotiales* order (*Eurotiomycetes* class) also decreased following the application of both types of amendment after 60 days (Fig. 7B). These changes could be attributed to a diminution in *Aspergillus terreus* (OTU 8), among others (Table S4). In the *Dothideomycetes* class, there was a particularly sharp decrease in *Pleosporales* [OTU 3 (*Preussia terricola*)] in the amended samples after 30 and 60 days. With regard to *Pezizomycetes* (*Pezizales* order), OTU 14 (*Ascobolus*), increased following treatment with CORDOR, especially after 60 days (Table S4). Finally, it is interesting to note that OTU 9, identified as *Cryptococcus* sp. (Basidiomycota), and OTU 13, identified as *Coprotus ochraceus* (*Leotiomycetes*, Ascomycota), also responded positively to DOR and CORDOR treatments after 30 and 60 days (Table S4).

Discussion

Soil microbial diversity

Human actions are causing a biodiversity crisis, with species extinction rates up to 1000 times higher than background rates [2]. Conservation strategies are therefore necessary, especially in vulnerable zones such as the Mediterranean biome, which is currently considered to be one of the most vulnerable of the Earth's thirteen terrestrial biomes [3]. In this context, acquiring

A

Glomeromycota (0.04) Blastocladiomycota (0.13%)

Chytridiomycota (5.58%)

Basidiomycota (5.87%)

Ascomycota (88.38%)

B

Others (12) (1.35%)

Agaricomycetes (4.52%)

Sordariomycetes (59.88%)

Eurotiomycetes (4.52%)

Chytridiomycetes (6.17%)

Dothideomycetes (10.60%)

Pezizomycetes (13.68%)

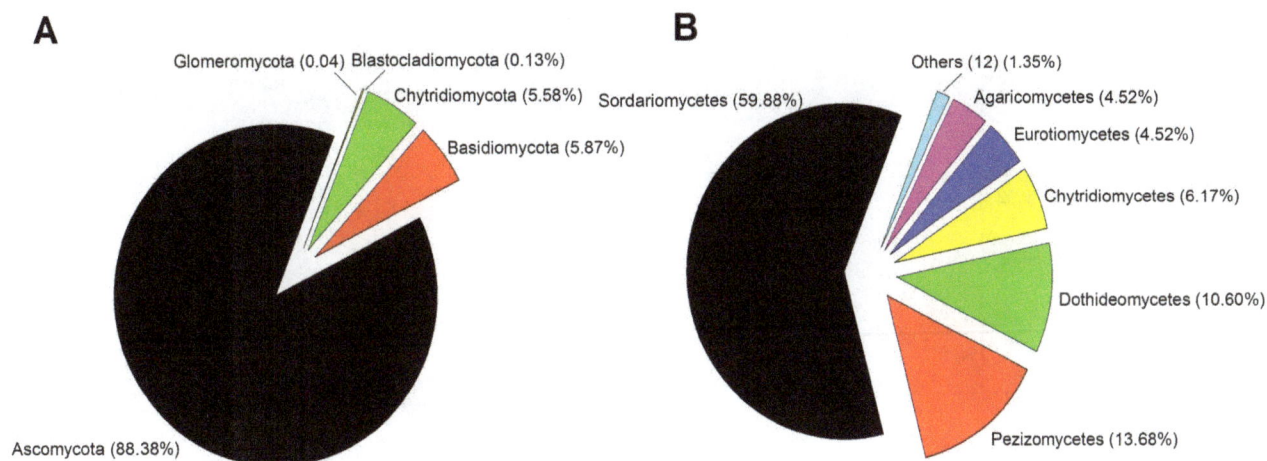

Figure 2. Composition of fungal community in the soil studied based on 28S rRNA gene pyrosequencing at phylum (A) and class (B) level.

knowledge of soil biodiversity would be a first step in the development of sustainable activities.

Our results demonstrated that microbial communities in this type of soil are extremely diverse, with high values for richness estimators recorded in each sample. This was corroborated by the high proportion of unclassified sequences at phylum level despite the use of a 50% threshold for read classification [40]. The phylum composition of bacterial communities in this soil was consistent with previous pyrosequencing surveys of soils in the Iberian Peninsula [41–43]. Although our study found a higher proportion of Acidobacteria, especially in the Gp6 class, than the aforementioned works, this could be related to the high alkaline pH (~8.4) of soil [44]. Higher relative abundance levels were found for *Gemmatimonadetes* (8.63%) than the normal levels reported in other soils [45], which could be due to the xeric conditions of Mediterranean basin soils [46]. On the other hand, it is not surprising that our results for bacterial diversity differed substantially from data obtained for the present experiment using culture-dependent techniques. To carry out the culture-dependent study, 900 strains were isolated, clustered by their fatty acid methyl ester

profiles and groups of isolates were identified by partial sequencing of 16S rRNA gene [23]. Culturable bacterial diversity was distributed between Actinobacteria (50.6%), Proteobacteria (40.4%), Firmicutes (4.5%) and Bacteroidetes (4.4%). Other studies have also shown that important differences in the bacterial diversity of an environment are observed when culture–dependent and culture–independent studies are carried out simultaneously [47–49]. These discrepancies are understandable given the difficulty in obtaining culturable members of certain bacterial groups. In this respect, it is interesting to note that, despite the high abundance levels of Acidobacteria in this soil and the proven effectiveness of VL 70 medium with regard to Acidobacteria isolation [50], it was not possible to obtain any strain belonging to this phylum. The absence of culturable Acidobacteria in this soil may be due to the slow growth of these bacteria or the inhibition in the development of these colonies caused by other culturable bacteria [47]. It is also worth noting that Actinobacteria was the most common culturable phylum, while pyrosequencing data showed that Proteobacteria dominated bacterial diversity in this soil. This bias may be a consequence of the copiotrophic lifestyle of

Table 3. Functional microbial diversity characteristics (mean±standard deviation) obtained from unamended soil (C) and soil amended with untransformed DOR (DOR) or *C. floccosa*–transformed DOR (CORDOR) at 0 (T0), 30 (T1) and 60 (T2) days.

Soil sample	S_f	H_f	J_f
C-T0	27.33±0.58 cd	3.15±0.01 d	0.95±0.01 a
DOR-T0	22.50±3.54 bcd	3.00±0.11 d	0.96±0.01 a
CORDOR-T0	24.00±1.41 cd	3.05±0.05 de	0.96±0.01 a
C-T1	24.00±1.41 de	3.00±0.01 e	0.95±0.01 a
DOR-T1	19.50±0.71 ab	2.84±0.01 c	0.94±0.01 a
CORDOR-T1	17.50±2.12 a	2.49±2.49 a	0.92±0.03 a
C-T2	24.50±0.71 de	3.12±0.02 e	0.97±0.01 a
DOR-T2	20.50±0.71 abc	2.79±0.02 c	0.94±0.01 a
CORDOR-T2	17.00±1.41 a	2.64±0.07 b	0.94±0.02 a

S_f –Functional richness.
H_f–Functional Shannon index.
J_f–Functional evenness.
For each variable, data followed by different letter are significantly different according to Tukey's HSD test (P≤0.05).

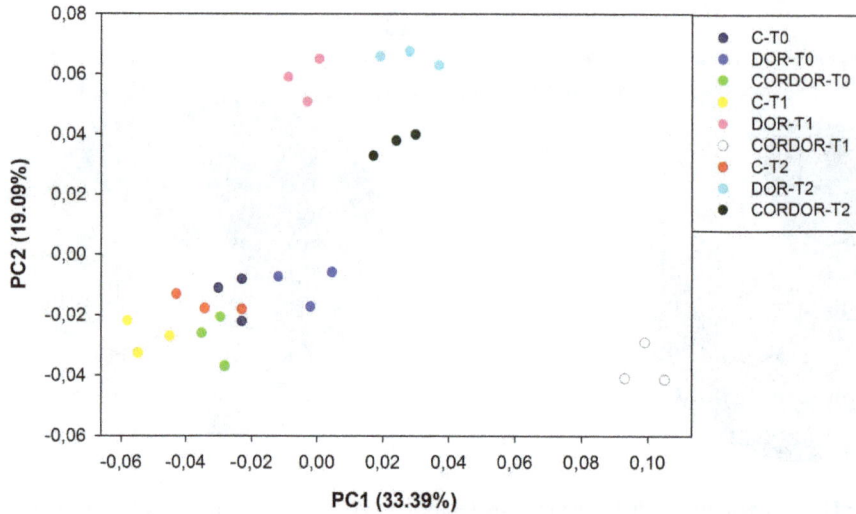

Figure 3. PCoA based on Euclidean distances of community level physiological profile (CLPP) dataset for unamended soil (C) and soil amended with untransformed DOR (DOR) or *C. floccosa*-transformed DOR (CORDOR) at 0 (T0), 30 (T1) and 60 (T2) days. Percent variability explained by each principal coordinate is shown in parentheses after each axis legend.

some Actinobacteria groups [51]. Pyrosequencing data showed that genera such as *Microvirga*, *Nocardioides* and *Rhizobium* were among the most abundant in this Mediterranean soil, and it is interesting to note that numerous isolates belonging to these genera were found in the previous culture-dependent survey [23]. In this way, VanInsberghe et al. [49], in a work assessing bacterial diversity using culture-dependent and pyrosequencing techniques, suggested that the combination of both approaches may be useful as the isolates obtained can be used for genomic and physiological research.

Ascomycota clearly dominated the fungal diversity of the soil studied, a finding which is in line with other pyrosequencing studies of Mediterranean soils [3,42,52]. This research also demonstrated that Basidiomycota and Chytridiomycota, though in lower concentrations than Ascomycota, may also be found in

these environments. The limited presence of Glomeromycota (phylum associated with arbuscular mycorrhiza) in the present work is worth noting, which may be due to the absence of plants in the plots when soil samples were collected. The fungal pyrosequencing-based diversity found in the present research was consistent with the diversity obtained by the application of culturable-dependent techniques to these soil samples. In this culture-dependent survey, 1,733 strains were obtained and characterized with the aid of morphological and molecular techniques [24], with the majority of isolates being distributed among the *Sordariomycetes*, *Eurotiomycetes* and *Dothideomycetes* classes [24]. It is interesting to note that some of the most abundant fungal species (*Chaetomium*, *Fusarium*, *Aspergillus*, *Alternatia* and *Cladosporium*) in pyrosequencing data were also the most common culturable fungi [24]. These findings are not surprising as soil fungi are dominated by readily culturable saprobic filamentous forms [53]. Similarly, Klaubauf et al. [54] have reported that data from both molecular-based and culture-dependent techniques correlate more closely in soil fungal communities than in bacterial communities.

Effects of DOR and CORDOR on soil microbial communities

The present survey assessed the impact of DOR and CORDOR as organic amendments on the soil microbial community. However, the application of raw DOR is not possible as it causes oxidative stress in plants and presents considerable phytotoxic activity due to its high phenol content and other substances such as fatty acids [55]. In our greenhouse experiment, the phytotoxicity of DOR at agronomic rates (150 Mg ha^{-1}) was confirmed; DOR produced a drastic reduction in the shoot dry weight of sorghum plants, while CORDOR reduced toxicity levels [23]. These results were also corroborated by a sorghum field-based experiment involving both amendments (unpublished data). Other studies have reported that the application of raw DOR to olive groves did not lead to a diminution in olive production over the long term, although only small doses of DOR were applied in this experiment (27 and 54 Mg ha^{-1}) [6]. However, other works have shown that olive oil wastewater (OMW), a liquid residue obtained from a

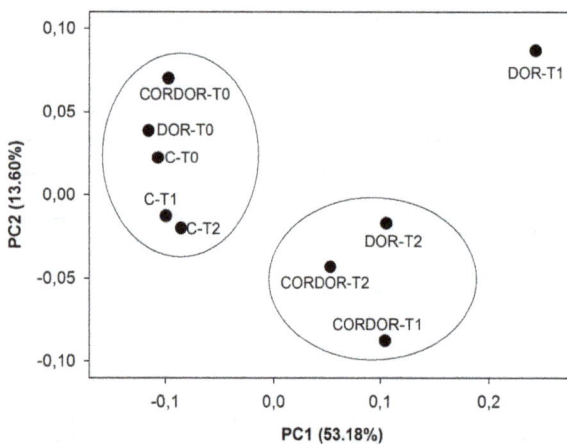

Figure 4. PCoA based on unweighted UniFrac distances of bacterial community found in unamended soil (C) and soil amended with untransformed DOR (DOR) or *C. floccosa*-transformed DOR (CORDOR) at 0 (T0), 30 (T1) and 60 (T2) days. Percent variability explained by each principal coordinate is shown in parentheses after each axis legend.

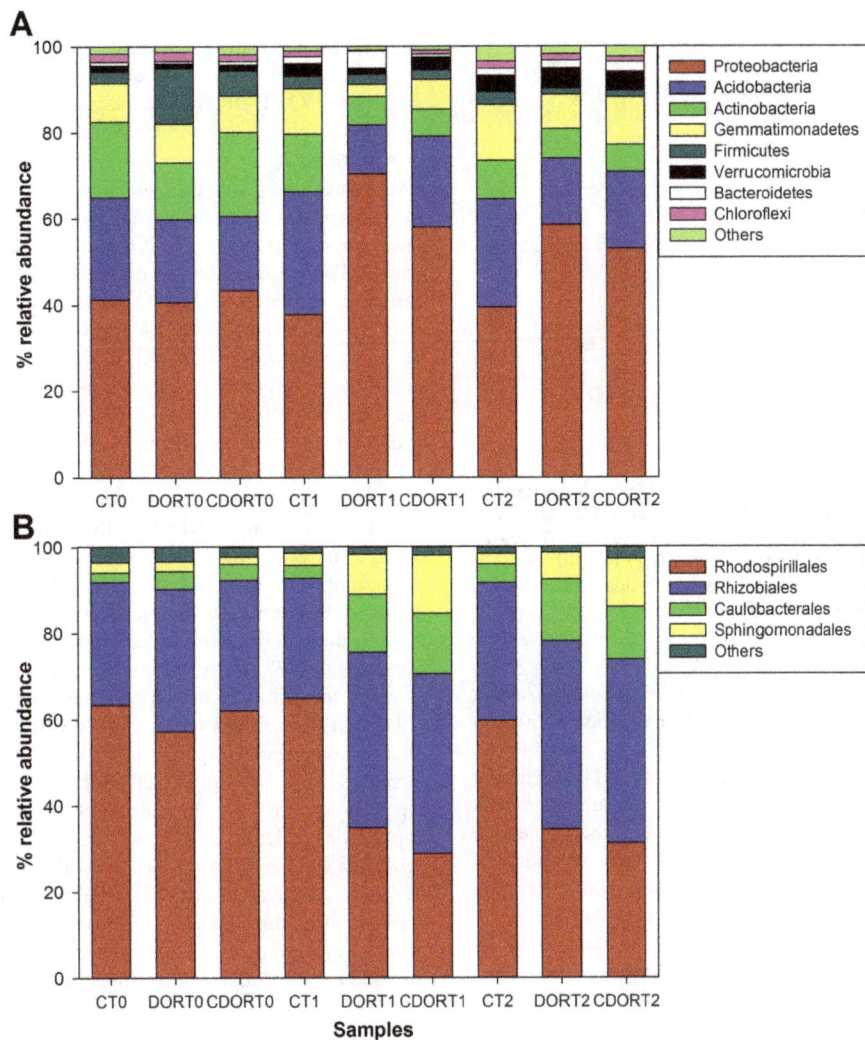

Figure 5. Relative abundance of the different bacterial phyla (A) and orders of *Alphaproteobacteria* (B) found in unamended soil (C) and soil amended with untransformed DOR (DOR) or *C. floccosa*-transformed DOR (CDOR) at 0 (T0), 30 (T1) and 60 (T2) days.

three-phase olive-mill extraction system with a composition similar to that of DOR and produced in other Mediterranean countries such as Italy and Greece, can cause olive grove death (800 m^3 ha^{-1}) and negatively affect the quality of olive oil at high doses [56,57]. Thus, the phytotoxic activity of DOR depends on the doses applied, with seasonal crops being especially susceptible to these effects.

Previous studies have assessed the impact of raw and transformed DOR as well as OMW on soil microbiology using DGGE, PLFA and colony forming unit (CFU) counts [21,58–61]. Thus, our understanding of the changes produced by DOR and OMW (olive mill waste) in the phylogenetic composition of microbial communities is limited, although these studies have demonstrated that OMW has a beneficial effect on microbial abundance. Previous surveys have found, through the use of culture-independent techniques, that DOR caused a rapid and marked increase in bacterial and fungal abundance in the soil analyzed in the present work after 30 and 60 days of treatment, while CORDOR resulted in slower and more moderate increases [23,26]. Thus, although microtoxic effects have been associated with raw olive mill waste [62,63], it was not possible to detect these

effects in relation to abundance levels in an environment as diverse and complex as soil. The beneficial impact of this waste on certain microbial groups, due to the input of easily degradable compounds, probably masked potential microtoxicity. In this respect, previous studies have established that the impact of olive wastes on soil microbiology is the result of complex, sometimes contradictory, effects, and depends on the relative amounts of beneficial and toxic organic and inorganic compounds added [57,64].

CLPP analyses showed a diminution in functional diversity (S_f and H_f) and changes in the functional structure of microbial communities after amendments application, especially in the case of CORDOR after 30 days. Some studies, using the Biolog system, have shown an increase or no change in functional diversity after the application of amendments [65], while others have demonstrated a reduction in this diversity [42]. These differences between works are probably due to variations in the kinds of organic amendments used. In the present study, the high level of functional diversity in the soil studied can be regarded as normal in relation to agronomical soil [66]. The subsequent diminution in diversity following the application of amendments could be due to the

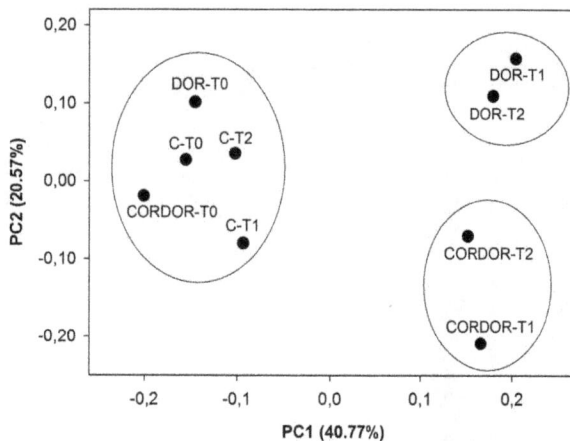

Figure 6. PCoA based on unweighted UniFrac distances of fungal community found in unamended soil (C) and soil amended with untransformed DOR (DOR) or *C. floccosa*-transformed DOR (CORDOR) at 0 (T0), 30 (T1) and 60 (T2) days. Percent variability explained by each principal coordinate is shown in parentheses after each axis legend.

adaptation and selective proliferation of certain microorganisms at the expense of the incorporated nutrients since functional diversity negatively correlated with total organic C ($R_{pearson}$ -0.814, P< 0.05) and total N ($R_{pearson}$ -0.729, P<0.05). The different functional community structure found in the soil amended with CORDOR after 30 days as compared to the other amended samples may be a consequence of the different nutrients added, as this type of treatment was especially characterised by the oxidation of C sources containing N (amino acids and amines/amides). The ability of some saprobic fungi to increase N content in DOR during bioremediation has been reported in previous studies [14].

However, no dramatic changes were caused by DOR and CORDOR in phylogenetic microbial diversity with respect to the unamended samples, with DOR reducing bacterial diversity after 30 days and CORDOR increasing fungal diversity after 30 and 60 days. Thus, it is possible to conclude that the most significant effects of the amendments on soil microbiology are related to alterations in the community structure. Similar conclusions were reached in the culture-dependent studies of this experiment. [23,24]. In the case of the bacterial community, the most important change mediated by amendments was an increase in Proteobacteria, which was more marked in the case of DOR at 30 days. A detailed analysis of this phylum found that the most significant changes occurred in the *Alphaproteobacteria* class. *Rhodospirillales* decreased with the addition of DOR and CORDOR after 30 and 60 days. In fact, a negative correlation between organic C and the relative abundance of *Rhodospirillales* ($R_{pearson}$ -0.953, P<0.001) was found, probably due to the adaptation of this bacterial group to oligotrophic nutritional conditions [67]. On the other hand, *Rhizobiales*, *Caulobacterales* and *Sphingomonadales* increased with the addition of DOR and CORDOR after 30 and 60 days due to their saprophytic lifestyle and their capacity to degrade recalcitrant compounds, even phenols [68,69]. There was a positive correlation between the relative abundance of these groups and total soil organic C ($R_{pearson}$ 0.945, P<0.001). It is remarkable the increase of *Rhizobiales* with both amendments, as this group contains species associated with nitrogen fixation and plant growth promotion [70], which may be a beneficial aspect of the application of these amendments to soil. By contrast, given that Acidobacteria have

been identified as oligotrophic bacteria [71] and alkaline soil inhabitants, especially subgroups 5, 6 and 7 [44], the increase in nutrients and decrease in pH observed in soil after applications of DOR and CORDOR could explain the diminution in the relative abundance of this group. Unlike the aforementioned bacterial groups, there are previous works assessing the impact of olive mill wastes on Actinobacteria, which reported a positive effect of OMW on this bacterial group over the short term [58,72]. However, with the aid of DGGE, Karpouzas et al. [73] suggested that OMW is responsible for dramatic alterations in this bacterial community. Siles et al. [23] found that culturable Actinobacteria responded differently to DOR and CORDOR depending on the phylogenetic group considered. In the present study, the changes in this community mediated by the amendments were limited with the exception of members of the *Propionibacterinae* suborder, which decreased with DOR and CORDOR after 30 and 60 days. Using a culturable-dependent approach, Siles et al. [23] also found that *Streptomyces* spp. were negatively affected by DOR and that the application of CORDOR offset this adverse effect. In the present survey, it was not possible to obtain conclusive results concerning this bacterial group as the number of sequences obtained belonging to *Streptomyces* sp. was very low. In this respect, these findings are in line with the research carried out by Shade et al. [48], where culture dependent studies were regarded as useful for assessing the soil rare biosphere.

For the fungal community, it is worth noting the findings obtained with respect to *Fusarium* spp., whose relative abundance increased with the addition of DOR. Siles et al. [24] have also reported an increase in culturable *Fusarium* spp. after the application of DOR, which is a reasonable finding given that *Fusarium* spp. have been associated with lignocellulosic wastes due to their saprophytic lifestyle [74]. This may be a drawback for the application of this residue in its raw state, which may adversely affect crop development as some species of *Fusarium* have been identified as potential phytopathogens [75]. On the other hand, CORDOR treatments led to a decrease in this fungal group. Previous studies have also demonstrated the suppressive effect of composted OMW on certain fungal phytopathogen species [76]. *Aspergillus terreus* and *Cryptoccocus* sp. were also observed to decrease and increase, respectively, following the application of amendments, which is in line with the culture-dependent study [24]. Curiously, Karpouzas et al. [59] observed an increment in *Cryptococcus* sp. after OMW application to soil, which was explained as the result of the ability of these microorganisms to metabolize a high variety of substrates. On the other hand, we observed a striking reduction in the relative abundance of *Chaetomium* spp., which are saprobic fungi and potential degraders of cellulosic material, with both amendments after 30 and 60 days [77]. This could be explained by the high sensitivity of this group's members to phenols [78]. In this respect, a negative correlation was found between soil phenol content and the number of *Chaetomium* sp. sequences ($R_{pearson}$ -0.759, P<0.05). However, CORDOR presented a lower phenol content than DOR due to its transformation by *C. floccosa* [27]. Thus, the decrease in *Chaetomium* sp., in amended samples, could also be explained by the ability of other microbial groups positively affected by inputs to inhibit their proliferation [79]. By contrast, other fungal groups such as *Podospora* sp. and *Cercophora* sp., which have been identified as coprophilous fungi [80] and have been associated with lignocellulosic material [81,82], were found to increase after 30 and 60 days following the addition of CORDOR, thus supporting the hypothesis that olive mill wastes affects soil microbiology in contradictory ways.

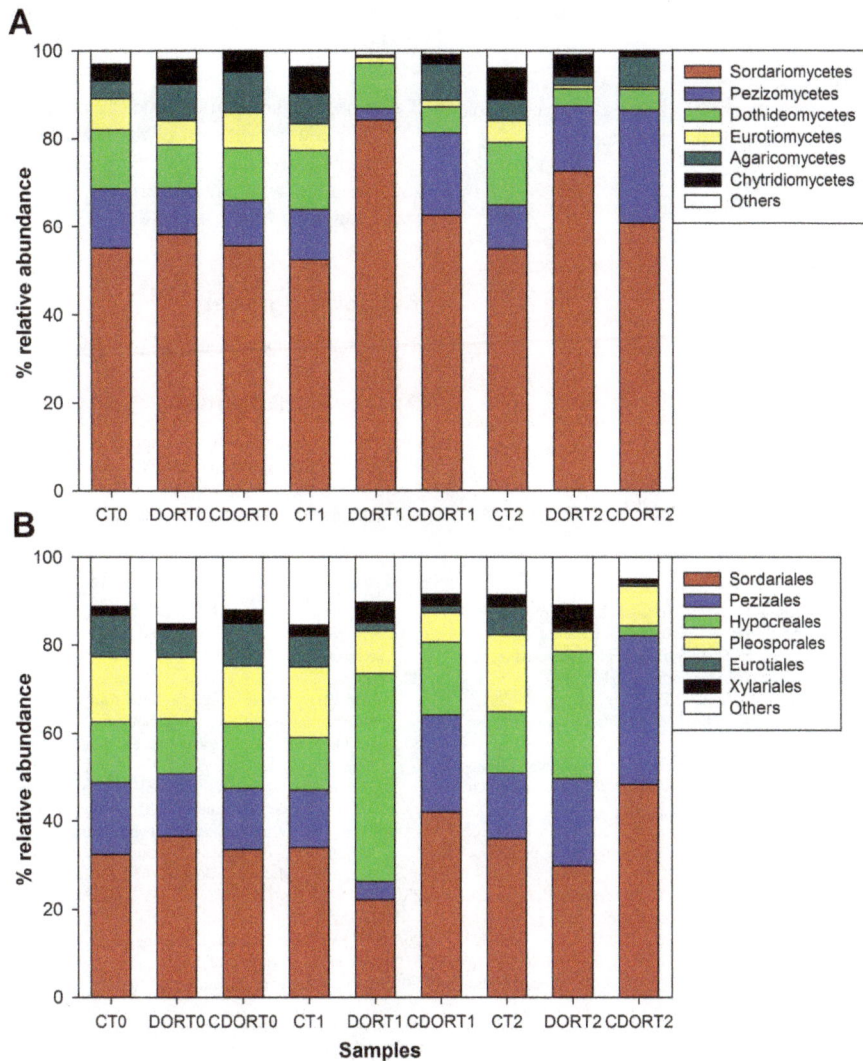

Figure 7. Relative abundance of the different fungal classes (A) and orders (B) found in unamended soil (C) and soil amended with untransformed DOR (DOR) or *C. floccosa*-transformed DOR (CDOR) at 0 (T0), 30 (T1) and 60 (T2) days.

In conclusion, this work has showed that the Mediterranean soil analyzed has an incredible microbial diversity. The application of DOR and CORDOR resulted in a diminution in functional diversity as well as changes in functional community structures depending on the kind of treatment applied given the different types of C sources provided. In addition to its phytotoxicity, DOR was shown to be more disruptive than CORDOR in relation to bacterial and fungal communities as the impact of olive mill wastes on soil microbial communities depends on the relative amounts of beneficial and inhibitory components added, which alter nutrient and toxic substance levels and chemical soil properties. Although a direct link cannot be established, the bacterial (*Rhizobiales*, *Caulobacterales* and *Sphingomonadales*) and fungal (*Fusarium* sp., *Cryptococcus* sp., *Podospora* sp. and *Cercophora* sp.) groups that benefited most from amendments, are probably responsible for changes in soil functionality.

Supporting Information

Figure S1 Bacterial rarefaction curves. Rarefaction curves for bacteria obtained from unamended soil (C) and soil amended with untransformed DOR (DOR) or *C. floccosa*-transformed DOR (CORDOR) at 0 (T0), 30 (T1) and 60 (T2) days.

Figure S2 Fungal rarefaction curves. Rarefaction curves for fungi obtained from unamended soil (C) and soil amended with untransformed DOR (DOR) or *C. floccosa*-transformed DOR (CORDOR) at 0 (T0), 30 (T1) and 60 (T2) days.

Figure S3 Changes in bacterial community mediated by amendments. Relative abundance of the different *Proteobacteria* classes (A), *Acidobacteria* classes (B) and *Actinobacteria* suborders (C) found in unamended soil (C) and soil amended with untransformed DOR (DOR) or *C. floccosa*-transformed DOR (CDOR) at 0 (T0), 30 (T1) and 60 (T2) days.

Figure S4 Changes in fungal community mediated by amendments. Relative abundance of the different fungal phyla found in unamended soil (C) and soil amended with untransformed DOR

(DOR) or *C. floccosa*-transformed DOR (CDOR) at 0 (T0), 30 (T1) and 60 (T2) days.

Table S1 Soil chemical properties. The chemical properties of the soil used in the study (mean±standard deviation).

Table S2 Correlation of carbon sources with principal coordinates. Correlation of carbon sources with the first (PC1) and second principal coordinates (PC2) after principal coordinate analysis (PCoA) of community level physiological profiles (CLPP) from unamended soil and soil amended with untransformed DOR or *C. floccosa*-transformed DOR at 0, 30 and 60 days.

Table S3 Identification and abundance of the major bacterial OTUs. Basic information of the bacterial 14 most abundant OTUs and their relative abundance (percent) in unamended soil (C) and soil amended with untransformed DOR (DOR) or *C.*

floccosa-transformed DOR (CORDOR) at 0 (T0), 30 (T1) and 60 (T2) days.

Table S4 Identification and abundance of the major fungal OTUs. Basic information of the fungal 14 most abundant OTUs and their relative abundance (percent) in unamended soil (C) and soil amended with untransformed DOR (DOR) or *C. floccosa*-transformed DOR (CORDOR) at 0 (T0), 30 (T1) and 60 (T2) days.

Acknowledgments

The authors thank John Quensen for providing comments on the manuscript.

Author Contributions

Conceived and designed the experiments: JAS CTCCR IS IG-R JMT. Performed the experiments: JAS. Analyzed the data: JAS CTCCR. Contributed reagents/materials/analysis tools: IG-R JMT. Contributed to the writing of the manuscript: JAS CTCCR IS IG-R JMT.

References

1. Myers N, Mittermeier RA, Mittermeier CG, da Fonseca GA, Kent J (2000) Biodiversity hotspots for conservation priorities. Nature 403: 853–858.
2. Brooks TM, Mittermeier RA, da Fonseca GA, Gerlach J, Hoffmann M, et al. (2006) Global biodiversity conservation priorities. Science 313: 58–61.
3. Orgiazzi A, Lumini E, Nilsson RH, Girlanda M, Vizzini A, et al. (2012) Unravelling soil fungal communities from different Mediterranean land-use backgrounds. PLoS One 7: e34847.
4. Lozano-García B, Parras-Alcántara L (2013) Short-term effects of olive mill by-products on soil organic carbon, total N, C:N ratio and stratification ratios in a Mediterranean olive grove. Agr Ecosyst Environ 165: 68–73.
5. Morillo JA, Antizar-Ladislao B, Monteoliva-Sánchez M, Ramos-Cormenzana A, Russell NJ (2009) Bioremediation and biovalorisation of olive-mill wastes. Appl Microbiol Biot 82: 25–39.
6. López-Piñeiro A, Albarrán A, Rato Nunes JM, Peña D, Cabrera D (2011) Long-term impacts of de-oiled two-phase olive mill waste on soil chemical properties, enzyme activities and productivity in an olive grove. Soil Till Res 114: 175–182.
7. Tortosa G, Alburquerque JA, Ait-Baddi G, Cegarra J (2012) The production of commercial organic amendments and fertilisers by composting of two-phase olive mill waste ("alperujo"). J Clean Prod 26: 48–55.
8. Sampedro I, Cajthaml T, Marinari S, Petruccioli M, Grego S, et al. (2009) Organic matter transformation and detoxification in dry olive mill residue by the saprophytic fungus *Paecilomyces farinosus*. Process Biochem 44: 216–225.
9. López-Piñeiro A, Albarrán A, Nunes JM, Barreto C (2008) Short and medium-term effects of two-phase olive mill waste application on olive grove production and soil properties under semiarid mediterranean conditions. Bioresource Technol 99: 7982–7987.
10. Diacono M, Montemurro F (2010) Long-term effects of organic amendments on soil fertility. A review. Agron Sustain Dev 30: 401–422.
11. Thangarajan R, Bolan NS, Tian G, Naidu R, Kunhikrishnan A (2013) Role of organic amendment application on greenhouse gas emission from soil. Sci Total Environ 465: 72–96.
12. Ntougias S, Bourtzis K, Tsiamis G (2013) The microbiology of olive mill wastes. Biomed Res Int 2013: 784591.
13. Sampedro I, D'Annibale A, Ocampo JA, Stazi SR, García-Romera I (2005) Bioconversion of olive-mill dry residue by *Fusarium lateritium* and subsequent impact on its phytotoxicity. Chemosphere 60: 1393–1400.
14. Sampedro I, Marinari S, D'Annibale A, Grego S, Ocampo JA, et al. (2007) Organic matter evolution and partial detoxification in two-phase olive mill waste colonized by white-rot fungi. Int Biodeter Biodegr 60: 116–125.
15. Reina R, Liers C, Ocampo JA, García-Romera I, Aranda E (2013) Solid state fermentation of olive mill residues by wood- and dung-dwelling Agaricomycetes: Effects on peroxidase production, biomass development and phenol phytotoxicity. Chemosphere 93: 1406–1412.
16. Kirk JL, Beaudette LA, Hart M, Moutoglis P, Klironomos JN, et al. (2004) Methods of studying soil microbial diversity. J Microbiol Methods 58: 169–188.
17. Trevors JT (2010) One gram of soil: a microbial biochemical gene library. Anton Leeuw Int J G 97: 99–106.
18. Rincon-Florez VA, Carvalhais LC, Schenk PM (2013) Culture-independent molecular tools for soil and rhizosphere microbiology. Diversity 5: 581–612.
19. Ranjard L, Richaume A (2001) Quantitative and qualitative microscale distribution of bacteria in soil. Res Microbiol 152: 707–716.
20. Ritter WF, Scarborough RW (1995) A review of bioremediation of contaminated soils and groundwater. J Environ Sci Heal A 30: 333–357.
21. Sampedro I, Giubilei M, Cajthaml T, Federici E, Federici F, et al. (2009) Short-term impact of dry olive mill residue addition to soil on the resident microbiota. Bioresource Technol 100: 6098–6106.
22. López-Piñeiro A, Albarrán A, Rato Nunes JM, Peña D, Cabrera D (2011) Cumulative and residual effects of two-phase olive mill waste on olive grove production and soil properties. Soil Sci Soc Am J 75: 1061–1069.
23. Siles JA, Pascual J, González-Menéndez V, Sampedro I, García-Romera I, et al. (2014) Short-term dynamics of culturable bacteria in a soil amended with biotransformed dry olive residue. Syst Appl Microbiol 37: 113–120.
24. Siles J, González-Menéndez V, Platas G, Sampedro I, García-Romera I, et al. (2014) Effects of dry olive residue transformed by *Coriolopsis floccosa* (Polyporaceae) on the distribution and dynamic of a culturable fungal soil community. Microbial Ecol 67: 648–658.
25. USDA-NRCS (1996) Soil survey laboratory methods manual. Soil Survey Investigations Report N. 42, Version 3.0. USDA. Washington, DC.
26. Siles JA, Pérez-Mendoza D, Ibáñez JA, Scervino JM, Ocampo JA, et al. (2014) Assessing the impact of biotransformed dry olive residue application to soil: Effects on enzyme activities and fungal community. Int Biodeter Biodegr 89: 15–22.
27. Siles JA, Cajthaml T, Hernández P, Pérez-Mendoza D, García-Romera I, et al. Shifts in soil chemical properties and bacterial communities responding to biotransformed dry olive residue used as organic amendment. Microbial Ecol: Submitted for publication.
28. Nair A, Ngouajio M (2012) Soil microbial biomass, functional microbial diversity, and nematode community structure as affected by cover crops and compost in an organic vegetable production system. Appl Soil Eco 58: 45–55.
29. Insam H, Goberna M, (2004) Use of Biolog for the community level physiological profiling (CLPP) of environmental samples. In: Kowalchuk, GA, de Brujin, FJ, Head, IM, Akkermans, AD, van Elsas, JD, editors. Molecular Microbial Ecology Manual. The Netherlands: Springer, pp. 853–860.
30. Hammer Ø, Harper DAT, Ryan PD (2001) Past: Paleontological statistics software package for education and data analysis. Palaeontol Electron 4: XIX-XX.
31. Rodrigues JLM, Pellizari VH, Mueller R, Baek K, Jesus EDC, et al. (2013) Conversion of the Amazon rainforest to agriculture results in biotic homogenization of soil bacterial communities. Proc Natl Acad Sci USA 110: 988–993.
32. Penton CR, Stlouis D, Cole JR, Luo Y, Wu L, et al. (2013) Fungal diversity in permafrost and tallgrass prairie soils under experimental warming conditions. Appl Environ Microbiol 79: 7063–7072.
33. Schloss PD, Westcott SL, Ryabin T, Hall JR, Hartmann M, et al. (2009) Introducing mothur: Open-source, platform-independent, community-supported software for describing and comparing microbial communities. Appl Environ Microb 75: 7537–7541.
34. Gottel NR, Castro HF, Kerley M, Yang Z, Pelletier DA, et al. (2011) Distinct microbial communities within the endosphere and rhizosphere of *Populus deltoides* roots across contrasting soil types. Appl Environ Microb 77: 5934–5944.
35. Edgar RC, Haas BJ, Clemente JC, Quince C, Knight R (2011) UCHIME improves sensitivity and speed of chimera detection. Bioinformatics 27: 2194–2200.
36. Magurran A (1988) Ecological diversity and its measurement. Princeton, Princeton University Press.

37. Kim O-S, Cho Y-J, Lee K, Yoon S-H, Kim M, et al. (2012) Introducing EzTaxon-e: a prokaryotic 16S rRNA gene sequence database with phylotypes that represent uncultured species. Int J Syst Evol Micr 62: 716–721.

38. Poulsen PHB, Al-Soud WA, Bergmark L, Magid J, Hansen LH, et al. (2013) Effects of fertilization with urban and agricultural organic wastes in a field trial – Prokaryotic diversity investigated by pyrosequencing. Soil Biol Biochem 57: 784–793.

39. Lee OO, Wang Y, Yang J, Lafi FF, Al-Suwailem A, et al. (2011) Pyrosequencing reveals highly diverse and species-specific microbial communities in sponges from the Red Sea. ISME J 5: 650–664.

40. Rachid CT, Santos AL, Piccolo MC, Balieiro FC, Coutinho HL, et al. (2013) Effect of sugarcane burning or green harvest methods on the Brazilian Cerrado soil bacterial community structure. PLoS One 8: e59342.

41. Yuste JC, Barba J, Fernandez-Gonzalez AJ, Fernández-López M, Mattana S, et al. (2012) Changes in soil bacterial community triggered by drought-induced gap succession preceded changes in soil C stocks and quality. Ecol Evol 2: 3016–3031.

42. Bastida F, Hernández T, Albaladejo J, García C (2013) Phylogenetic and functional changes in the microbial community of long-term restored soils under semiarid climate. Soil Biol Biochem 65: 12–21.

43. Curiel Yuste J, Fernandez-Gonzalez AJ, Fernández-López M, Ogaya R, Penuelas J, et al. (2014) Strong functional stability of soil microbial communities under semiarid Mediterranean conditions and subjected to long-term shifts in baseline precipitation. Soil Biol Biochem 69: 223–233.

44. Rousk J, Baath E, Brookes PC, Lauber CL, Lozupone C, et al. (2010) Soil bacterial and fungal communities across a pH gradient in an arable soil. ISME J 4: 1340–1351.

45. Janssen PH (2006) Identifying the dominant soil bacterial taxa in libraries of 16S rRNA and 16S rRNA genes. Appl Environ Microbiol 72: 1719–1728.

46. DeBruyn JM, Nixon LT, Fawaz MN, Johnson AM, Radosevich M (2011) Global biogeography and quantitative seasonal dynamics of *Gemmatimonadetes* in soil. Appl Environ Microb 77: 6295–6300.

47. Zhang L, Xu Z, Patel B (2009) Culture-dependent and culture-independent microbial investigation of pine litters and soil in subtropical Australia. J Soil Sediment 9: 148–160.

48. Shade A, Hogan CS, Klimowicz AK, Linske M, McManus PS, et al. (2012) Culturing captures members of the soil rare biosphere. Environ Microbiol 14: 2247–2252.

49. VanInsberghe D, Hartmann M, Stewart GR, Mohn WW (2013) Isolation of a substantial proportion of forest soil bacterial communities detected via pyrotag sequencing. Appl Environ Microb 79: 2096–2098.

50. Sait M, Hugenholtz P, Janssen PH (2002) Cultivation of globally distributed soil bacteria from phylogenetic lineages previously only detected in cultivation-independent surveys. Environ Microb 4: 654–666.

51. Ramirez KS, Craine JM, Fierer N (2012) Consistent effects of nitrogen amendments on soil microbial communities and processes across biomes. Glob Change Biol 18: 1918–1927.

52. Orgiazzi A, Bianciotto V, Bonfante P, Daghino S, Ghignone S, et al. (2013) 454 pyrosequencing analysis of fungal assemblages from geographically distant, disparate soils reveals spatial patterning and a core mycobiome. Diversity 5: 73–98.

53. Bills GF, Christensen M, Powell M, Thorn G (2004) Saprobic soil fungi. In: Mueller G, Bills GF, Foster MS, editors. Biodiversity of fungi, inventory and monitoring methods. Oxford: Elsevier Academic Press. pp. 271–302.

54. Klaubauf S, Inselsbacher E, Zechmeister-Boltenstern S, Wanek W, Gottsberger R, et al. (2010) Molecular diversity of fungal communities in agricultural soils from Lower Austria. Fungal Divers 44: 65–75.

55. García-Sánchez M, Garrido I, Casimiro IdJ, Casero PJ, Espinosa F, et al. (2012) Defence response of tomato seedlings to oxidative stress induced by phenolic compounds from dry olive mill residue. Chemosphere 89: 708–716.

56. Gioffré D, Cannavò S, Smorto D (2004) Risultati sugli effetti delle acque reflue olearie somministrate con diverse modalità su terreno ulivetato in pieno campo e su piante di olivo allevate in mastello. In: Zimbone SM, editor. Valorizzazione di acque reflue e sottoprodotti dell'industria agrumaria e olearia. Rome: Aracne. pp. 81–98.

57. Mechri B, Issaoui M, Echbili A, Chehab H, Mariem FB, et al. (2009) Olive orchard amended with olive mill wastewater: Effects on olive fruit and olive oil quality. J Hazard Mater 172: 1544–1550.

58. Mechri B, Echbili A, Issaoui M, Braham M, Elhadj SB, et al. (2007) Short-term effects in soil microbial community following agronomic application of olive mill wastewaters in a field of olive trees. Appl Soil Eco 36: 216–223.

59. Karpouzas DG, Rousidou C, Papadopoulou KK, Bekris F, Zervakis GI, et al. (2009) Effect of continuous olive mill wastewater applications, in the presence

60. Ipsilantis I, Karpouzas DG, Papadopoulou KK, Ehaliotis C (2009) Effects of soil application of olive mill wastewaters on the structure and function of the community of arbuscular mycorrhizal fungi. Soil Biol Biochem 41: 2466–2476.

61. Magdich S, Jarboui R, Rouina BB, Boukhris M, Ammar E (2012) A yearly spraying of olive mill wastewater on agricultural soil over six successive years: impact of different application rates on olive production, phenolic compounds, phytotoxicity and microbial counts. Sci Total Environ 430: 209–216.

62. Medina E, Romero C, de Los Santos B, de Castro A, Garcia A, et al. (2011) Antimicrobial activity of olive solutions from stored Alpeorujo against plant pathogenic microorganisms. J Agric Food Chem 59: 6927–6932.

63. Justino CI, Pereira R, Freitas AC, Rocha-Santos TA, Panteleitchouk TS, et al. (2012) Olive oil mill wastewaters before and after treatment: a critical review from the ecotoxicological point of view. Ecotoxicology 21: 615–629.

64. Piotrowska A, Rao MA, Scotti R, Gianfreda L (2011) Changes in soil chemical and biochemical properties following amendment with crude and dephenolized olive mill waste water (OMW). Geoderma 161: 8–17.

65. Frąc M, Oszust K, Lipiec J (2012) Community level physiological profiles (CLPP), characterization and microbial activity of soil amended with dairy sewage sludge. Sensors 12: 3253–3268.

66. Montes-Borrego M, Navas-Cortés JA, Landa BB (2013) Linking microbial functional diversity of olive rhizosphere soil to management systems in commercial orchards in southern Spain. Agr Ecosyst Environ 181: 169–178.

67. King AJ, Freeman KR, McCormick KF, Lynch RC, Lozupone C, et al. (2010) Biogeography and habitat modelling of high-alpine bacteria. Nature Commun 1: 53.

68. Kolvenbach BA, Corvini PFX (2012) The degradation of alkylphenols by Sphingomonas sp. strain TTNP3 - a review on seven years of research. New Biotechnol 30: 88–95.

69. Mahmoudi N, Porter TM, Zimmerman AR, Fulthorpe RR, Kasozi GN, et al. (2013) Rapid degradation of deepwater horizon spilled oil by indigenous microbial communities in louisiana saltmarsh sediments. Environ Sci Technol 47: 13303–13312.

70. Lin Y-T, Tang S-L, Pai C-W, Whitman W, Coleman D, et al. (2014) Changes in the soil bacterial communities in a cedar plantation invaded by moso bamboo. Microbial Ecol 67: 421–429.

71. Fierer N, Jackson JA, Vilgalys R, Jackson RB (2005) Assessment of soil microbial community structure by use of taxon-specific quantitative PCR assays. Appl Environ Microbiol 71: 4117–4120.

72. Mekki A, Dhouib A, Sayadi S (2006) Changes in microbial and soil properties following amendment with treated and untreated olive mill wastewater. Microbiol Res 161: 93–101.

73. Karpouzas DG, Ntougias S, Iskidou E, Rousidou C, Papadopoulou KK, et al. (2010) Olive mill wastewater affects the structure of soil bacterial communities. Appl Soil Eco 45: 101–111.

74. De Gannes V, Eudoxie G, Hickey WJ (2013) Insights into fungal communities in composts revealed by 454-pyrosequencing: Implications for human health and safety. Front Microbiol 13: 164.

75. Doohan FM, Brennan J, Cooke BM (2003) Influence of climatic factors on *Fusarium* species pathogenic to cereals. Eur J Plant Pathol 109: 755–768.

76. Aviani I, Laor Y, Medina S, Krassnovsky A, Raviv M (2010) Co-composting of solid and liquid olive mill wastes: Management aspects and the horticultural value of the resulting composts. Bioresource Technol 101: 6699–6706.

77. Soytong K, Kanokmedhakul S, Kukongviriyapa V, Isobe M (2001) Application of *Chaetomium* species (Ketomium) as a new broad spectrum biological fungicide for plant disease control: A review article. Fungal Divers 7: 1–15.

78. Asiegbu PO, Paterson A, Smith JE (1996) Inhibition of cellulose saccharification and glycolignin-attacking enzymes of five lignocellulose-degrading fungi by ferulic acid. World J Microb Biot 12: 16–21.

79. Zaccardelli M, De Nicola F, Villecco D, Scotti R (2013) The development and suppressive activity of soil microbial communities under compost amendment. J Plant Nutr Soil Sci 13: 730–742.

80. Chang JH, Kao HW, Wang YZ (2010) Molecular phylogeny of Cercophora, Podospora, and Schizothecium (Lasiosphaeriaceae, Pyrenomycetes). Taiwania 55: 110–116.

81. Bonito G, Isikhuemhen OS, Vilgalys R (2010) Identification of fungi associated with municipal compost using DNA-based techniques. Bioresource Technol 101: 1021–1027.

82. Souza RC, Cantão ME, Vasconcelos ATR, Nogueira MA, Hungria M (2013) Soil metagenomics reveals differences under conventional and no-tillage with crop rotation or succession. Appl Soil Eco 72: 49–61.

11

A 2-Year Field Study Shows Little Evidence That the Long-Term Planting of Transgenic Insect-Resistant Cotton Affects the Community Structure of Soil Nematodes

Xiaogang Li[1,2], Biao Liu[1]*

1 Nanjing Institute of Environmental Sciences, Ministry of Environmental Protection of China, Nanjing, Jiangsu Province, China, 2 Key Laboratory of Soil Environment and Pollution Remediation, Institute of Soil Science, Chinese Academy of Sciences, Nanjing, Jiangsu Province, China

Abstract

Transgenic insect-resistant cotton has been released into the environment for more than a decade in China to effectively control the cotton bollworm (*Helicoverpa armigera*) and other Lepidoptera. Because of concerns about undesirable ecological side-effects of transgenic crops, it is important to monitor the potential environmental impact of transgenic insect-resistant cotton after commercial release. Our 2-year study included 1 cotton field where non-transgenic cotton had been planted continuously and 2 other cotton fields where transgenic insect-resistant cotton had been planted for different lengths of time since 1997 and since 2002. In 2 consecutive years (2009 and 2010), we took soil samples from 3 cotton fields at 4 different growth stages (seedling, budding, boll-forming and boll-opening stages), collected soil nematodes from soil with the sugar flotation and centrifugation method and identified the soil nematodes to the genus level. The generic composition, individual densities and diversity indices of the soil nematodes did not differ significantly between the 2 transgenic cotton fields and the non-transgenic cotton field, but significant seasonal variation was found in the individual densities of the principal trophic groups and in the diversity indices of the nematodes in all 3 cotton fields. The study used a comparative perspective to monitor the impact of transgenic insect-resistant cotton grown in typical 'real world' conditions. The results of the study suggested that more than 10 years of cultivation of transgenic insect-resistant cotton had no significant effects–adverse or otherwise–on soil nematodes. This study provides a theoretical basis for ongoing environmental impact monitoring of transgenic plants.

Editor: Nicolas Desneux, French National Institute for Agricultural Research (INRA), France

Funding: This research was jointly supported by the National Special Transgenic Project (2011ZX08012-005; 2013ZX08012-005), the National Science and Technology Pillar Program (No. 2013BAD11B01), the Jiangsu Province Science Foundation for Youths (BK2012498). The funders had no role in study design, data collection and analysis, decision to publish, or preparation of the manuscript.

Competing Interests: The authors have declared that no competing interests exist.

* E-mail: 85287064@163.com

Introduction

Many crops have been transformed to provide enhanced resistance against pests and diseases. Crops expressing δ-endotoxins of *Bacillus thuringiensis* (*Bt*) active against Lepidopteran and Coleopteran insect pests are the most widely grown [1]. The global area of transgenic crops has increased from 1.7 million hectares in 1996 to 160 million hectares in 2011 [1]. Transgenic insect-resistant cotton expressing Cry1Ab/c and/or CpTI (Cowpea Trypsin Inhibitor) has been released into the environment for commercial cultivation for more than a decade in China. Its planted area currently represents 71.5% of the total cotton grown in China [1]. These lines effectively control cotton bollworm and other Lepidoptera, resulting in a significant reduction in the usage of chemical insecticides, thus protecting the environment and human health while yielding substantial socioeconomic benefits [2–4]. Nevertheless, as with any technology, there have been questions about the potential environmental risks associated with transgenic plants. One of the major ecological concerns about the

environmental risks of transgenic insect-resistant plants is the potential effects of these plants on non-target organisms [5–7].

The effects of transgenic insect-resistant cotton on non-target pests and natural enemies have been extensively assessed [8,9]. Most studies have found no convincing and meaningful negative effects of transgenic insect-resistant cotton on the population density, abundance, species richness and diversity of non-target arthropod natural enemies [9–13]. Pollinators are also important non-target organisms, and the impacts of transgenic crops on these organisms have also been evaluated. Feeding tests, as well as field surveys, have been extensively performed to evaluate the safety of *Bt* plants for honey bees or pollinating beetles, and no significant adverse effects on longevity, feeding and learning behavior, the development of the hypopharyngeal glands or superoxide dimutase activity have been observed in these insects [14–21].

Transgenic insect-resistant cotton could also affect soil organisms. Transgenic proteins, such as Cry1Ab and Cry1Ac, can be released into the soil from cotton residues, root exudates and pollen during growth and after harvest [22,23]. Once in the soil, the toxins can be bound to clay and humus particles [24]. This state protects them from biodegradation and preserves their insecticidal activity [25], and it may pose a potential, inadvertent risk to soil-dwelling organisms [5,26]. Free-living soil nematodes are the most abundant and species-rich metazoan group in soils [27]. Nematodes are useful indicators of soil quality because of their great diversity and participation in many functions at different levels of the food webs in the soil. They are relatively simple to separate and enumerate. Their populations, in contrast to those of bacteria, are stable in response to changes in soil moisture and temperature [28]. Impacts on soil nematodes are, therefore, an important aspect of the environmental risk assessment and post-release monitoring of transgenic insect-resistant plants.

Certain *Bt* toxins, e.g., Cry5B, Cry6A, Cry14A and Cry21A, have been found to have direct toxic effects on some nematode species [29]. Nevertheless, Cry1Ac and CpTI expressed simultaneously by transgenic crops have not been evaluated for their effects on nematodes. To date, studies on the effects of transgenic insect-resistant plants, such as those expressing Cry1Ab, on soil nematodes have produced contrasting results. Negative effects of Cry1Ab protein on the growth, number of eggs and reproduction of soil nematodes have been detected in the rhizosphere soil of transgenic Cry1Ab maize [30]. However, the results of laboratory and field studies have generally shown no consistent effects of transgenic insect-resistant plants on soil nematodes [31–34]. For example, no significant differences in the numbers, communities and biodiversity of nematodes have been found in the soil of maize expressing the Cry1Ab protein relative to non-*Bt* maize [31,32]. A significant but transient decrease in the numbers of nematodes in soil under *Bt* maize expressing the Cry1Ab protein at 3 different field sites was found in a comparison with non-*Bt* maize, whereas studies conducted in a greenhouse showed no toxic effects of the Cry1Ab protein on populations of nematodes [35,36]. The reasons for the differences between the 2 studies are unclear, but they may have resulted from different environmental conditions in the greenhouse and the field, which could affect the interactions between plants and soil organisms.

As the results of these studies only reflect short histories of transgenic cultivation, we still face considerable gaps in our scientific understanding of longer-term community-level impacts on soil nematodes from the cultivation of transgenic insect-resistant crops [37]. The biosafety regulations for genetically modified organisms (GMOs) in many jurisdictions, including both China and the European Union (EU), require monitoring of environmental impacts after the environmental release and commercial cultivation of transgenic crops [38,39]. Transgenic insect-resistant cottons have now been planted for more than a decade in China, a nation prominent in pioneering the use of this new technology. The cotton fields of China therefore offer a valuable opportunity to address scientific questions of longer-term

impacts and to fulfill the ongoing demands of biosafety regulations. The purpose of this study was to investigate the population density and community structure of soil nematodes over 2 years, comparing the soil nematodes of a conventional non-transgenic cotton plantation with those from soils that have been planted with transgenic cotton expressing Cry1Ab/c and CpTI for up to 10 years, and to provide a theoretical basis for environmental impact monitoring of transgenic plants.

Materials and Methods

Plant Material and Field Trial

This study was conducted in fields at a cotton farm in Baibi town, Anyang, Henan Province, China. The farm belongs to the Cotton Research Institute (CRI) of the Chinese Academy of Agricultural Sciences (CAAS). This field site is in the North Temperate Zone and has a continental monsoon climate. The annual mean temperature is 13.6 °C, and the annual mean precipitation is 606.1 mm. Three types of cotton fields were selected for this study (Table 1). Field T-1 was originally planted with non-transgenic cotton, then sown beginning in 2002 with the transgenic insect-resistant cotton line Zhong-41 expressing Cry1Ab/c and CpTI. Field T-2 was planted with 2 transgenic Cry1Ab/c cotton lines, Zhong-29 and Zhong-30, from 1999 to 2001 and subsequently planted with Zhong-41 beginning in 2002. Field CK has been planted with the conventional non-transgenic cotton line Zhong-35 since 1999. Zhong-29 and Zhong-30 were developed by CRI, CAAS, and approved by the Ministry of Agriculture (MOA) of China in 1998. Zhong-41 was developed jointly by CRI and the Biotechnology Research Institute of CAAS and approved by the MOA in 2002 [40]. The content of Cry1Ab/c expressed in the leaves of Zhong-41, Zhong-29 and Zhong-30 was determined, and the results indicated that the transgene expression remained stable during the survey period (Table 1).

The 3 cotton fields monitored were all located at 36° 7′ N and 116° 22′ E and were composed of a cambisol-type soil (FAO (1998) classification) with the following properties (on a dry mass basis): pH (soil: water ratio 1:2.5) 7.82, organic C 16.10 g•kg^{-1}, total N 0.84 g•kg^{-1}, total P 0.85 g•kg^{-1}, total K 7.61 g•kg^{-1}, available P 26.62 mg•kg^{-1}, available K 134.47 mg•kg^{-1} and soil clay (<0.002 mm) 9.29%. The fields CK, T-1 and T-2 were distributed side by side from south to north and separated by belts 50 m wide. The cotton growing season extended from April to November annually. The agricultural practice for the cotton in the 3 fields was the same as that used for conventional cottons. Fertilizer was applied at the seedling stage and at the budding stage. During the growing seasons, chemical pesticides were used for pest control as necessary (Fig. 1). In addition, all 3 fields lay fallow from November to the following April.

Table 1. Planting information for 3 cotton fields.

Treatments (Cotton fields)	Cotton lines	Transgenes	Contents of Cry1Ab/c (ng •g^{-1} fresh tissue)	Planting time
CK	Zhong-35	–	–	1999 to 2010
T-1	Zhong-41	Cry1Ab/c[#] & CpTI	150–600	2002 to 2010
T-2	Zhong-29 & Zhong-30	Cry1Ab/c	200–550	1999 to 2001
	Zhong-41	Cry1Ab/c & CpTI	150–600	2002 to 2010

[#]Cry1Ab/c represents a fusion gene of Cry1Ac and Cry1Ab.

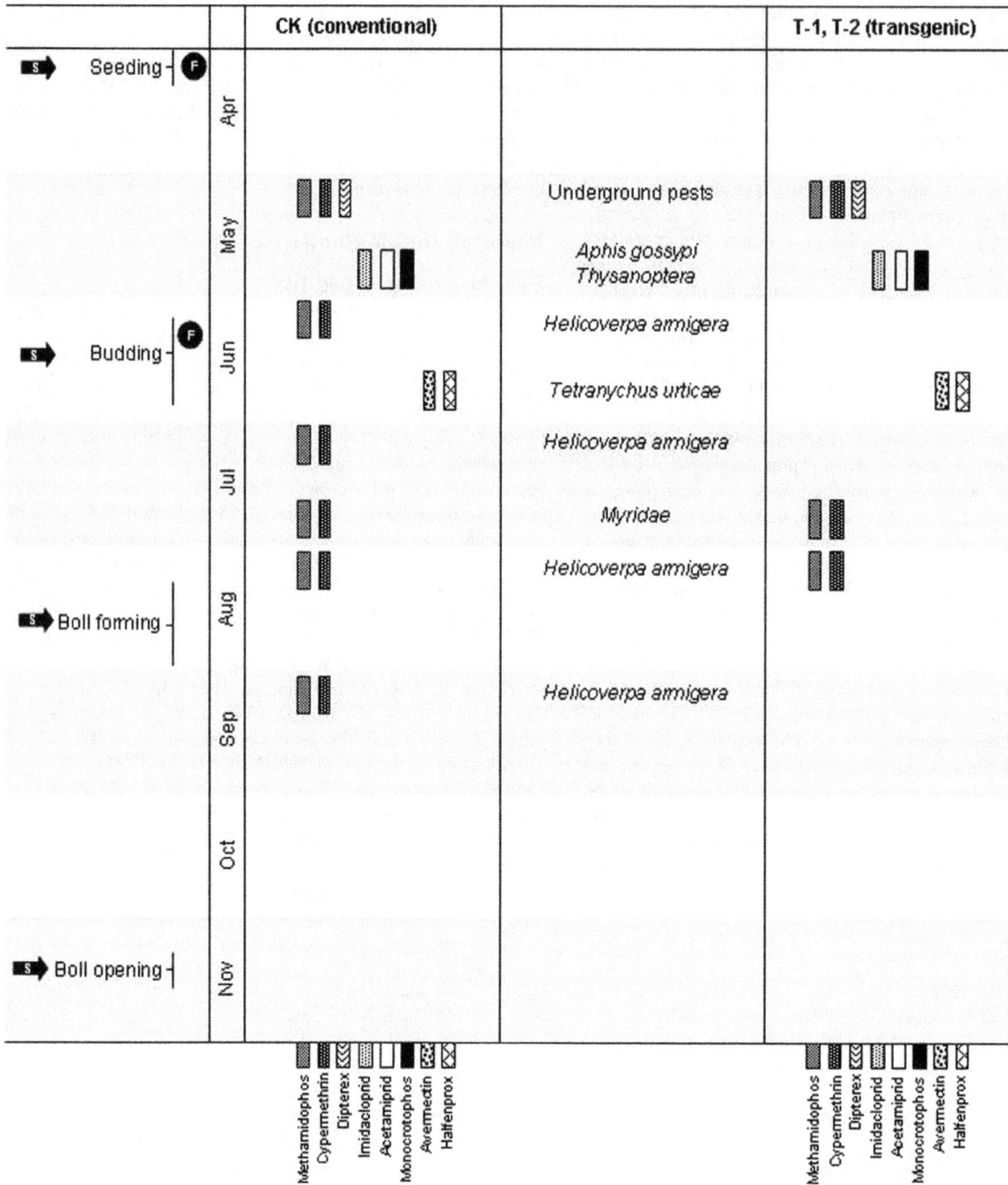

Figure 1. Schematic of agricultural practices applied to the 3 cotton fields during this study (2007–2010). Patterned bars represent different pesticides, with target pest listed in central column. Black circles with 'F' indicate fertilizer application, and black arrows with 'S' represent sampling times. Symbol placement is indicative of timing but is not precise.

Soil Sample Collection

Soil samples were collected 4 times per year from 2009 to 2010, inclusive, coinciding with the major growth stages of cotton, namely, seedling (April), budding (June), boll-forming (August) and boll-opening (November). Fifteen meters were left at both ends of every treatment to eliminate marginal field effects on soil sampling. Each type of treatment field was established in triplicate, and the plot size for each replication was 0.17 hectare. One soil sample was collected between 0 and 20 cm deep with 5 cores using a soil auger with a 4 cm diameter and then placed in a sterile plastic bag. Three soil samples were taken in each replication according to the checkerboard method [41]. The soil samples were immediately transported to the laboratory to isolate the soil nematode specimens. The data from 3 soil samples in each replication were pooled, and 3 replicates from each field were used in further statistical analyses.

Extraction and Identification of Nematodes

Nematodes were extracted from fresh soil equivalent to 100 g dried soil with a sieving process followed by sugar flotation [42].

Table 2. Composition and abundance (individual/100 g dry soil) of the soil nematodes extracted from different samples from 3 cotton fields during 2009–2010. P_i denotes the proportion of soil nematode individuals in each treatment.

Trophic groups	Genus	Treatment	2009				2010				Total	Percentage of total	Average abundance
			Seedling	Budding	Boll forming	Boll opening	Seedling	Budding	Boll forming	Boll opening			
Plant parasites	Tylenchus	CK	0.31	0.71	1.30	2.40	1.85	0.51	1.10	1.99	10.17	4.2%	
		T-1	0.00	0.34	2.02	2.10	1.47	0.29	1.39	1.54	9.15	3.7%	4.0%
		T-2	0.54	0.00	2.29	1.93	1.47	0.51	0.69	1.73	9.17	3.9%	
	Filenchus	CK	2.90	1.47	2.72	1.95	3.08	1.30	3.16	3.28	19.86	8.2%	
		T-1	2.43	1.41	1.55	2.99	2.93	1.67	3.22	3.12	19.32	7.9%	8.2%
		T-2	2.86	1.49	2.78	2.48	3.12	1.54	2.71	3.04	20.01	8.5%	
	Psilenchus	CK	0.75	0.51	0.00	0.00	0.00	0.00	0.00	0.00	1.26	0.5%	
		T-1	1.66	0.57	0.00	0.00	0.00	0.29	0.29	0.00	2.80	1.1%	1.1%
		T-2	0.00	0.31	0.59	0.96	0.98	0.00	0.29	0.85	3.97	1.7%	
	Helicotylenchus	CK	1.95	1.55	4.28	3.99	1.95	2.48	1.67	2.66	20.54	8.5%	
		T-1	2.39	2.08	4.22	4.12	1.99	2.37	2.04	2.30	21.51	8.8%	8.6%
		T-2	2.68	1.56	3.82	4.29	1.67	2.27	1.79	2.12	20.20	8.6%	
	Pratylenchus	CK	2.80	1.81	2.91	2.17	1.54	1.54	1.20	2.20	16.17	6.7%	
		T-1	2.25	1.20	2.77	2.21	1.39	1.39	1.39	2.20	14.79	6.0%	6.4%
		T-2	2.94	1.81	2.65	1.93	1.10	1.39	1.30	1.90	15.01	6.4%	
	Paratylenchus	CK	1.52	1.21	0.00	1.72	0.00	0.00	0.51	0.00	4.96	2.1%	
		T-1	2.67	1.41	1.13	1.99	0.00	0.00	0.69	0.00	7.90	3.2%	2.9%
		T-2	2.43	1.10	1.94	2.15	0.69	0.00	0.00	0.00	8.31	3.5%	
	Longidorella	CK	0.29	0.98	0.55	1.25	0.85	0.00	0.00	0.00	3.92	1.6%	
		T-1	0.00	0.93	1.05	1.48	0.69	0.00	0.51	0.29	4.95	2.0%	1.7%
		T-2	0.30	0.31	0.54	0.31	0.98	0.00	0.29	0.98	3.71	1.6%	
Bacterivores	Mesorhabditis	CK	1.66	1.47	2.74	2.61	0.85	0.85	0.29	0.85	11.31	4.7%	
		T-1	2.33	1.33	2.78	2.59	1.20	0.00	0.69	0.00	10.92	4.5%	4.9%
		T-2	2.18	1.47	2.76	2.88	1.30	0.29	0.51	1.47	12.86	5.5%	
	Protorhabditis	CK	2.09	1.00	1.89	3.32	1.61	1.47	2.20	1.95	15.53	6.4%	
		T-1	2.05	0.93	2.60	2.49	1.61	1.30	1.95	1.73	14.65	6.0%	6.2%
		T-2	1.80	1.12	1.70	2.74	1.90	1.85	1.39	1.67	14.16	6.0%	
	Eucephalobus	CK	0.55	0.98	1.09	1.17	1.20	0.85	1.47	0.69	8.01	3.3%	
		T-1	0.30	2.37	1.66	2.31	1.10	1.30	1.39	1.20	11.64	4.8%	3.6%
		T-2	0.00	0.72	0.35	0.41	0.69	1.39	1.20	1.67	6.43	2.7%	
	Heterocephalobus	CK	1.60	1.62	1.69	0.69	2.77	2.48	1.61	2.16	14.62	6.1%	
		T-1	2.25	1.83	0.42	0.69	2.54	2.46	1.67	2.16	14.04	5.7%	5.8%

Table 2. Cont.

Trophic groups	Genus	Treat-ment	2009				2010				Total	Percentage of total	Average abundance
			Seedling	Budding	Boll forming	Boll opening	Seedling	Budding	Boll forming	Boll opening			
		T-2	1.35	1.43	0.00	1.11	2.59	2.51	1.99	1.99	12.97	5.5%	
	Acrobeles	CK	1.94	0.87	1.30	0.81	0.29	0.51	0.00	0.00	5.72	2.4%	2.0%
		T-1	1.25	0.82	1.99	1.04	0.00	0.00	0.29	0.00	5.38	2.2%	
		T-2	1.44	0.29	0.35	0.00	0.51	0.00	0.51	0.00	3.10	1.3%	
	Acrobeloides	CK	2.85	2.83	2.35	2.47	2.75	2.40	2.98	3.28	21.92	9.1%	9.0%
		T-1	2.89	3.32	0.42	2.54	2.85	2.34	3.19	3.39	20.94	8.6%	
		T-2	2.68	2.54	1.95	2.52	2.83	2.30	3.26	3.50	21.58	9.2%	
	Panagrolaimus	CK	0.29	0.30	1.31	1.25	0.00	0.00	0.00	0.51	3.66	1.5%	1.5%
		T-1	0.53	0.34	1.44	0.69	0.00	0.00	0.29	0.29	3.58	1.5%	
		T-2	0.00	0.87	0.00	0.00	0.51	0.29	0.51	1.20	3.38	1.4%	
	Plectus	CK	0.72	0.87	1.09	0.00	0.00	0.29	0.00	0.00	2.97	1.2%	1.5%
		T-1	1.14	0.95	1.06	0.69	0.00	0.00	0.51	0.00	4.35	1.8%	
		T-2	1.16	0.87	0.59	0.00	0.51	0.00	0.29	0.00	3.42	1.5%	
	Chronogaster	CK	0.31	0.29	0.00	1.08	0.00	0.00	0.51	0.00	2.19	0.9%	1.1%
		T-1	0.00	0.31	1.06	0.69	0.29	0.85	0.69	0.00	3.89	1.6%	
		T-2	0.30	0.00	0.00	0.00	0.00	0.69	0.29	0.69	1.97	0.8%	
	Pseudoaulolaimus	CK	0.29	0.00	0.00	0.00	1.73	2.12	1.73	0.85	6.73	2.8%	2.5%
		T-1	0.30	0.00	0.86	0.00	0.98	1.85	1.39	0.29	5.66	2.3%	
		T-2	0.00	0.00	0.00	0.00	1.10	1.73	1.47	0.98	5.28	2.3%	
	Alaimus	CK	1.25	0.87	1.39	1.90	2.54	1.54	1.10	1.61	12.20	5.1%	4.9%
		T-1	1.43	1.20	1.05	1.92	2.54	1.47	1.30	0.98	11.90	4.9%	
Fungivores		T-2	1.44	1.32	0.57	0.92	2.12	1.47	1.73	1.30	10.88	4.6%	
	Ditylenchus	CK	0.74	1.21	1.09	0.00	0.51	0.69	0.85	0.00	5.10	2.1%	2.4%
		T-1	0.73	1.17	0.74	1.96	0.29	0.51	1.20	0.00	6.61	2.7%	
		T-2	0.54	0.00	0.99	2.18	0.51	0.00	1.20	0.29	5.71	2.4%	
	Aphelenchus	CK	1.14	1.39	1.96	1.79	0.00	0.00	0.00	0.00	6.27	2.6%	3.3%
		T-1	1.85	2.01	1.91	1.24	0.00	1.10	0.00	0.00	8.10	3.3%	
		T-2	2.05	2.74	2.18	1.63	0.00	0.85	0.00	0.00	9.46	4.0%	
	Aphelenchoides	CK	2.79	2.34	1.30	1.75	1.67	1.14	0.98	0.69	12.68	5.3%	4.9%
		T-1	2.60	2.84	1.28	1.55	1.47	0.29	0.29	0.29	10.59	4.3%	
		T-2	2.40	3.23	1.40	2.38	1.10	0.85	0.00	0.29	11.65	5.0%	
	Dorylaimoides	CK	0.87	0.00	0.80	0.69	0.29	0.00	0.00	0.00	2.64	1.1%	1.0%
		T-1	0.88	0.00	0.00	0.00	0.00	0.00	0.29	0.00	1.17	0.5%	

Table 2. Cont.

Trophic groups	Genus	Treat-ment	2009				2010				Total	Percentage of total	Average abundance
			Seedling	Budding	Boll forming	Boll opening	Seedling	Budding	Boll forming	Boll opening			
Predators		T-2	1.79	0.69	0.81	0.00	0.00	0.00	0.00	0.00	3.30	1.4%	
	Mesodorylaimus	CK	0.29	0.00	1.24	1.09	0.00	0.00	0.29	0.00	2.91	1.2%	
		T-1	0.72	0.00	1.31	0.00	0.29	0.29	0.98	0.00	3.59	1.5%	1.2%
		T-2	0.55	0.00	0.00	0.00	0.29	0.29	0.51	0.69	2.33	1.0%	
	Thonus	CK	0.31	1.54	1.54	1.38	0.85	0.29	0.98	1.47	8.36	3.5%	
		T-1	0.31	2.30	0.71	1.46	0.69	0.51	0.85	1.10	7.92	3.2%	3.3%
		T-2	0.72	1.42	1.72	1.09	0.29	0.98	1.10	0.51	7.83	3.3%	
Omnivores	Eudorylaimus	CK	0.31	0.00	1.09	0.85	0.00	0.00	0.00	0.00	2.25	0.9%	
		T-1	0.00	0.00	0.76	1.19	0.51	0.00	0.29	0.00	2.75	1.1%	0.9%
		T-2	0.31	0.29	0.54	0.00	0.29	0.00	0.00	0.29	1.72	0.7%	
	Epidorylaimus	CK	1.96	1.31	0.00	0.00	0.69	0.29	1.47	1.39	7.11	2.9%	
		T-1	1.52	0.34	0.00	0.00	0.29	1.10	1.30	0.98	5.52	2.3%	2.5%
		T-2	1.85	0.69	0.00	0.00	0.85	0.85	1.10	0.29	5.63	2.4%	
	Microdorylaimus	CK	0.72	0.86	1.54	0.00	2.04	1.10	1.39	1.67	9.32	3.9%	
		T-1	0.54	1.62	0.41	1.50	1.95	0.98	0.00	1.54	8.54	3.5%	3.5%
		T-2	0.72	1.12	0.59	0.00	1.73	0.98	0.69	1.20	7.05	3.0%	
	Aporcelaimellus	CK	0.26	0.14	0.49	0.42	6.06	0.36	0.20	0.69	0.60	1.1%	
		T-1	0.41	0.00	0.49	0.35	4.72	0.15	0.00	0.69	0.49	1.0%	1.2%
		T-2	0.29	0.42	0.68	0.35	7.35	0.01	0.60	0.43	0.09	1.5%	

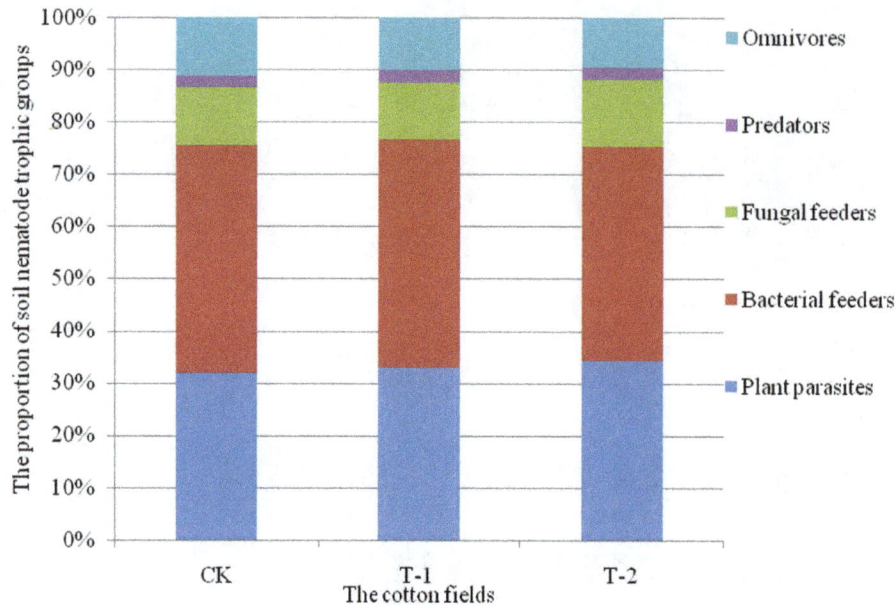

Figure 2. The soil nematode composition in 3 cotton fields during 2009–2010.

The nematodes were heat killed and fixed in 4% formaldehyde. They were then counted under a dissecting microscope at 25× magnification. A total of 100 specimens per sample were then randomly selected and identified to the genus level, as described in Liang et al. (2009), at 200× magnification using an inverted compound microscope [43].

Nematode Community Analysis

Nematode abundances were ln (x +1) transformed prior to statistical analysis and expressed as numbers per 100 g dry soil. Nematode biodiversity was measured with the Simpson and Shannon–Wiener diversity indices. Both indices are sensitive to the abundance of the most common/dominant species in a population [44]. Simpson's index is defined by the equation $C = 1 - \sum_{i=1} (P_i)^2$.

Table 3. Generalized Linear Mixed Model (GLMM) results for overall effects on the numbers of soil nematodes.

Effects	Degrees of freedom	F Value	p Value
Total soil nematodes			
Treatments	2;14	0.69	0.52
Sampling time	7;14	24.53	0.00
Filenchus			
Treatments	2;14	0.12	0.88
Sampling time	7;14	9.98	0.01
Helicotylenchus			
Treatments	2;14	1.00	0.39
Sampling time	7;14	49.47	0.00
Acrobeloides			
Treatments	2;14	0.18	0. 84
Sampling time	7;14	5.34	0.00

The Shannon–Wiener diversity index is defined by the equation $H = - \sum_{i=1} P_i \ln P_i$. In these equations, P_i denotes the proportion of soil nematode individuals in each treatment; $P_i = N_i/N$, where N_i is the abundance of the ith species and N is the overall total abundance in each treatment. $P_i \geq 10\%$ represents dominant groups, $10\% > P_i \geq 1\%$ represents common groups and $P_i < 1\%$ represents rare groups.

Nematode taxa were ranked along a colonizer–persister (c–p) scale of 1–5 according to Bongers and Ferris (1999) [45]. The Maturity Index (MI) was calculated using the equation of Bongers (1990) [46]: $MI = - \sum_{i=1}^{n} cp_i \times P_i$, where P_i is the frequency of the taxon in the sample and cp_i is the c-p value of taxon i.

Nematode taxa were classified into 5 main trophic groups: bacterial feeders, fungal feeders, plant parasites, omnivores and predators [47]. Based on the c–p and feeding type classification, nematode taxa were also categorized in functional guilds according to Ferris *et al.* (2001): Ba_n, Fu_n, Ca_n, and $Om_n =$ bacterial feeders, fungal feeders, predators, and omnivores, respectively, with n = c–p value [48]. The following indices were calculated to describe the enrichment and structure conditions as well as the predominant decomposition channels in the soil food webs: Enrichment index: $EI = 100 \times \frac{e}{e+b}$, Structure index: $SI = 100 \times \frac{s}{s+b}$, Channel index: $CI = 100(0.8 \ Fu_2)/(3.2 \ Ba_1 + 0.8 \ Fu_2)$, where $b = (Ba_2 + Fu_2) \times 0.8$, $e = Ba_1 \times 3.2 + Fu_2 \times 0.8$, and $s = Ca_2 \times 0.8 + (Ba_3 + Ca_3 + Fu_3 + Om_3) \times 1.8 + (Ba_4 + Ca_4 + Fu_4 + Om_4) \times 3.2 + (Ba_5 + Ca_5 + Fu_5 + Om_5) \times 5$.

The response of the soil nematode community to the factors 'treatment' and 'sampling time' was examined with a 2-way ANOVA (Proc GLM). Significance was measured at the alpha = 0.05 level. Principal component analysis (PCA), a repeated-measures multivariate ordination analysis, was performed to identify the influence of treatment and sampling time on community structure (SPSS 13.0 for Windows). The principal components whose eigenvalue exceeded 1 were selected for the analysis.

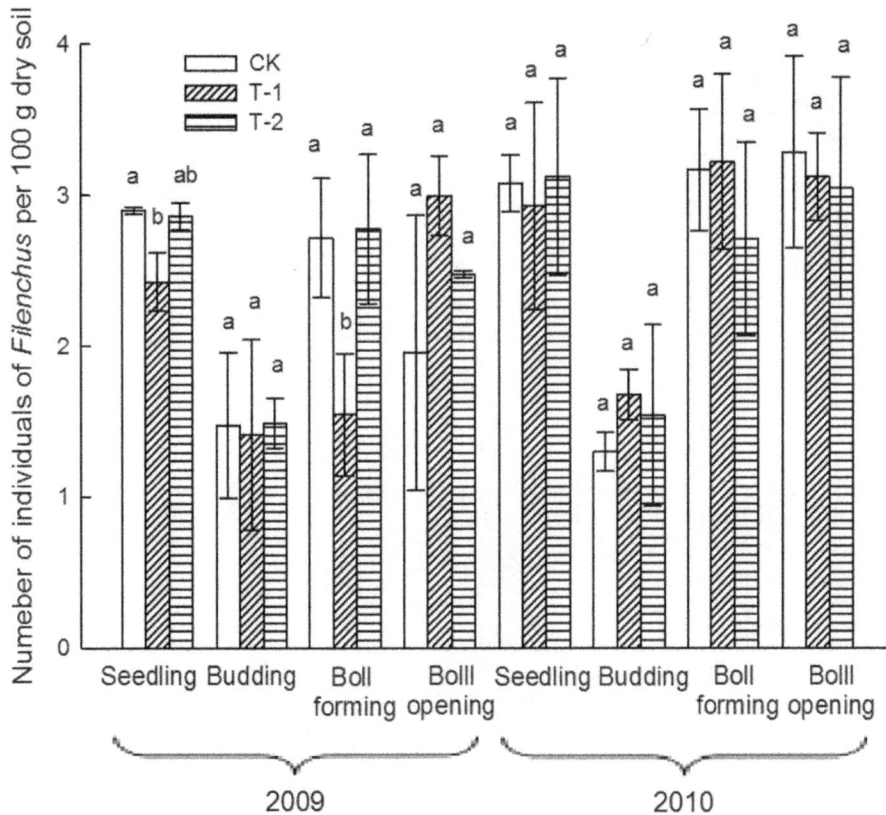

Figure 3. Number of *Filenchus* in 3 cotton fields at different sampling times. Error bars indicate standard errors (n = 3). Different letters above bars denote a statistically significant difference between the means of the fields.

Results

The Composition of the Soil Nematode Community in the 3 Cotton Fields

Twenty-eight genera of soil nematodes were identified in the 3 cotton fields during different cotton growth stages over 2 years (Table 2). The overall results showed that the most abundant common groups in CK, T-1 and T-2 were *Helicotylenchus*, *Filenchus* and *Acrobeloides*. Most of the genera collected from the soil, such as *Tylenchus*, *Pratylenchus*, *Paratylenchus*, *Mesorhabditis*, *Protorhabditis*, *Eucephalobus*, *Heterocephalobus*, *Acrobeles*, *Pseudoaulolaimus*, *Alaimus*, *Ditylenchus*, *Aphelenchus*, *Aphelenchoides*, *Thonus*, *Epidorylaimus* and *Microdorylaimus*, represented common groups. *Eudorylaimus* was a rare group. Over the 2-year field period, the composition of soil nematode communities was essentially uniform in the transgenic insect-resistant cotton fields and the non-transgenic cotton field (Fig. 2). In brief, the soil nematode communities in the 3 fields did not differ significantly.

Effect of Transgenic Insect-resistant Cotton on the Number of Soil Nematodes

Due to the particularly arid conditions occurring throughout 2010, the number of total soil nematodes for the 4 sampling times in 2010 was obviously less than that for the 2009 samples (Table 2). During the 2-year sampling period, the number of total soil nematodes and the most abundant nematodes, such as *Filenchus*, *Helicotylenchus* and *Acrobeloides*, in each cotton field varied significantly among different sampling times (i.e., different growth stages of the plants; $p < 0.01$; Table 3) but did not differ significantly

overall among the 3 cotton fields ($p > 0.05$; Table 3). At the seedling and boll-forming stages in 2009, the number of *Filenchus* in T-1 was significantly lower than that in CK ($p < 0.05$; Fig. 3). At the seedling stage in 2009, the number of *Helicotylenchus* in T-2 was significantly greater than that in CK ($p < 0.05$; Fig. 4). At boll-forming stage in 2009, the number of *Acrobeloides* in T-1 was significantly lower than that in CK and T-2 ($p < 0.05$; Fig. 5). For the other sampling times, the numbers of *Filenchus*, *Helicotylenchus* and *Acrobeloides* did not vary significantly among the 3 fields ($p > 0.05$; Figs. 3–5).

Effect of Transgenic Insect-resistant Cotton on Soil Nematode Trophic Groups

In all, 26–57% of the nematodes were bacterial feeders, 24–47% plant parasites, 1–25% fungal feeders, 3–15% omnivores and 1–8% predators. The feeding-type composition was relatively constant throughout the 2-year survey, with average proportions of 41–44% bacterial feeders, 32–34% plant parasites, 11–13% fungal feeders, 9–11% omnivores and 2% predators (Table 4). These values showed no significant differences in feeding-type composition among the 3 cotton fields at different sampling times ($p > 0.05$; one-way ANOVA). During the 2 years (2009 and 2010), the proportions of bacterial feeders, fungal feeders, plant parasites, omnivores and predators in each cotton field varied significantly among sampling times ($p < 0.05$; Table 5) but did not differ significantly overall among the 3 cotton fields ($p > 0.05$; Table 5).

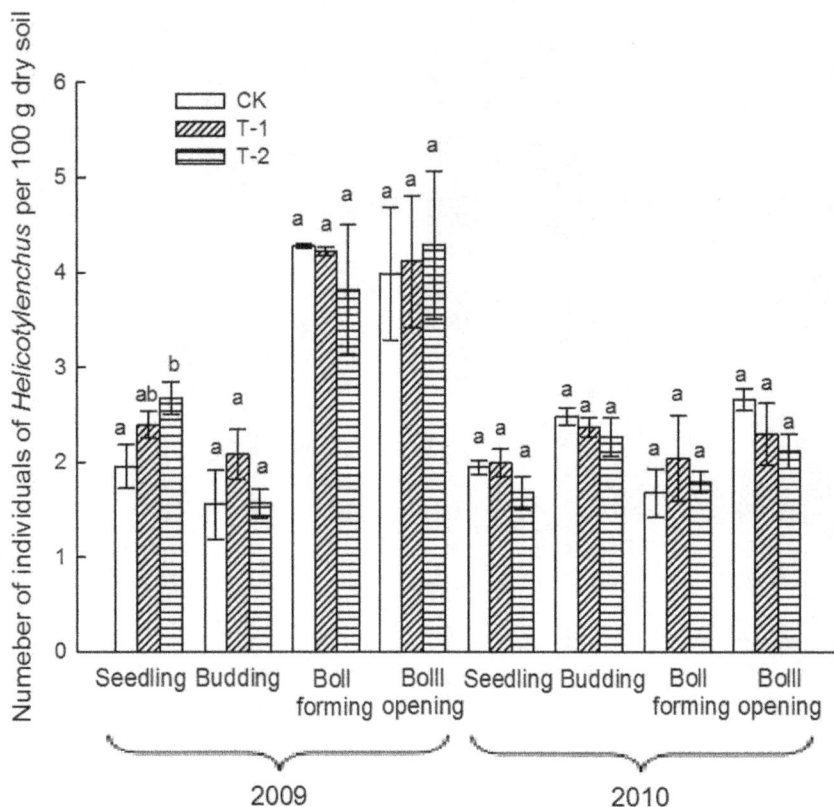

Figure 4. Number of *Helicotylenchus* in 3 cotton fields at different sampling times. Error bars indicate standard errors (n = 3). Different letters above bars denote a statistically significant difference between the means of the fields.

Effect of Transgenic Insect-resistant Cotton on the Ecological Indices of Soil Nematodes

The initial values of the Shannon–Wiener, Simpson, EI, SI, CI and MI ecological indices were obtained from the analysis of the nematode fauna at the seedling stage. Differences (Δ) from these initial values were obtained for each cotton field at the budding, boll-forming and boll-opening stages (Fig. 6). A univariate general linear model analysis suggested that these differences in the ecological indices varied significantly overall among different sampling times ($p < 0.05$) but did not differ significantly overall among the 3 cotton fields ($p > 0.05$). A further analysis used a one-way ANOVA with a priori contrasts to compare these ecological indices for the conventional cotton field with the corresponding indices for the 2 transgenic fields but detected no differences in the values of the ecological indices of the transgenic fields relative to the conventional cotton field at the 8 sampling times.

Principal Component Analysis of Soil Nematode Composition in Cotton Fields

Five principal components were selected for the analysis based on a cumulative contribution rate of 85% for the principal components extracted. The contribution rates of the first 2 principal components were 31.73% and 20.53%, respectively (Fig. 7). Different sampling times showed a distinct separation along the principal component axes, whereas different fields formed a cluster at the same sampling time. The first principal component axis clearly separated the budding, boll-forming and boll-opening stages. The second principal component axis clearly distinguished the seedling and budding stages (Fig. 7).

Discussion

With the cultivation of more varieties of transgenic insect-resistant plants and their large-scale planting, environmental impact monitoring after commercial release has attracted increasing attention from the scientific community [37,49–51]. High population densities and large numbers of species of nematodes occur in almost all soils. In this study, potential effects on soil nematodes at the community level were monitored during a 2-year survey to assess the environmental risks associated with transgenic fields planted with transgenic insect-resistant cotton for more than 10 years relative to the risks associated with a field planted with non-transgenic cotton. Based on the nematode communities examined at the genus level, the overall findings of the 2-year field study showed that the community structure of the soil nematodes was similar in the 3 cotton fields and that the most abundant common groups of soil nematodes all included *Helicotylenchus*, *Filenchus* and *Acrobeloides*. In soil planted with *Bt* maize and *Bt* eggplant expressing the Cry1Ac and Cry3Bb1 proteins, respectively, no effects were found on nematode community structure [32,52]. However, a distinct shift in community structure, i.e., a significantly higher proportion of mycophagous nematodes and a lower proportion of phytophagous nematodes, were observed in soil planted with *Bt* canola relative to soil planted with the respective non-*Bt* isoline [53]. In a study of the effects of Mon88017 cultivation on the community structure of the indigenous soil-inhabiting nematodes, no significant differences in the generic composition of the nematodes was found between a *Bt* maize plot and non-*Bt* maize plots, but a significant shift in generic composition was found at the final sampling date [48]. In a

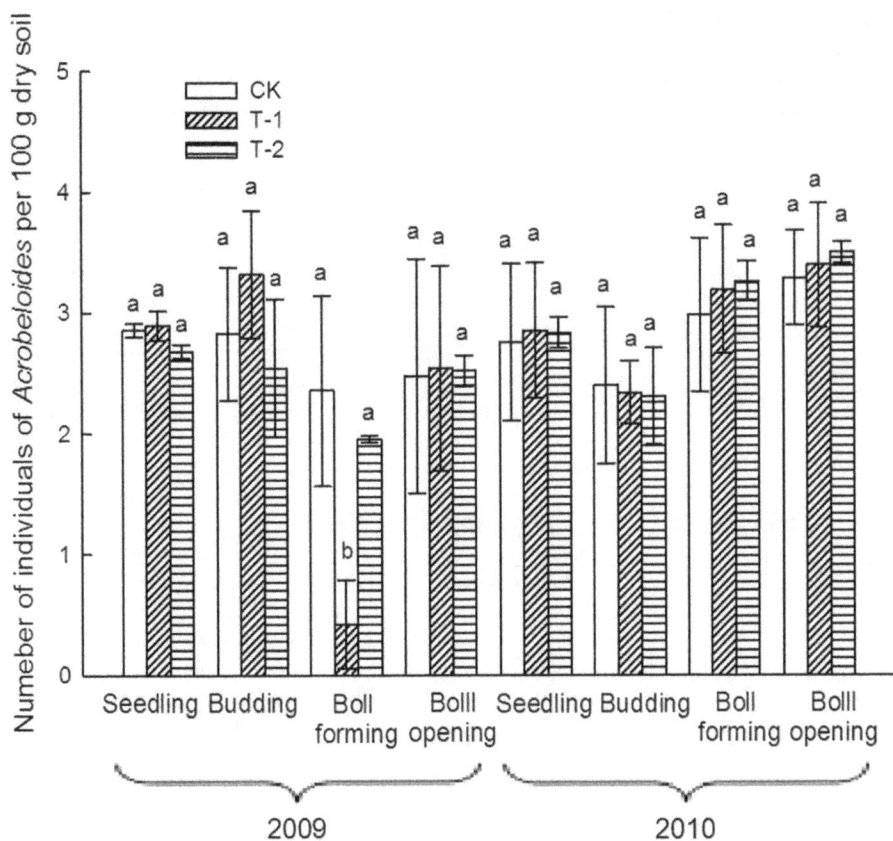

Figure 5. Number of *Acrobeloides* in 3 cotton fields at different sampling times. Error bars indicate standard errors (n = 3). Different letters above bars denote a statistically significant difference between the means of the fields.

Table 4. Changes in the composition of nematode trophic groups (%) among different cotton fields at different sampling times for the 2-year survey (mean±SE).

Trophic groups	Treatment	2009				2010			
		Seedling	Budding	Boll forming	Boll opening	Seedling	Budding	Boll forming	Boll opening
Plant parasites	CK	31.6±0.8[a]	29.5±2.1[a]	31.6±4.7[a]	37.1±2.9[a]	31.3±2.6[a]	26.4±3.7[a]	28.9±2.4[a]	36.1±6.7[a]
	T-1	32.3±1.9[a]	25.1±1.7[a]	36.2±3.8[a]	37.1±1.3[a]	30.7±3.2[a]	26.9±2.4[a]	32.8±3.9[a]	40.4±3.8[a]
	T-2	33.2±0.2[a]	24.0±2.3[a]	46.4±9.7[a]	43.5±7.6[a]	34.0±4.5[a]	23.9±4.3[a]	27.4±2.8[a]	36.7±4.2[a]
Bacterial feeders	CK	40.8±4.9[a]	39.6±11.7[a]	40.0±6.3[a]	42.1±5.8[a]	46.5±4.6[a]	56.5±7.6[a]	44.9±1.2[a]	42.3±4.7[a]
	T-1	41.0±5.8[a]	42.4±5.2[a]	43.6±8.5[a]	39.0±3.6[a]	47.6±5.5[a]	51.7±9.4[a]	46.0±2.7[a]	42.9±6.9[a]
	T-2	34.9±10.0[a]	38.8±9.6[a]	26.3±13.7[a]	32.7±7.4[a]	47.8±7.3[a]	52.5±3.7[a]	51.0±4.5[a]	50.0±5.6[a]
Fungal feeders	CK	16.7±5.9[a]	17.7±2.4[a]	13.9±3.2[a]	11.7±3.2[a]	8.4±2.7[a]	8.3±1.4[a]	6.9±1.3[a]	2.5±0.6[a]
	T-1	17.1±4.4[a]	19.0±2.3[a]	11.2±3.1[a]	11.8±2.3[a]	6.4±3.7[a]	8.5±1.7[a]	6.1±2.7[a]	1.2±0.7[a]
	T-2	19.2±7.3[a]	24.3±6.1[a]	17.1±4.7[a]	19.2±3.5[a]	5.5±1.6[a]	7.1±2.7[a]	4.7±1.9[a]	2.0±1.3[a]
Predators	CK	1.8±0.3[a]	5.5±1.5[a]	7.5±1.9[a]	6.8±2.1[a]	2.9±1.2[a]	1.3±0.9[a]	4.8±2.1[a]	5.2±1.8[a]
	T-1	2.9±0.2[a]	7.3±1.3[a]	5.7±2.2[a]	3.6±1.5[a]	3.6±1.8[a]	3.6±1.5[a]	6.3±1.8[a]	4.7±2.2[a]
	T-2	3.6±1.2[a]	5.2±2.8[a]	5.5±1.9[a]	3.4±1.7[a]	2.0±0.5[a]	5.3±2.8[a]	6.2±2.7[a]	4.2±1.7[a]
Omnivores	CK	9.0±1.3[a]	7.8±3.2[a]	7.1±2.7[a]	2.3±1.1[a]	11.0±2.5[a]	7.6±2.3[a]	14.5±1.9[a]	13.9±5.1[a]
	T-1	6.7±3.5[a]	6.2±1.8[a]	3.3±1.9[a]	8.4±3.7[b]	11.8±1.3[a]	9.3±3.8[a]	8.8±2.1[a]	10.8±2.3[a]
	T-2	9.0±0.3[a]	7.7±1.4[a]	4.8±2.3[a]	1.3±2.8[a]	10.7±2.9[a]	11.2±3.7[a]	10.7±2.9[a]	7.1±3.8[a]

Mean values and standard error of 3 replicates are presented. Use of the same letter as a superscript indicates that variable means did not differ significantly among different treatments at each sampling date (p>0.05).

Table 5. Generalized Linear Mixed Model (GLMM) results for the overall effects on soil nematode trophic groups.

Effects	Degrees of freedom	F Value	P Value
Plant parasites			
Treatments	2;14	0.62	0.55
Sampling time	7;14	6.28	0.00
Bacterial feeders			
Treatments	2;14	0.85	0.45
Sampling time	7;14	4.56	0.01
Fungal feeders			
Treatments	2;14	2.07	0.16
Sampling time	7;14	15.77	0.00
Predators			
Treatments	2;14	0.12	0.89
Sampling time	7;14	4.24	0.01
Omnivores			
Treatments	2;14	0.78	0.48
Sampling time	7;14	1.99	0.13

few cases, the studies cited reported some differences in the nematode communities among the different treatments at certain sampling times, but no consistent significant differences were found over the entire sampling period, and these studies only covered a short sampling time period. The current study observed no significant differences in soil nematode communities between fields of transgenic insect-resistant cotton and a field of non-transgenic cotton, and the study indicated that long-term cultivation of transgenic insect-resistant cotton had no significant impact on the composition and community structure of soil nematodes in agricultural soils.

Laboratory and field studies have shown no consistent effects of transgenic insect-resistant maize on the number of soil nematodes relative to non-Bt maize [31,33,34]. The cultivation of Bt maize expressing the Cry1Ab protein significantly decreased the number of soil nematodes at 3 different field sites relative to non-Bt maize, whereas a study conducted in a greenhouse showed no toxic effects of the Cry1Ab protein on populations of nematodes. In contrast, both the sampling site and the time had greater significant influences on the population density of soil nematodes than that of the maize lines [35,36]. Based on a GLMM analysis of data collected over 2 years of sampling, our results indicated that the number of functional guilds in the 3 cotton fields showed significant seasonal variation (necessarily following the progression of different cotton growth stages). However, the effect of long-term cultivation of transgenic insect-resistant cotton (T-1 and T-2) on the number of total soil nematodes and on certain dominant groups, such as Filenchus, Helicotylenchus and Acrobeloides, was not significant relative to the values found for conventional cotton cultivation. In only a few instances in our study did we find significant differences in the numbers of soil nematodes between fields of different cultivars. No consistent trend was found over the 2 study years. This general finding is in accord with work by other teams in other regions and on a variety of crops [31,33,35,36,54,55].

Parasitic nematodes can cause considerable economic damage worldwide to many types of crops, including cotton in the major growing areas of certain countries [56,57]. Several plant parasitic nematodes, such as Helicotylenchus, Tylenchorhynchus, Tylenchus, Pratylenchus and Filenchus, have been detected in the cotton fields of north China [58–60]. Our study found that Filenchus, Helicotylenchus, Tylenchus and Pratylenchus were the principal genera of plant parasitic nematodes in the 3 cotton fields. These results were consistent with the findings of other studies conducted in China. Certain genera, e.g., Meloidogyne, Rotylenchulus and Belonolaimus, that are known to infest cotton were not detected in the soils of the 3 cotton fields. The principal explanation for this result might be that these nematodes are of concern in the United States, India, Pakistan, Egypt and Brazil but not in China [61] and are seldom separated from soils sampled in the cotton-growing region of north China [58–60]. Moreover, these nematodes primarily parasitize the root tissues of the plant and are very rare in the soil around the plant roots. Finally, outbreaks of cotton nematode diseases have not been recorded in the 3 cotton fields for more than 20 years. Therefore, the nematode communities collected from the soil samples in the study were typical of China.

Functional analyses and indices such as those used in ecological studies have proven relatively useful in detecting the true effects of the cultivation of transgenic plants [18,34,62]. Over 2 years of sampling, we observed strong and significant seasonal variations in the ecological indices of the soil nematodes collected in all 3 cotton fields, in parallel with the distinct growth stages of cotton. However, no statistically significant effects of the long-term cultivation of transgenic insect-resistant cotton (in fields T-1 and T-2) on the ecological indices of the soil nematodes were evident relative to the non-Bt cotton (field CK). This result agreed with those reported by Manachini and Lozzia (2002) and Höss et al. (2011), who found that Bt maize expressing the Cry1Ab and Cry3Bb1 toxins had no significant effects on the diversity of soil nematodes [32,34].

The accumulation of Bt protein in soil has long been posited as one of the main putative mechanisms for the effects of transgenic Bt plants on soil organisms [9,68–70]. However, many studies have demonstrated that Cry proteins degrade rapidly in soil under laboratory conditions [70–72] and, hence, are unlikely to accumulate or persist in fields where Bt crops have been planted for years [73–75]. For example, the content of Cry1Ab protein was above the detection limit of an ELISA test in only half of the soil samples obtained from transgenic plots, ranging from 0.19 to 1.31 ng g^{-1} dry weight [34]. To address this possible but unlikely occurrence at our study site, we determined the residual levels of Cry1Ac protein in the soils of T-1 and T-2 using a QualiplateTM kit for Cry1Ab/Cry1Ac (EnviroLogix, USA) and found extremely low levels of Cry1Ac protein, below the quantitative limit of the kit [76]. These results agreed with the findings of others [73–75], indicating that Cry1Ab/c protein did not accumulate in cambisol soil with prolonged planting of transgenic cotton.

There was also no indication that the community structure of the soil nematodes was influenced by indirect effects of transgenic proteins, such as Cry1Ab and Cry1Ac, via the food web. Microbial communities such as bacteria, actinomycetes and fungi and soil invertebrate communities such as Collembola, Opisthophora and Acarina, which were studied in the same cotton fields investigated in the present study, showed no significant differences in abundance and diversity between transgenic insect-resistant cotton and non-Bt cultivars [76,77]. These findings are consistent with the results of another study examining the effect of transgenic insect-resistant plants on microbial populations [75,78,79] and other soil organisms, such as Collembola, mites and earthworms [62,80–82]. Clearly, soil organisms are not impacted or are only

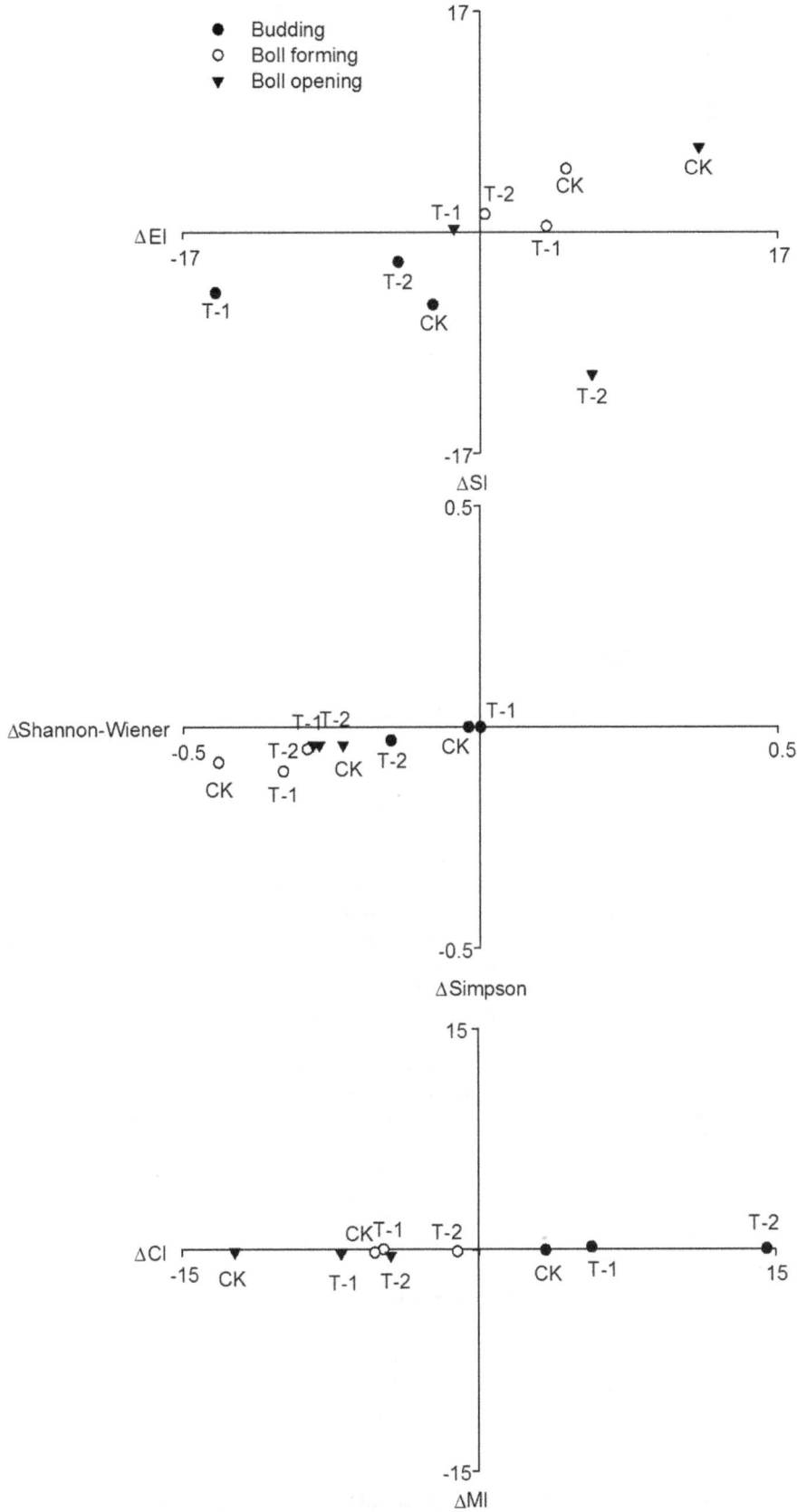

Figure 6. Changes in the values of ecological indices for soil nematode faunal analysis relative to seedling stage among different cotton fields.

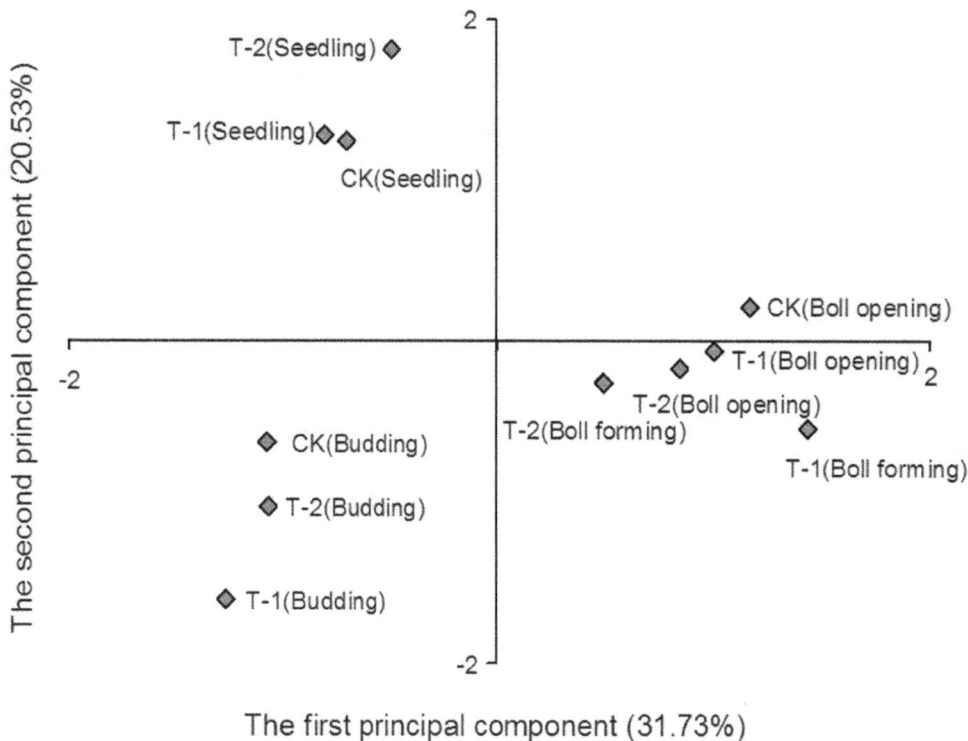

Figure 7. Factor loading graph of principal component analysis of soil nematode generic composition in cotton fields at different sampling times.

slightly impacted by the cultivation of transgenic insect-resistant cotton.

Many studies on the effects of pesticides (primarily nematicides and insecticides) on total nematode abundance and feeding groups under field or semi-field conditions have been conducted. The majority of these studies have indicated no negative effects of pesticides such as malathion, imicyafos and carbofuran on nematodes [63–66]. In one study, a decrease in abundance was observed as an effect of pesticides such as nemacur [67]. In the current study, the same spraying dose of chemical pesticides (i.e., avermectin and halfenprox) used to control insects with piercing-sucking mouthparts, such as the red spider, was applied to the 3 fields. However, the applications of chemical pesticides (i.e., methamidophos and cypermethrin) used to control Lepidoptera, such as the cotton bollworm, in the 2 transgenic cotton fields were fewer in number than those applied to the conventional cotton field (Fig. 1). Therefore, the differences in the spraying of chemical pesticides to control Lepidoptera in the 3 cotton fields might be another important influence on the soil nematode community, and the total pesticide applications could mask the effects of different cotton lines in the present study, or the effects of transgenic cotton lines on the soil nematode community might be smaller than those of the pesticide regimes.

Transgenic plants must be monitored for environmental risk after being commercially released [50,83]. The risks potentially posed by transgenic plants, especially *Bt* crops, to the environment have been extensively assessed worldwide over the past 10 years, and no scientific evidence has shown that the cultivation of *Bt* crops has caused sustained environmental harm to communities of soil organisms, such as nematodes, earthworms, collembolans or mites [33,35,50,62]. However, the soil environment is a very complex ecosystem in which many factors affect the soil biota. In field studies, high variability in biotic parameters is inherent and usually present, and this variability must be considered seriously if the ecological risks posed by transgenic plants are to be monitored. Moreover, the soil biota may be strongly stressed as a result of the influence of environmental factors (e.g., pH, salinity, redox potential, vegetation and water-holding capacity), which may cause higher or lower levels of sensitivity to transgenic plants [23,40,56]. Therefore, it is necessary to continue monitoring the effects of transgenic plants on the soil ecosystem in different environments and to define the ecological significance of the planting of transgenic crops.

Author Contributions

Conceived and designed the experiments: BL XGL. Performed the experiments: XGL. Analyzed the data: XGL. Contributed reagents/materials/analysis tools: XGL BL. Wrote the paper: XGL.

References

1. James C (2011) Global Status of Commercialized Biotech/GM Crops: 2011. ISAAA Brief No. 43. ISAAA, Ithaca, New York.
2. Lu YH, Wu KM, Jiang YY, Guo YY, Desneux N (2012) Widespread adoption of *Bt* cotton and insecticide decrease promotes biocontrol services. Nature 487: 362–365.
3. Choudhary B, Gaur K (2010) *Bt* cotton in india: a country profile. ISAAA Series of Biotech Crop Profiles. ISAAA, Ithaca, New York.
4. Hutchison WD, Burkness EC, Mitchell PD, Moon RD, Leslie TW, et al. (2010) Areawide suppression of European corn borer with *Bt* maize reaps savings to non-Bt maize growers. Science 330: 222–225.

5. Lilley AK, Bailey MJ, Cartwright C, Turner SL, Hirsch PR (2006) Life in earth: the impact of GM plants on soil ecology? Trends Biotech 24: 9–14.

6. Cortet J, Griffiths BS, Bohanec M, Demsar D, Andersen MN, et al. (2007) Evaluation of effects of transgenic Bt maize on microarthropods in a European multi-site experiment. Pedobiologia 51: 207–218.

7. Rose R, Dively GP (2007) Effects of insecticide treated and lepidopteran-active Bt transgenic sweet corn on the abundance and diversity of arthropods. Environ Entomol 36: 1254–1268.

8. Yu HL, Li YH, Wu KM (2011) Risk assessment and ecological effects of transgenic Bacillus thuringiensis crops on non-target organisms. J Integr Plant Biol 53: 520–538.

9. Duan JJ, Lundgren JG, Naranjo S, Marvier M (2010) Extrapolating non-target risk of Bt crops from laboratory to field. Biol Lett 6: 74–77.

10. Naranjo SE (2005) Long-term assessment of the effects of transgenic Bt cotton on the abundance of nontarget arthropod natural enemies. Environ Entomol 34: 1193–1210.

11. Marvier M, McCreedy C, Regetz J, Kareiva P (2007) A meta-analysis of effects of Bt cotton and maize on non-target invertebrates. Science 316: 1475–1477.

12. Li GP, Feng HQ, McNeil JN, Liu B, Chen PY, et al. (2011) Impacts of transgenic Bt cotton on a non-target pest, Apolygus lucorum (Meyer-Dür) (Hemiptera: Miridae), in northern China. Crop Prot 30: 1573–1578.

13. Xu Y, Wu KM, Li HB, Liu J, Ding RF, et al. (2012) Effects of transgenic Bt+CpTI cotton on field abundance of non-target pests and predators in Xinjiang, China. J Integr Agr 11: 1493–1499.

14. Liu B, Shu C, Xue K, Zhou KX, Li XG, et al. (2009) The oral toxicity of the transgenic Bt plus CpTI cotton pollen to honeybees (Apis mellifera). Ecotox Environ Safe 72: 1163–1169.

15. Han P, Niu CY, Lei CL, Cui JJ, Desneux N (2010) Quantification of toxins in a Cry1Ac+CpTI cotton cultivar and its potential effects on the honey bee Apis mellifera L. Ecotoxicology 19: 1452–1459.

16. Chen LZ, Cui JJ, Ma WH, Niu CY, Lei CL (2011) Pollen from Cry1Ac/CpTI-transgenic cotton does not affect the pollinating beetle Haptoncus luteolus. J Pest Sci 84: 9–14.

17. Han P, Niu CY, Biondi A, Desneux N (2012) Does transgenic Cry1Ac+CpTI cotton pollen affect hypopharyngeal gland development and midgut proteolytic enzyme activity in the honey bee Apis mellifera L. (Hymenoptera, Apidae)? Ecotoxicology 21: 2214–2221.

18. Wolfenbarger LL, Naranjo SE, Lundgren JG, Bitzer RJ, Watrud LS (2008) Bt crop effects on functional guilds of non-target arthropods: A meta-analysis. PloS ONE 3 (5): e2118. doi: 10.1371/journal.pone.0002118.

19. Duan JJ, Marvier M, Huesing J, Dively G, Huang ZY (2008) A meta-analysis of effects of Bt crops on honeybees (Hymenoptera: Apidae). PLoS ONE 3 (1): e1415. doi:10.1371/journal.pone.0001415.

20. Rose R, Dively GP, Pettis J (2007) Effects of Bt corn pollen on honey bees: emphasis on protocol development. Apidologie 38: 368–377.

21. Hofs JL, Schoeman AS, Pierre J (2008) Diversity and abundance of flower-visiting insects in Bt and non-Bt cotton fields of Maputaland (KwaZulu Natal Province, South Africa). Int J Trop Insect Sci 28: 211–219.

22. Gupta VVSR, Watson S (2004) Ecological impacts of GM cotton on soil biodiversity: Below-ground production of Bt by GM cotton and Bt cotton impacts on soil biological processes. Australian Government Department of the Environment and Heritage, CSIRO Land and Water.

23. Knox OGG, Gupta VVSR, Nehl DB, Stiller WN (2007) Constitutive expression of Cry proteins in roots and border cells of transgenic cotton. Euphytica 154: 83–90.

24. Tapp H, Stotzky G (1998) Persistence of the insecticidal toxin from Bacillus thuringiensis subsp. kurstaki in soil. Soil Biol Biochem 30: 471–476.

25. Koskella J, Stotzky G (1997) Microbial utilization of free and clay-bound insecticidal toxins from Bacillus thuringiensis and their retention of insecticidal activity after incubation with microbes. Appl Enviro Microbiol 63: 3561–3568.

26. Andrássy I (1992) A short census of free-living nematodes. Fund Appl Nematol 15: 187–188.

27. Blair JM, Bohlen PJ, Freckman DW (1996) Soil invertebrates as indicators of soil quality. In: Doran JW, Jones AJ, editors. Methods for Assessing Soil Quality. SSSA Special Publication No. 49. Soil Science Society of America, Madison, 283–301.

28. Sanvid O, Stark M, Romeis J, Bigler F (2006) Ecological impacts of genetically modified crops: experiences from ten years of experimental field research and commercial cultivation. Swiss Expert Committee for Biosafety. ART-Schriften-reihe 1: 1–84.

29. Wei JZ, Hale K, Carta L, Platzer E, Wong C, et al. (2003) Bacillus thuringiensis crystal proteins that target nematodes. PANS 100: 2760–2765.

30. Lang A, Arndt M, Beck R, Bauchhenss J, Pommer G (2006) Monitoring of the environmental effects of the Bt gene. Bayerische Landesanstalt für Land-wirtschaft. Available: http://www.lfl-neu.bayern.de/publikationen/daten/schriftenreihe_url_1_43.pdf. Accessed 2012 Nov 3.

31. Saxena D, Stotzky G (2001) Bacillus thuringiensis (Bt) toxin released from root exudates and biomass of Bt corn has no apparent effect on earthworms, nematodes, protozoa, bacteria, and fungi in soil. Soil Biol Biochem 33: 1225–1230.

32. Manachini B, Lozzia GC (2002) First investigations into the effects of Bt corn crop on Nematofauna. Boll Zool Agraria e di Bachicoltura Serie II 34: 85–96.

33. Al-Deeb MA, Wilde GE, Blair JM, Todd TC (2003) Effect of Bt corn for corn rootworm control on nontarget soil microarthropods and nematodes. Environ Entomol 32: 859–865.

34. Höss S, Nguyen HT, Menzel R, Pagel-Wieder S, Miethling-Graf R, et al. (2011) Assessing the risk posed to free-living soil nematodes by a genetically modified maize expressing the insecticidal Cry3Bb1 protein. Sci Total Envir 409: 2674–2684.

35. Griffiths BS, Caul S, Thompson J, Birch ANE, Scrimgeour C, et al. (2005) A comparison of soil microbial community structure, protozoa, and nematodes in field plots of conventional and genetically modified maize expressing the Bacillus thuringiensis Cry1Ab toxin. Plant Soil 275: 135–146.

36. Griffiths BS, Caul S, Thompson J, Birch ANE, Scrimgeour C, et al. (2006) Soil microbial and faunal community responses to Bt maize and insecticide in two soils. J Environ Qual 35: 734–741.

37. Bartsch D, Devos Y, Hails R, Kiss J, Krogh PH, et al. (2010) Environmental impact of genetically modified maize expressing cry1 proteins. Genet Modif Plants Biotechnol Agr For 64: 575–614.

38. The States Council of China (2001) Regulations on safety of agricultural genetically modified organisms. The Central People's Government of the People's Republic of China. Available: http://www.gov.cn/flfg/2005-08/06/content_21003.htm. Accessed 2013 Feb 7.

39. EFSA (2011) Guidance on the post-market environmental monitoring (PMEM) of genetically modified plants. J EFSA 9(8): 2316. doi:10.2903/j.efsa.2011.2316.

40. Guo XM, Liu JS (2003) The guidance for seed introduction of cotton. The JinDun Publishing House, Beijing.

41. Committee of handbook of soil fauna research methods (1998) Handbook of soil fauna research methods. China Forestry Publishing House, Beijing.

42. Jenkins WR (1964) A rapid centrifugal-floatation technique for separating nematodes from soil. Plant Dis Rep 48: 692.

43. Liang WJ, Lou YL, Li Q, Zhong S, Zhang XK, et al. (2009) Nematode faunal response to long-term application of nitrogen fertilizer and organic manure in Northeast China. Soil Biol Biochem 41: 883–890.

44. Magurran AE (2004) Measuring biological diversity. Blackwell Science, Oxford.

45. Bongers T, Ferris H (1999) Nematode community structure as a bioindicator in environmental monitoring. Trend Ecol Evolut 14: 224–228.

46. Bongers T (1990) The maturity index: an ecological measure of environmental disturbance based on nematode species composition. Oecologia 83: 14–19.

47. Yeates GW, Bongers T, de Goede RGM, Freckman DW, Georgieva SS (1993) Feeding habits in soil nematode families and genera – an outline for soil ecologists. J Nematol 25: 315–31.

48. Ferris H, Bongers T, de Goede RGM (2001) A framework for soil food web diagnostics: extension of the nematode faunal analysis concept. Appl Soil Ecol 18: 13–29.

49. Graef F, Züghart W, Hommel B, Heinrich U, Stachow U, et al. (2005) Methodological scheme for designing the monitoring of genetically modified crops at the regional scale. Environ Monit Assess 111: 1–26.

50. Sanvido O, Aviron S, Romeis J, Bigler F (2007) Challenges and perspectives in decision-making during post-market environmental monitoring of genetically modified crops. J Für Verbraucherschutz 2: 37–40.

51. Züghart W, Benzler A, Berhorn F, Sukopp U, Graef F (2008) Determining indicators, methods and sites for monitoring potential adverse effects of genetically modified plants to the environment: the legal and conceptual framework for implementation. Euphytica 164: 845–852.

52. Manachini B, Fiore MC, Landi S, Arpaia S (2003) Nematode species assemblage in Bt-expressing transgenic eggplants and their isogenic control. Biodiversity Implications of Genetically Modified Plants. September 7–13, Ascona, Switzerland.

53. Manachini B, Landi S, Fiore MC, Festa M, Arpaia S (2004) First investigations on the effects of Bt-transgenic Brassica napus L. on the trophic structure of the nematofauna. IOBC/WPRS Bull 27: 103–108.

54. Powell JR, Levy-Booth DJ, Gulden RH, Asbil WL, Campbell RG, et al. (2009) Effects of genetically modified, herbicide-tolerant crops and their management on soil food web properties and crop litter decomposition. J Appl Ecol 46: 388–396.

55. Green J, Wang D, Lilley CJ, Urwin PE, Atkinson HJ (2012) Transgenic potatoes for potato cyst nematode control can replace pesticide use without impact on soil quality. PloS ONE 7 (2): e30973. doi:10.1371/journal.pone.0030973.

56. Karuri HW, Amata R, Amugune N, Waturu C (2010) Occurrence and distribution of soil nematodes in cotton (Gossypium hirsutum L.) production areas of Kenya. Afr J Agr Res 5: 1889–1896.

57. Lingaraju S, Sonavane P, Jamadar MM, Harlapur SI, Bhat RS, et al. (2012) Plant parasitic nematodes associated with Bt cotton. Curr Sci India 103: 926–932.

58. Wang RX, Yang ZW, Pang HL, Cheng P (1989) Studies on the effect of major groups of nematode in the cotton fields of Shaanxi province on Fusarium wilt of cotton. Acta Phytopathologica Sinica 19: 205–209.

59. Dong DF, Chen YF, Steinberger Y, Cao ZP (2008) Effects of different soil management practices on soil free-living nematode community structure, Eastern China. Can J Soil Sci 88: 115–127.

60. Li KM, Liang Z, Xu WL, Zhang XB, Tunuhe, etal. (2009) The species of parasitic nematodes in cotton field of Xinjiang. Acta Agri Boreali-occidentalis Sinica 18: 273–275.

61. Robinson AF (2007) Nematode management in cotton. In: Ciancio A, Mukerji KG, editors. Integrated management and biocontrol of vegetable and grain crops nematodes. Springer: Springer Netherlands. 149–182.

62. Priestley AL, Brownbridge M (2009) Field trials to evaluate effects of *Bt*-transgenic silage corn expressing the Cry1Ab insecticidal toxin on non-target soil arthropods in northern New England, USA. Transgenic Res 18: 425–443.

63. Moser T, Schallnaß HJ, Joses SE, Gestel CAMV, Koolhaas JEE, et al. (2004) Ring-testing and field-validation of a terrestrial model ecosystem (TME)–an instrument for testing potentially harmful substances: effects of carbendazim on nematodes. Ecotoxicology 13: 61–74.

64. Parmelee RW, Phillips CT, Checkai RT, Bohlen PJ (1997) Determining the effects of pollutants on soil faunal communities and trophic structure using a refined microcosm system. Environ Toxicol Chem 16: 1212–1217.

65. Wada S, Toyota K (2008) Effect of three organophosphorous nematicides on non-target nematodes and soil microbial community. Microbes Environ 23: 331–336.

66. Chelinho S, Sautter KD, Cachada A, Abrantes I, Brown G, et al. (2011) Carbofuran effects in soil nematode communities: Using trait and taxonomic based approaches. Ecotox Environ Safe 74: 2002–2012.

67. Pen-Mouratov S, Steinberger Y (2005) Responses of nematode community structure to pesticide treatments in an arid ecosystem of the Negev Desert. Nematology 7: 179–191.

68. Lee L, Saxena D, Stotzky G (2003) Activity of free and clay-bound insecticidal proteins from *Bacillus thuringiensis* subsp. Israelensis against the mosquito *Culex pipiens*. Appl Enviro Microbiol 69: 4111–4115.

69. Stotzky G (2004) Persistence and biological activity in soil of the insecticidal proteins from *Bacillus thuringiensis*, especially from transgenic plants. Plant Soil 266: 77–89.

70. Icoz I, Stotzky G (2008) Cry3Bb1 protein from *Bacillus thuringiensis* in root exudates and biomass of transgenic corn does not persist in soil. Transgenic Res 17: 609–620.

71. Sims SR, Ream JE (1997) Soil inactivation of the *Bacillus thuringiensis* subsp. *kurstaki* Cry IIA insecticidal protein within transgenic cotton tissue: laboratory microcosm and field studies. J Agr Food Chem 45: 1502–1505.

72. Hopkins DW, Gregorich EG (2003) Detection and decay of the *Bt* endotoxin in soil from a field trial with genetically modified maize. Eur J Soil Sci 54: 793–800.

73. Head G, Surber JB, Watson JA, Martin JW, Duan JJ (2002) No detection of Cry1Ac protein in soil after multiple years of transgenic *Bt* cotton (Bollgard) use. Environ Entomol 31: 30–36.

74. Ahmad A, Wilde GE, Zhu KY (2005) Detectability of coleopteran-specific Cry3Bb1 protein in soil and its effect on nontarget surface and below-ground arthropods. Environ Entomol 34: 385–394.

75. Icoz I, Saxena D, Andow DA, Zwahlen C, Stotzky G (2008) Microbial populations and enzyme activities in soil in situ under transgenic corn expressing Cry proteins from *Bacillus thuringiensis*. J Environ Qual 37: 647–662.

76. Li XG, Liu B, Wang XX, Han ZM, Cui JJ, et al. (2012) Field trials to evaluate effects of continuously planted transgenic insect-resistant cottons on soil invertebrates. J Environ Monit 14: 1055–1063.

77. Li XG, Liu B, Cui JJ, Liu DD, Ding S, et al. (2011) No evidence of persistent effects of continuously planted transgenic insect-resistant cotton on soil microorganisms. Plant Soil 339: 247–257.

78. Devare M, Londoño-R LM, Thies JE (2007) Neither transgenic *Bt* maize (MON863) nor tefluthrin insecticide adversely affect soil microbial activity or biomass: A 3-year field analysis. Soil Biol Biochem 39: 2038–2047.

79. Fließbach A, Messmer M, Nietlispach B, Infante V, Mäder P (2012) Effects of conventionally bred and *Bacillus thuringiensis* (*Bt*) maize varieties on soil microbial biomass and activity. Biol Fert Soils 48: 315–324.

80. Bitzer RJ, Rice ME, Pilcher CD, Pilcher CL, Lam WF (2005) Biodiversity and community structure of epedaphic and euedaphic springtails (Collembola) in transgenic enrootworm *Bt* corn. Environ Entomol 34: 1346–1376.

81. Cortet J, Griffiths BS, Bohanec M, Demsar D, Andersen MN, et al. (2007) Evaluation of effects of transgenic *Bt* maize on microarthropods in a European multi-site experiment. Pedobiologia 51: 207–218.

82. Duc C, Nentwig W, Lindfeld A (2011) No adverse effect of genetically modified antifungal wheat on decomposition dynamics and the soil fauna community – a field study. PloS ONE 6(10): e25014. doi:10.1371/journal.pone.0025014.

83. Sanvido O, Widmer F, Winzeler M, Bigler F (2005) A conceptual framework for the design of environmental post-market monitoring of genetically modified plants. Environ Biosaf Res 4: 13–27.

12

Quantitative Trait Loci for Mercury Accumulation in Maize (*Zea mays* L.) Identified Using a RIL Population

Zhongjun Fu[1,2✪]**, Weihua Li**[1✪]**, Qinbin Zhang**[1]**, Long Wang**[1]**, Xiaoxiang Zhang**[1]**, Guiliang Song**[1]**, Zhiyuan Fu**[1]**, Dong Ding**[1]**, Zonghua Liu**[1]***, Jihua Tang**[1]***

1 National Key Laboratory of Wheat and Maize Crops Science, Collaborative Innovation Center of Henan Grain Crops, College of Agronomy, Henan Agricultural University, Zhengzhou, China, **2** Maize Research Institute, Chongqing Academy of Agricultural Sciences, Chongqing, China

Abstract

To investigate the genetic mechanism of mercury accumulation in maize (*Zea mays* L.), a population of 194 recombinant inbred lines derived from an elite hybrid Yuyu 22, was used to identify quantitative trait loci (QTLs) for mercury accumulation at two locations. The results showed that the average Hg concentration in the different tissues of maize followed the order: leaves > bracts > stems > axis > kernels. Twenty-three QTLs for mercury accumulation in five tissues were detected on chromosomes 1, 4, 7, 8, 9 and 10, which explained 6.44% to 26.60% of the phenotype variance. The QTLs included five QTLs for Hg concentration in kernels, three QTLs for Hg concentration in the axis, six QTLs for Hg concentration in stems, four QTLs for Hg concentration in bracts and five QTLs for Hg concentration in leaves. Interestingly, three QTLs, *qKHC9a*, *qKHC9b*, and *qBHC9* were in linkage with two QTLs for drought tolerance. In addition, *qLHC1* was in linkage with two QTLs for arsenic accumulation. The study demonstrated the concentration of Hg in Hg-contaminated paddy soil could be reduced, and maize production maintained simultaneously by selecting and breeding maize Hg pollution-safe cultivars (PSCs).

Editor: Ben Lehner, CRG, Spain

Funding: This work was supported by the State Key Basic Research and Development Plan of China (2014CB138203) and Scientific Personnel Innovation Fund of Henan Province in China. The funders had no role in study design, data collection and analysis, decision to publish, or preparation of the manuscript.

Competing Interests: The authors have declared that no competing interests exist.

* Email: zhliu4000@sohu.com (ZL); tangjihua1@163.com (JT)

✪ These authors contributed equally to this work.

Introduction

Mercury (Hg) is a non-essential element in higher plants, and is one of the most hazardous heavy metals; it can accumulate in living organisms and cause serious damage [1]. With the development of industry and modern agriculture, Hg pollution has become a worldwide environmental problem [2,3]. Numerous cases of mercury pollution in soils have been reported throughout the world [4]. Generally, Hg concentration in unpolluted arable soil is 20–150 μg kg^{-1} [5]. However, the Hg concentration in the soils of paddy fields is much higher, especially in the Philippines and Japan; the average Hg concentration is 24 mg kg^{-1} and 146 mg kg^{-1}, respectively, for land irrigated with water contaminated with Hg [6]. The concentration of Hg in paddy soil is as high as 45.9 mg kg^{-1} in some area of China [7].

At high concentrations in soil, Hg can poison plant cells and cause physiological disorders [8], producing detrimental effects on plant growth and metabolism [9,10]. Hg accumulation in roots prevents the uptake and transport of other mineral nutrients [11], and excess Hg in solution culture can inhibit biomass production [5]. Hg toxicity is not only associated with water uptake and transpiration [12], but also with the decrease in chlorophyll content and photosynthetic efficiency [13]. In addition, Hg stress is believed to trigger the production of reactive oxygen species (ROS), causing oxidative stress and membrane lipid peroxidation

in plants [14,15]. Hg in soil can accumulate in the edible parts of vegetables and crops, and is then transferred to humans via the food chain [16]. Hg is also toxic to humans, causing impaired health in adults. Hg has toxicological effects on the developing central nervous system, on the general physiological systems of children, and adverse effects on the cardiovascular system [17] and fetal brain development [18].

Hg is more easily absorbed and transported in straw and grain than other heavy metals [19]. Previous studies have shown that different genotypes in rice differ widely in their tolerance to Hg toxicity [14]. Wang et al. identified three quantitative trait loci (QTLs) for Hg tolerance in rice seedlings using a recombinant inbred line (RIL) population [20]. Using doubled haploid (DH) lines, three QTLs for Hg accumulation in rice were determined by Yu et al. [14]. Rugh et al. expressed the *merApe9* gene in *Arabidopsis thaliana*, and found that the transgenic seedlings showed low levels of toxicity during the growth and flowering stages because *merApe9* converted toxic Hg^{2+} to the less toxic Hg^{0} [21]. Shen et al. isolated three novel HO genes from rapeseed, and HO-1 has been tested for its ability to regulate plant tolerance to Hg-induced oxidative stress [22]. Wei et al. found that the overexpression of HO-1 confers algal tolerance to excess Hg, whereas silencing of HO-1 had adverse effects on algal growth [23]. Recently, Chen et al. reported that overexpression of the *MTH1745* gene could enhance mercury tolerance in transgenic rice [24]. In maize, there have been some studies on the physiological and biochemical responses to Hg tolerance. The net photosynthesis rate and carboxylation efficiency of maize leaves exposed to 50 ng m^{-3} Hg (in the air) were significantly reduced [25]. *In vivo*, HgCl$_2$ inhibited 5-aminolevulinic acid dehydratase activity in excised greening leaf segments of maize, and the inhibition could be alleviated by the addition of KNO$_3$ [26]. Additionally, Hg had a strong toxic effect in the roots of maize seedlings, as inferred from the observed higher proportion of oxidized glutathione (GSSG), enhanced carbonyl content and the negative effects on growth [27].

Maize is one of the most important staple crops in the world, and is used as feedstuff and raw-food material in many countries. Maize is used as a staple food for more than 1.2 billion people in sub-Saharan Africa and Latin America [28]. Maize planted in Hg-contaminated paddy soil can accumulate Hg in the grains, which is transferred to consumers by the food chain (soil-plant-human or soil-plant-animal-human), posing a human health risk. However the Hg concentration and distribution pattern in the different tissues of maize and the genetic basis for Hg accumulation remain unclear. Therefore, the objectives of the study were to: (i) examine the accumulation and distribution of Hg in different tissues of maize, (ii) detect QTLs for Hg accumulation in the tissues of maize under Hg-contaminated paddy soil treatment, and to dissect the genetic bases of Hg accumulation in maize.

Materials and Methods

The experimental population

A RIL population including 194 inbred lines was used in the study. This population was derived from a cross between two inbred lines, Zong3 and 87-1; the result was an elite hybrid, Yuyu22, which was grown on about 2.7 million hectares per year from 2001 to 2004 in China [29]. The parent Zong3 was selected from a synthetic population with Chinese domestic germplasm; while the inbred line 87-1 was selected from an exotic germplasm. In 2012, the RIL population, both parents and the hybrid were evaluated in Xixian and Changge, Henan Province, China. The experimental materials followed a randomized complete block design with three replications. Each plot included fifteen plants with one 4-m-long and 0.67-m-wide row. The density was 45,000 plants per hectare.

The mercury content in soil

The experimental materials were evaluated in Xixian country of Xinyang city in Henan province of China (E114°95′–114°72′, N32°35′–32°44′) and Changge country of Xuchang city in Henan province of China (E113°34′–114°08′, N34°09′–34°20′), with an annual average temperature of 15.2°C and 14.3°C; and annual average rainfall of 946 mm and 711.1 mm. The field studies did not involve endangered or protected species, and no specific permissions were required for these locations/activities. The average value of agricultural soil Hg exposure in the Xixian experimental area is 457.57±31.30 µg kg^{-1}Hg (pH 6.5), as a result of irrigation with mercury-rich surface water. The average value of agricultural soil Hg exposure in the Changge experimental area is 345.40±22.24 µg kg^{-1}Hg (pH 6.5), which was used as a control.

Quantification of mercury content in the different tissues of maize

Five consecutive plants per row were harvested at physiological maturity stage and at natural withering for mercury content measurements in the different tissues of maize. The whole plant

was separated into five tissues: kernels, axis, stems, bracts and leaves, which were ground into powders. A 0.5-g sample of each tissue was digested in 5 ml of HNO$_3$/HClO$_4$ (80/20 v/v) on a heating block (Digestion Systems of AIM500, A. I. Scientific, Australia). An atomic fluorescence spectrometry (AFS-3000, Beijing Haiguang Analytical Instrument Co, Beijing, China) was used to determine the Hg concentration. Data analyses using the PROC MIXED procedure were performed using SAS 8.0 statistical software.

Molecular Linkage Construction and QTL Mapping

A total of 263 SSR markers that covered the whole genome of maize were used to construct the genetic linkage map for the RIL population, which spanned 2.361 cM with an average interval of 9 cM between markers [29]. The composite interval mapping method in software QTL Cartographer 2.5 [30] was employed for QTL mapping of measured traits, using the average data of three replications. Model 6 of the Zmapqtl module was used, with scanning intervals of 2 cM between markers and putative QTLs and a 10-cM window. The number of marker cofactors for background control was set by forward–backward stepwise regression with five controlling markers. The logarithm of odds (LOD) threshold was set for each trait by randomly permuting 1,000 times at a significance level of P = 0.05.

Results

Hg concentrations in different maize tissues at two locations

The Hg concentration in the five tissues varied widely in the RIL population (Table 1, Fig 1). In Xixian, the Hg concentration in the kernels (KHC) for the parent Zong3 was higher (3.53 µg kg^{-1}) than that in the parent 87-1 (2.14 µg kg^{-1}); the same trend for the two parents was observed in Changge. The KHC was higher in the hybrid Yuyu22 (4.62 µg kg^{-1}, 2.74 µg kg^{-1}) than in the two parents in both locations. The average KHC of the RIL population was 2.99±1.25 µg kg^{-1} (range 0.73–7.47 µg kg^{-1}) and 2.37±1.52 µg kg^{-1} (range 0.26–7.18 µg kg^{-1}) in Xixian and Changge, respectively.

The Hg concentration in the axis (AHC) for the parent Zong3 was higher (2.89 µg kg^{-1}) than that of the parent 87-1(1.51 µg kg^{-1}) in Xixian. However, in Changge, the value of AHC for the parent Zong3 was lower (1.86 µg kg^{-1}) than that of the parent 87-1(3.71 µg kg^{-1}). The hybrid had a higher AHC (3.40 µg kg^{-1}) in Xixian and a mid-parent value (3.32 µg kg^{-1}) in Changge, compared with its parents. The average AHC for the RIL population was 3.84±1.49 µg kg^{-1} (range 1.08–8.37 µg kg^{-1}) in Xixian and 2.71±1.76 µg kg^{-1} (range 0.18–8.18 µg kg^{-1}) in Changge.

For the Hg concentration in the stem (SHC), the value of the parent Zong3 was lower (3.71 µg kg^{-1}) than that of the parent 87-1(6.02 µg kg^{-1}) in Xixian, but the opposite trend was found in Changge. The hybrid showed a mid-parent value (5.15 µg kg^{-1}, 3.67 µg kg^{-1}) at both locations. For the RIL population, the average SHC was 5.36±1.30 µg kg^{-1} (range 1.82–9.02 µg kg^{-1}) in Xixian and 4.26±2.32 µg kg^{-1} (range 0.47–11.83 µg kg^{-1}) in Changge.

For the Hg concentration in the bract (BHC), the parent Zong3 had higher values (8.80 µg kg^{-1}, 9.18 µg kg^{-1}) than the parent 87-1 (6.21 µg kg^{-1}, 8.23 µg kg^{-1}) at both locations. In addition, the BHC (5.95 µg kg^{-1}, 7.14 µg kg^{-1}) for the hybrid expressed the same low-parent performance at both locations. The average BHC for the RIL population was 7.59±1.81 µg kg^{-1} (range 4.07–

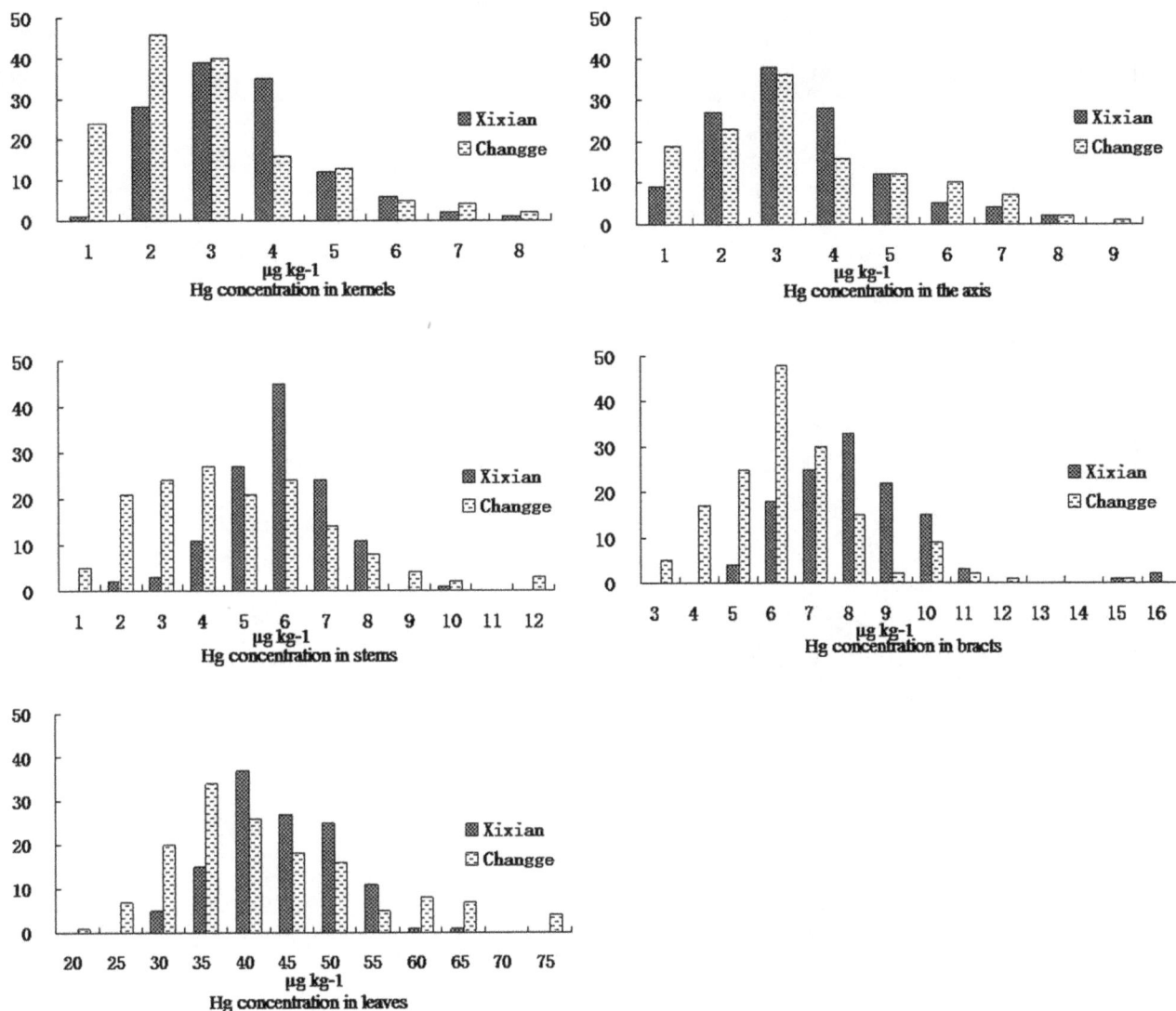

Figure 1. Histogram of Hg concentration in the five tissues of the RIL population.

15.56 µg kg^{-1}) in Xixian and 5.83±1.84 µg kg^{-1} (range 2.06–14.03 µg kg^{-1}) in Changge.

The parent Zong3 had a lower Hg concentration (30.20 µg kg^{-1}) than the parent 87-1 (43.20 µg kg^{-1}) in the leaves in Xixian, but the value of the parent Zong3 (43.14 µg kg^{-1}) was higher than the parent 87-1 (37.31 µg kg^{-1}) in Changge. The Hg concentration in the leaves (LHC) for the hybrid was 43.83 µg kg^{-1} in Xixian and 45.23 µg kg^{-1} in Changge. For the LHC of the RIL populations, the average value was 41.06±6.65 µg kg^{-1} (range 27.30–61.45 µg kg^{-1}) in Xixian and 39.60±11.79 µg kg^{-1} (range 18.64–74.50 µg kg^{-1}) in Changge.

Among the different tissues of maize, the kernels had the lowest Hg concentration, followed by the axis, stems, bracts and leaves, and both locations showed the same trend. According to variance analysis, the Hg concentration for the five measured tissues (kernels, axis, stems, bracts, and leaves) in the RIL population exhibited significant variations between genotypes and environments, as well as in the interaction of genotypes and environments (Table 2, P<0.01). For the Hg concentration in five measured tissues in Xixian, the KHC and AHC displayed significant positive relationships, yet the BHC positively correlated with LHC.

However, there were no significant relationships among the Hg concentrations in the five measured tissues in the RIL population in Changge by the phenotypic relationship analysis (Table 3, P< 0.01).

QTL analysis for Hg concentration in the five tissues of maize

Twenty-three QTLs were detected for Hg concentration in the five tissues of maize in the RIL population at the two locations (Table 4, Fig 2). These QTLs were distributed on chromosomes 1, 4, 7, 8, 9 and 10. Five QTLs for Hg concentration in the kernels were identified in the two environments. In Xixian, there were three QTLs; two of them, qKHC9a and qKHC9b was adjacent and explained 10.75% and 10.85% of the phenotypic variance, respectively. These two were derived from the parent 87-1. In Changge, two QTL qKHC8 and qKHC10 were identified, with 13.00% and 8.42% contribution rate to total phenotypic variance, which came from the parent Zong3 and 87-1, respectively.

For the Hg concentration in the axis, two QTLs, qAHC4 and qAHC7 were detected in Xixian and qAHC8 was detected in

Table 1. Hg concentration in the five measured tissues in the RIL population.

Location	Population	Trait	KHC (µg kg⁻¹)	AHC (µg kg⁻¹)	SHC (µg kg⁻¹)	BHC (µg kg⁻¹)	LHC (µg kg⁻¹)
Xixian	P_1	Mean	3.53	2.89	3.71	8.80	30.20
	P_2	Mean	2.14	1.51	6.02	6.21	43.20
	F_1	Mean	4.62	3.40	5.15	5.95	43.83
	RIL	Mean	2.99±1.25	3.84±1.49	5.36±1.30	7.59±1.81	41.06±6.65
		Range	0.73–7.47	1.08–8.37	1.82–9.02	4.07–15.56	27.30–61.45
		Skewness	1.05	0.83	−0.14	1.54	0.28
		Kurtosis	1.51	0.66	0.34	4.99	−0.06
Changge	P_1	Mean	1.96	1.86	4.70	9.18	43.14
	P_2	Mean	0.91	3.71	2.83	8.23	37.31
	F_1	Mean	2.74	3.32	3.67	7.14	45.23
	RIL	Mean	2.37±1.52	2.71±1.76	4.26±2.32	5.83±1.84	39.60±11.79
		Range	0.26–7.18	0.18–8.18	0.47–11.83	2.06–14.03	18.64–74.50
		Skewness	1.11	0.94	0.76	1.11	0.92
		Kurtosis	0.90	0.25	0.56	2.51	0.46

Note: KHC, Hg concentration in kernels; AHC, Hg concentration in the axis; SHC, Hg concentration in stems; BHC, Hg concentration in bracts; LHC, Hg concentration in leaves.

Table 2. Variance analysis of the five measured tissues in the RIL population.

	Kernel	Axis	Stem	Bract	Leave
L	135.49**	161.05**	59.1**	205.2**	5.43*
G	7.03**	2.68**	7.17**	3.07**	6.71**
L×G	7.37**	2.83**	6.46**	3.88**	8.48**

Note: *significant at $P<0.05$, **significant at $P<0.01$.

Changge, which contributed 8.18%, 22.10% and 6.69% of the phenotypic variance in the Hg concentration in the axis, respectively. The alleles of *qAHC4* and *qAHC7* were derived from the parent Zong3, and the *qAHC8* allele came from the parent 87-1.

Six QTLs associated with SHC were detected at the two locations, which were located on chromosomes 1, 4 and 8. The QTLs contributed 6.70%–20.63% of the total phenotypic variance. Among the QTLs, *qSHC8b* was found at both environments simultaneously, accounting for 20.63% and 6.70% of the total variance, and the effects resulted from the parent Zong3.

Four QTLs were associated with BHC in this study. In Xixian, two QTLs, *qBHC8a* and *qBHC9*, contributed 7.80% and 8.30% of phenotypic variance, respectively. In Changge, the QTLs *qBHC4* and *qBHC8b* explained 7.34% and 6.44% of the phenotypic variance of BHC, respectively. All these QTLs were derived from the high performance parent, Zong3.

For the Hg concentration in the leaves, there were five QTLs at both locations, which explained 9.22%–16.46% of the phenotypic variance. QTL *qLHC4a* was found at both locations simultaneously, and was responsible for 16.46% and 14.38% of the total variance, respectively.

Discussion

Hg pollution has aroused global concern because of its toxicity to organisms, persistence in the environment and long-range transport [25]. In some contaminated areas, food chain transfer is very common [31–34]. In rice, the Hg concentration is much higher in roots than in shoots [14]. Similar results were obtained from other plant species [35,36]. Meng et al. reported that Hg levels in rice tissues followed the trend: root > stalk> leaf > husk > seed [37], which is consistent with previous studies [38–40]. Liu et al. found that the Hg content in the different tissues of maize had features as follows: root > leaf > stalk > grain [41]. In this study, we observed a similar distribution in Hg concentration in different maize tissues (leaves > bracts > stems >axis > kernels).

These results demonstrated that the Hg concentration in kernels was lower than that in the other tissues, and the mechanism of Hg accumulation and distribution in different tissues of maize was possibly related to the plant detoxification mechanism. In this study, the Hg concentration in soil in Xixian was higher (457.57 ± 31.30 µg kg^{-1}) than that in Changge (345.40 ± 22.24 µg kg^{-1}). Compared with the average Hg concentration for the five measured tissues in the two locations, the average value in the RIL population was higher in Xixian than in Changge for all tissues. These phenomena indicated that the Hg concentrations in the different tissues of maize were mainly reflected in the Hg concentration in the soil.

Heavy metals are increasingly becoming environmental concerns because of their release into ecosystems, which pose a threat to human health [23]. Low-cost and ecologically sustainable strategies are needed to restore heavy metal-contaminated soils [28]. However, physical or chemical methods to address heavy metal contamination are environmentally invasive, expensive and inefficient, especially for large scale clean up [42]. Phytoremediation is considered a cost-effective and environmentally friendly approach to remove heavy metals from contaminated soils [43]; however, its application is limited because it is time consuming, there is a lack of suitable plant species and production during remediation is unprofitable [44]. Yu et al. raised the concept of pollution-safe cultivars (PSCs), in which a specific pollutant was accumulated by the edible part at a low level for safe consumption, even when planted in contaminated soil [45]. As an alternative choice, there has been increasing interest in selecting and breeding PSCs [46]. In this study, we found that Hg accumulated in high concentrations in the axis, bracts, stems and leaves, which represent the main biomass products in maize, while the kernels contained the lowest concentration among the five tested tissues. For KHC in the RIL population, two parents and the hybrid, the highest value was 4.62 µg kg^{-1} in Xixian and 2.74 µg kg^{-1} in Changge; however, the values were much lower than 20 µg kg^{-1}, which is the maximum recommended by the Chinese National Standard Agency [47]. Maize is the most commonly planted crop

Table 3. Correlation coefficients among five measured tissues in the RIL population.

	Kernel	Axis	Stem	Bract	Leave
Kernel		0.25**	0.16	−0.07	−0.03
Axis	0.12		0.04	0.09	0.01
Stem	−0.07	−0.01		0.05	0.05
Bract	−0.05	−0.11	−0.03		0.27**
Leave	−0.05	−0.04	0.09	0.15	

Note: **significant at $P<0.01$.
Correlation coefficients in Xixian are above the diagonal, while those in Changge are below the diagonal.

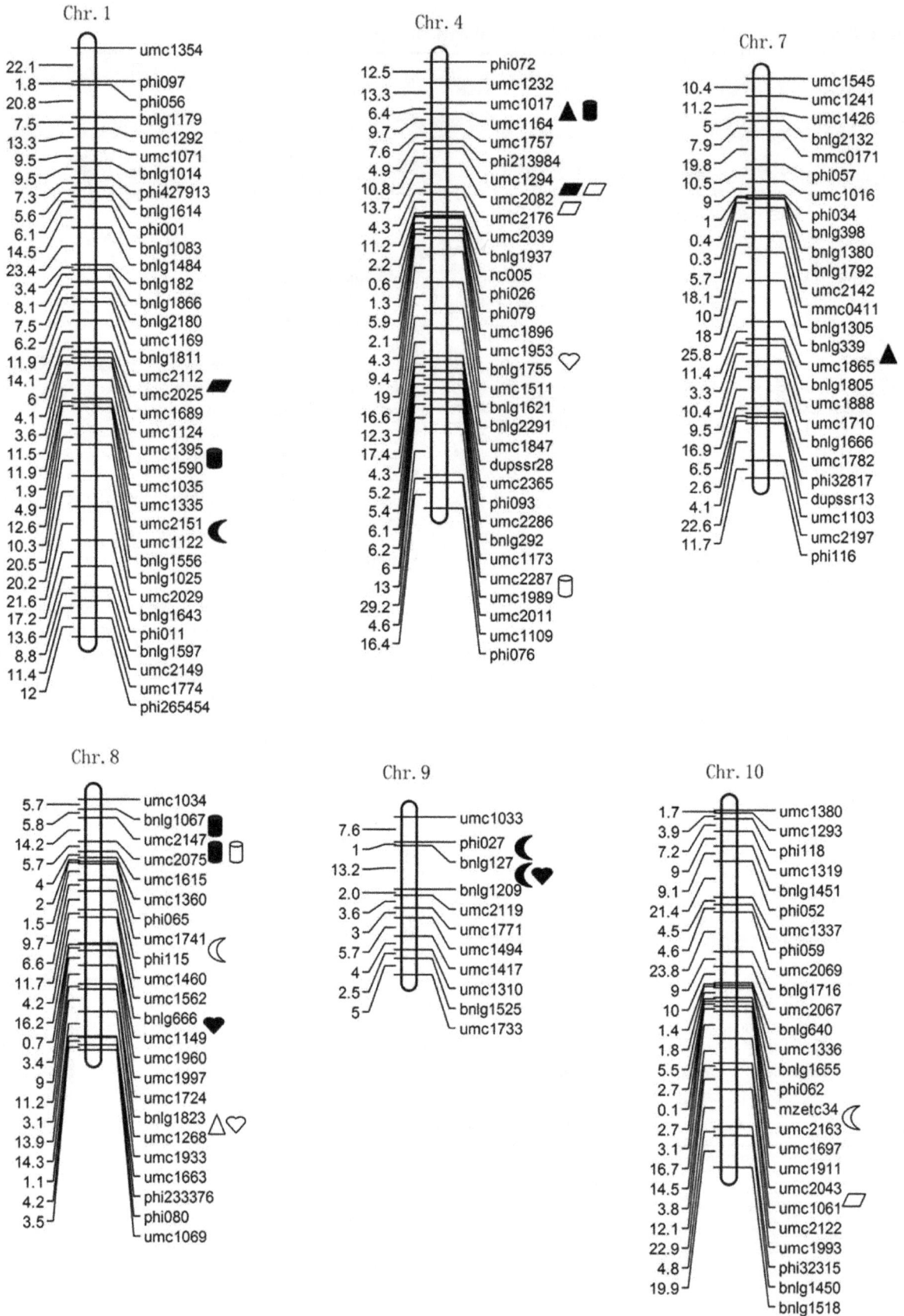

Figure 2. Chromosomal locations of QTLs for Hg concentration for five maize tissues. Note: *Moon* QTL detected for Hg content in kernels, *Triangle* QTL detected for Hg content in the axis, *Cylinder* QTL detected for Hg content in stems, *Heart* QTL detected for Hg content in bracts, *Quadrangle* QTL for Hg content in leaves. Black indicates a QTL detected in Xixian, and lucidity indicates a QTL detected in Changge.

Table 4. QTLs detected for Hg concentration in five maize tissues.

Location	Tissue	QTL[a]	Location	Flanking-markers	LOD[b]	A[c]	R[2d]
Xixian	kernel	qKHC1	238.51	umc2151-umc1122	3.66	0.51	15.11
		qKHC9a	9.61	phi027-bnlg127	3.06	−0.42	10.75
		qKHC9b	18.91	bnlg127-bnlg1209	2.50	−0.42	10.85
	axis	qAHC4	25.81	umc1017-umc1164	2.86	0.44	8.18
		qAHC7	145.31	bnlg339-umc1865	3.08	0.70	22.10
	stem	qSHC1	211.31	umc1395-umc1590	2.50	0.47	11.76
		qSHC4a	30.81	umc1017-umc1164	3.62	0.43	10.94
		qSHC8a	6.71	bnlg10067-umc2147	5.88	0.56	18.24
		qSHC8b	16.51	umc2147-umc2075	4.20	0.59	20.63
	bract	qBHC8a	71.11	bnlg666-umc1149	2.65	0.52	7.80
		qBHC9	14.91	bnlg127-bnlg1209	2.85	0.53	8.30
	leave	qLHC1	178.51	umc2112-umc2025	2.64	−2.07	9.30
		qLHC4a	57.41	umc1294-umc2082	3.19	−3.08	16.46
Changge	kernel	qKHC8	46.91	umc1741-phi115	3.18	0.69	13.00
		qKHC10	115.61	mzetc34-umc2163	3.16	−0.46	8.42
	axis	qAHC8	111.61	bnlg1823-umc1268	2.55	−0.52	6.69
	stem	qSHC4b	221.71	umc2287-umc1989	3.76	−0.88	13.94
		qSHC8b	25.51	umc2147-umc2075	2.71	0.61	6.70
	bract	qBHC4	109.51	umc1953-bnlg1755	2.81	0.53	7.34
		qBHC8b	114.61	bnlg1823-umc1268	2.66	0.48	6.44
	leave	qLHC4a	59.41	umc1294-umc2082	3.55	4.63	14.38
		qLHC4b	69.21	umc2082-umc2176	2.51	3.94	10.39
		qLHC10	138.21	umc2043-umc1061	3.38	3.73	9.22

Notes: [a]QTL detected for Hg concentration in the five maize tissues; [b]LOD for each QTL; [c]Additive effect; positive values indicate that Zong3 alleles increase rates; [d]R^2, contribution ratio.

worldwide, with broad adaptability. Thus, there is the potential to decrease the concentration of Hg in Hg-contaminated paddy soil, and maintain maize production simultaneously by selecting and breeding maize Hg PSCs.

There are many forms of Hg in soil, including the elemental (Hg^0), ionic (Hg^{2+}), hydroxide ($Hg(OH)_2$), methyl (MeHg), and sulfide (HgS) [48,49]. The environment and the genotype control the uptake of Hg species in rice [50], and rice varieties differ widely in their absorption and tolerance of Hg toxicity [14]. In peas, six Hg^{2+} responsive genes were identified using suppression subtractive hybridization [51]. Their gene products were involved in the salicylic acid (SA) biological defense system, the biosynthetic pathway of isoflavonoids, antioxidant reactions, sulfur metabolism, and cell wall rigidity [51]. Heidenreich et al. profiled the transcriptome of *A. thaliana* exposed to Hg^{2+}, and identified Hg-induced genes that encoded proteins involved in chlorophyll synthesis, cell wall metabolism, and P450-mediated biosynthesis of secondary metabolites [52]. In addition, Yamaguchi et al. found that Hg and other heavy metals could induce certain genes at the same time [53]. Recently, several other genes related to Hg accumulation have been identified [54,55]. In rice, 20 proteins that were differentially expressed in roots under Hg treatment

were identified using 2-D electrophoresis. They were involved in cellular functions, including redox and hormone homeostasis, chaperone activity, metabolism, and transcription regulation [56]. Wang et al. identified three QTLs for relative root length in rice under Hg stress, which coincided with QTLs for drought, Zn^{2+}, Fe^{2+}, Cd^{2+}, Cu^{2+} and Al^{3+} toxicity tolerance in rice [19]. In this study, out of the 23 QTLs for Hg concentration in different maize tissues, three (*qKHC9a, qKHC9b, qBHC9*) were adjacent to two QTLs for drought tolerance using the $F_{2:3}$ populations derived from the same parents Zong3 and 87-1 [57]; one QTL, *qLHC1*, was in linkage with two QTLs for arsenic accumulation using RIL populations derived from a cross between two maize inbred lines, Huang-C and Xu178 [58]. These results suggested that the QTLs related to Hg accumulation had pleiotropic effects.

Author Contributions

Conceived and designed the experiments: WL ZL JT. Performed the experiments: Zhongjun Fu LW QZ XZ GS. Analyzed the data: Zhongjun Fu Zhiyuan Fu DD. Contributed to the writing of the manuscript: Zhongjun Fu JT.

References

1. Zhang L, Wong MH (2007) Environmental mercury contamination in China: Sources and impacts. Environment International 33: 108–121.
2. Chen J, Shiyab S, Han FX, Monts DL, Waggoner CA, et al. (2009) Bioaccumulation and physiological effects of mercury in Pteris vittata and Nephrolepis exaltata. Ecotoxicology 18: 110–121.
3. Elbaz A, Wei YY, Meng Q, Zheng Q, Yang ZM (2010) Mercury induced oxidative stress and impact on antioxidant enzymes in Chlamydomonas reinhardtii. Ecotoxicology 9: 1285–1293.
4. Luis R, Jesusa R, Isaac A, Laura RC (2007) Capability of selected crop plants for shoot mercury accumulation from polluted soils: Phytoremediation Perspectives. International Journal of Phytoremediation 9: 1–13.
5. Du X, Zhu YG, Liu WJ, Zhao XS (2005) Uptake of mercury (Hg) by seedlings of rice (Oryza sativa L.) grown in solution culture and interactions with arsenate uptake. Environmental and Experimental Botany 54: 1–7.
6. Li P, Feng XB, Qiu GL, Shang LH, Li ZG (2009) Mercury pollution in Asia: A review of the contaminated sites. Journal of Hazardous Materials 168: 591–601.
7. China National Environmental Monitoring Station (1990) Background elemental concentrations of Chinese soils. China Environmental Science Press, Beijing, pp. 8–10.
8. Zhou ZS, Huang SQ, Guo K, Mehta SK, Zhang PC, et al. (2007) Metabolic adaptations to mercury-induced oxidative stress in roots of Medicago sativa L. Journal of Inorganic Biochemistry 101: 1–9.
9. Patra M, Sharma A (2000) Mercury toxicity in plants. Botanical Review 66: 379–422.
10. Tamas L, Mistrik I, Huttova J, Haluskova L, Valentovicova K, et al. (2010) Role of reactive oxygen species-generating enzymes and hydrogen peroxide during cadmium, mercury and osmotic stresses in barley root tip. Planta 231: 221–231.
11. Boening DW (2000) Ecological effects, transport, and fate of mercury: A general review. Chemosphere 40: 1335–1351.
12. Zhang WH, Tyerman SD (1999) Inhibition of water channels by HgCl2 in intact wheat root cells. Plant Physiology 120: 849–857.
13. Xylander M, Hagen C, Braune W (1996) Mercury increases light susceptibility in the green alga Haematococcus lacustris. Bot. Acta 109: 222–228.
14. Yu JY, Hu HT, Wang CC, Y L (2011) QTL analysis of mercury tolerance and accumulation at the seedling stage in rice (Oryza sativa L.). Journal of Food Agriculture and Environment 9(2): 748–752.
15. Han Y, Xuan W, Yu T, Fang WB, Lou TL, et al. (2007) Exogenous hematin alleviates mercury-induced oxidative damage in the roots of Medicago sativa. Journal of Integrative Plant Biology 49: 1703–1713.
16. Zheng N, Wang QC, Zheng DM (2007) Mercury contamination and health risk to crops around the zinc smelting plant in Huludao City, northeastern China. Environmental Geochemistry and Health 29: 385–393.
17. Abbas ES, Esmail A, Sharif JS, Seyed MG (2012) Hair mercury levels in six Iranian sub-populations for estimation of methylmercury exposure: A mini-review. Iranian Journal of Toxicology 15: 541–547.
18. Drum DA (2009) Are toxic biometals destroying your children's future? Biometals 22: 697–700.
19. Liu WX, Shen LF, Liu JW, Wang YW, Li SR (2007) Uptake of toxic heavy metals by rice (Oryza sativa L.) cultivated in the agricultural soil near Zhengzhou city, People's Republic of China. Bulletin of Environmental Contamination and Toxicology 79: 209–213.
20. Wang CC, Wang T, Mu P, Li ZC, Yang L (2013) Quantitative trait loci for mercury tolerance in rice seedlings. Rice Science 20(3):238–242.
21. Rugh CL, Wilde HD, Stack NM, Thompson DM, Meagher R B (1996) Mercuric ion reduction and resistance in transgenic Arabidopsis thaliana plants expressing a modified bacterial merA gene. PNAS 93: 3182–3187.
22. Shen Q, Jiang M, Li H, Che LL, Yang ZM (2011) Expression of a Brassica napus heme oxygenase confers plant tolerance to mercury toxicity. Plant Cell and Environment 34: 752–763.
23. Wei YY, Zheng Q, Liu ZP, Yang ZM (2011) Regulation of tolerance of Chlamydomonas reinhardtii to heavy metal toxicity by heme oxygenase-1 and carbon monoxide. Plant and Cell Physiology 52(9): 1665–1675
24. Chen Z, Pan YH, Wang SS, Ding YF, Yang Wj, et al. (2012) Overexpression of a protein disulfide isomerase-like protein from Methanothermobacter thermo-autotrophicum enhances mercury tolerance in transgenic rice. Plant Science 197: 10–20.
25. Niu ZC, Zhang XS, Wang ZW, Ci ZJ (2011) Field controlled experiments on the physiological responses of maize (Zea mays L.) leaves to low-level air and soil mercury exposures. Enviroment Pollution 159: 2684–2689.
26. Priyanka G, Meeta J, Juliana S, Rekha G (2013) Inhibition of 5-aminolevulinic acid dehydratase by mercury in excised greening maize leaf segments. Plant Physiology and Biochemistry 62: 63–69.
27. Rubén RÁ, Cristina OV, Ana ÁF, Francisca FC, Luis EH (2006) A stress response of Zea mays to cadmium and mercury. Plant and Soil 279: 41–50.
28. Wuana RA, Okieimen FE (2010) Phytoremediation Potential of Maize (Zea mays L.) A Review. African Journal of General Agriculture 6 (4): 275–287.
29. Tang JH, Teng WT, Yan JB, Ma XQ, Meng YJ, et al. (2007) Genetic dissection of plant height by molecular markers using a population of recombinant inbred lines in maize. Euphytica 155: 117–124.
30. Zeng ZB (1994) Precision mapping of quantitative trait loci. Genetics 136: 1457–1468.
31. Qiu GL, Feng XB, Li P, Wang SF, Li GH, et al. (2008) Methylmercury accumulation in rice (Oryza sativa L.) grown at abandoned mercury mines in Guizhou, China. Journal of Agricultural and Food Chemistry 56: 2465–2468.
32. Horvat M, Nolde N, Fajon V, Jereb V, Logar M, et al. (2003) Total mercury, methylmercury and selenium in mercury polluted areas in the province Guizhou, China. Science of the Total Enviroment 304: 231–256.
33. Meng B, Feng XB, Qiu GL, Cai Y, Wang DY, et al. (2010) Distribution patterns of inorganic mercury and methylmercury in tissues of rice (Oryza sativa L.) plants and possible bioaccumulation pathways. Journal of Agricultural and Food Chemistry 58: 4951–4958.
34. Zhang H, Feng XB, Larssen T, Shang LH, Li P (2010) Bioaccumulation of methylmercury versus inorganic mercury in rice (Oryza sativa L.) grain. Environmental Science & Technology 44: 4499–4504.
35. Israr M, Sahi S, Datta R, Sarkar D (2006) Bioaccumulation and physiological effects of mercury in Sesbania drummondii. Chemosphere 65: 591–598.
36. Wang Y, Greger M (2004) Clonal differences in mercury tolerance, accumulation and distribution in willow. Journal of Environmental Quality 33: 1779–1785.
37. Meng M, LI B, Shao JJ, Wang TH, He Bin, et al. (2014) Accumulation of total mercury and methylmercury in rice plants collected from different mining areas in China. Environmental Pollution 184: 179–186.

38. Sierra MJ, Millán R, Esteban E (2008) Potential use of Solanum melongena in agricultural areas with high mercury background concentrations. Food and Chemical Toxicology 46: 2143–2149.

39. Sierra MJ, Millán R, Cardona AI, Schmid T (2011) Potential cultivation of Hordeum Vulgare L. in soils with high mercury background concentrations. International Journal of Phytoremediation 13: 765–778.

40. Zornoza P, Millán R, Sierra MJ, Seco A, Esteban E (2009) Efficiency of the white lupin in the removal of mercury from contaminated soils: soil and hydroponic experiments. Journal of Environmental Sciences 22(3): 421–427.

41. Liu R, Wang Q, Lv X, Li Z, Wang Y (2004) Distribution and stock of mercury in typical wetland plant in the Sanjiang Plain. Chinese Journal of Applied Ecology 15: 287–290.

42. Kärenlampi S, Schat H, Vangronsveld J, Verkleij JAC, Lelie D, et al. (2000) Genetic engineering in the improvement of plants for phytoremediation of metal polluted soil. Environmental Pollution 107: 225–231.

43. Kramer U (2005) Phytoremediation: novel approaches to cleaning up polluted soils. Current Opinion in Biotechnology 16: 133–141.

44. Pilon-Smits EAH, Freeman JL (2006) Environmental cleanup using plants: biotechnological advances and ecological considerations. Frontiers in Ecology and the Environment 4 (4): 203–210.

45. Yu H, Wang JL, Fang W, Yuan J, Yang ZY (2006) Cadmium accumulation in different rice cultivars and screening for pollution-safe cultivars of rice. Science of the Total Enviroment 370(2–3): 302–309.

46. Grant CA, Clarke JM, Duguid S, Chaney RL (2008) Selection and breeding of plant cultivars to minimize cadmium accumulation. Science of the Total Environment 390(2–3): 301–310.

47. Chinese National Standard Agency (1994) Tolerance limit of mercury in foods. GB 2762–94. Beijing, China.

48. Heaton ACP, Rugh CL, Wang NJ, Meagher RB (2005) Physiological responses of transgenic merA-tobacco (Nicotiana tabacum) to foliar and root mercury exposure. Water Air and Soil Pollution 161: 137–155.

49. Han FX, Su Y, Monts DL, Waggoner CA, Plodinec MJ (2006) Binding, distribution, and plant uptake of mercury in a soil from Oak Ridge, Tennessee, USA. Science of the Total Enviroment 368: 753–768.

50. Rothenberg SE, Feng XB, Zhou WJ, Tu M, Jin BW, et al. (2012) Environment and genotype controls on mercury accumulation in rice (Oryza sativa L.) cultivated along a contamination gradient in Guizhou, China. Science of the Total Enviroment 426: 272–280.

51. Sävenstrand H, Strid Å (2004) Six genes strongly regulated by mercury in Pisum sativum roots. Plant Physiology and Biochemistry 42: 135–142.

52. Heidenreich B, Mayer K, Sandermann JR, Ernst D (2001) Mercury-induced genes in Arabidopsis thaliana: identification of induced genes upon long-term mercuric ion exposure. Plant Cell and Environment 24: 1227–1234.

53. Yamaguchi H, Fukuoka H, Arao T, Ohyama A, Nunome T, et al. (2010) Gene expression analysis in cadmium-stressed roots of a low cadmium-accumulating solanaceous plant, Solanum torvum. Journal of Experimental Botany 61: 423–437.

54. Hsieh JL, Chen CY, Chiu MH, Chein MF, Chang JS, et al. (2009) Expressing a bacterial mercuric ion binding protein in plant for phytoremediation of heavy metals. Journal of Hazardous Materials 161: 920–925.

55. Ruiz ON, Alvarez D, Torres C, Roman L, Daniell H (2011) Metallothionein expression in chloroplasts enhances mercury accumulation and phytoremediation capability. Plant Biotechnology Journal 9: 609–617.

56. Chen YA, Chi WC, Huang LT, Lin CY, Quynh Nguyeh TT, et al. (2012) Mercury-induced biochemical and proteomic changes in rice roots. Plant Physiology and Biochemistry 55: 23–32.

57. Liu Y, Subhash C, Yan JB, Song CP, Zhao JR, et al. (2011) Maize leaf temperature responses to drought: Thermal imaging and quantitative trait loci (QTL) mapping. Enviroment and Experimental Botany 71: 158–165.

58. Ding D, Li WH, Song GL, Qi HY, Liu JB, et al. (2011) Identification of QTLs for arsenic accumulation in maize (Zea mays L.) using a RIL population. PLOS ONE 6: e25646–25652.

13

Organic Fertilization and Sufficient Nutrient Status in Prehistoric Agriculture? – Indications from Multi-Proxy Analyses of Archaeological Topsoil Relicts

Franziska Lauer[1], Katharina Prost[1]*, Renate Gerlach[2], Stefan Pätzold[1], Mareike Wolf[1], Sarah Urmersbach[1], Eva Lehndorff[1], Eileen Eckmeier[1,3], Wulf Amelung[1]

1 Institute of Crop Science and Resource Conservation – Soil Science and Soil Ecology, University of Bonn, Bonn, Germany, **2** Archaeological Heritage Management Rhineland (LVR-Amt für Bodendenkmalpflege im Rheinland), Bonn, Germany, **3** Department of Geography, RWTH Aachen University, Aachen, Germany

Abstract

Neolithic and Bronze Age topsoil relicts revealed enhanced extractable phosphorus (P) and plant available inorganic P fractions, thus raising the question whether there was targeted soil amelioration in prehistoric times. This study aimed (i) at assessing the overall nutrient status and the soil organic matter content of these arable topsoil relicts, and (ii) at tracing ancient soil fertilizing practices by respective stable isotope and biomarker analyses. Prehistoric arable topsoils were preserved in archaeological pit fillings, whereas adjacent subsoils served as controls. One Early Weichselian humic zone represented the soil status before the introduction of agriculture. Recent topsoils served as an additional reference. The applied multi-proxy approach comprised total P and micronutrient contents, stable N isotope ratios, amino acid, steroid, and black carbon analyses as well as soil color measurements. Total contents of P and selected micronutrients (I, Cu, Mn, Mo, Se, Zn) of the arable soil relicts were above the limits for which nutrient deficiencies could be assumed. All pit fillings exhibited elevated $\delta^{15}N$ values close to those of recent topsoils ($\delta^{15}N > 6$ to 7‰), giving first hints for prehistoric organic N-input. Ancient legume cultivation as a potential source for N input could not be verified by means of amino acid analysis. In contrast, bile acids as markers for faecal input exhibited larger concentrations in the pit fillings compared with the reference and control soils indicating faeces (i.e. manure) input to Neolithic arable topsoils. Also black carbon contents were elevated, amounting up to 38% of soil organic carbon, therewith explaining the dark soil color in the pit fillings and pointing to inputs of burned biomass. The combination of different geochemical analyses revealed a sufficient nutrient status of prehistoric arable soils, as well as signs of amelioration (inputs of organic material like charcoal and faeces-containing manure).

Editor: Liping Zhu, Institute of Tibetan Plateau Research, China

Funding: This study was funded by the German Research Foundation (DFG; SFB 806, D2) and the Stiftung zur Förderung der Archäologie im rheinischen Braunkohlenrevier (216). The funders had no role in study design, data collection and analysis, decision to publish, or preparation of the manuscript.

Competing Interests: The authors have declared that no competing interests exist.

* Email: kprost@uni-bonn.de

Introduction

Human activities had an impact on landscapes since the Neolithic, e.g. by the construction of settlements, arable cropping and animal husbandry. As a result, also the chemical composition of the soils changed [1–3]. These chemical changes of ancient soils may still be preserved in buried soils and, therefore, may be used to elucidate human activities inside and outside prehistoric settlements, for instance by analyzing prehistoric topsoil relicts. Buried ancient topsoil relicts can be preserved as fillings of pits that have been constructed outside prehistoric settlements. These pits are defined as off-site features that do not contain any characteristic anthropogenic artefacts, settlement material, or settlement related nutrient inputs. They have been part of the prehistoric agrarian landscape and therefore their dark humic soil filling presumably consists to a large extent of prehistoric topsoil from an open landscape that was most likely arable land [4–6]. Some of these arable topsoil relicts were buried in deep pit features, and hence have largely been preserved from the influence of roots, bioturbation and weathering processes by overlying sediments [6]. Examples for deep off-site features are man-made slot pits ("Schlitzgruben") and man-made pit alignments located outside prehistoric settlements [4,7] in which relocated topsoil material that bears information on prehistoric nutrient status and fertilization measures has been conserved as the infilling of those pits [8]. It was shown that pit fillings investigated in German archaeological excavation sites usually exhibited larger contents of soil organic carbon, total nitrogen, as well as extractable and organic phosphorus compared with the adjacent subsoils [4,8]. However, the origin of this organic matter remained unclear.

Geochemical analyses of ancient topsoils have been conducted since the 1930s as means of archaeological prospection – mainly in settlement areas [9–11]. In recent years new methods have enlarged the spectrum of analyses for archaeological soil material. The mapping of activity areas within settlements on the basis of soil phosphorus (P) concentrations is an established method in archaeology (for general overviews see [12–15]). However, the differentiation between the various P fractions has scarcely been explored so far [16], especially in off-site features like slot pits and pit alignments [8]. Elevated extractable P (i.e. sum of all sequentially extracted inorganic and organic P fractions) and plant available inorganic P contents (i.e. sum of resin-P_i, $NaHCO_3$-P_i and NaOH-P_i) in ancient topsoil relicts compared with control samples can be seen as first hints for soil amelioration in prehistoric times [8]. Yet, it remained uncertain to which degree elevated contents of organic P were the result of a P-cycling from mineral phases by plants or microorganisms, or an indication of organic fertilization. For the differentiation between both sources, w more information on other nutrients that are less abundant in the organic phase (like micronutrients) and on the origin of soil organic matter (SOM) preserved within prehistoric topsoils are required.

Several geochemical methods for characterizing soil micronutrient contents and for detecting traces of fertilization are nowadays available. Therefore, we combined the analyses of different chemical markers in archaeological soil samples. Some micronutrients possess a low mobility in soils, and are therefore suitable to track past human activities in settlement areas, e.g. to separate areas of food preparation, hearths, manuring or craft working [17,18]. Depending on the soil properties in distinct regions, even human health can be affected by micronutrient deficiency [19], e.g. the iodine content of food depends on the iodine content of the soil in which it is grown. Prehistoric fertilization using nitrogen fixing plants might be evidenced by measurements of stable nitrogen isotope and the amino acid composition in ancient topsoil material. This is possible because legumes fix relatively more of the "light" ^{14}N isotope, leading to a changed ^{14}N to ^{15}N ratio [20–22], and the amino acid composition may be specific for distinct crops like, e.g. legumes [23–25]. Moreover, compound-specific $\delta^{15}N$ analysis of amino acids in soil has been used to identify land use and manure application in archaeological contexts (i.e. in Medieval and Bronze Age time periods) [26]. Beside the cultivation of legumes, application of livestock manure or human faeces might have been practiced to preserve and enhance soil fertility in prehistory. Measurements of specific steroids, i.e. 5β-stanols and bile acids, provide a reliable tool to identify ancient faecal deposition. Among the steroids, 5β-stanols such as coprostanol and 5β-stigmastanol were found to be indicators for the faecal input of omnivores and herbivores species, respectively [27–30]. However, there are studies that could also show a contribution of microbial sterol degradation to an enhanced 5β-stanol signal in sediments and sewage sludge [31,32]. Therefore, there is most likely a natural background concentration of 5β-stanols in soils. Similarly to 5β-stanols, stanones are also thought to be indicative for the input of faeces because they are generated both via intestinal and sedimentary processes of sterol reduction [33–35]. Additionally, bile acids serve as faecal markers because they are exclusively formed by vertebrates, are presumed to exhibit an greater resistance to degradation than 5β-stanols [36], and even allow to distinguish between different faecal sources (human, ruminant, and porcine) [28–30,37]. Up to now there are only few studies that have used stanones as biomarker [35,38,39]. Even fewer studies investigated bile acids to determine manuring in prehistory, because of the requirement of a complex chemical isolation procedure [40].

The burning of the fields was another activity connected to prehistoric farming, as it allowed to clean the soils from native vegetation, to fertilize them and to fight weeds [41,42]. These former slash-and burn practices can be evidenced by the combination of black carbon analysis and soil color measurements [7,43–45]. The objective of this study was to contribute to a better understanding of the nutrient status and organic matter properties of ancient topsoil relicts in archaeological pit fillings using a combination of different geochemical analyses. We hypothesized that prehistoric agriculture left fingerprints of altered soil nutrient contents and fertilization. Therefore, we (i) elucidated the total P and micronutrient status of prehistoric arable soils, and ii) characterized the origin of soil organic matter using natural $\delta^{15}N$ abundance as markers for legume cultivation and manure application [46,47], amino acid composition as markers for the cultivation of legumes [23–25], benzene polycarboxylic acids and color measurements as markers for charred remains in soils [7,43–45], and sterols and bile acids as markers for faecal residues [1,30,48,49].

Materials and Methods

For the described study all necessary permits were obtained from the Museum and Archaeological State Service of Saxony-Anhalt and the Archaeological Heritage Management Rhineland, which complied with all relevant regulations.

1 Sites and archaeological topsoil relicts

Selected archaeological topsoil relicts (i.e. fillings of slot pits and pit alignments) were investigated from two prehistoric settlement areas in Germany: Central (now comprising Chernozems and Phaeozems) and Western Germany (now comprising Luvisols) (Table 1; see also Lauer et al. [8]). Both regions had been settled since Early Neolithic time by sedentary farmers. The studied soil relicts were preserved in deep features such as slot pits found in the surroundings of Middle and Younger Neolithic sites (i.e. around 7,000 until 5,500 before present (BP); [50,51]) and pit alignments dating to the Younger Bronze Age or Early Iron Age (between 3,200 and 2,700 BP; [52]). Both slot pits and pit alignments are enigmatic features, situated outside prehistoric settlements presumably in the areas of prehistoric arable fields (off-site features) and normally do not contain any archaeological artefacts. Until now, little is known about the function of the slot pits [7,50,53]. They have a slot-like appearance, being up to 3 m deep, 2 m long and narrow (0.2–1 m), and they are defined as anthropogenic due to their regular and specific shape and because they often appear in regularly arranged groups [50,51]. With respect to their narrow shape it is presumed that they remained open for only a few days to several weeks and were refilled with autochthonous topsoil and subsoil materials, either on purpose, via erosion, or both. Humic soil material (= ancient topsoil material) in archaeological features like pits or ditches is regarded as prominent archaeological finding [4–6,54].

The Bronze Age pit alignments are evenly spaced pits arranged in lines which run for large distances in straight or curving lines [52]. Again, little is known about their original function [52]. Like in the case of the slot pits, also the pit alignments are clearly anthropogenic as proved by their shape [52]. Similarly to the slot pits the humic filling of these pits represents the prehistoric topsoil in off-site positions [50,51]. More detailed information about the archaeological topsoil relicts and the sampling strategy can be found in Lauer et al. [8].

Because of the complexity of the investigated proxies, this study focused on ten selected slot pits (n = 14 soil samples), six selected pits of pit alignments (n = 6 soil samples) as well as one Early Weichselian humic zone (Table 1). The Early Weichselian humic zone (i.e. "Humuszone": dark humic-rich paleosol, which is considered to be formed in an interstadial period during the Early

Table 1. Characterization of the study sites.

Region/ excavation site	Number of pits	Type of soil archive	Age of topsoil relicts	Geographical coordinates	MAP[a] mm	MAT[b] °C	Soil parent material
Central Germany							
Chernozem region							
Kleingräfendorf	2	Pit alignment	Bronze Age	51° 22′ N, 11° 51′ E	<500[c]	8.5–9[c]	loess
Oechlitz	1	Pit alignment	Bronze Age	51° 19′ N, 11° 45′ E			loess
Jüdendorf	3	Pit alignment	Bronze Age	51° 18′ N, 11° 42′ E			loess
Phaeozem region							
Prießnitz	2	Slot pit	Neolithic	51° 6′ N, 11° 46′ E	560–580[c]	8–8.5[c]	loess
Western Germany							
Luvisol region							
Merzenich[e]	6	Slot pit	Neolithic	50° 49′ N, 6° 32′ E	650–700[d]	10–11[d]	loess
Pulheim	2	Slot pit	Neolithic	50° 59′ N, 6° 48′ E			loess
Düren Arnoldsweiler (humic zone)[e]	1	Humic zone	Early Weichselian	50° 51′ N, 6° 30′ E			loess

[a]MAP = mean annual precipitation;
[b]MAT = mean annual temperature;
[c]Kropp et al. [133];
[d]Genßler et al. [134],
[e]excavation sites were selected for steroid analyses.

Weichselian; [55]), was located in Düren Arnoldsweiler (Lower Rhine Basin, Western Germany). Thus, this paleosol is assumed to represent the soil status before the onset of farming activity. It may therefore be assumed that the respective sample received if at all, only sparse amounts of wild-life faeces but no significant amounts of animal manure, like manured fields do. Thus, this paleosol served as a reference for the comparison of arable vs. natural soil and is designated as "reference soil" in the following. A comparable humic zone in Garzweiler/Elsbachtal also located in the Lower Rhine Basin (ca. 30 km distance apart) was dated to 87,100±8,300 BP [56]. Hence, the investigated humic zone of Düren Arnoldsweiler is considered to be of the same age.

The subsoils (n = 20 soil samples) were sampled in the same depth, but at short lateral distances from the pit fillings (<0.5 m). Moreover, we sampled recent topsoils (n = 7 soil samples). These samples served as control against post-depositional translocation processes. It should be noted, though, that along small distances we cannot fully discount the possibility that there was a lateral transport of any of the substances under study from the pit filling into the adjacent subsoil as well as vice versa (e.g. via water or soil dwelling animals). For the same reasons, also a vertical transport into the subsoil from upper soil layers by the same mechanisms cannot be ruled out. Hence, we did not expect to find zero biomarker and element contents for all parameters in the subsoils. The pit fillings and adjacent subsoils, however, were deep buried features (on average with a depth of 1.60 m). Additionally, we assumed that vertical leaching with convective-dispersive flow from the recent topsoil is unlikely for the majority of the studied, more or less immobile soil constituents (P, some micronutrients, peptide-bound amino acids, steroids, black carbon; see also Lauer et al. [8]).

2 Basic soil properties

Prior to laboratory analyses, all samples were dried and sieved to <2 mm. Sub-samples were ball-milled. For all samples the total carbon, nitrogen, and carbonate contents, as well as pH-values (CaCl$_2$) were determined according to [57–59]. The grain size distribution was determined in selected samples (n = 24) on the basis of the sedimentation and pipette method by Köhn [60].

3 Micronutrient analysis

The micronutrient contents were determined after *aqua regia* digestion [61] of dried and sieved soil samples with subsequent quantification via inductively coupled plasma optical emission spectrometry (ULTIMA 2 ICP-OES spectrometer, HORIBA, Japan). Besides the total P content, the present analyses concentrated on the following elements: iodine (I), copper (Cu), manganese (Mn), molybdenum (Mo), selenium (Se) and zinc (Zn), in addition we assessed the contents of Ca, Mg, S, Fe, as well as heavy metals like Co, Ni, Ti, Sr, Zr (see Table S1). The selected micronutrients also are typical essential trace elements for human nutrition (I), crops (Mo) or both (Cu, Mn, Se, Zn) [62,63].

4 Stable N isotope analysis

Stable isotope analysis of milled bulk soil samples was carried out by dry combustion in an elemental analyzer (Flash EA, 1112 Series, Thermo Fisher Scientific GmbH, Bremen, Germany) coupled with a Delta V Advantage isotope ratio mass spectrometer (Thermo Fisher Scientific GmbH, Bremen, Germany). Nitrogen isotopic values are expressed in δ^{15}N relative to the isotopic composition of air (reference standard):

$$\delta^{15}N = \left(\frac{^{15}N_{sample} \div ^{14}N_{sample}}{^{15}N_{air} \div ^{14}N_{air}} - 1 \right) \times 10^3, \, where \, ^{15}N / ^{14}N_{air} = 0.003663.$$

Calibration was performed using acetanilide (δ^{15}N: 1.15‰) and ammonium sulfate (δ^{15}N: 20.3‰) as certified standards (purchased from the International Atomic Energy Agency, Vienna, Austria). Precision of analyses was ±0.15‰ for duplicate measurements.

5 Amino acid analysis

Amino acid enantiomers were determined in two replicates by the method of Amelung and Zhang [64]. As microorganisms are known to synthesize a variety of D-amino acids in free and water-soluble forms [65], free amino acids were removed with 1 M HCl (12 h, 25°C; [66]). The remaining soil was hydrolyzed with 6 M HCl (12 h, 105°C), the solution was filtered, purified via cation exchange resins using 0.1 M oxalic acid for metal complexing, eluted with 2.5 M NH$_4$OH, and dried on a rotary evaporator. Amino acid enantiomers were converted into N-pentafluoropropionyl-amino acid isopropyl esters as described by Frank et al. [67]. Gas chromatographic separation of the amino acid derivatives was carried out on an Agilent 6890 gas chromatograph with mass-spectrometer, using a chiral column (Chirasil-L-Val, 25 m×0.25 mm; Agilent Technologies GmbH, Böblingen, Germany) and electron impact ionization. For recovery assessment L-norvaline (internal standard 1) was added after hydrolysis and calculated relative to D-methionine (internal standard 2), which was added before derivatization. Recovery of L-norvaline averaged 83±18% (mean ± standard deviation). Arginine, cysteine, histidine, and tryptophan cannot be measured with this method. Hydrolysis transforms glutamine and asparagine, if present, into their carboxylic acids; so we report concentrations of glutamic acid and aspartic acid. The total amino acid concentration was calculated as the sum of the D- and L-enantiomers.

6 Steroid analysis

The steroid contents were determined according to the method of Birk et al. [49] with slight modifications (Fig. S1; see following section for method details). Due to the complex method only selected samples (n = 13 soil samples) of two excavation sites in Western Germany (i.e. Merzenich and Düren Arnoldsweiler) were analyzed - in order to compare samples from one archive with same history of pedogenesis.

6.1 Steroid extraction, separation and derivatization. Steroid analyses followed the protocol of Birk et al. [49] (see also Fig. S1) with modifications concerning sample extraction and quantification. In brief, for each sample 10 g of milled soil were spiked with 5β-pregnan-3α-ol-20-one, 5β-pregnan-3α-ol, and isodeoxycholic acid as recovery standards and extracted subsequently with dichloromethane/methanol (2:1, v/v) and dichloromethane/methanol (1:3, v/v), using accelerated solvent extraction (ASE Dionex 350; at 100°C). The dried total lipid extracts (TLE) were saponified with 5% KOH in methanol (10–14 h). Afterwards, the extracts were separated into a neutral fraction (including the sterols, stanols and stanones) and an acidic fraction (including the bile acids) by a repeated liquid-liquid extraction with chloroform (3×15 mL), an acidification with 1 M HCl (pH≤2), and a further liquid-liquid extraction with chloroform (3×15 mL).

After evaporation and drying of the extracts, the neutral fraction was separated by solid phase extraction (SPE) using 5% deactivated silica gel and (i) 5 mL hexane (for preconditioning), (ii) 5 mL hexane, (iii) 3 mL dichloromethane and (iv) 2 mL dichloromethane/acetone (2:1, v/v). The second fraction eluted with hexane was discarded; the third and fourth fraction were combined and dried. The acidic fraction was methylated (addition of 1 mL dry 1.25 M HCl in methanol and heating at 80°C for 2 h; then addition of 1 mL Millipore water with following repeated liquid-liquid-extraction with 3×1 mL hexane) and separated in a fatty acid and a bile acid fraction by SPE. Therefore, activated silica gel was used with (i) 5 mL hexane/dichloromethane (2:1, v/v) (for preconditioning), (ii) 4 mL dichloromethane/hexane (2:1, v/v) and (iii) 5 mL dichloromethane/methanol (2:1, v/v). The second eluted fraction was discarded, the third fraction, containing the bile acid methyl esters, was dried.

The sterol, stanole, and stanone extracts were silylated adding Sylon as derivatization reagents and heating at 70°C for 1 h, the bile acid methyl ester extracts by adding 50 μl toluene and N,O-bis(trimethylsilyl)trifluoroacetamide (BSTFA) containing N-trimethylsilylimidazole (TSIM) (98:2, v/v) as derivatization reagents and heating at 80°C for 1 h.

6.2 Steroid measurements and quantification. As a second internal standard α-cholestane (in toluene) was added to both fractions (the bile acid methyl ester as well as the sterol, stanol, and stanone fraction) and finally the samples were analyzed using gas chromatography-mass spectrometry (GC/MS) with an Agilent 5973 quadrupole mass spectrometer coupled to an Agilent 6890 gas chromatograph. Gas chromatographic separation of the steroids was carried out with an Optima-5 MS column, including a 10 m pre-column (40 m×0.25 mm×0.25 μm; Macherey-Nagel, Düren, Germany) and electron ionization. The injection port was set to 250°C and samples were injected in splitless mode. For the sterol, stanol and stanone separation the column temperature program was: 80°C (1.5 min) to 265°C at 12°C min^{-1}, to 288°C at 0.75°C min^{-1}, to 300°C at 10°C min^{-1} (held 12 min), and to 340°C at 25°C min^{-1} (held 5 min). For the bile acid separation the column temperature program was: 80°C (1.5 min) to 265°C at 12°C min^{-1}, to 288°C at 0.75°C min^{-1}, to 300°C at 10°C min^{-1} (held 12 min), and to 340°C at 25°C min^{-1} (held 5 min). Scan mode and the comparison with external standards were used to verify peak identity; measurements in selected ion monitoring mode (SIM) were carried out for quantification. Table S2 shows the steroid structures, the retention times, and the selected characteristic ion fragments.

The quantification of steroids was done using an external standard series (with the relevant, commercially available steroids in five concentrations) with sample matrix for each sample. For sterols, stanols, and stanones the recovery of the first internal standard pregnanolone (5β-pregnan-3α-ol-20-one) averaged 83±21% (mean ± standard deviation), and for bile acids the recovery of isodeoxycholic acid averaged 56±20% (mean ± standard deviation). The limit of quantification was 5 μg kg^{-1} soil for coprostanol, epi-coprostanol, and deoxycholic acid and 10 μg kg^{-1} soil for all other steroids.

7 Black carbon analysis

Black carbon analyses were conducted using the benzene polycarboxylic acid (BPCA) method as described by Glaser et al. [68] with revisions from Brodowski et al. [69]. The samples (two replicates) were first treated with 4 M trifluoroacetic acid (TFA) to remove polyvalent cations and then digested with HNO$_3$ at 170°C for 8 h to yield benzene acids with different degree of carboxylation (for the structure of black carbon and benzene polycarboxylic acid degradation products see scheme in Kögel-Knaber and Amelung [70]). After cleanup via cation exchange resin (Dowex 50 W×8, 200–400 mesh, Fluka, Steinheim, Germany), the samples were silylated and BPCAs were measured using gas

chromatography with flame ionization detection (GC-FID; Agilent 6890 gas chromatograph; Optima-5 column; 30 m×0.25 mm; Supelco, Steinheim, Germany). For recovery assessment citric acid (internal standard 1) was added prior to the cleanup step and calculated to biphenyl dicarboxylic acid (internal standard 2), which was added to the samples prior to derivatization. Carefully monitoring the pH avoided decomposition of citric acid during sample processing (as criticized by Schneider et al. [44]); the recovery of the first internal standard averaged 67±1% (mean ± standard deviation; optimal recovery is >70%). Results are given in the following as the carbon content of BC (BC-C) corrected for loss during sample preparation via internal standard 1 and a conversion factor of 2.27 and normalized to SOC (g BC-C per kg SOC).

8 Color measurements

Dried and homogenized fine-earth samples (<2 mm) were measured in triplicates using a spectrophotometer (CM-5; Konica Minolta, Japan). Reflected light was detected under standardized observation conditions (2° Standard Observer, Illuminant C), and color spectra were obtained in the 360 to 740 nm range, in 10 nm increments. The spectral information was converted into the CIELAB Color Space (L*a*b*; CIE 1976) using the Software SpectraMagic NX (Konica Minolta, Japan). The L* values indicate lightness as the extinction of light, or luminance, on a scale from L* 0 (absolute black) to L* 100 (absolute white).

9 Statistical evaluation

Differences in geochemical parameters between pit filling, adjacent subsoil and recent topsoil were evaluated using STATISTICA (8.0 for Windows; Statsoft Europe GmbH, Hamburg, Germany). For the comparison of normally distributed data, we applied t-tests for paired samples, since the properties of the pit filling were directly related to those of the adjacent subsoil and recent topsoil. The Wilcoxon-Rank-test was used for data that did not pass the test of normal distribution (Shapiro-Wilk-Test). The differences between the soil regions were evaluated by a Student's t-test. The correlation between the amino acid contents and the SOC and total N as well as the correlation between the pH-values and D-lysine contents were determined using the Spearman rank correlation. Also the correlations between the SOC contents, the black carbon contents and the L* values were determined using the Spearman rank correlation. Significance was set at $p < 0.05$ unless otherwise stated.

Results and Discussion

1 Nutrient status of archaeological topsoil relics

In this study, we analyzed key elements in arable soil relics outside ancient settlements in order to gain more information on nutrient contents as they potentially affect plant growth (e.g. P, micronutrients) and human nutrition (e.g. I and Se). We presumed that if a nutrient was deficient, it should be depleted in the prehistoric topsoil relative to its adjacent subsoil, because of significant, continued root uptake of nutrients from topsoils and less intensively from subsoils. The results the *aqua regia* extractable contents of total P and of the micronutrients I, Cu, Mn, Se, and Zn of the pit fillings in the Luvisol and Phaeozem regions could not show a depletion compared with the respective adjacent subsoils, with one exception for the P content in one site in the Chernozem region (i.e. Jüdendorf; Table 2). The pit fillings of this site were also not enriched in extractable P when compared with the adjacent subsoils [8].

Table 2. Contents of selected *aqua regia* extractable elements (in mg kg^{-1} soil) in the pit fillings, adjacent subsoils and recent topsoils of the Chernozem and Phaeozem region.

Region/excavation site	Sample type	n^a	pH^b	P	I	Cu	Mn	Mo	Se	Zn
				[mg kg^{-1} soil]						
Central Germany										
Chernozem region										
Kleingräfendorf	pit filling	2	8.1	388 (348–428)	76 (0–151)	9 (5–13)	399 (388–410)	1 (0–2)	4 (0–8)	31 (25–37)
	subsoil	2	8.2	377 (337–416)	75 (0–150)	6 (3–9)	282 (246–318)	1 (0–2)	4 (0–7)	20 (19–21)
	topsoil	1	7.6	683	216	16	490	1	9	44
Oechlitz	pit filling	1	8.1	873	667	12	421	2	8	37
	subsoil	1	8.2	843	398	9	316	2	8	27
	topsoil	1	7.7	1339	645	15	476	3	10	42
Jüdendorf	pit filling	3	7.8	354 (37)	96 (30)	15 (6)	453 (42)	1 (0.6)	3 (3)	39 (5)
	subsoil	3	7.8	386 (9)	65 (9)	11 (4)	397 (48)	0 (0.1)	0 (5)	38 (5)
	topsoil	3	7.3	423 (21)	44 (21)	17 (5)	512 (41)	0 (0.2)	0	57 (5)
Mean	*pit filling*	*6*	*7.9 ab*	*452 (87) a*	*185 (99) a*	*13 (3) a*	*430 (22) a*	*1 (0.5) a*	*4 (2) a*	*36 (3) a*
	subsoil	*6*	*8.0 ab*	*459 (78) a*	*124 (58) b*	*9 (2) b*	*345 (33) a*	*1 (0.4) a*	*2 (2) a*	*31 (4) a*
	topsoil	*5*	*7.5 b*	*641 (178) a*	*211 (117) ab*	*16 (3) ab*	*505 (24) b*	*1 (0.5) a*	*4 (2) a*	*50 (4) b*
Phaeozem region										
Prießnitz	pit filling	2	6.5 a	593 (513–674) a	229 (138–320) a	13 (13–13) a	859 (511–1207) a	1 (0–1) a	10 (10–11) a	58 (54–62) a
	subsoil	2	7.5 a	375 (322–427) a	133 (76–190) a	14 (13–16) a	1220 (468–1972) a	2 (1–2) a	10 (9–11) a	41 (35–47) a
	topsoil	1	5.3 b	587 a	270 a	13 a	652 a	1 a	10 a	43 a

Means with standard errors in parentheses and for n = 2 each value is listed in parentheses. Within one region, different letters designate significant (*p*<0.05) differences between sample types.

It is noteworthy that the contents of some micronutrients in the pit fillings of all studied soil regions were similar to those in recent fertilized topsoils (Table 2 and Table 3). Similarly, also the contents of other nutrients like Ca, Mg, S, and Co did not point to significantly lowered nutrient stocks in the prehistoric topsoils when compared with recent topsoils (Table S1). Generally, the plant availability of many micronutrients depends on the soil's pH. The pH values of prehistoric topsoils were probably higher than today, because decalcification with subsequent chemical weathering on the primary calcareous loess (Upper Weichselian) did not start before the Late Glacial (see also Schlummer *et al.*, [71]). The micronutrients Cu, Mn, and Zn are immobilized in calcareous soils [72,73], therefore, their available portion in soil might have been small in prehistory despite of their elevated total contents. These micronutrients possess a low mobility under these conditions of relatively high pH values, because they form carbonate complexes or insoluble oxides. Thus, these elements may have been stored for long periods in a soil depth, where acid root exudates may not have found their way to mobilize them [74–76]. On the contrary, the availability of I, Mo, and Se usually increases with increasing soil pH, suggesting that these elements were likely even more available to plants in prehistory than they are nowadays.

In summary, the analyses do not support the hypothesis that there might have been ancient P or micronutrient deficiencies in the studied Neolithic and Bronze Age arable soils due to agricultural land use. Yet, it remains unclear whether there have been nutrient replacements, e.g. in form of N fixation or organic fertilization.

2 Proxies for prehistoric nitrogen input

Stable N isotope signature. Neolithic agriculture included cropping of legumes as shown by archaeobotanical analyses [77]. Nitrogen fixing plants, like legumes, adapt to the isotopic signature of N_2 in the air, and their $\delta^{15}N$ value approaches zero [20–22]. However, the increased abundance of legumes usually showed little if any bulk soil $\delta^{15}N$ discrimination during N_2 fixation [46,47,78]. Here, the recent topsoil of one excavation site (Merzenich; Luvisol region in Western Germany) showed $\delta^{15}N$ values of 5.9‰ (Fig. 1A). The slot pit was covered by a colluvial soil horizon ("M"; 38 cm thick), which exhibited larger $\delta^{15}N$ values (6.7‰; Fig. 1A). The $\delta^{15}N$ values of the slot pits ranged from 4.7‰ to 6.7‰. These values are far off typical values that were found in fields that were cropped with N fixing plants – hence, if there has been a legume cropping, it did not occur long enough to significantly influence the soil $\delta^{15}N$ values.

Recent native grassland whose vegetation frequently contains N fixing clover, usually also shows low soil $\delta^{15}N$ values close to zero, as well (in the range of −1‰ to +2‰, but not larger than 5‰) [47,79–81]. According to these $\delta^{15}N$ values, the prominence of a native grassland vegetation in the studied regions and time periods can be excluded. Also recent forests usually exhibit $\delta^{15}N$ values between −2.8 and +2‰ under comparable climatic conditions [82,83]. Hence, the large $\delta^{15}N$ values found in our study give strong support to the assumption that the pit filling material was neither grassland nor forest soil but, in consequence, arable land. However, also the Early Weichselian humic zone located in boreal vegetation [84] before farming activity exhibited $\delta^{15}N$ of 5.3‰. But as it is reported that also the soil age exerts an influence on ecosystem $\delta^{15}N$ values over longer timescales with increasing $\delta^{15}N$ values in older soils [82,85]. Thus, the humic zone as the oldest investigated soil horizon (i.e. Early Weichselian) might have been altered during the time of deposition. This aging effect might have

also been the cause for a slight increase of the $\delta^{15}N$ values of the pit fillings.

The subsoil outside the pit, exhibited significantly smaller $\delta^{15}N$ values (2.4–5.9‰) than the pit filling material (4.7–6.7‰), showing even lower values in the deeper loess dominated horizons (i.e. C-horizons) than the overlying reference soil horizons (Fig. 1a). This is remarkable, since $\delta^{15}N$ values of soils usually increase with soil depth or with decreasing organic N contents because the processes of soil organic nitrogen formation and loss result in an enrichment of the heavy isotope and hence, in larger $\delta^{15}N$ values [47,78]. Hence, we have to assume that the overlying material, particularly the recent Ap horizon as well as the prehistoric A horizon-material within the pits, was influenced by external N input. Typical inputs showing elevated $\delta^{15}N$ values are organic manures. During storage of organic manures usually large losses of the light ^{14}N occur by ammonia volatilization, resulting in $\delta^{15}N$ values of 5‰ or even larger [26,47,81,86]. The elevated ^{15}N ratios in the pit fillings could thus be seen as a first indicator of external nitrogen input caused by prehistoric manuring [47,86–88]. However, isotope ratios are usually not used as direct biomarkers for manure, hence, we added analyses of steroids, for a more specific indication (see below).

When comparing the three soil regions (i.e. Chernozems, Phaeozems and Luvisols) no change in $\delta^{15}N$ pattern can be observed: all pit fillings exhibited large $\delta^{15}N$ values (5.8–7.6‰) with no significant differences to the values of the respective recent topsoils (5.3–7.3‰). The material of the pit fillings was thus also enriched in ^{15}N compared with that of the adjacent subsoils (Fig. 1B). Hence, the $\delta^{15}N$ values reported for all archaeological soils of this study appear to be typical for larger soil areas.

Nevertheless, the level of the $\delta^{15}N$ values differed: Chernozem pit fillings in Central Germany exhibited on average significantly larger $\delta^{15}N$ values (7.6±0.3‰ standard error) than the Luvisol pit fillings in Western Germany (5.6±0.2‰) (Fig. 1B). The reasons for these divergences can be manifold: either there were different N-inputs [81,89], varying attributes of the soil (e.g. microbial population, mineralogical composition), or different pedogenetic processes. All of them probably contribute to a different degree of $\delta^{15}N$ fractionation [78,90–92].

Amino acid signature. The amino acid composition may be specific for distinct crops [23–25], hence, we analyzed their composition in archaeological soil samples to elucidate the sources of organic N inputs in prehistory. The total contents of amino acids (i.e. the sum of D- and L-enantiomers of the respective amino acids under study) correlated closely with SOC (r = 0.72) and total N contents (r = 0.87), which is in accordance with the results of other studies [93–95]. For a better understanding of the role of amino acids in SOM dynamics, the amino acids were expressed in grams per kilogram total N. The element-normalized amino acid contents then amounted to 334–1734 g kg^{-1} N (Table 4). Generally, the soil amino acids contain approximately 7–21% N. In the recent topsoils of our study, the amino acids explained 20–22% of total soil N irrespective of the study region (Table 4). Other studies reported similar proportions (i.e. 20–40% of total N; [96,97]. Contrastingly, the content of the amino acid-N of the pit fillings and the adjacent subsoils corresponded only to 5–16% of the contents of soil N (Table 4), which is significantly lower than the amino acid-N to soil N contribution in the recent topsoils. With increasing soil depth and increasing soil age the proportion of non-hydrolysable N and undefined N increased. This indicates either that older SOM in the pit fillings and in the subsoils comprised a larger fraction of chemically stable peptide N or a selective decomposition of hydrolysable amino acids in the pit fillings and subsoils. Similarly, Mikutta *et al.* [98] found a larger

Table 3. Contents of selected *aqua regia* extractable elements (in mg kg^{-1} soil) in the pit fillings, adjacent subsoils, recent topsoils and Early Weichselian humic zone of the Luvisol region.

Region/excavation site	Sample type	n[a]	pH[b]	P	I	Mn	Cu	Mo	Se	Zn
				[mg kg^{-1} soil]						
Western Germany										
Luvisol region										
Merzenich	pit filling	9	7.0	831 (38)	345 (47)	780 (55)	17 (1)	2 (0.2)	13 (1.1)	66 (7)
	subsoil	11	7.3	669 (69)	51 (28)	862 (125)	17 (1)	3 (0.2)	9 (0.8)	46 (3)
	topsoil	1	6.8	1225	12	652	15	3	7	66
Pulheim	pit filling	2	6.4	836 (817–856)	118 (0–236)	505 (427–583)	20 (14–27)	1 (0–3)	5 (0–11)	52 (43–61)
	subsoil	2	6.4	733 (725–740)	97 (0–194)	516 (508–524)	18 (12–24)	1 (0–3)	5 (0–11)	60 (40–80)
	topsoil	1	5.3	800	207	627	11	2	7	52
Mean	*pit filling*	*11*	*6.9 a*	*832 (36) a*	*310 (52) a*	*737 (52) a*	*18 (1) a*	*2 (0.3) a*	*12 (1) a*	*64 (6) ab*
	subsoil	*13*	*7.2 b*	*680 (63) b*	*59 (28) b*	*799 (120) a*	*17 (1) ab*	*3 (0.3) ab*	*8 (1) b*	*49 (4) ab*
	topsoil	*2*	*6.0 c*	*1013 (213) c*	*110 (98) b*	*639 (12) a*	*13 (2) b*	*3 (0.3) b*	*7 (0.1) c*	*59 (7) b*
Düren Arnoldsweiler	humic zone	1	6.6	257	109	767	13	1	10	39

Means with standard errors in parentheses and for n=2 each value is listed in parentheses. Within one region, different letters designate significant (p<0.05) differences between sample types.
[a] number of samples,
[b] in 0.01 CaCl$_2$.

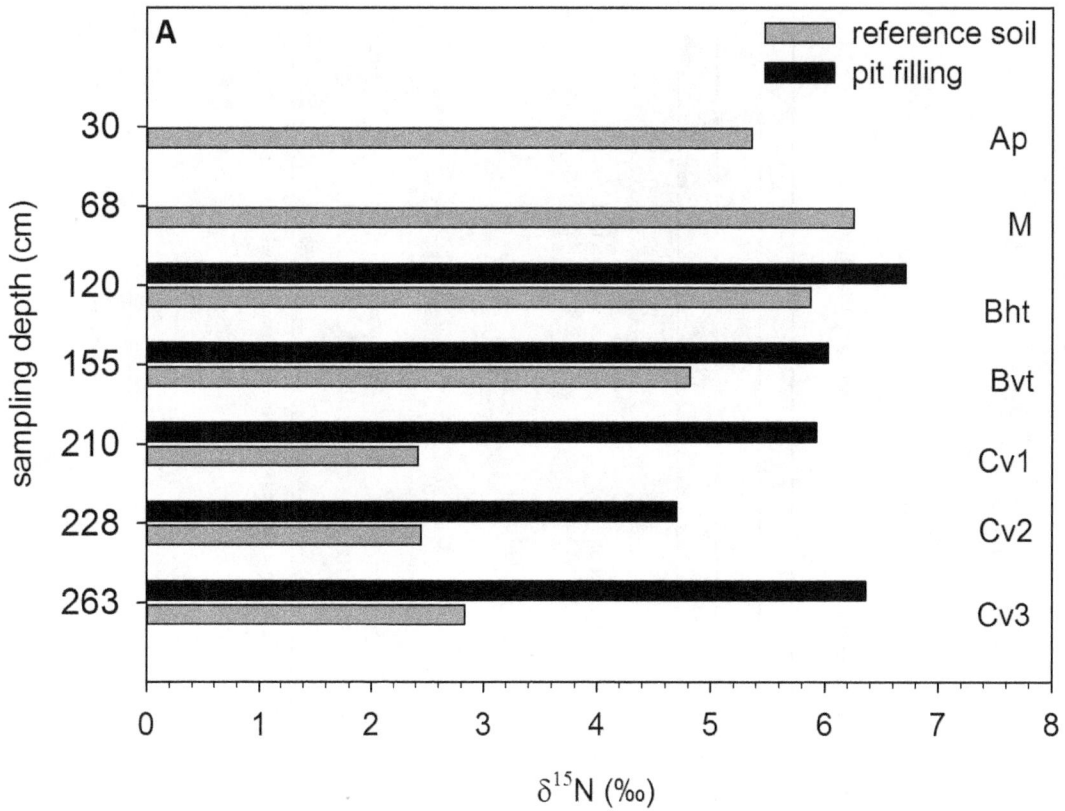

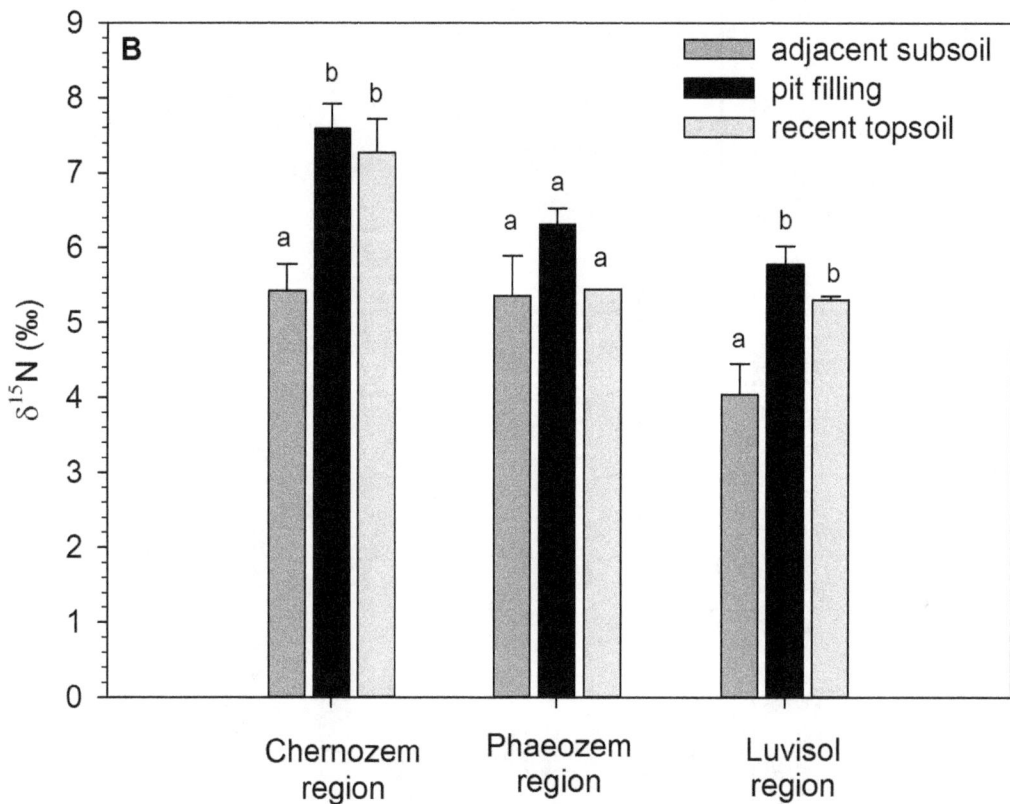

Figure 1. Depth distribution of δ^{15}N (in ‰) at the excavation site Merzenich in the Luvisol region in the reference soil outside the pit, and in the pit filling (A). Soil horizon designation of reference soil follows the German soil classification [132]. Ap = ploughed humic horizon, M = colluvial deposit, Bht = humic, argic horizon, Bvt cambic, argic horizon, Cv = parent loess, slightly weathered. And δ^{15}N values (in ‰) in the

adjacent subsoil, the pit filling representing the prehistoric topsoil, and recent topsoil in the Chernozem, Phaeozem, and Luvisol region (B). The error bars represent the standard error. Within one region, different letters designate significant ($p<0.05$) differences between sample sets.

proportion of non-hydrolysable amino acids across a long-term chronosequence in Hawaii, indicating a preferential accumulation of mineral-associated organic N with time (up to 4,100 years) and with soil depth. Equally, Glaser [99] revealed that the amino acid-N pool of Amazonian dark earths only contributed to 18–25% of total N, and assumed that a majority of these unknown N-pools were present in heterocyclic compounds. In any case, due to the low amount of amino acid-N recovered in the ancient pit fillings, the usefulness of the amino acids in identifying different soil organic nitrogen (SON) sources from prehistoric crops was limited, irrespective of possible transformations of the amino acid pool during SON genesis. Nitrogen that is introduced into the soil by legume cultivation can be detected due to the involved increase of hydrolysable amino acid contents [96,100]. The amino acid contents of the pit fillings were small compared to the recent topsoils, therefore, the amino acid analyses could not reveal an evidence for legume cultivation in the prehistoric arable topsoils.

Yet, the amino acid contents differed between the sites. The Neolithic pit fillings of the Luvisol and Phaeozem region showed significantly smaller contents of amino acid-N compared to the Bronze Age pit fillings of the Chernozem region (Table 4). On the one hand this finding might reflect the higher age of SON that – similarly to the subsoils – resulted in smaller amino acid-N proportions. On the other hand this finding may be a result of significant larger clay contents in the Neolithic pit fillings of the Phaeozem region (averagely 31% clay in the pit fillings) and Luvisol region (averagely 24% clay), compared with the Bronze Age pit fillings (averagely 6% clay) located in the Chernozem region as higher clay contents can have negative impact on the extractability of amino acids [93,101]. In both cases, the properties of the SON pool were apparently influenced by pedogenesis, suggesting that the amino acid method was not suitable biomarker for the origin of amino acids in our arable sites, as, e.g. previously found for compound-specific ^{15}N amino acid signals in Bronze Age grassland sites [26].

Lacking clear evidence of legume traces does not mean that legumes were not cropped at all; such a conclusion would also have contradicted archaeobotanical evidence [77]. Looking into the amino acid composition of different crops, for instance, showed that the threonine to lysine ratio is typically around 0.6 for legumes but 1.0 for cereals [23–25]. Hence, smaller threonine-to-lysine ratios in soils might reflect an increasing cropping of legumes. We tested this by analyzing an arable soil that had been recently cropped for three years with clover (*Trifolium pratense*). Indeed this soil exhibited an enrichment of lysine relative to threonine with a L-threonine to L-lysine ratio of 0.6 (data not shown). Since D-threonine is also produced by microbes [102], the amino acid ratios were based on the respective L-types of these two amino acids. The recent topsoils of the study regions showed L-threonine to L-lysine ratios of 1.1 to 1.5 (Table 4), thus reflecting the dominance of cereal production in modern crop rotations. In contrast, the prehistoric topsoils (i.e. pit fillings) revealed L-threonine to L-lysine ratios of 0.5 in the Luvisol region, 0.6 in the Phaeozem region, and 1.0 in the Chernozem region (Table 4). These results would have been in line with observations of archaeologists that there was legume cultivation in prehistoric agriculture, even if it was not dominant [77,103]. However, these data are not supported by low δ^{15}N ratios and elevated amino acid N contents (see above). Moreover, also the adjacent subsoils showed L-threonine to L-lysine ratios of 0.6 to 0.9, indicating that

this ratio might as well have been influenced by SOM genesis in the soil profiles besides legume-N input.

Amino acid enantiomers have been proposed as bacterial and age markers [94]. D-glutamic acid and D-alanine, being part of the bacterial cell wall, are enriched during SOM formation and transformation [94,95,98]. And indeed, we found more bacterial derived D-amino acids in the pit fillings and in adjacent subsoils than in the recent topsoils (Table 4; data for alanine). This finding suggests that larger proportions of the total SON pool had been cycled by bacteria in the pit fillings and subsoils than in the recent surface soils, i.e. the SOM of the prehistoric topsoils had been altered relative to the surface soils. Such processes may have contributed to changes in the amino acid composition as mentioned above, and they may in part explain the elevated δ^{15}N values in the pit fillings above those in the recent surface soils. These findings do explain the general large soil δ^{15}N values in comparison to the deeper subsoils.

Apart from using the contents of the D-enantiomers of glutamic acid and alanine as markers for the contribution of bacterial cell walls to SON, the D/L ratio of lysine has been discussed as a marker for an age assessment of the respective protein pool and of total soil organic matter because D-lysine seems to originate solely from abiotic cell aging [94]. Increasing D/L ratios of lysine therewith correlated with SOM age in other soils [94]. This marker confirmed our assumption that the SOM age of the pit fillings and the adjacent subsoils was significantly older than that of the respective recent topsoils (Table 4). This finding was valid for all studied soil regions. The Early Weichselian humic zone exhibited the highest D/L ratios indicating the oldest investigated soil sample. Comparing the studied geoarchives, the Bronze Age pits in the Chernozem region revealed significantly smaller D/L ratios of lysine, and therefore a younger age than the Neolithic pit fillings in the Phaeozem and Luvisol region (Table 4). However, a negative correlation of the D-lysine content and the pH value of the investigated soils ($r = -0.76$) could be determined. That is why the production of D-lysine might be hindered in the more alkaline soils in Central Germany in comparison to the more neutral soils in Western Germany. Thus, this finding is consistent with those of Amelung [94] and Amelung *et al.* [104], who reported that amino acid racemization only allows a relative SON dating for a given environment but no estimation of the absolute age of proteins.

In summary, there was a significant accumulation of total N in the Neolithic and Bronze Age pit fillings. This SON contained amino acids, i.e. peptide-like structures were preserved. The composition of amino acids was typical for legume cropping, however, also for advanced stages of microbial SON transformation and SOM genesis. Overall, a significant cropping of legumes in prehistoric agriculture could not be assured on the bases of geochemical proxies. Noteworthy, the prehistoric topsoils had significantly elevated δ^{15}N values compared with the adjacent subsoils, and therefore pointed to the use of manure in prehistoric agriculture. As the soil δ^{15}N values do not provide a direct proxy for the origin of manure, we analyzed steroids as specific markers for faecal residues.

3 Proxies for prehistoric faecal input

Sterols, stanols, and stanones. The use of manure can be elucidated by steroid biomarkers, including sterols, stanols, and stanones ([30,49,105] see also Table S2).

Table 4. Contents of total N and amino acid as well as respective contributions and selected amino acid ratios in the pit fillings, adjacent subsoils, and recent topsoils.

Region/excavation site	Sample type	n^a	total N [g kg^{-1}]	amino acids [g kg^{-1} N]	AA-N/N [%]	D/L alanine	D/L lysine	L-threonine/L-lysine
Central Germany								
Chernozem region	pit filling	6	0.5 (0.0) a	1121 (64) a	14.1 (0.8) a	0.18 (0.01) a	0.04 (0.003) a	1.0 (0.10) a
	subsoil	6	0.2 (0.1) a	1238 (107) a	15.5 (1.4) a	0.15 (0.01) ab	0.04 (0.004) a	0.9 (0.12) a
	topsoil	5	1.1 (0.2) b	1616 (67) b	20.2 (0.8) b	0.14 (0.01) b	0.03 (0.004) b	1.1 (0.13) b
Phaeozem region	pit filling	2	0.5 (0.1) ab	635 (11) a	8.1 (0.1) a	0.19 (0.01) a	0.06 (0.01) ab	0.7 (0.12) a
	subsoil	2	0.2 (0.1) a	635 (24) a	8.1 (0.4) a	0.18 (0.02) a	0.06 (0.002) a	0.6 (0.23) a
	topsoil	1	1.2 b	1694 b	21.1 b	0.12 a	0.03 b	1.4 a
Western Germany								
Luvisol region	pit filling	11	0.4 (0.0) a	334 (29) a	4.5 (0.4) a	0.20 (0.01) a	0.05 (0.002) a	0.5 (0.05) a
	subsoil	13	0.2 (0.0) b	345 (48) a	4.6 (0.6) a	0.17 (0.01) b	0.05 (0.002) a	0.6 (0.06) b
	topsoil	2	1.4 (0.6) c	1734 (183) b	22.0 (2.1) b	0.11 (0.02) c	0.02 (0.003) b	1.5 (0.05) c
Düren Arnoldsweiler	humic zone	1	0.4	362	4.9	0.21	0.06	0.3

Means with the standard errors in parentheses. Within one region, different letters designate significant ($p<0.05$) differences between sample types.
[a] number of samples;
[b] contribution of amino acid-N to total N.

All selected Neolithic soil relics of the Lower Rhine Basin and the respective adjacent subsoils exhibited only little if any quantifiable amounts of cholesterol (16–94 µg kg^{-1}) and the plant-derived sterol β-sitosterol (11–40 µg kg^{-1}; Table 5). Stigmasterol, also a plant-derived sterol, could not be quantified or even not be determined in all studied samples. 5β-stigmastanol could only be detected in small amounts in one pit filling (11 µg kg^{-1}). Similarly, coprostanol was not found in any of the studied samples. All reference soils including the adjacent subsoils (= control) and the Early Weichselian humic zone (representing the soil before any farming activity) did not exhibit these faecal markers (5β-stanols) (Table 5).

Without a positive proof of detectable amounts of 5β-stanols (coprostanol or β-stigmastanol) in the studied Neolithic arable soils, except for pit filling 4, stanol analyses did not provide evidence that faeces had been deposited on the prehistoric topsoils. This is in contrast to earlier studies, which found 5β-stanols in approximately 3,500 years old Minoan terraces [106] and even in 10,000 years old stratified shell middens [107]. In this study, relatively large contents of α-stigmastanol could be determined in the pit fillings as well as in the adjacent subsoils (Table 5). This finding points to the larger natural abundance of α-stanols in soils due to their production in course of the transformation of their precursors, like – in the case of α-stigmastanol – the plant sterols β-sitosterol and stigmasterol (see Fig. S2) [108,109].

In the studied soil relics and the reference soils, stanones (5α-cholestan-3-one and 5β-cholestan-3-one) could only be detected below the limit of quantification (i.e. 10 µg kg^{-1}; data not shown). Hence, also this third steroid group did not help to unambiguously detect prehistoric faecal inputs into the soils.

Bile acids. Bile acids are additional markers for faecal matter from vertebrates as they are formed from cholesterol in their liver and excreted in small amounts [30]. Combining bile acid and β-stanol analyses may even help to distinguish between the source of faecal matter (human, ruminant, and porcine) [28,30,37]. Additionally, bile acids are presumed to be more stable against degradation in soils than stanols and stanones [29,36,110] and, thus, can be detected in soils thousands of years after the application of the faeces [110–112].

All studied Neolithic pit fillings revealed larger contents of deoxycholic acid (DCA) and of hyodeoxycholic acid (HDCA) than the adjacent subsoils, in which only small contents of DCA and HDCA were detected (Fig. 2; see Table S3 for all data of the bile acids). There were little if any detectable amounts of other bile acids (i.e. lithocholic acid, chenodeoxycholic acid and ursodeoxycholic acid) in the pit fillings and the adjacent subsoils. The Early Weichselian humic zone which represented the soil before any human farming and manuring activity, therefore, served as a reference sample without any anthropogenic inputs of bile acids and indeed showed no contents of bile acids above the quantification limit. These findings confirm the current assumption that beside faeces of most vertebrates there are no other significant sources of bile acids in the environment [40]. Therefore, the very small natural background occurrence of bile acids in the Early Weichselian humic zone (here below quantification limit) can only originate from the sparse input of faeces from wildlife animals, and the low contents support the absence of manuring. Yet, we cannot fully exclude a degradation or aging of the bile acids since their input into the humic zone in view of the long period of time (90.000 years), as there are no studies that have examined bile acids in soil samples of this high age.

Different contents of DCA and HDCA have been used to differentiate between faeces of ruminants and omnivores: HDCA is a biomarker of pig faeces, with DCA being absent [30,36]. In contrast, human and cow faeces are dominated by DCA and lithocholic acid (LCA), but with cow faeces containing only small amounts of LCA [30,36,37]. Provided that all bile acids were decomposed equally quickly, the signal in the pits comprising DCA and HDCA indicated a mixture of pig, human and/or ruminant faeces. Pig faeces contain HDCA (and nearly no DCA), but human and ruminant faeces both contain DCA. The remarkably low contents of LCA in all pit fillings exclude an input of human faeces, because these faeces generally contain high amounts of LCA [30,36]. In contrast, an input of ruminant faeces is more likely as these faeces generally contain low amounts of LCA [30,36]. Hence, the enrichment of bile acids in the prehistoric arable topsoils - relative to the adjacent soil samples and the Early Weichselian humic zone - reflect that these soils had received faeces, likely produced by pigs and ruminant animals. It is very unlikely that these compounds were leached from the above-lying surface soils, because (i) there are indications that faecal residues are not prone to leaching [113], and (ii) because we could detect these compounds even in very deep pit fillings (>2.5 m), but hardly in the subsoils and not at all in quantifiable amounts in the Early Weichselian humic zone. For other rather immobile substances like phosphates, there was also no indication for a leaching into this depth [8]. We postulate, therefore, that the prehistoric surface soils had received a mixture of different manures.

Earlier archaeological studies that had used bile acids as tracers for faecal inputs into soils and sediments had mainly focused on Roman or younger soil relics, such as drainage channels for latrines or sediment-filled sewers [38,105,110]. The detection of bile acids in the studied Neolithic soil relics, being approx. 4,000 to 6,000 years old, confirms the long-lasting stability of these compounds as suggested by Bull *et al.* [110].

All in all, the studied sites in the Lower Rhine Basin contained indicators for a faecal input into the Neolithic arable soils, mainly because all pit fillings exhibited larger bile acid contents than the respective adjacent subsoils and the Early Weichselian humic zone (representing the soil before any farming activity). Here, the data suggests that Neolithic arable fields had been fertilized with faeces or livestock manure. Whether this has been done intentionally - by manuring or by hazard during grazing - remains unclear.

4 Proxies for prehistoric burning events

Soil color and black carbon. All pit fillings were characterized by a dark color, similar to the color of the recent topsoil but different from that of the adjacent subsoils [8]. This optical feature initially pointed to the origin of the pit fillings from ancient topsoil. Beyond the optical impression, color measurements allowed the quantitative comparison of the different sample types. The L* values, indicating the extinction of light on a scale from L* 0 (absolute black) to L* 100 (absolute white), were similar for the topsoils and pit fillings (Table 6). This is in line with the elevated contents of SOC in the pit filling material (Table 6). When relating color values (L* values) to black carbon (BC) and SOC contents (see Table S4 for single site data), there was a stronger correlation to BC (r = −0.76; $p<0.01$) than to SOC contents (r = −0.52; $p< 0.01$), suggesting that the soil color was mainly related to burned, aromatic C compounds represented by the BC contents [43,114].

In all samples the BC contents correlated significantly with the SOC contents (r = 0.71) with the contribution of BC to SOC being significantly larger in the pit fillings (24–38% of SOC was BC) than in the recent topsoils (8–25% of SOC was BC). Also in absolute amounts, the pit fillings revealed significantly larger BC contents (in g kg^{-1} fine earth) than the adjacent subsoils but

Table 5. Contents of sterols and stanols (in µg kg⁻¹ soil) of selected pit fillings, the respective adjacent subsoils (= control) and the Early Weichselian humic zone (= reference).

Region/excavation site	Sample type	Soil depth [cm]	Sterols [µg kg⁻¹ soil]			β-Stanols		α-Stanols	Epi-5β-Stanols		Sum
			Chole-sterol*	β-Sito-sterol	Stigma-sterol	Copro-stanol*	β-stigma-stanol	α-Cholestanol*	α-Stigma-stanol	Epi-coprostanol	Σ sterols and stanols
Western Germany/ Luvisol region											
Merzenich	pit filling 1	123	n.q.	15.8	n.q.	n.q.	n.q.	n.d.	71.3	n.d.	87.1
	adjacent subsoil 1	123	61.8 (19.8)	26.8 (8.9)	n.q.	n.q.	n.d.	n.d.	95.5 (15.4)	n.d.	184.1
	pit filling 2	125	41.1 (7.0)	10.5	n.q.	n.q.	n.q.	n.d.	72.2	n.d.	123.8
	adjacent subsoil 2	125	51.2	37.6	11.1	n.d.	n.d.	n.q.	14.1	n.d.	102.9
	pit filling 3	130	43.0	40.0	10.1	n.q.	n.q.	n.d.	64.1	n.d.	157.2
	adjacent subsoil 3	130	16.3 (2.3)	20.7 (5.9)	n.q.	n.q.	n.q.	n.d.	67.1 (17.3)	n.d.	104.0
	pit filling 4	175	93.5 (31.5)	16.2 (7.0)	n.q.	n.q.	10.9	n.d.	62.0 (5.0)	n.d.	182.6
	adjacent subsoil 4	175	10.7 (8.8)	n.q. (1.2)	n.q.	n.d.	n.d.	n.d.	13.7 (3.4)	n.d.	24.4
	pit filling 5	180	n.q. (6.5)	14.0 (0.2)	n.q.	n.d.	n.d.	n.q.	30.3 (6.6)	n.d.	44.3
	adjacent subsoil 5	180	n.q.	n.q.	n.q.	n.d.	n.q.	n.q.	30.8	n.d.	30.8
	pit filling 6	263	10.9 (8.3)	n.q. (2.3)	n.d.	n.q.	n.d.	n.d.	15.3 (11.7)	n.d.	26.3
	adjacent subsoil 6	263	n.q.	n.q.	n.q.	n.d.	n.d.	n.d.	16.9 (6.3)	n.d.	16.9
Düren Arnoldsweiler	humic zone	170	n.q.	n.q.	n.d.	n.d.	n.d.	n.d.	n.q.	n.d.	0.0

Standard deviation of the laboratory replicates is given in parentheses.

n.q. detectable, but not quantifiable (under quantification limit; 5 µg kg⁻¹ soil for coprostanol and epi-coprostanol, and 10 µg kg⁻¹ soil for all other steroids); n.d. not detectable (under detection limit);

*mark mainly animal-derived steroids. The sources of the other unmarked steroids are plant litter, root exudates, and faeces (of herbivores and omnivores).

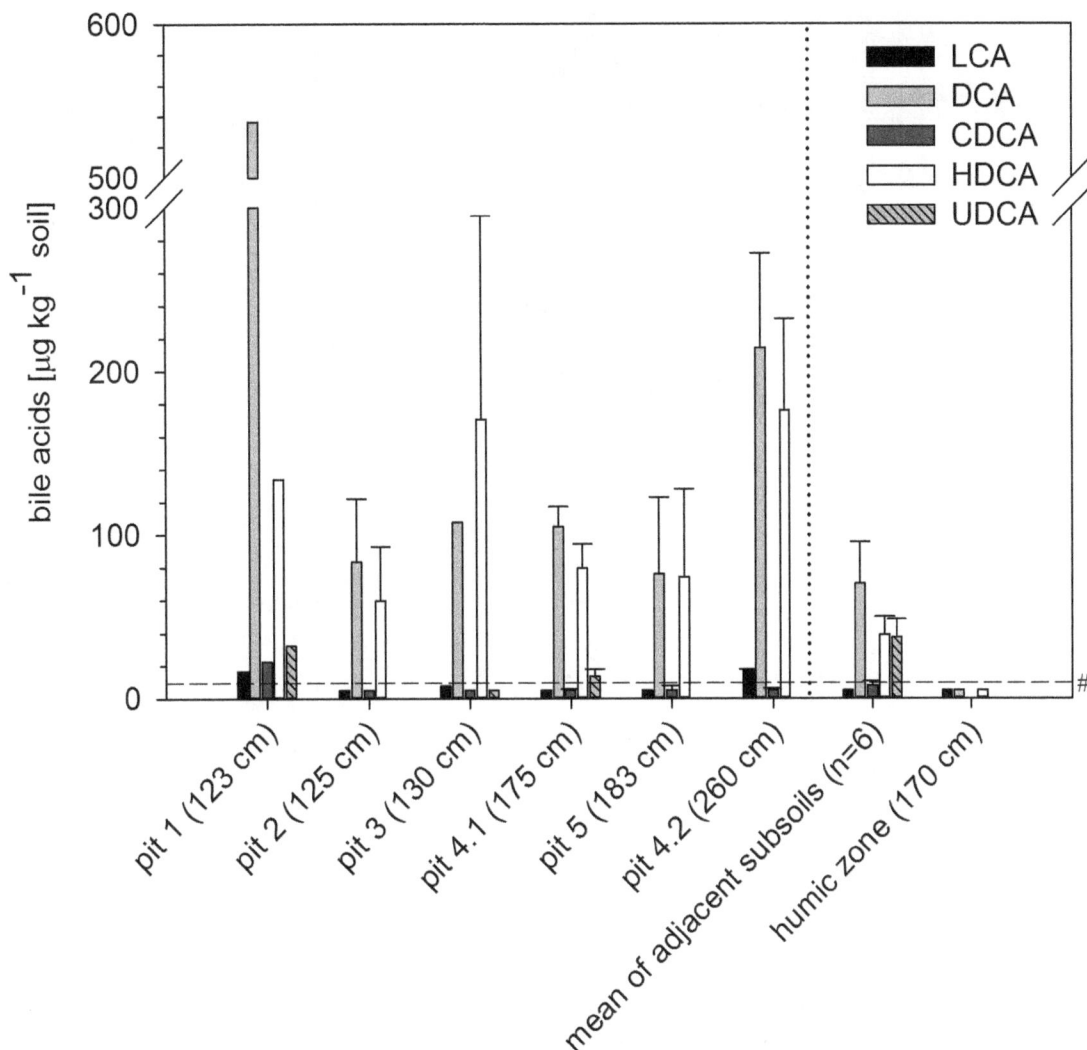

Figure 2. Bile acid contents of the selected pit fillings, the adjacent subsoils and the Early Weichselian humic zone at various soil depths as indicated in parentheses. The dashed line marks the limit of quantification. LCA = lithocholic acid, DCA = deoxycholic acid, CDCA = chenodeoxycholic acid, HDCA = hyodeoxycholic acid, UDCA = ursodeoxycholic acid. Error bars represent the standard deviation of laboratory replicates, whereas error bars of the mean of adjacent subsoils show the standard error of the six averaged adjacent subsoils.

similar BC contents compared with the respective recent topsoils (Table 6).

Among the individual soil regions, the Luvisol pit fillings revealed the largest contribution of BC to SOC (38%), whereas BC proportions in the Chernozem pit fillings were significantly smaller (25% BC of SOC; Table 6). Nevertheless, the recent topsoils of the Chernozem region revealed significantly larger BC contributions to SOC than the Luvisol topsoils, similar to those of other regions of the world [106]. Apparently, there was a preferred input of BC into the material of the pit fillings (the prehistoric topsoils) of the Luvisol region, which supports the idea of external BC inputs by slash- and burn managing practices [4,115]. Higher clay contents in the Luvisol region might have also intensified stabilization processes - like BC interactions with mineral phases – leading to larger BC to SOC contributions of the pit filling material of the Luvisol region compared with that of the Chernozem region [116–118].

In the reference sample (i.e. the Early Weichselian humic zone), BC comprised 22% of the SOC, although the absolute BC

contents were relatively small compared to those of the pit filling samples (Table 6). The BC stored in the Early Weichselian humic zone was likely produced by natural fires during this epoch [119,120], which was expected because fire occurs as a frequent natural disturbance in boreal forests [121]. Also Early to Middle Weichselian paleosol relicts from adjacent sites revealed BC to SOC proportions between 15–35%, which could be related to natural fires during this epochs, as well [122]. Eckmeier et al. [7] who investigated Neolithic slot pits in the Lower Rhine Basin concluded that the large BC proportions of SOC (36% BC-C of SOC) could be a result of vegetation fires ignited by man, as a burning of prevalent temperate deciduous forests without human impact is shown to be very unlikely [123–125]. Similarly, Kleber et al. [126] supposed anthropogenic fire in prehistoric agriculture due to BC contents of up to 13% of SOC in Neolithic soil relicts in Central Germany. Hence, an attribution to agricultural fires cannot be achieved alone via BC quantification, but needs to be supported by environmental (e.g. archaeobotanical) and chronological data.

Table 6. Contents of soil organic carbon and black carbon as well as respective contributions and soil color expressed as lightness (L*) in the pit fillings, adjacent subsoils, recent topsoils and Early Weichselian humic zone.

Region/excavation site	Sample type	n^a	C_{org} [g kg^{-1}]		BC [g kg^{-1} soil]		BC-C [% of C_{org}]		Color L*		B3CA [% of total BC]	B4CA		B5CA		B6CA	
Central Germany																	
Chernozem region	pit filling	6	8.5	(0.5) a	2.2	(0.3) a	25.2	(2.1) a	52	(1.5) ab	3.5 (0.3) a	20.1	(0.6) a	36.5	(1.1) a	39.9	(0.9) a
	subsoil	6	5.3	(0.8) a	0.6	(0.3) a	10.0	(2.8) a	61	(2.5) a	20.2 (16.0) a	34.0	(7.9) a	21.8	(4.7) b	24.0	(5.5) b
	topsoil	5	13.0	(1.8) b	2.4	(0.4) a	19.8	(2.7) a	47	(1.5) b	3.8 (0.3) a	20.3	(2.0) a	35.0	(1.0) ab	40.9	(2.2) a
Phaeozem region	pit filling	2	5.7	(1.8) ab	1.1	(0.1) a	29.1	(2.3) a	53	(2.4) ab	4.6 (0.3) a	24.2	(2.2) a	35.7	(1.0) a	35.6	(3.5) a
	subsoil	2	3.5	(1.0) a	0.2	(0.1) a	7.5	(5.2) a	64	(0.3) a	5.3 (0.2) a	42.0	(6.8) a	29.6	(8.5) a	23.1	(1.5) a
	topsoil	1	11.9	b	0.3	a	7.7	a	58	b	3.4 a	24.9	a	32.9	a	38.8	a
Western Germany																	
Luvisol region	pit filling	11	4.1	(0.5) a	1.6	(0.2) a	36.6	(4.2) a	54	(0.8) a	4.5 (0.3) a	26.5	(1.7) a	35.2	(1.4) a	33.7	(1.0) a
	subsoil	13	3.5	(0.5) a	0.3	(0.1) b	8.7	(2.9) b	57	(0.5) b	2.7 (0.2) b	38.2	(2.9) a	29.8	(1.9) a	29.3	(1.0) b
	topsoil	2	14.2	(5.0) b	1.4	(0.3) a	10.4	(1.9) b	55	(1.1) a	3.5 (0.4) b	20.6	(1.0) b	32.1	(3.8) a	43.8	(2.4) c
Düren Arnoldsweiler	humic zone	1	2.4		0.5		22.7		62		3.7	37.0		28.9		30.4	

Means with the standard errors in parentheses. Within one region, different letters designate significant ($p<0.05$) differences between sample types.
[a] number of samples;
[b] in 0.01 M CaCl$_2$.

Furthermore, the quality of BC can be used to gain information about the type of fire. Generally, BC from high temperature fire is characterized by large proportions of five- and six-times carboxylated benzene polycarboxylic acids (benzene pentacarboxylic acid = B5CA and mellitic acid = B6CA) [68,69]. In this study, B6CA contributed up to 40% and 44% of total BC, respectively, in the pit fillings and recent topsoils (Table 6). Such large B6CA proportions (>40% B6CA of total BC) are typical for high burning temperatures in charcoals (i.e. 600 °C) [44,45]. In the recent topsoils, the B6CA enrichment is most likely a result of modern fossil fuel combustion [117]. Yet, also the large contribution of B6CA to total BC in the pit fillings indicates human-induced fires on ancient arable fields, as these kinds of fires usually burn at higher temperatures compared with natural grass and forest fires ([45]; 23–33% B6CA of total BC in natural and forest fires). Indeed, in our study both, the subsoil samples and the humic zone, revealed elevated proportions of B4CA but only small proportions of B6CA to total BC (averagely 28% and 30%, respectively; Table 6) likely representing the "low temperature" natural fire background.

Summarizing, all studied arable topsoil relicts (pit filling material) were characterized by elevated amounts of fire-derived carbon, pointing to a human use of fire on prehistoric arable land. These fires were controlled burning events, which are also indicated by a change in BC quality. The early farmers might have used fire to clear the fields from understorey vegetation after cutting the trees [127]. They could have been aware of the benefits of periodically burning biomass on arable fields, i.e. albedo effect, weed suppression, liming, soil amelioration and fertilization [41,42,128]. However, in the studied regions with neutral to alkaline soil reaction, the liming effect was less important than the fertilizer effect due to the addition of plant available nutrients like calcium, magnesium, potassium and phosphorus from the burned biomass (mainly from the ash) [108,109,129] and the amelioration effects on physical, chemical, and biological soil properties [70,130].

Conclusions

We used a multi-proxy approach for studying the (micro-) nutrient status and the organic matter of Neolithic and Bronze Age arable topsoil relicts. The results showed:

- i.) that topsoil relicts did not hint at any nutrient deficiencies in Neolithic and Bronze Age arable soils, similar to findings for prehistoric phosphorus [8].
- ii.) that it was not possible to identify legumes by amino acid measurements; yet, D/L amino acid patterns pointed to the preservation of aged and microbially altered N forms in the prehistoric pit fillings.
- iii.) an enrichment of heavy ^{15}N isotope that give first indications of prehistoric manuring.
- iv.) manure- (faeces-) derived C in the pit fillings that was verified by the detection of stable bile acids. They proved to be better indicators of prehistoric faeces additions than coprostanol and 5β-stigmastanol.
- v.) that arable fields had received combustion residues (black carbon), most likely derived from human-induced biomass burning.

References

1. Bull ID, Simpson IA, van Bergen PF, Evershed RP (1999) Muck 'n' molecules: organic geochemical methods for detecting ancient manuring. Antiquity 73: 86–96.

Overall, the studied soils were fertile in prehistory. On the one hand, this reflects the favorable characteristics of the fertile loess soils as well as the low prehistoric population density. On the other hand, there were several hints that the fertility of the soils was sustained or even enhanced in prehistory, notably by additions of organic materials like charcoal and manure (faeces). Whether this was done intentionally remains unsure.

Supporting Information

Figure S1 **Flow chart of steroid analyses according to Birk et al.** [49].

Figure S2 **Products of sterol reduction to 5β-stanols in the gut of mammals, the subsequent conversion to epi-5β-stanols under anaerobic conditions, and the reduction products of sterols in soils (i.e. 5α-stanols and only to a minor extent 5β-stanols) (modified after Bull et al.** [30] **and Birk et al.** [131]**).** * If an input of algae can be excluded these steroids are animal-derived. The sources of the other unmarked steroids are e.g. plant litter, root exudates, and faeces (of herbivores and omnivores).

Table S1 **Contents of *aqua regia* extractable elements (in mg kg^{-1}) of all pit filling soil samples.**

Table S2 **Molecular structures, the retention times, and the selected characteristic ion fragments of the relevant steroids.**

Table S3 **Bile acid contents (in µg kg^{-1} soil) (average and standard deviation in parentheses) in all investigated samples.**

Table S4 **BPCA analysis (in duplicate) of all soil samples (average ± standard deviation; SD<15% for BPCAs; n.d., not detectable) and L* value.**

Acknowledgments

We thank the Archaeological State Service of Saxony-Anhalt for their help: Thorsten Schunke, Susanne Friederich, Matthias Becker, Andrea Moser, Karin Schwerdtfeger, Christian Bogen, Klaus Powroznik, the excavation team at the railway track construction site, and the excavation team at Prießnitz (Ildiko Bösze). We also thank Rainer Lubberich, Julia Gerz, and Andreas Folkers from the Archaeological Heritage Management of the Rhineland. Kirsten Unger is thanked for the stable N isotope measurements. Furthermore, we thank Maike Rotard for the English proofreading of the manuscript and the anonymous reviewers for helpful comments.

Author Contributions

Conceived and designed the experiments: RG SP EL WA. Performed the experiments: FL KP MW SU EE. Analyzed the data: FL KP MW SU. Contributed reagents/materials/analysis tools: RG EE WA. Contributed to the writing of the manuscript: FL KP RG EE SP EL WA.

2. Holliday VT, Gartner WG (2007) Methods of soil P analysis in archaeology. Journal of Archaeological Science (34): 301–333.

3. Eckmeier E, Wiesenberg GLB (2009) Short-chain n-alkanes (C16–C20) in ancient soil are useful molecular markers for prehistoric biomass burning. Journal of Archaeological Science 36 (7): 1590–1596.

4. Gerlach R, Baumewerd-Schmidt H, den Borg K, Eckmeier E, Schmidt MWI (2006) Prehistoric alteration of soil in the Lower Rhine Basin, Northwest Germany – archaeological, 14C and geochemical evidence. Geoderma 136 (1–2): 38–50.

5. Gerlach R, Hilgers A (2011) Grubenfüllungen als archäologische Quelle. In: Bork H, Meller H, Gerlach R, editors.Umweltarchäologie-Naturkatastrophen und Umweltwandel im archäologischen Befund: Tagungen des Landesmuseums für Vorgeschichte Halle (Saale). pp. 27–36.

6. Leopold M, Hürkamp K, Völkel J, Schmotz K (2011) Black soils, sediments and brown calcic luvisols: A pedological description of a newly discovered neolithic ring ditch system at Stephansposching, Eastern Bavaria, Germany. Quaternary International: 1–12.

7. Eckmeier E, Gerlach R, Tegtmeier U, Schmidt MWI (2008) Charred organic matter and phosphorus in black soils in the Lower Rhine Basin (Northwest Germany) indicate prehistoric agricultural burning. Charcoals from the past.

8. Lauer F, Pätzold S, Gerlach R, Protze J, Willbold S, et al. (2013) Phosphorus status in archaeological arable topsoil relicts—Is it possible to reconstruct conditions for prehistoric agriculture in Germany. Geoderma 207–208: 111–120.

9. Arrhenius O (1931) Die Bodenanalyse im Dienst der Archäologie. Zeitschrift für Pflanzenernährung, Düngung und Bodenkunde 10: 185–190.

10. Lorch W (1940) Die siedlungsgeographische Phosphatmethode. Naturwissenschaften 40/41: 633–640.

11. Eidt RC (1977) Detection and examination of anthrosols by phosphate analysis. Science 197: 1327–1333.

12. Dauncey KDM (1952) Phosphate Content of Soils on Archaeological Sites. Advancement of Science 9 (33): 33–37.

13. Cook SF, Heizer RF (1965) Studies in the Chemical Analysis of Archaeological Sites. Berkeley.

14. Proudfoot B (1976) The Analysis and Interpretation of Soil Phosphorus in Archaeological Contexts. In: Davidson DASML, editor. Geoarchaeology. London: Duckworth. pp. 93–113.

15. Moore PD (1983) Mapping human activity in soil. Nature 304: 122.

16. Lehmann J, Campos CV, Vasconselos de Macêdo JL, German L (2004) Sequential P Fractionation of Relict Anthropogenic Dark Earths of Amazonia. In: Glaser B, Woods WI, editors.Amazonian dark earths. Explorations in space and time. Berlin: Springer.

17. Terry RE, Fernández FG, Parnell JJ, Inomata T (2004) The story in the floors: chemical signatures of ancient and modern Maya activities at Aguateca, Guatemala. Journal of Archaeological Science (31): 1237–1250.

18. Wilson CA, Davidson A, Cresser MS (2008) Multi-element soil analysis: an assessment of its potential as an aid to archaeological interpretation. Journal of Archaeological Science 35 (2): 1–13.

19. Oliver MA (1997) Soil and human health: a review. European Journal of Soil Science 48: 573–592.

20. Virginia RA, Delwiche CC (1982) Natural 15N abundance of presumed N2-fixing and non-N2-fixing plants from selected ecosystems. Oecologia 54: 317–325.

21. Shearer G, Kohl D (1986) N2 fixation in field settings, estimations based on natural 15N abundance. Australian Journal of Plant Physiology 13: 699–757.

22. Marshall JD, Brooks JR, Lajtha K (2007) Sources of variation in the stable isotopic composition of plants. In: Michener R, Lajtha K, editors.Stable Isotopes in Ecology and Environmental Science.Blackwell Publishing: Blackwell Publishing. pp. 22–60.

23. Kircher M, Leuchtenberger W (1998) Aminosäuren - ein Beitrag zur Welternährung. Biologie in unserer Zeit 28 (5): 281–293.

24. Abel H, Sommer W, Weiß J (2004) Inhaltsstoffe, Futterwert und Einsatz von Ackerbohnen in der Nutztierfütterung. Union zur Förderung von Oel- und Proteinpflanzen e.V. Praxisinformation. Available: http://www.ufop.de/files/3013/4080/9202/RZ_Praxisinfo_Ackerbohne_100604.pdf Accessed 11 November 2013.

25. Bellhoff G, Spann B, Weiß J (2004) Inhaltsstoffe, Futterwert und Einsatz von Erbsen in der Nutztierfütterung. Union zur Förderung von Oel- und Proteinpflanzen e.V. Praxisinformation. Available: http://www.ufop.de/files/3613/4080/8200/RZ_Praxisinfo_Erbsen_100604.pdf Accessed 11 November 2013.

26. Simpson I, Bol R, Bull ID, Evershed RP, Petzke KJ, et al. (1999) Interpreting Early Land Management Through Compound Specific Stable Isotope Analyses of Archaeological Soils. Rapid Communications in mass spectrometry 13: 1315–1319.

27. Bethell PH, Goad LJ, Evershed RP, Ottaway J (1994) The study of molecular markers of human activity: The use of coprostanol in the soil as an indicator of human faecal material. Journal of Archaeological Science 21: 619–632.

28. Leeming R, Ball A, Ashbolt N, Nichols P (1996) Using faecal sterols from humans and animals to distinguish faecal pollution in recieving waters. Water Research 30: 2893–2900.

29. Elhmmali MM, Roberts DJ, Evershed RP (2000) Combined Analysis of Bile Acids and Sterols/Stanols from Riverine Particulates To Assess Sewage Discharges and Other Fecal Sources. Environmental Science and Technology 34: 39–46.

30. Bull ID, Lockheart MJ, Elhmmali MM, Roberts DJ, Evershed RP (2002) The origin of faeces by means of biomarker detection. Environment International 27: 647–654.

31. Gaskell S, Eglinton G (1975) Rapid hydrogenation of sterols in a contemporary lacustrine sediment. Nature 254: 209–211.

32. Taylor C, Smith SO, Gagosian RB (1981) Use of microbial enrichments for the study of the anaerobic degradation of cholesterol. Geochimica et Cosmochimica Acta 45: 2161–2168.

33. Björkhem I, Gustafsson J, Wrange Ö (1973) Microbial Transformation of Cholesterol into Coprostanol- Properties of a 3-oxo-D4-Steroid-5b-Reductase. European journal of biochemistry 37: 143–147.

34. Eyssen HJ, Parmentier GG, Compernolle FC, Pauw G, Piessens-Denef M (1973) Biohydrogenation of sterols by Eubacterium ATCC 21,408—nova species. European journal of biochemistry 36 (2): 411–421.

35. Grimalt JO, Fernandez P, Bayona JM, Albaiges J (1990) Assessment of fecal sterols and ketones as indicators of urban sewage inputs to coastal waters. Environmental Science & Technology 24 (3): 357–363.

36. Elhmmali MM, Roberts DJ, Evershed RP (1997) Bile acids as a new class of sewage indicator. Environmental Science and Technology 31: 3663–3668.

37. Tyagi P, Edwards DR, Coyne MS (2008) Use of Sterol and Bile Acid Biomarkers to Identify Domesticated Animal Sources of Fecal Pollution. Water, Air and Soil Pollution 187: 263–274.

38. Knights BA, Dickson CA, Dickson JH, Breeze DJ (1983) Evidence concerning the Roman military diet at Bearsden, Scotland, in the 2nd century A.D. Journal of Archaeological Science 10: 139–152.

39. Rogge WF, Medeiros PM, Simoneit BR (2006) Organic marker compounds for surface soil and fugitive dust from open lot dairies and cattle feedlots. Atmospheric Environment 40 (1): 27–49.

40. Bull ID, Evershed RP (2012) Organic geochemical signatures of ancient manure use. In: Jones R, editor. Manure matters.Historical, archaeological and ethnographic perspectives. Farnham Surrey, Burlington, VT: Ashgate Pub. Co.pp. 61–79.

41. Rösch M, Ehrmann O, Herrmann L, Schulz E, Bogenrieder A, et al. (2002) An experimental approach to Neolithic shifting cultivation. Vegetation History and Archaeobotany 11: 143–154.

42. Schier W (2009) Extensiver Brandfeldbau und die Ausbreitung der neolithischen Wirtschaftsweise in Mitteleuropa und Südskandinavien am Ende des 5. Jahrtausends v. Chr. Prähistorische Zeitschrift (84): 15–43.

43. Eckmeier E, Gerlach R, Skjemstadt JO, Ehrmann, Schmidt MWI (2007) Minor changes in soil organic carbon and charcoal concentrations detected in a temperate deciduous forest a year after an experimental slash-and-burn. Biogeosciences 4: 377–383.

44. Schneider MPW, Hilf M, Vogt UF, Schmidt MWI (2010) The benzenepolycarboxylic acid (BPCA) pattern of wood pyrolyzed between 200°C and 1000°C. Organic Geochemistry 41: 1082–1088.

45. Wolf M, Lehndorff E, Wiesenberg GLB, Stockhausen M, Schwark L, et al. (2013) Towards reconstruction of past fire regimes from geochemical analysis of charcoal. Organic Geochemistry 55: 11–21.

46. Högberg P (1997) 15N natural abundance in soil-plant systems. New Phytology 137: 179–203.

47. Kriszan M, Amelung W, Schellberg J, Gebbing T, Kühbauch W (2009) Long-term changes of the delta 15N natural abundance of plants and soil in a temperate grassland. Plant and Soil 325: 157–169.

48. Evershed R, Bethell PH (1996) Application of multimolecular biomarker techniques to the identification of faecal material in archaeological soils and sediments. ACS Symposium Series 625: 157–172.

49. Birk JJ, Dippold M, Wiesenberg GLB, Glaser B (2012) Combined quantification of faecal sterols, stanols, stanones and bile acids in soils and terrestrial sediments by gas chromatography-mass spectrometry. Journal of Chromatography A (1242): 1–10.

50. Struck W (1984) Schlitzgräbchen im Kaiserstuhlgebiet. Archäologische Informationen 7: 13–17.

51. Friederich S (2011) Bad Friedrichshall-Kochendorf und Heilbronn-Neckargartach. Studie zum mittelneolithischen Siedlungswesen im Mittleren Neckar. Stuttgart: K. Theiss. 896 p.

52. Stäuble H (2002) Lineare Gräben und Grubenreihen in Nordwestsachsen: Eine Übersicht. Arbeits- und Forschungsberichte zur Sächsischen Bodendenkmalpflege 44: 9–49.

53. Döhle H, Hüser A (2010) Hirschkälber in bronzezeitlichen Schlitzgruben: zwei nicht alltägliche Befunde bei Halle (Saale). Beiträge zur Archäozoologie und Prähistorischen Anthropologie 8: 35–44.

54. Schmid EM, Knickerm H., Bäumler R&KI (2001) Chemical composition of the organic matter in Neolithic soil material as revelaed by CPMAS 13C spectroscopy, polysaccharide analysis, and CuO oxidation. Soil Science 166 (9): 539–584.

55. Semmel A (1968) Studien über den Verlauf jungpleistozäner Formung in Hessen. Frankfurter geographische Hefte 45: 1–133.

56. Fischer P, Hilgers A, Protze J, Kels H, Lehmkuhl F, et al. (2012) Formation and geochronology of Last Interglacial to Lower Weichselian loess/palaeosol sequences–case studies from the Lower Rhine Embayment, Germany. Eiszeitalter und Gegenwart Quaternary Science Journal 61 (48–63).

57. ISO 10694 (1995) Soil Quality – Determination of organic and total carbon after dry combustion (elementary analysis).

58. ISO 10693 (1994) Soil Quality – Determination of carbonate content - Volumetric method.

59. ISO 10390 (1994) Soil Quality - Determination of pH.

60. ISO 11277 (2009) Soil quality - Determination of particle size distribution in mineral soil material - Method by sieving and sedimentation.

61. ISO 11466 (1995) Soil quality - Extraction of trace elements soluble in aqua regia.

62. Stevenson FJ, Cole MA (1999) Mirconutrients and toxic metals. In: Stevenson FJ, Cole MA, editors. Cycles of soil.Carbon, nitrogen, phosphorus, sulfur, micronutrients.New York, NY: Wiley. pp. 368–418.

63. Ekmekcioglu C (2006) Essentielle Spurenelemente: Klinik und Ernährungsmedizin. Berlin: Springer. 1 online resource (205 p.

64. Amelung W, Zhang X (2001) Determination of amino acid enantiomers in soils. Soil Biology and Biochemistry 33: 553–562.

65. Nagata Y, Fujiwara T, Kawaguchi-Nagata K, Fukumori Y, Yamanaka T (1998) Occurrence of peptidyl D-amino acids in soluble fractions of several eubacteria, archaea and eukaryotes. Biochimica Et Biophysica Acta-General Subjects 1379: 76–82.

66. Kvenvolden K, Peterson E, Brown FS (1970) Racemization of Amino Acids in Sediments from Saanich-Inlet, British-Columbia. Science 169: 1079–1082.

67. Frank H, Rettenmeier A, Weicker F, Nicholsen GJ, Bayer E (1982) Determination of enantiomer-labeled amino acids in small volumes of blood by gas chromatography. Analytical Chemistry 54: 715–719.

68. Glaser B, Haumaier L, Guggenberger G, Zech W (1998) Black carbon in soils: the use of benzenecarboxylic acids as specific markers. Organic Geochemistry 29 (4): 911.

69. Brodowski S, Rodionov A, Haumaier L, Glaser B, Amelung W (2005) Revised black carbon assessment using benzene polycarboxylic acids. Organic Geochemistry 36: 1299–1310.

70. Lehmann J, Rillig MC, Thies J, Masiello CA, Hockaday WC, et al. (2011) Biochar effects on soil biota – A review. Soil Biology & Biochemistry 43: 1812–1836.

71. Schlummer M, Hoffmann T, Dikau R, Eickmeier M, Fischer P, et al. (2014) From Point to Area: Upscaling Approaches for Late Quaternary Archaeological and Environmental Data. Earth Science Reviews 131: 22–48.

72. Liu Z, Zhu Q, Tang L (1983) Microelements in the main soils of China. Soil Science 135: 40–46.

73. Johnston AE (2004) Micronutrients in soil and agrosystems: occurrence and availability. Occurrence and availability. York: International Fertiliser Society. 31 p.

74. Lindsay WL (1979) Chemical Equilibria in Soils. New York: Wiley Interscience; Blackburn Press; Wiley. xix, 449 p.

75. Wells EC, Terry RE, Parnell JJ, Hardin JP, Jackson MW, et al. (2000) Chemical Analyses of Ancient Anthrosols in Residential Areas at Piedras Negras, Guatemala. Journal of Archaeological Science 27: 449–462.

76. Bais HP, Weir TL, Perry LG, Gilroy S, Vivanco JM (2006) The role of root exudates in rhizosphere interactions with plants and other organisms. Annual Review of Plant Biology 57: 233–266.

77. Lüning J (2000) Steinzeitliche Bauern in Deutschland. Die Landwirtschaft im Neolithikum. Bonn: Habelt. 285 p.

78. Yoneyama T (1996) Characterization of natural 15N abundance of soils. In: Boutton TW, Yamasaki S, editors.Mass Spectrometry of Soils. New York: Marcel Dekker, Inc. pp. 206–223.

79. Steele KW, Wilson AT, Saunders WMH (1981) Nitrogen isotope ratios in surface and sub-surface horizons of New Zealand improved grassland soils. New Zealand Journal of Agricultural Research 24 (2): 167–170.

80. Shearer G, Kohl D (1989) Estimations of N2 Fixation in Ecosystems: The Need for and Basis of the 15N Natural Abundance Method. In: Rundel PW, Ehleringer JR, Nagym KA, editors. Stable isotopes in ecological research. New York: Springer-Verlag. pp. 342–374.

81. Watzka M, Buchgraber K, Wanek W (2006) Natural 15N abundance of plants and soils under different management practices in a montane grassland. Soil Biology and Biochemistry 38: 1564–1576.

82. Martinelli LA, Piccolo MC, Townsend AR, Vitousek PM, Cuevas E, et al. (1999) Nitrogen isotopic composition of leaves and soil: Tropical versus temperate forests. Biogeochemistry 46: 45–65.

83. Peri PL, Ladd BPDA, Bonser SP, Laffan W, Amelung W (2012) Carbon (d13C) andnitrogen (d15N) stable isotope composition in plant and soil in Southern Patagonia's native forests. Global Change Biology 18: 311–321.

84. Lang G (1994) Quartäre Vegetationsgeschichte Europas: Methoden und Ergebnisse: Spektrum Akademischer Verlag.

85. Brenner DL, Amundson R, Baisden WT, Kendall C, Harden J (2001) Soil N and 15N variation with time in a California annual grassland ecosystem. Geochimica et Cosmochimica Acta 65: 4171–4186.

86. Bol R, Eriksen J, Smith P, Garnett MH, Coleman K, et al. (2005) The natural abundance of C-13, N-15, S-34 and C-14 in archived (1923-2000) plant and soil samples from the Askov long-term experiments on animal manure and mineral fertilizer. Rapid Communications in mass spectrometry 19: 3216–3226.

87. Senbayram M, Dixon L, Goulding KWT, Bol R (2008) Long-term influence of manure and mineral nitrogen applications on plant and soil 15N and 13C values from the Broadbalk Wheat Experiment. Rapid Communications in mass spectrometry 22: 1735–1740.

88. Kriszan M, Schellberg J, Amelung W, Gebbing T, Pötsch EM, et al. (2014) Revealing N management intensity on grassland farms based on natural δ15N abundance. Agriculture, Ecosystems and Environment 184: 158–167.

89. Högberg P, Johannisson C (1993) 15N abundance of forests is correlated with losses of nitrogen. Plant and Soil 157: 147–150.

90. Amundson R, Austin AT, Schuur EAG, Yoo K, Matzek V, et al. (2003) Global patterns of the isotopic composition of soil and plant nitrogen. Global Biogeochemical Cycles 17 (1): 1031.

91. Krull ES, Skjemstad JO (2003) d13C and d15N profiles in C-14-dated Oxisol and Vertisols as a function of soil chemistry and mineralogy. Geoderma 112: 1–29.

92. Bol R, Ostle NJ, Chenu C, Petzke KJ, Werner RA, et al. (2004) Long term changes in the distribution and delta 15N values of individual soil amino acids in the absence of plant and fertilizer inputs. Isotopes in Environmental and Health Studies 40 (4): 243–256.

93. Friedel JK, Scheller E (2002) Composition of hydrolysable amino acids in soil organic matter and soil microbial biomass. Soil Biology and Biochemistry: 315–325.

94. Amelung W (2003) Nitrogen biomarkers and their fate in soil. Journal of Plant Nutrition and Soil Science 166: 207–217.

95. Brodowski S, Amelung W, Lobe I, Du Preez CC (2004) Losses and biogeochemical cycling of soil organic nitrogen with prolonged arable cropping in the South African Highveld- evidence from D- and L-amino acids. Biogeochemistry 71: 17–42.

96. Stevenson FJ (1994) Organic forms of soil nitrogen. Stevenson (Hg.) – Humus chemistry. pp. 59–92.

97. Schulten HR, Schnitzer M (1998) The chemistry of soil organic nitrogen: a review. Biology and Fertility of Soils 26: 1–15.

98. Mikutta R, Kaiser K, Dörr N, Vollmer A, Chadwick OA, et al. (2010) Mineralogical impact on organic nitrogen across a long-term soil chronosequence (0.3–4100 kyr). Geochimica et Cosmochimica Acta 74: 2142–2164.

99. Glaser B (1999) Eigenschaften und Stabilität des Humuskörpers der "Indianerschwarzerde" Amazoniens. Bayreuther bodenkundliche Berichte 68: 196.

100. Campbell CA, Schnitzer M, Lafond GP, Zentner RP, Knipfel JE (1991) Thirty-year crop rotation and management practices effects on soil and amino nitrogen. Soil Sci. Soc. Am. J. 55: 739–745.

101. Rejsek K, Formanek P, Vranova V (2010) Amino acids in soil hydrolysates. In: Rejsek K, Formanek P, Vranova V, editors. The soil amino acids. Quality, distribution and site ecology; environmental science, engineering and technology. New York, N.Y: Nova Science Publ. pp. 15–31.

102. Vranova V, Zahradnickova H, Janous D, Skene KR, Matharu AS, et al. (2012) The significance of D-amino acids in soil, fate and utilization by microbes and plants: review and identification of knowledge gaps. Plant and Soil 354: 21–39.

103. Zimmermann A, Meurers-Balke J, Kalis AJ (2005) Das Neolithikum im Rheinland. Bonner Jahrbücher 205: 1–63.

104. Amelung W, Brodowski S, Sandhage-Hofmann A, Bol R (2008) Combining biomarker with stable isotope analyses for assessing the transformation and turnover of soil organic matter. Advances in Agronomy: 155–250.

105. Simpson IA, van Bergen PF, Perret V, Elhmmali MM, Roberts DJ, et al. (1999) Lipid biomarkers of manuring practice in relict anthropogenic soils. The Holocene 9 (2): 223–229.

106. Bull ID, Betancourt PP, Evershed RP (2001) An Organic Geochemical Investigation of the Practice of Manuring at a Minoan Site on Pseira Island, Crete. Geoarchaeology 16: 223–242.

107. Lombardo U, Szabo K, Caprilles J, Ma J, Amelung W, et al. (2013) Early and Middle Holocene Hunter-Gatherer Occupations in Western Amazonia: The Hidden Shell Middens. Plos one 8: e72746.

108. Singh B, Singh BP, Cowie AL (2010) Characterisation and evaluation of biochars for their application as a soil amendment. Australian Journal of Soil Research 48: 516–525.

109. Prost K, Borchard N, Siemens J, Kautz T, Séquaris J, et al. (2013) Biochar affected by composting with farmyard manure. Journal of Environmental Quality 42 (1): 164–172.

110. Bull ID, Elhmmali MM, Roberts DJ, Evershed RP (2003) The application of steroidal biomarkers to track the abandonment of a Roman wastewater course at the Agora (Athens, Greece). Archaeometry 45: 149–161.

111. Shillito LBID, Matthews W, Almond MJ, Williams JM, Evershed RP (2011) Biomolecular and micromorphological analysis of suspected faecal deposits at Neolithic Çatalhöyük, Turkey. Journal of Archaeological Science 38: 1869–1877.

112. Taube PS, Hansel FA, dos Santos Madureira LA, Teixeira WG (2013) Organic geochemical evaluation of organic acids to assess anthropogenic soil deposits of Central Amazon, Brazil. Organic Geochemistry 58: 96–106.

113. Lloyed CEM, Michaelides K, Chadwick DR, Dungait JAJ, Evershed RP (2012) Tracing the flow-driven vertical transport of livestock-derived organic matter through soil using biomarkers. Organic Geochemistry 43: 56–66.

114. Spielvogel S, Knicker H, Kögel-Knabner I (2004) Soil organic matter composition and soil lightness. Journal of Plant Nutrition and Soil Science 167: 545–555.

115. Gerlach R, Fischer P, Eckmeier E, Hilgers A (2012) Buried dark soil horizons and archaeological features in the Neolithic settlement region of the Lower Rhine area, NW Germany: Formation, geochemistry and chronostratigraphy. Quarternary International (265): 191–204.

116. Glaser B, Balashov E, Haumaier L, Guggenberger G, Zech W (2000) Black carbon in density fractions of anthropogenic soils of the Brazilian Amazon region. Organic Geochemistry 31: 669–678.
117. Brodowski S, Amelung W, Haumaier L, Zech W (2007) Black carbon contribution to stable humus in German arable soils. Geoderma 139: 220–228.
118. Eckmeier E, Egli M, Schmidt MWI, Schlumpf N, Nötzli M, et al. (2010) Preservation of fire-derived carbon compounds and sorptive stabilisation promote the accumulation of organic matter in black soils of the Southern Alps. Geoderma 159: 147–155.
119. Kolstrup E (1994) Examples of Weichselian environments: local versus regional developments. Eiszeitalter und Gegenwart Quaternary Science Journal 44: 16–19.
120. Daniau A, d'Errico F, Goñi MFS (2010) Testing the Hypothesis of Fire Use for Ecosystem Management by Neanderthal and Upper Palaeolithic Modern Human Populations. Plos one 5 (2): e9157.
121. Goldammer JG, Furyaev VV (1996) Fire in Ecosystems of Boreal Eurasia. Dordrecht: Springer Netherlands. 1 online resource (xii, 531 p.
122. Wolf M, Lehndorff E, Mrowald M, Eckmeier E, Kehl M, et al. (2014) Black Carbon: fire fingerprints in Pleistocene loess-palaeosol archives in Germany. Organic Geochemistry 70: 44–52.
123. Pyne SJ, Andrews PL, Laven RD (1996) Introduction to wildland fire. New York: John Wiley and Sons.
124. Kalis AJ, Tegtmeier U (1999) Gehölze als Nutzpflanzen. In: Knörzer KH, editor. PflanzenSpuren, Archäobotanik im Rheinland: Agrarlandschaft und Nutzpflanzen im Wandel der Zeiten. Materialien zur Bodendenkmalpflege im Rheinland (No. 10). Köln: Rheinland-Verlag. pp. 129–172.
125. Tinner W, Hubschmid P, Wehrli M, Ammann B, Conedera M (1999) Long-term forest fire ecology and dynamics in southern Switzerland. Journal of Ecology 87: 273–289.

126. Kleber M, Rößner J, Chenu C, Glaser B, Knicker H, et al. (2003) Prehistoric alteration of soil properties in a central german chernozemic soil: in search of pedologic indicators for prehistoric activity. Soil Science 168: 292–306.
127. Eckmeier E, Rösch M, Ehrmann, Schmidt MWI, Schier W, et al. (2007) Conversion of biomass to charcoal and the carbon mass balance from a slash-and-burn experiment in a temperate deciduous forest. The Holocene 17 (4): 539–542.
128. Ehrmann O, Rösch M, Schier W (2009) Experimentelle Rekonstruktion eine jungneolithischen Wald-Feldbaus mit Feuereinsatz- ein multidisziplinäres Forschungsprojekt zur Wirtschaftsarchäologie und Landschaftsökologie. Prähistorische Zeitschrift 84 (44–72).
129. Lehmann J, da Silva Jr JP, Steiner C, Nehls T, Zech W, et al. (2003) Nutrient availability and leaching in an archaeological Anthrosol and a Ferralsol of the Central Amazon basin: fertilizer, manure and charcoal amendments. Plant and Soil 249 (2): 343–357.
130. Atkinson CJ, Fitzgerald JD, Hipps NA (2010) Potential mechanisms for achieving agricultural benefits from biochar application to temperate soils: a review. Plant and Soil 337: 1–18.
131. Birk JJ, Teixeira WG, Neves EG, Glaser B (2011) Faeces deposition on Amazonian Anthrosols as assessed from 5β-stanols. Journal of Archaeological Science 38 (6): 1209–1220.
132. Ad-hoc-AG Boden (2005) Bodenkundliche Kartieranleitung. 438 p. Hannover.
133. Kropp J, Roithmeier O, Hattermann F, Rachimow C, Lüttger A, et al. (2009) Klimawandel in Sachsen-Anhalt - Verletzlichkeiten gegenüber den Folgen des Klimawandels. Final Report.
134. Genßler L, Hädicke A, Hübner T, Jacob S, König H, et al. (2010) Klima und Klimawandel in Nordrhein-Westfalen- Daten und Hintergründe. LANUV-Fachbericht, 27. Landesamt für Natur, Umwelt und Verbraucherschutz Nordrhein-Westfalen, Recklinghausen.

14

Scaling Up Stomatal Conductance from Leaf to Canopy Using a Dual-Leaf Model for Estimating Crop Evapotranspiration

Risheng Ding[1], Shaozhong Kang[1]*, Taisheng Du[1], Xinmei Hao[1], Yanqun Zhang[2]

1 Center for Agricultural Water Research in China, China Agricultural University, Beijing, China, 2 National Center of Efficient Irrigation Engineering and Technology Research-Beijing, China Institute of Water Resources and Hydropower Research, Beijing, China

Abstract

The dual-source Shuttleworth-Wallace model has been widely used to estimate and partition crop evapotranspiration (λET). Canopy stomatal conductance (G_{sc}), an essential parameter of the model, is often calculated by scaling up leaf stomatal conductance, considering the canopy as one single leaf in a so-called "big-leaf" model. However, G_{sc} can be overestimated or underestimated depending on leaf area index level in the big-leaf model, due to a non-linear stomatal response to light. A dual-leaf model, scaling up G_{sc} from leaf to canopy, was developed in this study. The non-linear stomata-light relationship was incorporated by dividing the canopy into sunlit and shaded fractions and calculating each fraction separately according to absorbed irradiances. The model includes: (1) the absorbed irradiance, determined by separately integrating the sunlit and shaded leaves with consideration of both beam and diffuse radiation; (2) leaf area for the sunlit and shaded fractions; and (3) a leaf conductance model that accounts for the response of stomata to PAR, vapor pressure deficit and available soil water. In contrast to the significant errors of G_{sc} in the big-leaf model, the predicted G_{sc} using the dual-leaf model had a high degree of data-model agreement; the slope of the linear regression between daytime predictions and measurements was 1.01 ($R^2 = 0.98$), with RMSE of 0.6120 mm s^{-1} for four clear-sky days in different growth stages. The estimates of half-hourly λET using the dual-source dual-leaf model (DSDL) agreed well with measurements and the error was within 5% during two growing seasons of maize with differing hydrometeorological and management strategies. Moreover, the estimates of soil evaporation using the DSDL model closely matched actual measurements. Our results indicate that the DSDL model can produce more accurate estimation of G_{sc} and λET, compared to the big-leaf model, and thus is an effective alternative approach for estimating and partitioning λET.

Editor: Ben Bond-Lamberty, DOE Pacific Northwest National Laboratory, United States of America

Funding: This research was partially supported by the National Natural Science Foundation of China (51321001 and 51309223), the National High-Tech Research and Development Program (2011AA100502 and 2013AA102904) and the Ph.D. Programs Foundation of Ministry of Education of China (20130008120007). The research is also supported by the Program of Introducing Talents of Discipline to Universities (B14002). The funders had no role in study design, data collection and analysis, decision to publish, or preparation of the manuscript.

Competing Interests: The authors have declared that no competing interests exist.

* E-mail: kangsz@cau.edu.cn

Introduction

Accurate estimation of evapotranspiration (λET) is important in understanding terrestrial hydrological cycles because λET is the largest component in the terrestrial water balance after precipitation [1]. In agricultural production, improved estimation of crop λET is also needed to develop precise irrigation scheduling and enhance water use efficiency, as soil water depletion is mostly determined by the rate of λET [2,3,4]. However, direct measurement of λET is often difficult, costly and not available in many regions [5,6]. Therefore, mathematical models are needed to estimate λET using readily measurable meteorological and environmental variables.

Vegetation transpiration (T_r) and soil evaporation (E_s), which are controlled by different biotic and physical processes, are the two major components of λET. Transpiration is strongly linked to crop productivity since it occurs concurrently with photosynthetic gas exchange [7]. Quantifying T_r is also critical to accurately predict the response of crop functioning and physiology to

changing climate [8]. Because the two separate processes occur simultaneously, there is no simple way to distinguish between them [9,10].

Several models have been developed to calculate λET and separately estimate soil evaporation and transpiration [11,12,13]. Shuttleworth and Wallace [14] described a dual-source model with a resistance-energy combination, which could separately predict T_r and E_s, and is also sufficiently simple [8]. This model has been widely used and can also be used to gain an understanding of the interaction of biophysical and hydrological processes in the crop canopy [14,15]. Determination of different resistances or conductances (the reciprocal of resistance) is necessary for its practical application. Specifically, canopy stomatal conductance (G_{sc}) is often calculated by scaling up leaf stomatal conductance of the leaves acting in parallel while treating the canopy as one big-leaf, hereafter the big-leaf approach [8,16].

The weakness of using the big-leaf approach is that the use of mean absorbed radiation can significantly overestimate G_{sc}, especially in dense canopies, because the light response of stomata

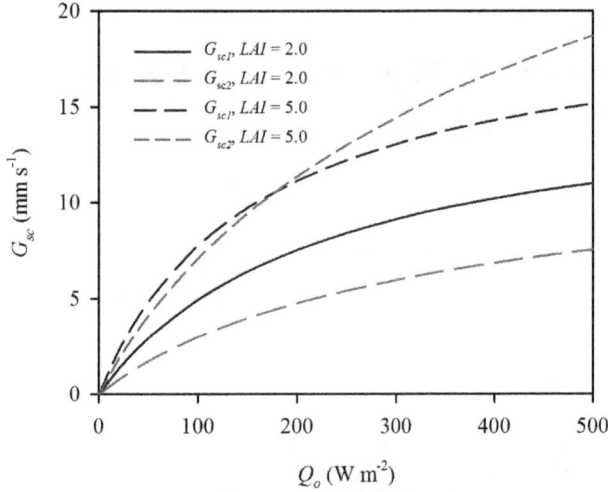

Figure 1. Response of canopy stomatal conductance (G_{sc}) to incident PAR above the canopy (Q_o) at different leaf area indices (*LAI*). G_{sc1} and G_{sc2} are calculated by the big-leaf and dual-leaf canopy stomatal resistance models, respectively.

is non-linear [17,18]. Moreover, the overestimation of G_{sc} by the big-leaf approach often occurred when using total leaf area index (*LAI*) to scale up stomatal conductance. To mitigate the overestimation of G_{sc}, researchers have used the effective or sunlit *LAI* (*LAI_e*) instead of total *LAI* [16,18]. But, the relationship between *LAI_e* and *LAI* is empirical and varies with the vegetation type and solar radiation angle. Therefore, models of G_{sc} should consider the non-linear response of stomata to irradiance as well as heterogeneous radiation profiles in the canopy. Beam and diffuse radiation penetrating the canopy must be considered separately due to differential attenuation in canopies, as should visible and near-infrared wavebands due to differential absorption by leaves [12,19]. A multilayers model could account for the mechanism of radiation penetration [20,21], but it is useless in practical applications [22]. Some studies indicate that radiation penetration within the canopy could be simplified by splitting the canopy into two fractions of leaves: sunlit and shaded [18,23,24]. Leuning et al. [25] used a multilayer model to show that photosynthesis from a canopy is closely approximated when calculated as the separate sums of sunlit and shaded fractions, weighted by their respective leaf area within the canopy. However, few studies have calculated G_{sc} using the approach of sunlit and shaded leaves (hereafter, the dual-leaf approach) in the dual-source S-W model for estimating λET.

Soil surface resistance (r_{ss}) is another key parameter for partitioning λET in the dual-source model because above-canopy λET is contaminated by flux from the soil substrate as variations in leaf area affect soil exposure, soil evaporation and absorbed radiation [14,26]. To reduce soil evaporation in the field, the ground is often mulched with plastic film, and this technique is widely used in northwest China [10,27]. But the effect of ground mulching on soil evaporation has not been taken into account when parameterizing r_{ss} in the dual-source model. In this study, a dual-leaf model of G_{sc} was developed by scaling up leaf stomatal conductance using the corresponding absorbed irradiance for the sunlit and shaded leaves separately, which overcomes the limitations of the big-leaf model. The dual-leaf model developed here was then incorporated into the dual-source model to estimate and partition λET. We evaluated the dual-source dual-leaf model (DSDL) by comparing the model estimations of λET with

measurements taken in an irrigated maize field mulched with plastic film.

Model Development for Evapotranspiration and Canopy Stomatal Conductance

1 Dual-source Evapotranspiration Model

The λET in the dual-source model was partitioned into two components, canopy transpiration (λT_c) and soil evaporation (λE_s) with a resistance network [14].

$$\lambda ET = \lambda T_c + \lambda E_s = \omega_c PM_c + \omega_s PM_s \tag{1}$$

$$PM_c = \frac{\Delta A + (\rho C_p VPD - \Delta r_{ac} A_s)/(r_{aa} + r_{ac})}{\Delta + \gamma[1 + r_{sc}/(r_{aa} + r_{ac})]} \tag{2}$$

$$PM_s = \frac{\Delta A + [\rho C_p VPD - \Delta r_{as}(A - A_s)]/(r_{aa} + r_{as})}{\Delta + \gamma[1 + r_{ss}/(r_{aa} + r_{as})]} \tag{3}$$

$$\omega_c = \frac{1}{1 + R_c R_a/[R_s(R_c + R_a)]} \tag{4}$$

$$\omega_s = \frac{1}{1 + R_s R_a/[R_c(R_s + R_a)]} \tag{5}$$

$$R_s = (\Delta + \gamma)r_{as} + \gamma r_{ss} \tag{6}$$

$$R_c = (\Delta + \gamma)r_{ac} + \gamma r_{sc} \tag{7}$$

$$R_a = (\Delta + \gamma)r_{aa} \tag{8}$$

where PM_c and PM_s are the terms similar to those in Penman–Monteith model for canopy transpiration and soil evaporation, respectively, and ω_c and ω_s are the weighting factors for the crop canopy and soil components, respectively. λ is the heat of water vaporization, ρ is air density, C_p is the specific heat of dry air at constant pressure, Δ is the slope of the saturation vapor pressure curve, γ is the psychrometric constant, VPD is vapor pressure deficit, and A and A_s are the total available energy and available energy for soil, respectively. r_{sc} is canopy stomatal resistance, r_{ss} is soil surface resistance, r_{ac} is canopy boundary layer resistance, r_{as} is soil boundary layer resistance between soil and vegetative canopy, and r_{aa} is aerodynamic resistance between canopy source and reference height, respectively. The calculation procedures of the other resistances except r_{sc} and r_{ss} are given in Appendix S1.

$$A = R_n - G \tag{9}$$

$$A_s = R_{ns} - G \tag{10}$$

$$R_{ns} = R_n \exp(-\kappa_R LAI) \tag{11}$$

where R_n and R_{ns} are net radiation above the canopy and at the soil surface, respectively, and G is the soil heat flux. The canopy extinction coefficient of net radiation, κ_R, is dependent on leaf orientation and solar zenith angle (ζ) [17].

$$\kappa_R = \frac{G_L}{\cos(\zeta)} \tag{12}$$

where G_L is 0.5 for a spherical leaf angle distribution. ζ, the angle subtended by the sun at the center of the earth, is perpendicular to the surface of the earth and calculated as in Appendix S1.

The two components, λT_c and λE_s were now calculated along with the VPD at the canopy source height (D_o).

$$D_o = VPD + [\Delta A - (\Delta + \gamma) \lambda ET] r_{aa} / \rho C_p \tag{13}$$

$$\lambda T_c = \frac{\Delta(A - A_s) + \rho C_p D_o / r_{ac}}{\Delta + \gamma(1 + r_{sc}/r_{ac})} \tag{14}$$

$$\lambda E_s = \frac{\Delta A_s + \rho C_p D_o / r_{as}}{\Delta + \gamma(1 + r_{ss}/r_{as})} \tag{15}$$

The measured r_{sc} was obtained by inverting Eq. (14), with λT_c calculated by the known or measured λET and λE_s.

$$\lambda T_c = \lambda ET - \lambda E_s \tag{16}$$

$$r_{sc} = \frac{\Delta(A - A_s) + \rho_a C_p D_o / r_{ac} - \lambda T_c(\Delta + \gamma)}{\gamma \lambda T_c / r_{ac}} \tag{17}$$

2 Irradiance within Crop Canopy

Incident PAR light above the canopy (Q_o) was divided into diffuse (Q_{od}) and beam irradiance (Q_{ob}) through the fraction of diffuse radiation (f_d).

$$Q_{od} = f_d Q_o \tag{18}$$

$$Q_{ob} = (1 - f_d) Q_o \tag{19}$$

The f_d was calculated from a simple model of atmospheric attenuation of radiation [17,28].

$$f_d = \frac{1 - \tau_a^{ma}}{1 + \tau_a^{ma}(1/f_a - 1)} \tag{20}$$

where τ_a is the atmospheric transmittance, f_a is the forward scattering coefficient of PAR in atmosphere, and m_a is the optical air mass, which can be calculated as follows.

$$m_a = \frac{P/P_o}{\cos \zeta} \tag{21}$$

where P is local atmospheric pressure and P_0 is atmospheric pressure at sea level.

At a depth ξ in the canopy, three types of irradiance can be calculated: the total beam, $Q_{\ell,bt}$ (unintercepted beam plus down scattered beam), direct beam, $Q_{\ell,b}$ (unintercepted beam) and the diffuse flux, $Q_{\ell,d}$ [17,29].

$$Q_{\ell,bt}(\xi) = Q_{ob}(1 - \rho_{cb})\sqrt{\alpha}\kappa_b \exp(-\sqrt{\alpha}\kappa_b \xi) \tag{22}$$

$$Q_{\ell,b}(\xi) = Q_{ob}(1 - \rho_{cb})\kappa_b \exp(-\kappa_b \xi) \tag{23}$$

$$Q_{\ell,d}(\xi) = Q_{od}(1 - \rho_{cd})\sqrt{\alpha}\kappa_d \exp(-\sqrt{\alpha}\kappa_d \xi) \tag{24}$$

$$\rho_{cb} = 1 - \exp(2\rho_h \kappa_b / (1 + \kappa_b)) \tag{25}$$

$$\rho_h = \frac{1 - \sqrt{\alpha}}{1 + \sqrt{\alpha}} \tag{26}$$

$$\rho_{cd} = \frac{2\kappa_d \rho_h}{\kappa_d + 1} \tag{27}$$

where α is absorptivity of leaves for irradiation, ρ_{cb} and ρ_{cd} are canopy reflectance for beam and diffuse irradiance respectively with a randomly spherical leaf-angle distribution, ρ_h is canopy reflectance for beam irradiance with a horizontal leaf-angle distribution, and κ_d is an extinction coefficient for diffuse radiation.

The absorbed irradiance in a canopy height (Q_ℓ) consists of the total beam radiation ($Q_{\ell,bt}$) and the diffuse radiation ($Q_{\ell,d}$).

$$Q_\ell(\xi) = Q_{\ell,bt}(\xi) + Q_{\ell,d}(\xi) \tag{28}$$

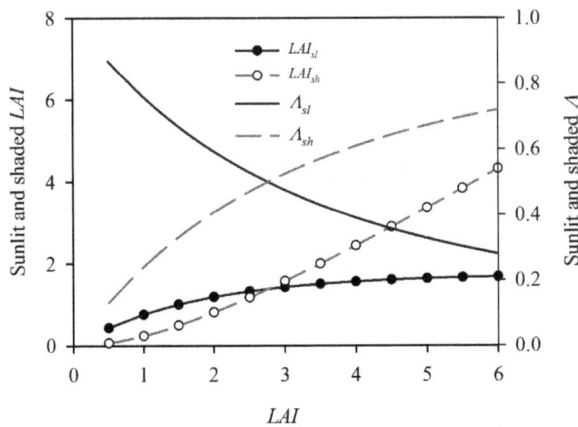

Figure 2. Variation in the sunlit and shaded leaf area indices (LAI_{sl} and LAI_{sh}) versus different LAI. The ratios of LAI_{sl} and LAI_{sh} to LAI (Λ_{sl} and Λ_{sh}) are also shown.

The total irradiance absorbed by the entire canopy (Q_c) per unit ground area was determined by integrating Q_ℓ over the total LAI.

$$Q_c = \int_0^{LAI} Q_l(\xi)d\xi = Q_{ob}(1-\rho_{cb})\left[1-\exp\left(-\sqrt{\alpha}\kappa_b LAI\right)\right]$$
$$+ Q_{od}(1-\rho_{cd})\left[1-\exp\left(-\sqrt{\alpha}\kappa_d LAI\right)\right] \tag{29}$$

The irradiance absorbed by the sunlit fraction in a specific canopy height ($Q_{\ell,sl}$) can be given as the sum of direct-beam ($Q_{\ell,b}$), diffuse ($Q_{\ell,d}$) and scattered-beam components ($Q_{\ell,s}$).

$$Q_{\ell,sl}(\xi) = Q_{\ell,b}(\xi) + Q_{\ell,d}(\xi) + Q_{\ell,s}(\xi) \tag{30}$$

$$Q_{\ell,s}(\xi)$$
$$= Q_{0b}\left[(1-\rho_{cb})\sqrt{\alpha}\kappa_b \exp\left(-\sqrt{\alpha}\kappa_b\xi\right) - \alpha\kappa_b \exp\left(-\kappa_b\xi\right)\right] \tag{31}$$

The irradiance absorbed by the sunlit fraction in the entire canopy (Q_{sl}) was obtained by integrating $Q_{\ell,sl}$ over the total LAI.

$$Q_{sl} = \int_0^{LAI} Q_{\ell,sl}(\xi)f_{\ell,sl}(\xi)d\xi$$
$$= \int_0^{LAI} Q_{\ell,b}(\xi)f_{\ell,sl}(\xi)d\xi + \int_0^{LAI} Q_{\ell,d}(\xi)f_{\ell,sl}(\xi)d\xi \tag{32}$$
$$+ \int_0^{LAI} Q_{\ell,s}(\xi)f_{\ell,sl}(\xi)d\xi$$

$$\int_0^{LAI} Q_{\ell,b}(\xi)f_{\ell,sl}(\xi)d\xi = Q_{ob}\alpha[1-\exp(-\kappa_b LAI)] \tag{32a}$$

$$\int_0^{LAI} Q_{\ell,d}(\xi)f_{\ell,sl}(\xi)d\xi$$
$$= Q_{od}(1-\rho_{cd})\frac{\sqrt{\alpha}\kappa_d}{\sqrt{\alpha}\kappa_d+\kappa_b}\left[1-\exp\left(-\left(\sqrt{\alpha}\kappa_d+\kappa_b\right)LAI\right)\right] \tag{32b}$$

$$\int_0^{LAI} Q_{\ell,s}(\xi)f_{\ell,sl}(\xi)d\xi$$
$$= Q_{ob}\left[(1-\rho_{cb})\frac{\sqrt{\alpha}}{\sqrt{\alpha}+1}\left(1-\exp\left(-\left(\sqrt{\alpha}\kappa_b+\kappa_b\right)LAI\right)\right)\right.$$
$$\left. -\alpha(1-\exp(-2\kappa_b LAI))/2\right] \tag{32c}$$

The total irradiance absorbed (Q_c) is the sum of the two parts, irradiance absorbed by the separate sunlit (Q_{sl}) and shade fractions (Q_{sh}) of the canopy. Thus, Q_{sh} was calculated as the difference between Q_c and Q_{sl}.

$$Q_{sh} = Q_c - Q_{sl} \tag{33}$$

3 Leaf and Canopy Stomatal Conductance

The stomatal conductance is represented by g_s for a single leaf and G_{sc} for the entire canopy.

3.1. Leaf stomatal conductance. The leaf g_s can be calculated using the Jarvis-Stewart type multiple formulae [30,31,32].

$$g_s = \frac{1}{r_s} = g_{smax} \Pi_i F_{X_i} \tag{34}$$

where g_{smax} is the maximum value of the leaf stomatal conductance and Fx_i is the stress function of the specific environmental variables (x_i), $0\leq Fx_i\leq1$. The original model used short-wave radiation as the light variable. Here we have used the photosynthetically active radiation absorbed by canopy leaves (Q_a) because stomatal aperture is determined by the received visible wavelength radiation, rather than short-wave radiation [17,18,33]. In addition, we incorporated the environmental stress impact on g_s by VPD and available soil water as follows.

$$F_Q = \frac{500+k_Q}{500}\frac{Q_a}{Q_a+k_Q} \tag{35}$$

$$F_D = e^{-k_D VPD} \tag{36}$$

$$F_w = \frac{1-\exp(-k_w\theta_E)}{1-\exp(-k_w)} \tag{37}$$

$$\theta_E = \frac{\theta-\theta_W}{\theta_F-\theta_W} \tag{38}$$

where the k_Q, k_D and k_w are the stress coefficients of Q_a, VPD and extractable soil water in the root zone (θ_E), and θ, θ_F and θ_W are the measured soil moisture, field capacity and wilting point in the root zone, respectively.

3.2. Big-leaf model of canopy stomatal conductance. The canopy stomatal conductance in the big-leaf model (G_{sc1}) is estimated by scaling up g_s weighing by the effective LAI (LAI_e) as if the canopy is a single big-leaf [16,18,32].

$$G_{sc1} = \frac{1}{r_{sc1}} = g_{s1} LAI_e \tag{39}$$

where g_{s1} is the mean leaf stomatal conductance for the entire big-leaf and can be calculated by Eq. (34) based on the mean absorbed irradiance of the entire canopy using Eq. (29); LAI_e is empirically equal to the actual LAI for $LAI\leq2$, $LAI/2$ for $LAI\geq4$, and 2.0 for others [16].

Table 1. The parameters used in the dual-source dual-leaf model.

Symbol	Description	Value	Units	Sources
b_1	Parameter in soil resistance model	15.2	s m^{-1}	Fitted in this study
b_2	Parameter in soil resistance model	−5.8	-	Fitted in this study
b_3	Parameter in soil resistance model	88.7	s m^{-1}	Fitted in this study
c_d	Mean drag coefficient for the individual vegetative elements	0.1	-	Meyers and Paw [42]
d_l	Characteristic leaf dimension	0.068	m	Measured in this study
G_L	Spherical leaf angle distribution	0.5	-	Campbell and Norman [17]
g_{smax}	Maximum value of the leaf stomatal conductance	7.5	mm s^{-1}	Fitted in this study
k	von Karman's constant	0.41	-	Brutsaert [41]
k_Q	Stress coefficients of the photosynthetically active radiation in the stomatal conductance model	150	W m^{-2}	Fitted in this study
k_D	Stress coefficients of the vapor pressure deficit in the stomatal conductance model	0.2	kPa^{-1}	Fitted in this study
k_w	Stress coefficients of the extractable soil water in the root zone in the stomatal conductance model	7.5	-	Fitted in this study
z_{0s}	Effective roughness length of the soil substrate	0.01	m	Shuttleworth and Wallace [14]
α	Absorptivity of leaves of irradiation	0.8	-	Monteith and Unsworth [12]
κ_d	Extinction coefficient for diffuse radiation	0.7	-	Campbell and Norman [17]
τ_a	Atmospheric transmittance	0.72	-	Brutsaert [41]
f_a	Forward scattering coefficient of PAR in the atmosphere	0.43	-	Brutsaert [41]
f_m	Fraction of ground mulched by plastic film	0.5/0.6[a]	-	Measured in this study

[a]0.5/0.6 is the fraction of ground mulched by plastic film in 2009 and 2010, respectively.

3.3. Dual-leaf model of canopy stomatal conductance. In the dual-leaf model, G_{sc} (G_{sc2}) is calculated by summing the contributions of sunlit and shaded fractions, G_{sl} and G_{sh}, respectively, which are scaled up using the associated g_s weighted by their respective fractions of LAI [18,24].

$$G_{sc2} = \frac{1}{r_{sc2}} = G_{sl} + G_{sh} = g_{sl} LAI_{sl} + g_{sh} LAI_{sh} \qquad (40)$$

where g_{sl} and g_{sh} are the mean leaf stomatal conductance for sunlit and shaded leaves, respectively, and can be calculated by Eq. (34) based on the separate absorbed irradiance using Eqs. (32) and (33). LAI_{sl} and LAI_{sh} are LAI for sunlit and shaded leaves in the entire canopy, respectively.

Assuming that all leaves in a canopy are randomly distributed, the fraction of sunlit leaves ($f_{\ell,sl}$) in a specific canopy depth declines exponentially with cumulative leaf area (ξ) [29,34].

$$f_{\ell,sl}(\xi) = \exp(-\kappa_b \xi) \qquad (41)$$

$$\xi(z) = \int_z^{h_c} L_d(z) dz \qquad (42)$$

where κ_b is an extinction coefficient for beam radiation, L_d is leaf area density, z is height above ground, and h_c is canopy height. LAI_{sl} is therefore calculated by integrating $f_{\ell,sl}$ for the entire canopy.

$$LAI_{sl} = \int_0^{LAI} f_{\ell,sl}(\xi) d\xi = \frac{1 - \exp(-\kappa_b LAI)}{\kappa_b} \qquad (43)$$

$$LAI_{sh} = LAI - LAI_{sl} \qquad (44)$$

4 Soil Surface Resistance

In this study, r_{ss} was directly calculated with a function dependent on surface soil water content [35], accounting for the effect of plastic mulching on reduction of soil evaporation by introducing a term for fraction of plastic mulch, f_m [i.e. r_{ss} is divided by the area of exposed substrate per unit ground area ($1 - f_m$)].

$$r_{ss} = \frac{1}{1 - f_m} \left[b_1 \left(\frac{\theta_g}{\theta_s} \right)^{b_2} + b_3 \right] \qquad (45)$$

where θ_g is the average soil water content between 0–0.1 m, θ_s is the saturated water content of surface soil, and b_1, b_2, and b_3 are the empirical coefficients.

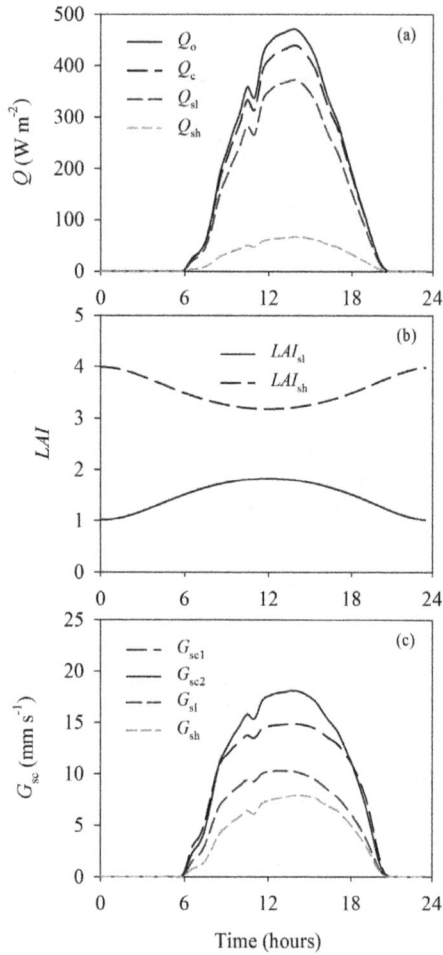

Figure 3. Diurnal variations of (a) modeled irradiance, (b) leaf area index and (c) canopy stomatal conductance at leaf area index (LAI) = 5.0. Q_o is the total irradiance above the canopy, Q_c is the irradiance absorbed by the entire canopy, and is separated into the irradiance absorbed by the sunlit leaves of the canopy (Q_{sl}) and irradiance absorbed by the shaded leaves of the canopy (Q_{sh}). LAI_{sl} and LAI_{sh} are the sunlit and shaded fractions of LAI, respectively. G_{sc1} and G_{sc2} are the canopy stomatal conductance calculated by the big-leaf and dual-leaf models, respectively; G_{sc2} is separated into the sunlit and shaded canopy stomatal resistance, G_{sl} and G_{sh}, respectively.

Materials and Methods

1 Experimental Arrangement

The experiments were conducted in an irrigated maize field in 2009 and 2010 at Shiyanghe Experimental Station for Water-saving in Agriculture and Ecology of the China Agricultural University in Gansu Province in northwest China (N 37°52′, E 102°50′, Altitude 1581 m). Grain maize was sown on April 21 and May 2, and harvested on September 28 and September 26 in 2009 and 2010, respectively. The ground was partly mulched with plastic films with a width of 100 cm, and there was bare soil of 65 and 45 cm in width between two film rows in 2009 and 2010, respectively. Maize seed was sown in holes of 5.0 cm diameter under the plastic films, with a row spacing of 50 and 45 cm and a plant spacing of 24 and 23 cm for 2 years. Maize was not sown in the bare soil. This planting scheme had an actual density of 76,300 and 82,500 plants ha^{-1} for 2009 and 2010, respectively. Actual fractions of ground-mulching were ~0.5 and 0.6, respectively,

which were defined as one minus the ratio of the summed surface areas of bare soil and holes to ground area. For the 0–1.0 m soil depth, the soil type was silt loam, with a bulk density of 1.38 g cm^{-3}, a field capacity of 0.30 m^3 m^{-3}, and a wilting point of 0.12 m^3 m^{-3}. Over the entire growing season, maize was border-irrigated four times, with a total irrigation water amount of 420 mm for both years. The amount of each irrigation event was measured by a water meter. Each irrigation amount was 105 mm on June 15, July 6, July 29 and August 20, 2009, and 105, 120, 90, and 105 mm on June 22, July 27, August 5 and August 29, 2010, respectively.

2 Measurements of Evapotranspiration and Soil Evaporation

λET was measured using an eddy covariance (EC) system installed in the center of the maize field. The EC consists of a fast response 3D sonic anemometer (CSAT3, Campbell Scientific Inc., UT, USA), a Krypton hygrometer (KH20, Campbell Scientific Inc.) and a temperature and humidity sensor (HMP45C, Vaisala Inc., Helsinki, Finland). All sensors were connected to a data logger (CR5000, Campbell Scientific, Inc.). The sonic anemometer and Krypton hygrometer were installed at a height of 1.0 m over the crop canopy. Net radiation (R_n) was measured by a net radiometer (NR-LITE, Kipp & Zonen, Delft, Netherlands), installed at a height of 3.5 m. Two soil heat fluxes (HFP01, Hukseflux, Delft, Netherlands) were installed at a soil depth of 8.0 cm under the plastic film and bare soil. Soil temperature above each soil heat flux plate was measured using thermocouples at depths of 0.0 cm, 2.0 cm and 6.0 cm. Soil water content from 0–10.0 cm was measured using a soil moisture reflectometer (EnviroSMART, Sentek Sensor Technologies, SA, Australia). Ground heat flux (G) was estimated by correcting heat fluxes at 8.0 cm for heat storage above the transducers. The heat storage was determined from changes in soil temperature and moisture above the transducers. Based on the covariance of the 10 Hz air temperature and specific humidity with vertical wind velocity, the latent heat flux in 30 min durations was computed using the eddy covariance methodology with the CarboEurope recommendations [36]. Daytime λET was adjusted by the Bowen-ratio forced closure method, and nighttime λET was adjusted using the filtering interpolation method as proposed by Ding et al. [37].

Soil evaporation (E_s) was measured by the micro-lysimeter. Eight micro-lysimeter cylinders, made from PVC tubes with a diameter of 10 cm and height of 20 cm, were installed in bare soil between two plastic film rows. The cylinders were weighed at 20:00 every day by an electric scale with a precision of 0.1 g. The micro-lysimeters were reinstalled within one day after each irrigation and heavy rain. E_s at the field scale can be calculated by weighting the fraction of ground-mulching (f_m) from the following equation.

$$E_s = (1 - f_m) 10 \frac{\overline{\Delta M_s}}{\rho_w A_e} \tag{46}$$

where $\overline{\Delta M_s}$ is the mean weight change of micro-lysimeter every day, A_e is the cross sectional area of the micro-lysimeters (78.5 cm^2 here), ρ_w is water density (1.0 g cm^{-3}) and 10 is a conversion factor for changing units from cm to mm.

3 Other Measurements

Solar radiation, precipitation, air temperature, relative humidity and wind speed were measured with a standard automatic weather station (Hobo, Onset Computer Corp., USA) at a height of 2.0 m

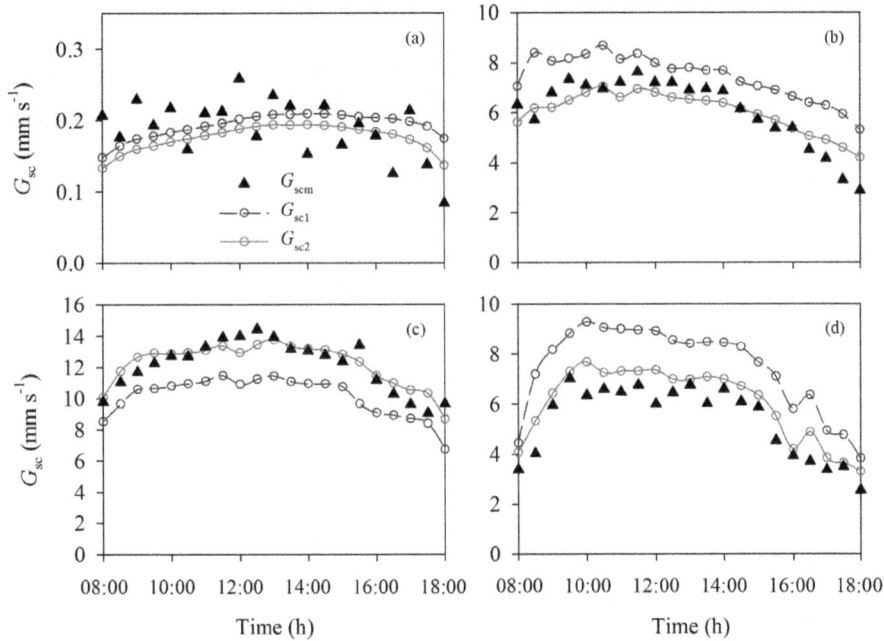

Figure 4. Diurnal variations of half-hourly canopy stomatal conductance (G_{sc}) calculated by the big-leaf (G_{sc1}) and dual-leaf models (G_{sc2}), and measured G_{sc} (G_{scm}) inverted by the S-W model, respectively for four typical clear-sky days in different growth stages of maize in 2009. Leaf area index (*LAI*) was 0.16, 2.62, 5.38 and 2.99, on (a) May 22, (b) June 23, (c) July 27, and (d) September 17, respectively.

above the ground. Volumetric soil water content in the root zone (θ_{rz}) was measured with PVC access tubes using the portable device Diviner 2000 (Sentek Sensor Technologies). Measurements were made at intervals of 0.1 m with a maximal soil depth of 1.0 m at intervals of 3–5 days. Additional samplings were conducted before and after irrigation events, as well as after rainfall events. The measurements were calibrated by oven drying of soil samples. Interpolation was applied between consecutive irrigations to determine θ_r for each day. In addition, two sets of ECH2O probes (Decagon Devices Inc., Pullman, WA, USA) were added to monitor soil moisture at 30 min intervals in 2010.

Ten maize plants were randomly selected to measure leaf length and width, and height at intervals of ~10 days during the growing season. Leaf area was calculated by summing the rectangular area of each leaf (leaf length×maximum width) multiplied by a factor of 0.74, a conversion factor obtained by analyzing the ratio of the rectangular area to the real area measured by an AM300 (ADC BioScientific Ltd., UK). *LAI* is defined as maize green leaf area per unit ground area. The daily *LAI* was obtained by linear interpolation.

Leaf-scale physiological measurements were performed to derive the stomatal conductance model parameters. A LI-6400 portable photosynthesis system (Li-Cor Inc., Lincoln, NE, USA) was used to measure leaf stomatal conductance on the first fully expanded leaf, which is the fourth leaf counted from the top of the shoot. The diurnal measurements of leaf gas exchange were performed once every 2 h from 8:00 to 18:00 on six sunny days in 10 randomly selected maize plants. Care was taken to keep leaves in their natural positions during measurement. The response of leaf stomatal conductance to varying PAR was measured at 30°C and at a CO_2 concentration of 400 mol mol^{-1} on 29 August, 2009. Measurements were taken at PAR levels of 2000, 1600, 1300, 1000, 800, 600, 400, 200, 100, 50, 20, and 0 μmol m^{-2} s^{-1}. The stomatal light-response curve was fit by a rectangular

hyperbola to obtain the parameter values of k_Q using the Jarvis-Stewart model.

4 Model Performance

Half-hourly G_{sc} and λET were calculated using the big-leaf and dual-leaf models Eqs. (39) and (40) with the dual-source equation based on the half-hourly measured meteorological data. The *LAI* and soil water were set as constants at the half-hourly time scale. Daily λET was calculated using Eqs. (39) and (40) with the dual-source equation based on the measured average daily meteorological data. We evaluated the two models by comparing with measurements taken over an irrigated maize field.

The parameters in the Jarvis-Stewart model were obtained using measurements of the stomatal light-response curve and the diurnal leaf gas exchange calculated by non-linear least-squares analysis (SPSS 13.0, SPSS Inc., Chicago, IL, USA). There were ~15 days with no crop cover before the emergence of maize, providing the opportunity to parameterize the empirical coefficients in the soil surface resistance model using the flux observations.

The coefficient of determination (R^2), root mean square error (RMSE) and the Willmott's index of agreement (d) were used to evaluate model performance [38].

Results

1 Model Parameter Estimation and Sensitivity

From the stomatal light-response curve, we obtained best-fitting estimates of k_Q by non-linear least-squares analysis using the Jarvis-Stewart model (Table 1). The stress coefficients of *VPD* and θ_E, k_D and k_w, were optimized using the diurnal measurements of leaf gas exchange (Table 1). Soil surface resistance (r_{ss}) was calculated by inverting the flux-resistance equation for the case of no crops [12]. Based on the relationship between r_{ss} and relative soil water

Table 2. Comparison of measured and estimated evapotranspiration (λET) and soil evaporation (E_s) during the growth periods of maize in 2009 and 2010.

Years	Time-scales	Models	Average values		Linear regression equation with zero intercept	R^2	RMSE (mm s^{-1})	d
			Measurements (x)	Estimates (y)				
2009	Half-hourly λET	Big-leaf	102.4	79.3	y=0.94x	0.83	72.22	0.9521
		Dual-leaf	102.4	87.6	y=1.02x	0.90	58.06	0.9706
	Daily E_s	Big-leaf	0.44	0.43	y=0.95x	0.63	0.1198	0.9036
		Dual-leaf	0.44	0.46	y=1.02x	0.68	0.1220	0.9129
2010	Half-hourly λET	Big-leaf	107.0	75.2	y=0.93x	0.82	70.97	0.9472
		Dual-leaf	107.0	82.9	y=1.03x	0.88	62.31	0.9626
	Daily E_s	Big-leaf	0.45	0.44	y=0.93x	0.64	0.1593	0.9054
		Dual-leaf	0.45	0.46	y=1.01x	0.73	0.1565	0.9218

Note: R^2 is the coefficient of determination, RMSE is the root mean square error, and d is the Willmott's index of agreement. The units of half-hourly and daily values are W m^{-2} and mm d^{-1}, respectively.

content (θ_g/θ_s) of the top soil, the best-fitting parameters of b_1, b_2 and b_3 were obtained (Table 1).

The big-leaf model estimates did not closely match the response of G_{sc} to incident PAR above the canopy (Q_o) predicted by the dual-leaf model (Fig. 1). At lower LAI (2.0), the big-leaf model overestimated G_{sc} by up to 47.4% at an intermediate irradiance (300 W m^{-2}). At higher LAI (5.0), the big-leaf model overestimated G_{sc} when $Q_o<200$ W m^{-2} and underestimated G_{sc} when $Q_o>200$ W m^{-2}. The sensitivity of the dual-leaf model was further analyzed by investigating the variations of the sunlit and shaded leaf area index (LAI_{sl} and LAI_{sh}) against the different LAI (Fig. 2). LAI_{sl} approached a maximum of 1.6 when $LAI \geq 3.0$, while LAI_{sh} almost linearly increased as LAI increased. As a result, the ratios of LAI_{sl} and LAI_{sh} to LAI (Λ_{sl} and Λ_{sh}) nonlinearly decreased and increased, respectively, as LAI increased.

The diurnal variations of modeled irradiance absorbed by the entire canopy (Q_c) and its separation into sunlit and shaded fractions (Q_{sl} and Q_{sh}) are shown in Fig. 3a at higher $LAI = 5.0$, using the measured diurnal courses of meteorological and environmental variables from June 23, 2009. Q_c, Q_{sl} and Q_{sh} exhibited the typical diurnal patterns, while Q_{sh} had a lower magnitude than Q_{sl} throughout the day. The average Q_{sl} accounted for 84.2% of the Q_c. The partitioning of leaves into sunlit and shaded fractions continually changes throughout the day (Fig. 3b). LAI_{sl} was a convex parabola, while LAI_{sh} was a concave parabola. The magnitude of LAI_{sh} was greater than LAI_{sl} even during midday when the solar zenith angle is lowest at the higher LAI, which is consistent with the result in Fig. 2. The diurnal distributions of G_{sc} calculated by the big-leaf model (G_{sc1}) and dual-leaf model (G_{sc2}) and its separation between sunlit and shaded fractions (G_{sl} and G_{sh}) are presented in Fig. 3c. G_{sc1} was significantly lower than G_{sc2} through most of the day at the higher LAI. G_{sh} showed a pattern similar to G_{sl}, while the magnitude of G_{sh} was lower than G_{sl}. The maximum difference between them occurred at the midday around 11:00.

2 Comparisons of Canopy Stomatal Conductance by Big-leaf and Dual-leaf Models

Diurnal patterns of G_{sc} calculated by the big-leaf and dual-leaf models (G_{sc1} and G_{sc2}, respectively) and the measured G_{sc} (G_{scm}) calculated by using Eq. (17) are shown in Fig. 4 for four typical clear-sky days in different growth stages of maize in 2009. The diurnal courses of the G_{sc1} and G_{sc2} were similar to the G_{scm}, but there were differences in magnitude. G_{sc1} was significantly lower than G_{scm}, while G_{sc2} closely matched G_{scm} at higher LAI (Fig. 4c). G_{sc1} was higher than G_{scm} at the lower LAI where there was still good agreement between G_{sc2} and G_{scm} values (Fig. 4b and d). Although the slope of the linear regression between G_{sc1} and G_{scm} was 0.97, G_{sc1} was overestimated at the lower G_{scm} and underestimated at the higher G_{scm} (Fig. 5a), as shown by the lower R^2 of 0.81, greater RMSE of 1.7047 mm s^{-1} and lower d of 0.9563, indicating that the big-leaf model yielded large errors for estimating G_{sc}. In contrast, the slope of the linear regression between G_{sc2} and G_{scm} was 1.01, with an R^2 of 0.98, RMSE of 0.6120 mm s^{-1} and d of 0.9951, indicating that there was good data-model agreement between predictions and measurements (Fig. 5b).

3 Comparisons of Crop Evapotranspiration by Big-leaf and Dual-leaf Models

Diurnal variations of half-hourly λET calculated by the dual-source big-leaf (λET_1) and dual-leaf models (λET_2), and measured λET (λET_m), respectively are presented for four typical clear-sky

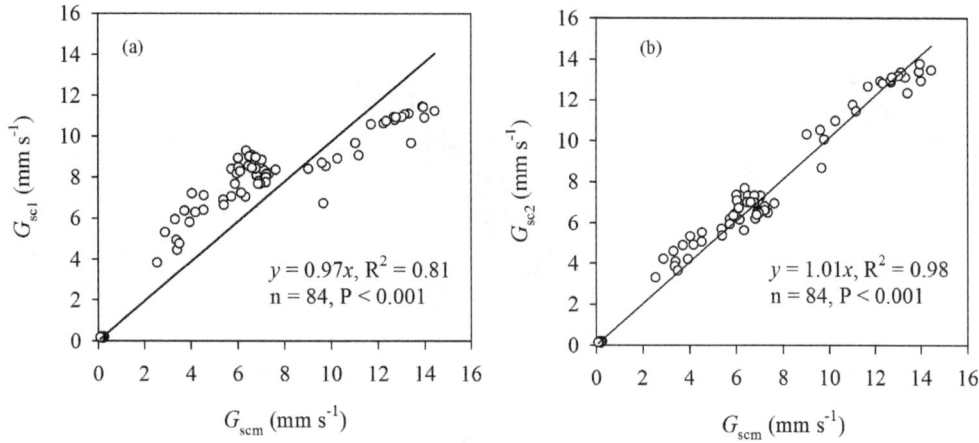

Figure 5. Relationship between canopy stomatal conductance (G_{sc}) estimated by the big-leaf (G_{sc1}) and dual-leaf models (G_{sc2}), and measured G_{sc} (G_{scm}) inverted by the S-W model, respectively for four typical clear-sky days in different growth stages of maize in 2009.

days in different growth stages of maize in 2009 (Fig. 6). The diurnal patterns of estimated λET were similar to the measurements. λET_1 was overestimated at lower *LAI*, and underestimated at higher *LAI*. The linear regression presented that λET_1 was overestimated by 8.7% ($R^2 = 0.97$) and 19.7% ($R^2 = 0.96$) for *LAI* = 2.62 and 2.99, respectively (Fig. 6b and d). λET_1 was underestimated by 13.9% ($R^2 = 0.97$) for *LAI* = 5.38 (Fig. 6c). In

contrast, λET_2 had a good agreement with measurements for differing *LAI*, with a linear slope of 1.01 and an R^2 of 0.97.

The irrigation scheduling and mulching fractions in 2009 and 2010 were different (See section 3.1), which yielded different *LAI* and soil water regimes for the two years (See Figure 1 and Table 1 in Ding et al.[39]). The maximum and averaged values of *LAI* respectively were 5.4 and 3.1 for 2009, and 4.7 and 2.7 for 2010.

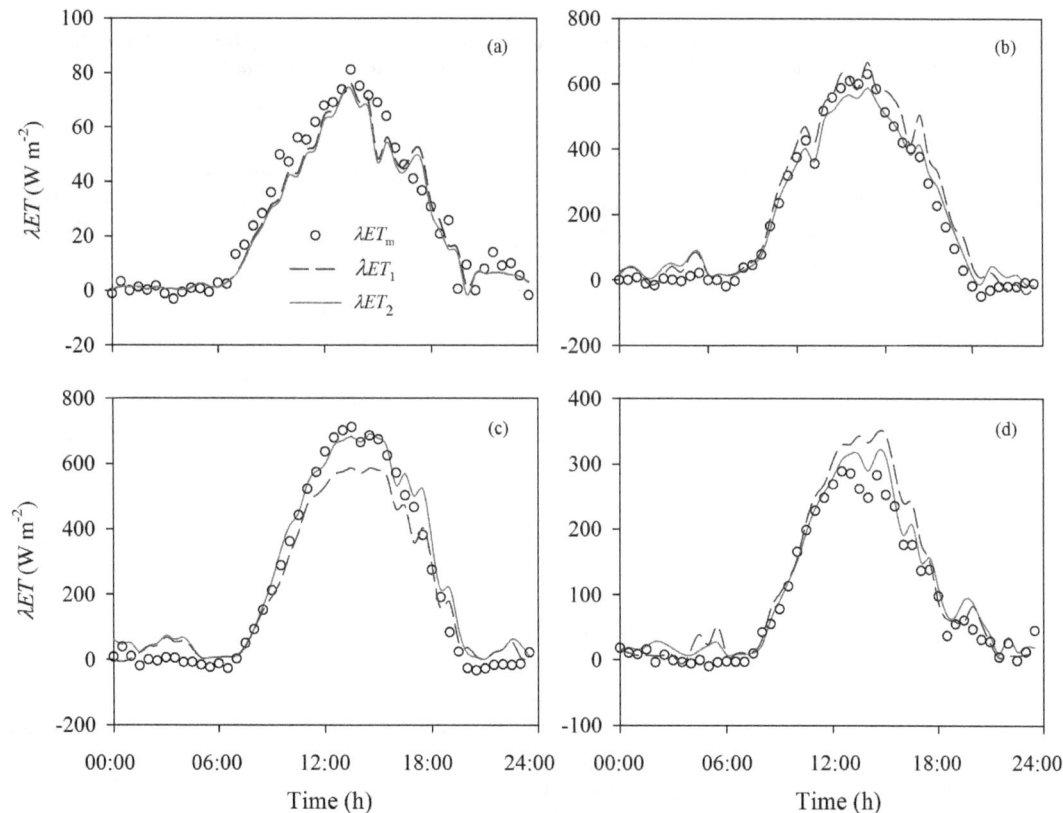

Figure 6. Diurnal variations of half-hourly crop evapotranspiration (λET) calculated by the dual-source with the big-leaf (λET_1) and dual-leaf models (λET_2), and measured λET (λET_m), respectively for four typical clear-sky days in different growth stages of maize in 2009. Leaf area index (*LAI*) was 0.16, 2.62, 5.38 and 2.99, on (a) May 22, (b) June 23, (c) July 27, and (d) September 17, respectively.

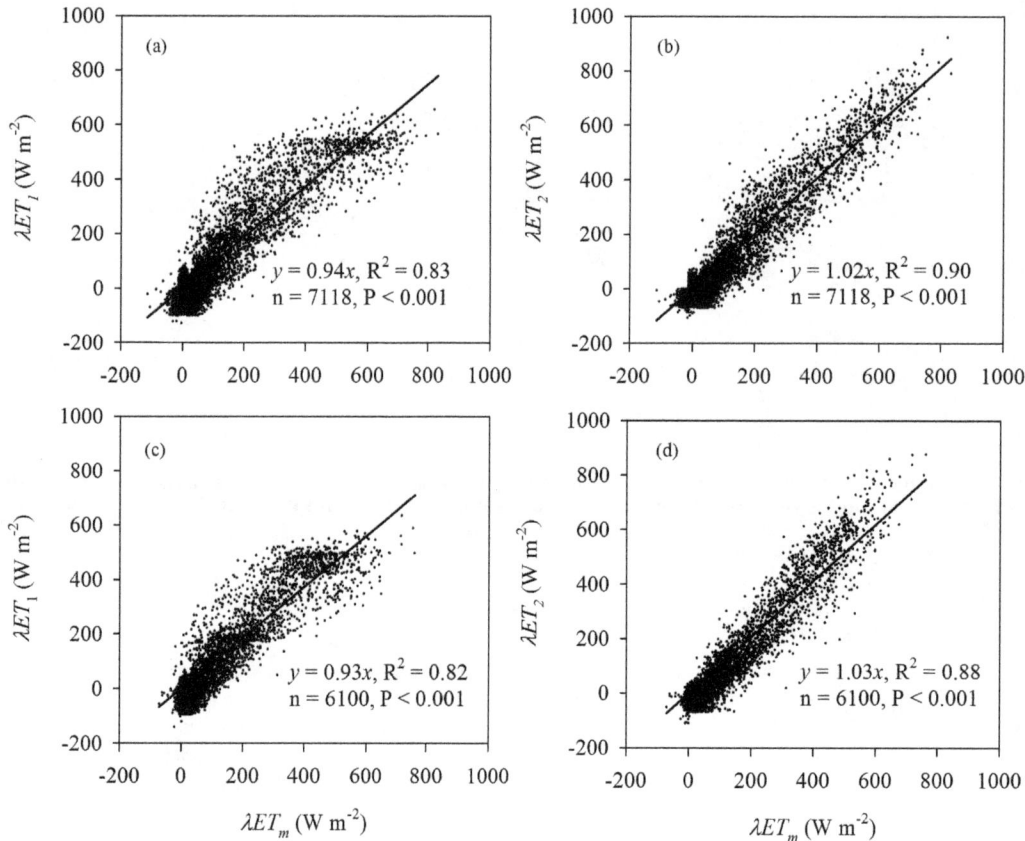

Figure 7. Comparison between half-hourly estimated evapotranspiration by the dual-source with the big-leaf (λET_1) and dual-leaf models (λET_2) versus measured λET by eddy covariance (λET_m) during the entire growth period of maize in (a and b) 2009 and (c and d) 2010.

The extractable soil water in the root zone (θ_E) was significantly different between the two years. Before the first and second irrigation events in 2010, there were 9 and 12 days of θ_E below 50% total available water in the root zone (TAW), which was regarded as a threshold of crop water stress [13]. Conversely, most θ_E was higher than 50% of TAW in 2009. All of these differences led to differing λET and its components, which provided a good dataset to test the big-leaf and dual-leaf models over two different hydrometeorological and management strategies.

The scatterplots of half-hourly λET exhibited that λET_1 was overestimated for lower values and underestimated for higher values, respectively (Fig. 7a and 7c). Total λET_1 was underestimated, with a slope of linear regression of 0.94 ($R^2 = 0.83$) and 0.93 ($R^2 = 0.82$), respectively, for 2009 and 2010. RMSE was 72.22 and 70.97 W m^{-2}, and d was 0.9521 and 0.9472 for 2009 and 2010, respectively (Table 2). In contrast, there was good data-model agreement between measurements and estimated half-hourly λET_2 in 2009 and 2010 (Fig. 7b and 7d). The slopes of linear regressions between the estimates and measurements were 1.02 and 1.03, with R^2 of 0.90 and 0.88, RMSE of 58.06 and 62.31 W m^{-2} and d of 0.9706 and 0.9626 for 2009 and 2010, respectively (Table 2). Daily estimated λET_2 enhanced data-model agreement, with R^2 of 0.91 for the 2 years despite the linear slopes were the same as those of half-hourly values (data not shown). The statistical test showed that the slopes were not significantly different with one (P = 0.114 and 0.092), and the intercepts were not significantly different with zero (P = 0.215 and 0.174) for the 2 years.

Seasonal variations of daily estimated and measured E_s using Eq. (15) combined with Eqs. (39) and (40) are presented in Fig. 8 for 2009 and 2010. Both the dual-source big-leaf and dual-leaf models could capture the variability of E_s when irrigation or precipitation occurred and when the canopy partially covered the ground during the initial growth stages. However, daily values of E_{s1} were underestimated at the early and late stages and overestimated at the middle stage (Fig. 8a and b). In general, total E_{s1} was underestimated by 5% and 7% for 2009 and 2010, respectively (Fig. 8c and d). In contrast, there was satisfactory data-model agreement between predicted and measured E_s using the DSDL model for the two years. The slopes of linear regressions between estimates and measurements were 1.02 ($R^2 = 0.68$) and 1.01 ($R^2 = 0.73$), with RMSE of 0.1220 and 0.1565 mm d^{-1} and d of 0.9129 and 0.9218 for 2009 and 2010, respectively (Table 2).

Discussion

In this paper, we have extended the big-leaf model by developing a dual-leaf model. The dual-leaf model presented here is an improvement over the previous big-leaf model, as more realistic non-uniform vertical profiles of radiation and stomatal conductance are now incorporated into the model. The penetration of beam radiation, its variation and the dynamics of sunlit and shaded LAI throughout the day all affect the ability of the big-leaf model to simulate diurnal changes in G_{sc} [18,29]. In the dual-leaf model, these canopy features can be explicitly incorporated by dividing the canopy into sunlit and shaded fractions and modeling

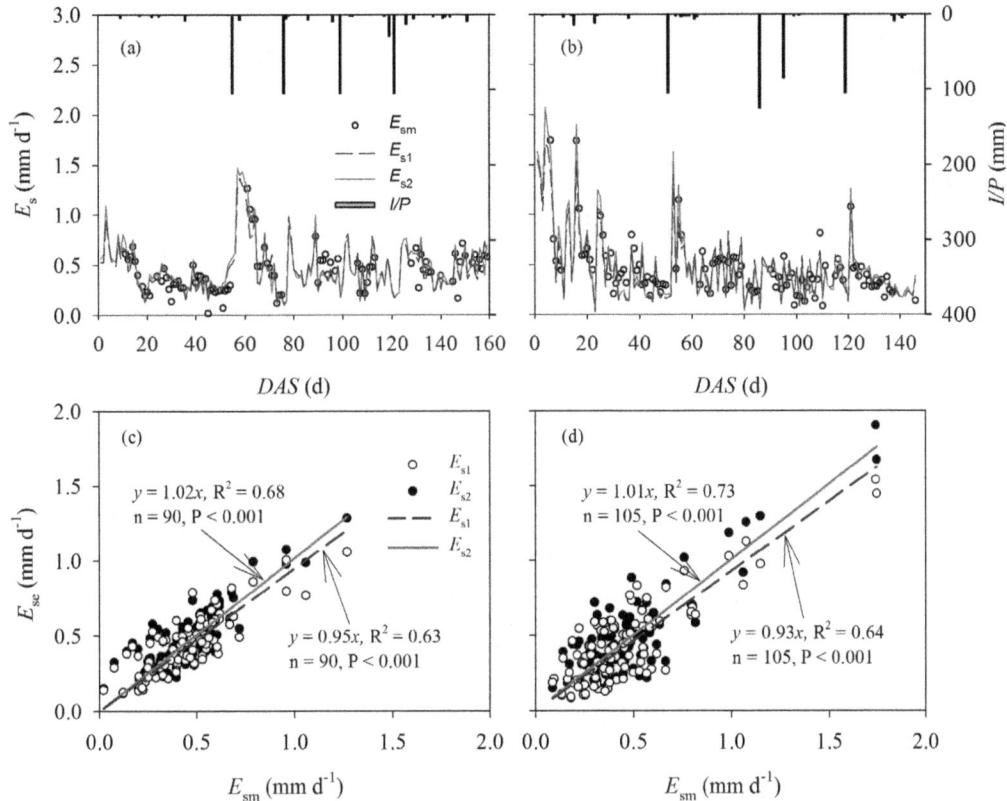

Figure 8. Comparison between daily estimated soil evaporation (E_{se}) by the dual-source with the big-leaf (E_{s1}) and dual-leaf models (E_{s2}) versus measurements (E_{sm}) during the entire growth period of maize in 2009 and 2010. Seasonal variations of the estimated and measured E_s against days after sowing (DAS) are presented in (a) and (b). The linear regressions between them are presented in (c) and (d).

each fraction of G_{sc} by scaling up the respective stomatal conductance separately. It is more complex than the big-leaf model, but the dynamic partitioning of LAI and irradiance between sunlit and shaded leaves has further reduced the errors associated with simplifying the leaves to only a big-leaf using either the total or empirically effective LAI.

The ability of the dual-leaf model was examined by comparing the estimated values and actual measurements. In contrast to significant errors by the big-leaf model, the dual-leaf model accurately reproduced the variation of G_{sc} (Fig. 4 and 5). One reason the dual-leaf model works so well is that it accommodates the nonlinear response of stomata to light [25,40]. Stomata-light responses of leaves can vary with depth in the canopy and this variation can be incorporated by partitioning the canopy into several layers and estimating the sunlit and shaded leaf fractions in each layer [20,34]. However, usually this is not necessary and a single, representative light response curve can be used for the entire canopy [17,40]. In the dual-leaf model, the entire G_{sc} may be calculated by summing contributions of sunlit and shaded leaves. These two contributions are added separately because sunlit leaves will be light-saturated while shaded leaves will be in the linear portion of the light-stomata relationship [18,33]; thus G_{sc} is not proportional to average light levels [18]. Because of the nonlinear relationship between stomatal conductance and PAR, the predicted G_{sc} will be overestimated when the average absorbed PAR is used to scale up the leaf stomatal conductance for the entire canopy as in the big-leaf model [17,40]. On the other hand, the G_{sc} is underestimated when the effective LAI is used to scale up the leaf stomatal conductance at higher LAI ((Fig. 4c). Since the

nonlinear relationship was considered and the sunlit-shaded method was introduced in the dual-leaf model, estimates using the dual-leaf model closely match the measured G_{sc} (Fig. 4 and 5).

The performance of the DSDL for estimating λET was investigated. The DSDL is robust for estimating λET over a range of canopy leaf areas and environmental variables (Fig. 6, Fig. 7 and Table 2). The good data-model agreement indicated the strengths of the DSDL model as a model framework and the reference for validating other approaches of calculating the G_{sc} using the dual-leaf model. Our framework of modeling λET also provided a soil evaporation estimation model Eq.(15) with a modified soil surface resistance term, which is useful for enhancing crop production by reducing the E_s fraction of λET [10,15].

The DSDL model is physically process-based, yet sufficiently simple to be effectively parameterized. The dual-leaf model requires only four additional equations, Eqs (32), (33), (43) and (44), beyond those required in the model of leaf stomatal conductance, to calculate the LAI and absorbed irradiance of the sunlit and shaded leaves. This simplicity makes it attractive for incorporation into in crop models, land surface schemes, and regional or global water cycle studies [29,40]. This model can also be used to assess effects of climate change on crop ecophysiology.

Conclusions

In this paper, a dual-leaf model for scaling-up stomatal conductance from the leaf to the canopy level was developed through the dynamic partitioning of the leaf area index and irradiance between sunlit and shaded leaves. In the model, canopy stomatal conductance was calculated by dividing the canopy into sunlit and shaded fractions and each fraction was modeled separately based on the absorbed irradiances. The dual-leaf model provided estimates of G_{sc} which were nearly the same as measurements, and were significantly more accurate than those of the big-leaf model. Our results showed excellent agreements between λET measurements gathered by the eddy covariance technique over an irrigated maize field during 2009 and 2010 under two different hydrometeorological and management conditions, and estimates of λET using the DSDL. The framework of the model can also satisfactorily estimate soil evaporation. Our proposed model provides an alternative approach to calculate λET, which is simple and attractive for incorporation into other comprehensive crop models.

Acknowledgments

We thank the two anonymous reviewers for their constructive comments that have helped to improve the manuscript.

Author Contributions

Conceived and designed the experiments: RD SK TD. Performed the experiments: RD YZ. Analyzed the data: RD YZ. Contributed reagents/materials/analysis tools: SK TD. Wrote the paper: RD SK TD XH YZ.

References

1. Leuning R, Zhang YQ, Rajaud A, Cleugh H, Tu K (2008) A simple surface conductance model to estimate regional evaporation using MODIS leaf area index and the Penman-Monteith equation. Water Resources Research 44: W10419, doi:10.1029/2007WR006562.
2. Pereira LS, Perrier A, Allen RG, Alves I (1999) Evapotranspiration: concepts and future trends. Journal of Irrigation and Drainage Engineering 125: 45–51.
3. Ding R, Kang S, Li F, Zhang Y, Tong L (2013) Evapotranspiration measurement and estimation using modified Priestley–Taylor model in an irrigated maize field with mulching. Agricultural and Forest Meteorology 168: 140–148.
4. Zhang X, Chen S, Sun H, Shao L, Wang Y (2011) Changes in evapotranspiration over irrigated winter wheat and maize in North China Plain over three decades. Agricultural Water Management 98: 1097–1104.
5. Katerji N, Rana G (2006) Modelling evapotranspiration of six irrigated crops under Mediterranean climate conditions. Agricultural and Forest Meteorology 138: 142–155.
6. Irmak S, Mutiibwa D (2010) On the dynamics of canopy resistance: Generalized linear estimation and relationships with primary micrometeorological variables. Water Resources Research 46: W8526, doi:10.1029/2009WR008484.
7. Pieruschka R, Huber G, Berry JA (2010) Control of transpiration by radiation. Proceedings of the National Academy of Sciences 107: 13372–13377.
8. Hu Z, Yu G, Zhou Y, Sun X, Li Y, et al. (2009) Partitioning of evapotranspiration and its controls in four grassland ecosystems: Application of a two-source model. Agricultural and Forest Meteorology 149: 1410–1420.
9. Er-Raki S, Chehbouni A, Boulet G, Williams DG (2010) Using the dual approach of FAO-56 for partitioning ET into soil and plant components for olive orchards in a semi-arid region. Agricultural Water Management 97: 1769–1778.
10. Ding R, Kang S, Zhang Y, Hao X, Tong L, et al. (2013) Partitioning evapotranspiration into soil evaporation and transpiration using a modified dual crop coefficient model in irrigated maize field with ground-mulching. Agricultural Water Management 127: 85–96.
11. Shuttleworth WJ (2007) Putting the "vap" into evaporation. Hydrology and Earth System Sciences 11: 210–244.
12. Monteith JL, Unsworth MH (2008) Principles of Environmental Physics. New York, USA: Academic Press. 418 p.
13. Allen RG, Pereira LS, Raes D, Smith M (1998) Crop Evapotranspiration: Guidelines for computing crop water requirements. Rome, Italy: FAO Irrigation and Drainage Paper, No. 56. 300 p.
14. Shuttleworth WJ, Wallace JS (1985) Evaporation from sparse crops-an energy combination theory. Quarterly Journal of the Royal Meteorological Society 111: 839–855.
15. Zhu G, Su Y, Li X, Zhang K, Li C (2013) Estimating actual evapotranspiration from an alpine grassland on Qinghai-Tibetan plateau using a two-source model and parameter uncertainty analysis by Bayesian approach. Journal of Hydrology 476: 42–51.
16. Zhang B, Kang S, Li F, Zhang L (2008) Comparison of three evapotranspiration models to Bowen ratio-energy balance method for a vineyard in an arid desert region of northwest China. Agricultural and Forest Meteorology 148: 1629–1640.
17. Campbell GS, Norman JM (1998) An Introduction to Environmental Biophysics. New York, USA: Springer-Verlag. 286 p.
18. Irmak S, Mutiibwa D, Irmak A, Arkebauer TJ, Weiss A, et al. (2008) On the scaling up leaf stomatal resistance to canopy resistance using photosynthetic photon flux density. Agricultural and Forest Meteorology 148: 1034–1044.
19. Ross J (1976) Radiative transfer in plant communities. In: Monteith JL, edito. Vegetation and the Atmosphere: Principles. New York, USA: Academic Press. 13–55.
20. Pyles RD, Weare BC, Pawu KT (2000) The UCD Advanced Canopy Atmosphere Soil Algorithm: Comparisons with observations from different climate and vegetation regimes. Quarterly Journal of the Royal Meteorological Society 126: 2951–2980.
21. Baldocchi D, Meyers T (1998) On using eco-physiological, micrometeorological and biogeochemical theory to evaluate carbon dioxide, water vapor and trace gas fluxes over vegetation: a perspective. Agricultural and Forest Meteorology 90: 1–25.
22. Raupach MR, Finnigan JJ (1988) 'Single-layer models of evaporation from plant canopies are incorrect but useful, whereas multilayer models are correct but useless': Discuss. Functional Plant Biology 15: 705–716.
23. Sinclair TR, Murphy CE, Knoerr KR (1976) Development and evaluation of simplified models for simulating canopy photosynthesis and transpiration. Journal of Applied Ecology: 813–829.
24. Zhang B, Liu Y, Xu D, Cai J, Li F (2011) Evapotranspiraton estimation based on scaling up from leaf stomatal conductance to canopy conductance. Agricultural and Forest Meteorology 151: 1086–1095.
25. Leuning R, Kelliher FM, De Pury DGG, Schulze ED (1995) Leaf nitrogen, photosynthesis, conductance and transpiration: scaling from leaves to canopies. Plant, Cell and Environment 18: 1183–1200.
26. Shuttleworth WJ, Gurney RJ (1990) The theoretical relationship between foliage temperature and canopy resistance in sparse crops. Quarterly Journal of the Royal Meteorological Society 116: 497–519.
27. Hou X, Wang F, Han J, Kang S, Feng S (2010) Duration of plastic mulch for potato growth under drip irrigation in an arid region of Northwest China. Agricultural and Forest Meteorology 150: 115–121.
28. Weiss A, Norman JM (1985) Partitioning solar radiation into direct and diffuse, visible and near-infrared components. Agricultural and Forest meteorology 34: 205–213.
29. De Pury D, Farquhar GD (1997) Simple scaling of photosynthesis from leaves to canopies without the errors of big-leaf models. Plant Cell and Environment 20: 537–557.
30. Jarvis PG (1976) The interpretation of the variations in leaf water potential and stomatal conductance found in canopies in the field. Philosophical Transactions of the Royal Society of London. Series B, Biological Sciences 273: 593–610.
31. Whitley R, Medlyn B, Zeppel M, Macinnis-Ng C, Eamus D (2009) Comparing the Penman-Monteith equation and a modified Jarvis-Stewart model with an artificial neural network to estimate stand-scale transpiration and canopy conductance. Journal of Hydrology 373: 256–266.
32. Stewart JB (1988) Modelling surface conductance of pine forest. Agricultural and Forest Meteorology 43: 19–35.
33. Farquhar GD, Sharkey TD (1982) Stomatal conductance and photosynthesis. Annual Review of Plant Physiology 33: 317–345.
34. Norman JM (1979) Modeling the complete crop canopy. In: Barfield G, edito. Modification of the Aerial Environment of Plants: American Society of Agricultural Engineers. 249–280.
35. Sun SF (1982) Moisture and heat transport in a soil layer forced by atmospheric conditions. Storrs: University of Connecticut.
36. Mauder M, Foken T (2004) Documentation and instruction manual of the eddy covariance software package TK2. Work Report University of Bayreuth, Dept. of Micrometeorology 26: 1614–8916.
37. Ding R, Kang S, Li F, Zhang Y, Tong L, et al. (2010) Evaluating eddy covariance method by large-scale weighing lysimeter in a maize field of northwest China. Agricultural Water Management 98: 87–95.
38. Willmott CJ (1982) Some comments on the evaluation of model performance. Bulletin of the American Meteorological Society 63: 1309–1313.
39. Ding R, Tong L, Li F, Zhang Y, Hao X, et al. (2014) Variations of crop coefficient and its influencing factors in an arid advective cropland of northwest China. Hydrological Processes, DOI:10.1002/hyp.10146.
40. Wang YP, Leuning R (1998) A two-leaf model for canopy conductance, photosynthesis and partitioning of available energy I: Model description and comparison with a multi-layered model. Agricultural and Forest Meteorology 91: 89–111.
41. Brutsaert W (1982) Evaporation into the Atmosphere: Theory, History, and Applications. Dordrecht, Netherlands: Kluwer Academic Publishers. 299 p.
42. Meyers TP, Paw U (1987) Modelling the plant canopy micrometeorology with higher-order closure principles. Agricultural and Forest Meteorology 41: 143–163.

15

Soil Organic Carbon Redistribution by Water Erosion – The Role of CO$_2$ Emissions for the Carbon Budget

Xiang Wang*, Erik L. H. Cammeraat, Paul Romeijn, Karsten Kalbitz

Earth Surface Science, Institute for Biodiversity and Ecosystem Dynamics, University of Amsterdam, Amsterdam, The Netherlands

Abstract

A better process understanding of how water erosion influences the redistribution of soil organic carbon (SOC) is sorely needed to unravel the role of soil erosion for the carbon (C) budget from local to global scales. The main objective of this study was to determine SOC redistribution and the complete C budget of a loess soil affected by water erosion. We measured fluxes of SOC, dissolved organic C (DOC) and CO$_2$ in a pseudo-replicated rainfall-simulation experiment. We characterized different C fractions in soils and redistributed sediments using density fractionation and determined C enrichment ratios (CER) in the transported sediments. Erosion, transport and subsequent deposition resulted in significantly higher CER of the sediments exported ranging between 1.3 and 4.0. In the exported sediments, C contents (mg per g soil) of particulate organic C (POC, C not bound to soil minerals) and mineral-associated organic C (MOC) were both significantly higher than those of non-eroded soils indicating that water erosion resulted in losses of C-enriched material both in forms of POC and MOC. The averaged SOC fluxes as particles (4.7 g C m^{-2} yr^{-1}) were 18 times larger than DOC fluxes. Cumulative emission of soil CO$_2$ slightly decreased at the erosion zone while increased by 56% and 27% at the transport and depositional zone, respectively, in comparison to non-eroded soil. Overall, CO$_2$ emission is the predominant form of C loss contributing to about 90.5% of total erosion-induced C losses in our 4-month experiment, which were equal to 18 g C m^{-2}. Nevertheless, only 1.5% of the total redistributed C was mineralized to CO$_2$ indicating a large stabilization after deposition. Our study also underlines the importance of C losses by particles and as DOC for understanding the effects of water erosion on the C balance at the interface of terrestrial and aquatic ecosystems.

Editor: Ben Bond-Lamberty, DOE Pacific Northwest National Laboratory, United States of America

Funding: This work was supported by China Scholarship Council (CSC) and University of Amsterdam (UvA). The funders had no role in study design, data collection and analysis, decision to publish, or preparation of the manuscript.

Competing Interests: The authors have declared that no competing interests exist.

* E-mail: X.Wang@uva.nl

Introduction

Climate change will likely modify current precipitation regimes influencing the global carbon (C) cycle in relation to erosion processes [1,2]. The length and intensity of droughts and the intensity of more sporadic rainfall events are predicted to increase for Western Europe [3], which will accelerate soil erosion. Soil erosion has significant impacts on the redistribution and transformation of soil organic carbon (SOC) within a landscape [4,5]. Even now, there is no consensus whether soil erosion is acting as a net C sink [5,6] or source [7] of atmospheric CO$_2$. Therefore, quantitative assessments of soil organic C redistribution along geomorphic gradients and the processes involved become increasingly important in a changing climate to resolve this controversy [8]. It is crucial that such studies comprise the different processes associated with the redistribution of C along the slope including CO$_2$ emissions as a result of changes in C mineralization upon erosion, transport and subsequent deposition. Based on such studies, complete C budgets of soils affected by erosion processes can be determined.

Soil erosion seems to preferentially remove fresh and more labile materials from C rich topsoils in upslope eroding positions, i.e. SOC with low density (e.g. free light fraction) and dissolved organic C (DOC) [7–10]. However, the fate of this organic C has rarely been studied. It is well known that most of the eroded

sediments are re-deposited close to the source areas and in the catchment (e.g. [4,11]). Deposition of C enriched sediments lead to accumulation of SOC in the downslope positions. The eroded and deposited C can be stabilized by interaction with minerals thereby decreasing mineralization of deposited C in soil profiles [12]. In addition, soil erosion could affect dissolved organic carbon (DOC) dynamics in soils. Wang et al. [12] found higher DOC concentration at eroding sites in comparison to depositional sites.

Soil erosion drastically influences not only lateral SOC distribution within a landscape but also vertical CO$_2$ fluxes into the atmosphere [7,10]. Van Oost et al. [5] summarized at least three key mechanisms controlling the net flux of C between the soil and atmosphere: 1) dynamic replacement of SOC at the eroding sites [6]; 2) deep burial of SOC rich topsoils at depositional sites [4,13]; 3) enhanced decomposition of SOC because of the chemical or physical breakdown of soil during detachment and transport [7]. Particularly, the second and the third mechanisms should be susceptible to changes in the precipitation regime.

A key uncertainty of erosion-induced C loss is C mineralization resulting from the breakdown of soil aggregates as a direct response to extreme precipitation [7,14,15]. During a given erosion event, rainfall leads to breakdown of aggregates and releases the encapsulated C due to flow shear and raindrop impact [14]. Some studies suggest that aggregates breakdown by raindrop impact and wetting is mainly caused by initial fast slaking [16] or welding [17]. However, the extent of additional CO_2 fluxes from breakdown of aggregates due to erosion is still largely unknown. Franzluebbers [18] estimated a 10–60% increase in CO_2 evolution from various soils after breakdown of aggregates during 0–3 days. Polyakov and Lal [14] suggested that mainly the breakup of initial soil aggregates by erosive forces is responsible for increased CO_2 emission. However, conducting a set of rainfall simulation experiments, Bremenfeld et al. [19] recently suggested that interill erosion and associated soil aggregates breakdown have no prominent effect on soil respiration *in situ*. Therefore, effects of erosion-induced breakdown of aggregates on CO_2 evolution need to be further assessed.

Estimates of soil and SOC redistribution and associated CO_2 emissions show a large spatial and temporal variability. As field SOC and CO_2 fluxes of soils under erosion strongly depend on temporal variability of environmental conditions (e.g. location, soil management, initial soil moisture, and rainfall event characteristics) rainfall simulations under controlled laboratory conditions may help to shed light on C flux processes. Several rainfall simulation experiments have attempted to investigate soil erosion and associated SOC dynamics [20–24]. Jacinthe et al. [24] determined mineralization of SOC in runoff under no-till, chisel till and moldboard plow conditions with rainfall simulation approach. Van Hemelryck et al. [23] experimentally simulated three typical agriculture erosion events to quantify CO_2 emission. So far, however, there is no direct process assessment on combining effects of erosion, transport and subsequent deposition on C redistribution including vertical CO_2 fluxes. Changes in SOC pools indicative for important mechanisms of SOC redistribution and differing in their stability against microbial decay are not well known.

To get a better process understanding of soil erosion, transport and deposition on the redistribution and mineralization of SOC, the main objective of the present study was to determine SOC redistribution and a complete C budget of a loess soil affected by water erosion using a pseudo-replicated rainfall simulation experiment under standardized conditions. The following processes were studied and considered in our C budget:

(i)   We determined SOC mineralization by measuring CO_2 emissions at different slope positions.

(ii) We analyzed soil and C redistribution along the slope including potential export into aquatic ecosystems. We measured C enrichment in the redistributed sediment. In order to test the hypothesis that POC is preferentially eroded and exported into aquatic ecosystem we fractionated SOC by density into particulate organic C (free POC, C not bound to minerals) and mineral associated organic C (MOC).

(iii) Finally, we analyzed concentrations of DOC in soil solutions at different positions of the slope and in runoff and determined above and belowground lateral DOC fluxes.

Materials and Methods

Ethics Statement

The experimental station 'Proefboerderij Wijnandsrade' (The Netherlands) permitted access to their land and allowed for taking soil sample material from their cereal fields for the research carried out.

Site Description and Sampling

The loess soil was collected from an agricultural field with winter wheat in South Limburg (50°53′58. 42″N, 5°53′16. 23″E), The Netherlands in May 2011. South Limburg is part of the European loess belt and has a temperate maritime climate. This region has a mean annual precipitation of 825–850 mm [25] and a mean annual temperature of 10.2°C. The sampled soil has a silty loam texture, and is classified as a Haplic Luvisol [26]. In the present study, the top 10 cm of the *Ap* horizon was collected and sieved over an 8 mm mesh to homogenize the soils and to keep aggregates intact as much as possible. Agricultural management at the sampled site is characterized by a potato-winter wheat-beet-winter wheat rotation. Soils are plowed 30 cm by a cultivator in spring and conventional tillage was applied in winter (including 30 cm plowing). The basic physical and chemical properties of the used soil are shown in Table 1.

Soil Analysis

Field bulk density was estimated from undisturbed 100 cm^{-3} cores that were oven-dried at 105°C for 24 hours [27]. Grain size distribution of soils was obtained using a particle size analyser (Micromeritics, SediGraph 5100, Norcross, USA). Soil pH (1:2.5 in H_2O) was measured with a multi-parameter analyser (CON-SORT C832, Abcoude, The Netherlands). Soil water content was continually determined by a multi-channel Metallic TDR cable tester system [28]. Carbon and nitrogen (N) contents in bulk soils, sediments and density fractions were determined using a C and N analyser (Elementar VarioEL, Hanau, Germany).

Experimental Design

The erosion experiment was carried out using a 1.25 m×3.75 m experimental stainless steel flume (Figure 1). The upper 1.75 m had a slope of 15° (upslope position) and the lower 2 m had a slope of 2° (downslope position). To assess the effects of erosion, transport and subsequent deposition on redistribution of soils and C along the erosion slope, the experimental flume was divided into three zones according to the positions of the slope and observed results of sediments redistribution (Figure 1): 1) the eroding zone, at the upper half of the upslope position; 2) the transport zone, at the lower half of the upslope position and the upper half of the downslope position; 3) the depositional zone, at the lower half of the downslope position of the flume. We used a static definition of the different zones as dynamic measurement locations would have disrupted the soil surface. We recognize that these zones can change during the event and between events and that during events in every zone also local deposition and re-entrainment will occur.

The entire flume was subdivided into three parallel replicates of 40±2 cm wide. The soil was laid on top of a 2 cm thick layer of inert quartz sand to allow water to drain away. On top there was a 20 cm layer of soil on the upper (erosion) section where soil was supposed to erode and a thinner (5 cm) soil layer on the lower deposition section to allow for material deposition. On the transport section there was a gradual transition from 20 to 5 cm soil layer. While placing the air-dried soil it was compacted for every 2 cm, using a hammer and wooden piece of board (30×30 cm) to distribute the applied force. The compaction was such that it approached bulk density under field conditions (1.28 g cm^{-3}). In addition, there were three controls. Three control buckets (diameter 34 cm) were filled with a 20 cm loess soil layer on top of a 2 cm quartz sand layer, similar to the main flume. These control buckets were also compacted to the same bulk density. The buckets were placed next to the flume so that they

Table 1. Basic properties of the loess soil used in the experiment. Results are shown as mean and standard error of three replicates.

Depth (cm)	Bulk density (g cm^{-3})	pH	SOC^a (%)	TN^b (%)	C/N	Soil texture (%)		
						Sand	Silt	Clay
0–10	1.28 (0.05)	6.5 (0.06)	1.07 (0.06)	0.11 (0.01)	10.3 (0.7)	8.6	82.2	9.2

^{a}: Soil organic carbon.
^{b}: Total nitrogen.

received the same rainfall as well, but no lateral displacement of soil material took place.

The soil layer was pre-wetted to an initial standard moisture contents (Table 2) in 10–15 min to initiate runoff generation prior to commencing the real rainfall experiment. Four 18-minutes rainfall events were carried out at a monthly time interval. Measurements were carried out every 2 minutes during rainfall simulation. Rainfall was simulated with two nozzles (Lechler 460 788) applying at 1600 hPa demineralized water using an average rainfall intensity of 41.8 ± 1.9 mm h^{-1}. A rainfall event with this intensity and duration of 18 minutes has a return period of about 2 years [29]. Mean drop size of the applied rainfall was 2.0 mm ($D_{50} = 2.0$). With an average falling height of 1.8 m, the kinetic energy applied on the soil surface was 12.5 J m^{-2} mm^{-1}. Demineralized water was used instead of tap water to prevent flocculation problems with dispersible soil material [30,31]. As the total load of ions in rainwater is very low (the annual average electrical conductivity EC_{25} was below 20 μS cm^{-1} at the official Dutch sampling site Beek [32], about 10 km from the soil sampling site) the physico-chemical impact of demineralized water on soil particles is considered to be the same as for rain water. The temperature was kept as constant as possible ($18.1 \pm 0.9°C$).

Sampling during Erosion Experiments

Sediment traps were installed in the middle of the eroding, transport and depositional zones respectively with entrance of the traps at the upslope side and at the same level as the soil surface to capture mobilized sediment in overland flow (Figure 1). The sediment traps were modified 12 ml Polypropylene screw cap tubes (Greiner Bio-One GmbH, Frickenhausen, Germany). The traps had a small diameter to minimize disturbance to overland flow and resulting erosion patterns. An opening in the side was made to collect mobilized sediments. The sediment traps were sampled every two minutes and the collected materials were transferred to containers, oven-dried at 35°C, weighed and later analysed for C and N contents.

Runoff and sediments were collected from weirs at the end of the flumes at 2-min intervals once continuous runoff had developed. Total runoff was collected using a polystyrene gutter that was installed at the lowest part of the experimental flume. The contents of the flume were then pumped into V-notched bottles to measure flow rates using a simple siphon pump made of Tygon R-3603 tubes (Saint-Gobin, Courbevoie, France). The lower end was constrained to 4 mm diameter to provide a constant flow velocity, without risking clogging by larger soil particles and keeping effects on the aggregation of the sediments limited. The V-notched bottles overflowed into sampling boxes which were replaced every two minutes or when the sampling box was full.

At the lowest end of the flume, three holes per replicate flume were present at the level of the sand drainage layer to collect through flow. Through flow was defined as the lateral underground flow in contrast to the overland flow. The holes were covered from the inside by a 63-μm stainless steel mesh allowing water to pass through, but to prevent clogging up. On the outside of the walls attached tubes drained into bottles, similar to the runoff setup.

Sampling after Erosion Experiments

Density Fractionation of Bulk Soils and Exported Sediments. After four rainfall events the 0–2 mm topsoils at the eroding, transport and depositional zones and sediments exported during the first and fourth events were fractionated into three fractions by a sodium polytungstate (NaPT) solution with a density of 1.6 g cm^{-3}: the free light fraction (fLF) which consisted

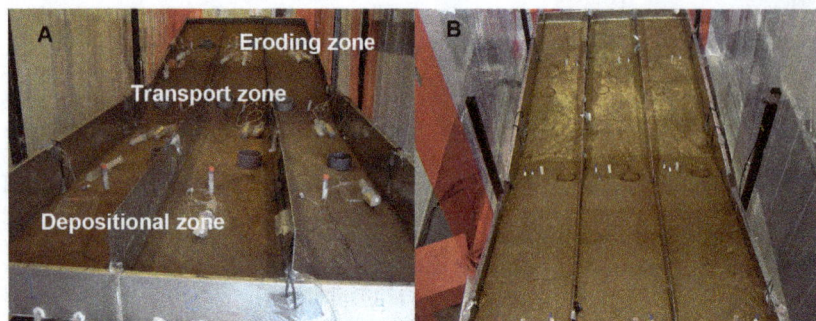

Figure 1. Photographs of the experimental setup and sampling locations along the experimental flume. It included the eroding, transport and depositional zones of the flume. A shows the lateral view; B shows the vertical view.

of large, undecomposed or partly decomposed root and plant fragments, the light fraction occluded in aggregates (oLF) and the heavy fraction (HF), which was associated with minerals [33,34]. Soil organic C in fLF, oLF and HF are defined as fPOC, oPOC and MOC, respectively. Particulate organic C (POC) is the C not bound to soil minerals including both fPOC and oPOC. The oPOC represents C sequestered in aggregates. Methods and procedures were followed as described in Cerli et al. [34]. All fractions were freeze-dried, homogenized and later analysed for C and N contents. Density fractionation was done in triplicate.

Dissolved Organic Carbon (DOC)

To investigate dynamics of dissolved organic C in different soil depths and positions as affected by soil redistribution, soil moisture samplers (MACRO RHIZON 19.21.35, 9 cm porous, 4.5 mm OD, 0.2 μm, Wageningen, The Netherlands) were inserted in the eroding, transport and depositional zones of the flume. Each sampler was connected to a syringe (50 mL) to collect the soil solution. At the eroding and transport zones of the flume, soil solutions were collected at 4 cm and 9 cm depths. At the depositional zones soil solutions were sampled at 4 cm only because of the thinner soil layer on the lower deposition section. Soil solutions were sampled twice per week during the first week immediately after one rainfall event because of higher soil water moisture. As the soil dried, soil solutions were collected once per week. Concentrations of DOC were determined by a TOC analyser (TOC-V CPH, Shimadzu, Kyoto, Japan).

Soil CO$_2$ Efflux Measurements

Soil respiration was measured using a Portable Gas Exchange and Fluorescence System (LI- 6400XT; LICOR Biosciences, Lincoln, NE USA). In order to enhance the comparability of data, most CO$_2$ efflux measurements were conducted in the afternoon between 17:00 and 19:00 at local time in PVC collars (10.2 cm in diameter and 7 cm in height). Soil CO$_2$ efflux was determined before and after each rainfall simulation event. As the 7 cm high collars, necessary for the CO$_2$ efflux measurements, would strongly affect the overland flow and erosion patterns during the rainfall event, the 7 cm high collars were replaced by smaller collars (same diameter but 1.5 cm tall). These were inserted at exactly the same place, to temporary fill the imprint of the high collar in the soil surface. The top of the collar was placed exactly equal to the soil surface, to minimize the disturbance of the sampling location by the CO$_2$ measurements but still enabling to measure the CO$_2$ efflux exactly at the same position later on. Overland flow was possible and erosion, transport and deposition processes at the surface of the area used for measuring CO$_2$ were hardly affected

by this strategy. Two to three measurements per site (i.e. per collar) were carried out each time. The number of replicated measurements per collar depended on the variation after the first two analysis with an additional measurement if the relative deviation of the second one was larger than 10%. Additionally, pre-experiments were carried out using the same loess soil to test impacts of soil depth on soil CO$_2$ efflux. In these experiments the CO$_2$ efflux was measured in columns with increasing soil thickness under constant soil moisture and temperature conditions. Results showed that soil depth did not have significant effect on soil respiration per soil weight up to a depth of 30 cm (data not published). Based on these results, all data measured in different experimental zones, and control soils, having different soil depths, were corrected to 20 cm soil layers in order to directly compare effects of erosion, transport and deposition on CO$_2$ effluxes.

Erosion-induced Carbon Budget

Fluxes of SOC and DOC were calculated by multiplying concentrations of SOC and DOC with the volume of the overland flow. Other parameters were calculated as follows:

$$Carbon\ enrichment\ ratio\ (CER) = \frac{C_{sediment}}{C_{control\ soils}}$$

$$Total\ C\ losses = Lateral\ C\ exported$$
$$+ Vertical\ CO_2\ emission$$

$$Lateral\ C\ exported = SOC\ exported\ in\ overland\ flow$$
$$+ DOC\ exported\ in\ overland\ flow$$
$$+ DOC\ exported\ in\ through\ flow$$

$$Net\ additional\ CO_2\ emission$$
$$= CO_2\ emission\ from\ soil\ in\ flume$$
$$- CO_2\ emission\ from\ soil\ in\ control\ treatment$$

Based on the 4-month data we calculated annual C fluxes by linear extrapolation making comparisons with the literature easier. However, the shortcomings of such budgets based on short-term laboratory experiments only are obvious.

The definition of C source and sink areas for calculating the C budget was based on two experimental observations. After the fourth rainfall event, soil layers with relocated materials were clearly visible in the flume, particularly in the downslope part of

Table 2. Initial soil water contents (m^3/m^3) before and after pre-wetting before starting the rainfall simulation.

Zones	Event 1		Event 2		Event 3		Event 4	
	Before	After	Before	After	Before	After	Before	After
Eroding	0.25	0.32	0.27	0.33	0.26	0.31	0.26	0.31
Transport	0.39	0.44	0.36	0.42	0.33	0.39	0.30	0.37
Depositional	0.36	0.48	0.33	0.46	0.30	0.42	0.23	0.44

the depositional zone (Figure 1B). In addition, we found that SOC was significantly depleted in the transport zone comparing with controls soils (cf. section results). Based on these two observations, the eroding and transport zones were defined as the C source area and the depositional zone and the runoff leaving the flume (exported into aquatic system) were defined as the C sink area. We calculated an erosion-induced SOC budget for the four rainfall events over the entire period using a mass balance approach (i.e. source = sink area). Changes in C distribution between the density fractions were appropriately considered by using the data of the original soil for the source area. This approach enabled us to include any changes in C redistribution between density fractions induced by erosion.

Statistical Analyses

Differences in C enrichment ratios, amounts of sediment exported and DOC concentrations in overland flow were tested with one-way ANOVA and the Post-hoc Duncan test to differentiate between individual differences. The difference of CO_2 effluxes measured in the 4-month period at eroding, transport and depositional zones of the gutter was tested by repeated measurement ANOVA. Averaged CO_2 efflux in different experimental zones was compared using a one-way ANOVA. For all tests, a significance level of $P = 0.05$ was set using the Post-hoc Duncan test, unless otherwise indicated. The relationship between cumulative CO_2 emission and DOC concentration was tested by two-tailed Pearson test. All statistical tests were performed using SAS software (Version 8.1) and SPSS (IBM Statistics 20).

Results

Loss of Sediment and Carbon Enrichment Ratios in Overland Flow

Total sediment losses in the overland flow increased during the course of the experiment from 9.5 g m^{-2} in the first event to 31.0 g m^{-2} in the fourth event (Figure 2). During the first rainfall event the average sediment concentration was 1.1 ± 0.2 g L^{-1} and doubled to 2.3 ± 0.8 g L^{-1} in the fourth event.

Carbon enrichment ratios (CER) of sediment loads of overland flow trapped at the eroding, transport and depositional zones of the flume ranged from 0.8 to 2.9. The CER was significantly higher at the depositional zone compared to those of the eroding and transport zones (Table 3). Carbon enrichment was even stronger in the sediments of the runoff with CER between 1.3 and 4.0. Carbon enrichment ratios decreased with increasing concentrations of suspended solids in the overland flow (Figure 3). Concentrations of suspended solids were smaller at the beginning of each rainfall event, resulting in larger C enrichment but also in larger variation of the data.

Preferential Erosion and Deposition of Organic Carbon at the Soil Surface

After four rainfall events, a thin sedimentation layer was present in the depositional zone (approximately 2 mm thick) without any layering. However the depositional zone clearly showed patterns of deposition of finer grained materials along the flow lines of overland flow and the whole lower part of the gutter. Soil organic C concentration (mg^{-1} g soil) of the surface soil (2 mm) decreased by 6.0% at the eroding zone and increased by 3.9% at the depositional zone if compared to control soils (Table 3). Nevertheless, soil organic C concentration did not differ significantly between the control, the eroding, transport and depositional zones of the gutter. Also the relative distribution of C in density fractions of the soil was not affected by soil erosion. Most of the C (86% to

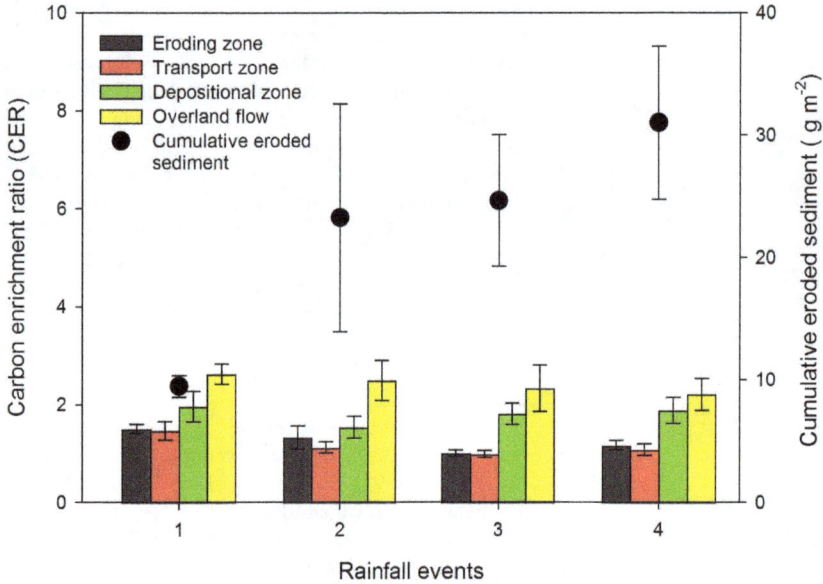

Figure 2. Average total eroded sediment per rainfall event exported by overland flow and carbon enrichment ratios (CER) during four rainfall events.

91%) was found in the heavy fraction, i.e. mineral associated organic C (MOC; Table 3). The rest was almost equally distributed between the free light fraction (particulate organic C in free light fraction = fPOC) and the fraction occluded within aggregates (oPOC). The free light fraction was significantly enriched in C at the transport and the depositional zone whereas the occluded light fraction (oLF) was depleted in C at the eroding zone (Table 3). The heavy fractions of surface soils in the flume did not significantly change in C contents.

In the sediments, all fractions were strongly enriched in C with the largest enrichment in the free light fraction. This C enrichment was smaller in the occluded and smallest in the heavy fraction and also decreased from the first to the last event (Table 3). However, the C content of the heavy fraction of the sediments (first event) was more than double the C content of the heavy fraction of the control soil (Table 3).

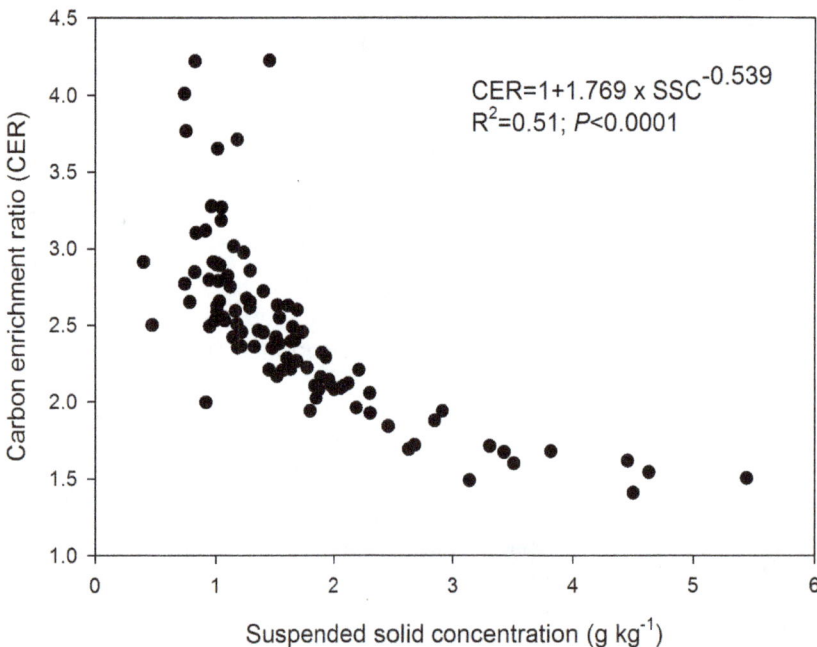

Figure 3. Relationship between carbon enrichment ratio (CER) and suspended solid concentration (SSC) in the overland flow.

Table 3. Carbon concentrations and specific carbon fractions of soils and sediments for different zones and events.

Zones	C concentration[a] (mg g^{-1} soil)	C concentration[b] (mg C g^{-1} specific density fraction)			C enrichment ratio (CER)[c] (-)				Relative proportion of MOC[d] (% SOC)
	Bulk soils	fPOC	oPOC	MOC	Bulk soils	fPOC	oPOC	MOC	MOC
Control	10.0 (0.5)	134.3 (28.9)	162.3 (24.4)	8.0 (0.1)					91
Eroding	9.4 (0.2)	189.8 (26.3)	175.3 (20.4)	8.0 (0.4)	0.94	1.1	0.8	0.9	91
Transport	9.7 (0.2)	220.3 (60.3)	143.1 (31.5)	7.7 (0.2)	0.97	1.9	1.0	0.9	87
Depositional	10.4 (0.5)	205.0 (61.8)	175.5 (25.1)	7.9 (0.3)	1.04	1.6	0.9	1.0	90
Overland flow 1	22.9 (0.9)	151.2 (42.2)	345.5 (20.1)	17.3 (0.6)	2.30	3.9	3.2	2.2	86
Overland flow 4	16.6 (1.6)	219.1 (54.3)	296.4 (39.3)	13.9 (0.9)	1.67	2.2	2.3	1.6	88

C in free light fraction = free particulate organic C, fPOC; C in occluded light fraction = occluded particulate organic C, oPOC; C in heavy fraction = mineral-associated organic C, MOC. Results are shown as mean and standard error of three replicates.
a. Carbon concentration of bulk soils (mg C g^{-1} soil).
b. Carbon concentration of the three density fractions fPOC, oPOC and MOC in relation to the total weight of that specific soil fraction (mineral + C parts) (mg C g^{-1} soil fraction).
c. Carbon enrichment ratios, calculated on the basis of mg C soil fraction g^{-1} soil organic C.
d. Relative proportion of MOC (%SOC) in bulk soils, density fractions and sediments of overland flow for the first (Overland flow 1) and fourth rainfall event (Overland flow 4).

Soil CO$_2$ Efflux

All measured CO$_2$ efflux rates for the whole experiment ranged from 0.12 to 4.34 g C m^{-2} day^{-1} (Figure 4). During the entire experimental period, rates of CO$_2$ emissions exhibited a similar behaviour in the eroding, transport and depositional zones and the non-eroded control with a sharp initial increase immediately after each rainfall event, followed by continuously decreasing rates thereafter. Rates of CO$_2$ efflux significantly decreased with time during the four events ($P = 0.001$). The spatial and temporal variability of CO$_2$ efflux rates was larger in the first rainfall event than during the other events.

The largest mean CO$_2$ efflux was observed in the transport zone during the first three rainfall events (Figure 5). In the fourth event, however, the depositional zone had the largest mean CO$_2$ efflux. The relative differences of the mean CO$_2$ efflux between the depositional and the eroding zones increased during the course of the whole experiment and became significant in the fourth event.

Cumulative CO$_2$ fluxes in the eroding, transport and depositional zones ranged from 80 to 180, 116 to 317, and 146 to 204 g C m^{-2} yr^{-1}, respectively. The largest mean cumulative CO$_2$ fluxes (221 g C m^{-2} yr^{-1}) were observed in the transport zone. Mean cumulative CO$_2$ fluxes in the depositional zone (181 g C m^{-2} yr^{-1}) were significantly larger than those in the control soils ($P = 0.02$) while CO$_2$ fluxes in the eroding zone were similar in comparison to the control. The total losses of C as CO$_2$ emission during the entire experiment accumulated to 1.8 to 2.9% of total soil organic C stocks.

Dissolved Organic Carbon (DOC)

Concentrations of DOC in soil solutions at eroding, transport and depositional zones ranged from 7.1 to 25.9 mg L^{-1} during four rainfall events (Figure 6). In the shallow soil (4 cm depth), mean concentration of DOC decreased in the following order: transport zone (15.1 mg L^{-1}) > control soils (14.3 mg L^{-1}) > depositional zone (12.3 mg L^{-1}) > eroding zone (11.8 mg L^{-1}). However, only DOC concentrations in the depositional and eroding zones were significantly lower than those in the transport zone and the control. Mean concentrations of DOC in the deeper soil (10 cm) were almost equal as in the shallow soil and decreased in the following order (not statistically significant): control soils (16.8 mg L^{-1}) > transport zone (15.2 mg L^{-1}) > eroding zones (12.3 mg L^{-1}). Concentrations of DOC in soil solutions of both depths showed distinct temporal patterns in all zones of the gutter. They increased at the beginning of each rainfall event, then decreased and increased again with time. This trend was less obvious during the first rainfall.

Concentration of DOC in overland flow remained constant during each single event, ranging from 0.3 to 8.3 mg L^{-1} and significantly decreased from the first to the third rainfall event (means of the four rainfall events: 7.2±0.4, 2.6±0.4, 0.9±0.7, 0.7±0.4 mg L^{-1}, no further data shown). Cumulated DOC fluxes transported by overland flow were on average 0.23 g C m^{-2} yr^{-1} (Figure 7). The amount of C exported as DOC by overland flow was small, accounting for 0.014% of the total SOC stocks in the flume. Fluxes of DOC in through flow (i.e. 0.002% of total SOC stocks) were significantly smaller than overland flow.

Discussion

Preferential Transport and Deposition of Organic Carbon

As expected from the literature [8,12], the soil of the eroding zone was depleted in C whereas the soil of the depositional zone and the sediments of the overland flow were enriched in C after the four rainfall events (Table 3). The results of the density

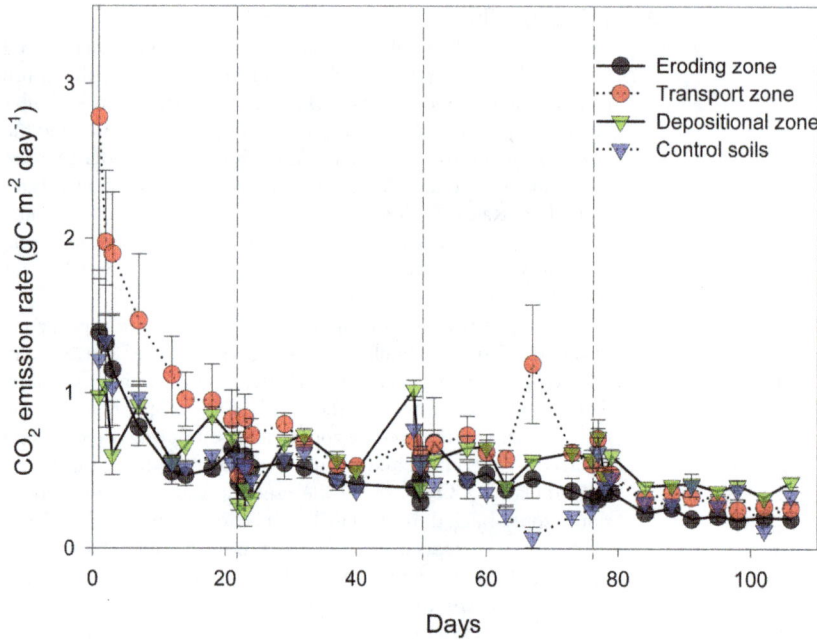

Figure 4. CO$_2$ efflux at different zones of the gutter and control soil during four rainfall events. Solid line + circle represents the eroding zone; dotted line + circle represents the transport zone; solid line + triangle represents the depositional zones; and dotted line + triangle represents control soils. Values are mean± standard error of three replicates.

fractionation clearly showed a large loss of C occluded in aggregates in the eroding zone, which was accompanied by an enrichment of C in the fPOC fraction in the other zones of the flume and in overland flow (Table 3). We assume that the disruption of macro-aggregates by raindrop peeling [35] and aggregate welding and development of a structural crust [17] resulted in the liberation of fPOC, which was preferentially

transported [8]. The disruption of macro-aggregates will result in the release of micro-aggregates (smaller than 250 μm) from the macro-aggregates too. The C content of micro-aggregates within macro-aggregates is usually larger than that of macro-aggregates [36–38]. The release of such small aggregates and selective transport of small aggregates with low density [39] could be the reasons for the observed significant C enrichment of oPOC in

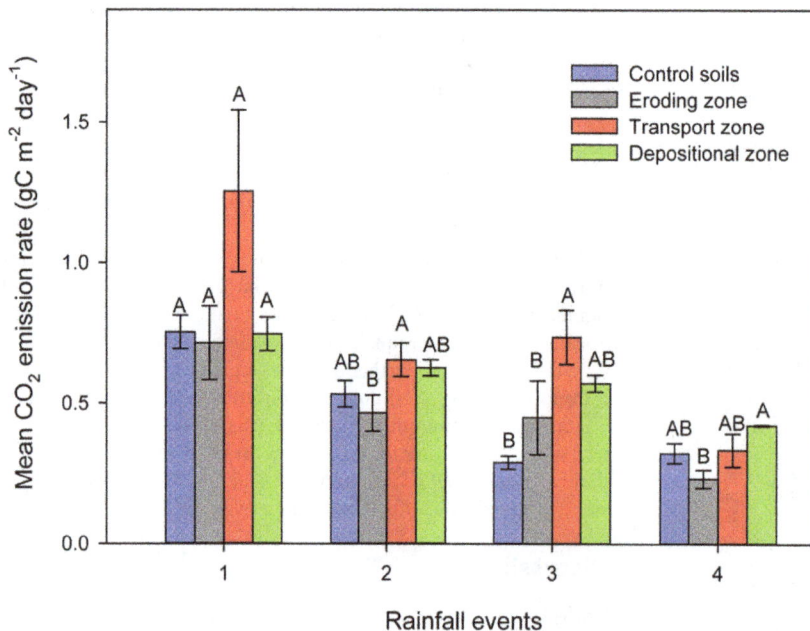

Figure 5. Mean cumulative CO$_2$ emission at the eroding, transport and depositional zones and control soil. Different capital letters mean significant difference at a single rainfall event between the different zones. Values are mean± standard error of three replicates.

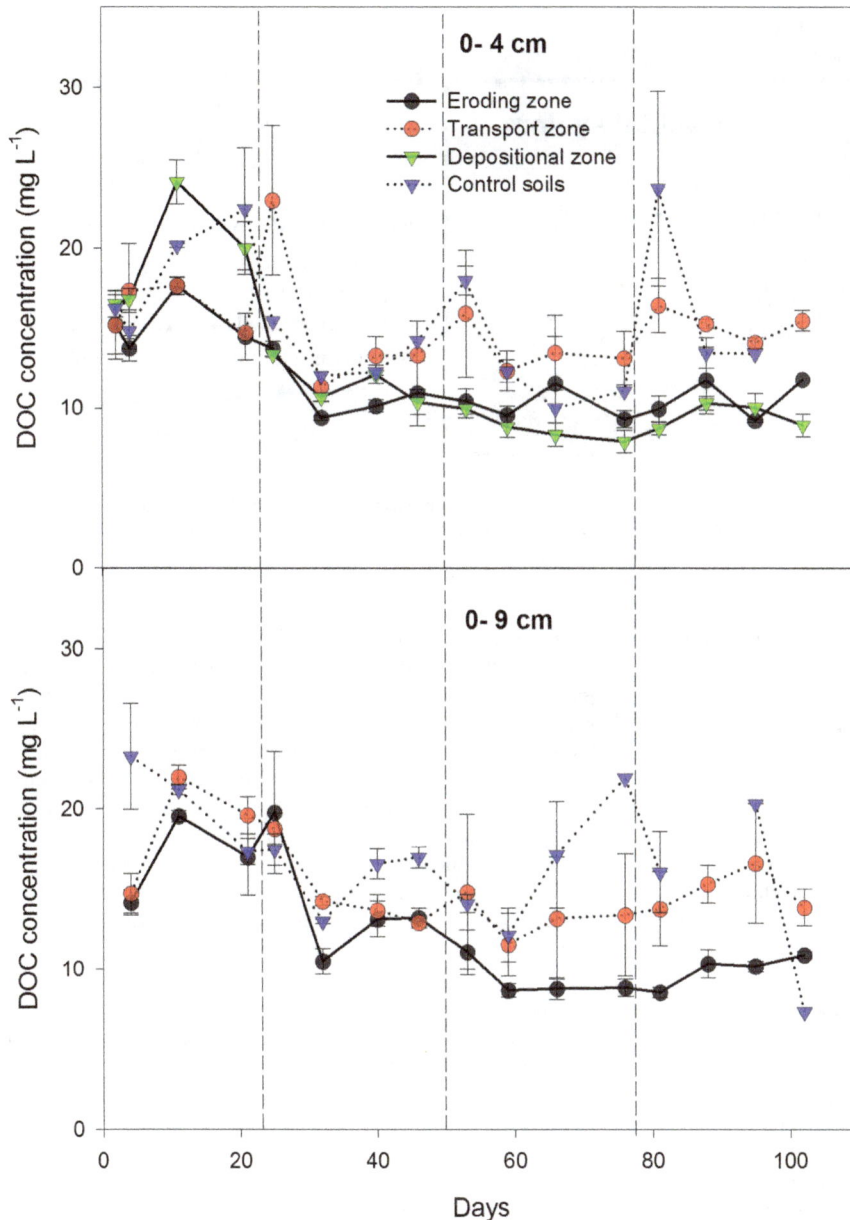

Figure 6. Dissolved organic carbon (DOC) concentrations. DOC solutions were collected at 0–4 cm and 0–9 cm depths of the eroding, transport and depositional zones of the flume during four rainfall events. Solid line + circle represents the eroding zone; dotted line + circle represents the transport zone; solid line + triangle represents the depositional zones; and dotted line + triangle represents control soils. Values are mean± standard error.

sediments ranging from 2.3 to 3.2 (Table 3). However, we did not study aggregate stability and the detailed processes resulting in breakdown of the aggregates neither the related preferential erosion, transport and deposition of different sizes of aggregates and particles. That should be done in follow-up experiments.

The calculated mass balance of the experiment illustrates the disruption of aggregates in the eroding zone and the redistribution of C from aggregates to fPOC with an erosion-induced accumulation of fPOC in the sink area of 0.24 g C (Table 4). This accumulation is equal to an increase in fPOC by 48% comparing the source and the sink area. One logical source of this additional POC would be C occluded within aggregates in the eroding zone at the beginning of the experiment. This large

accumulation of fPOC in the sink area contributed to the observed relative increase by 6% (Table 3) in the SOC content of the first two mm layer of the depositional zone.

Particulate organic C already present in the soils and formed by disruption of aggregates (cf. above) was preferentially eroded and transported by overland flow as indicated by the largest CER ratio of the density fractions in any of the sampled soils and sediments. Per definition, fPOC is the lightest fraction, not associated with minerals and therefore easier to be translocated by water than soil particles with a higher density [23,40,41]. The high C enrichment of mineral-associated organic C (MOC) in the sediments of the overland flow suggested that water erosion separated the whole soil particles according to their density (Table 3). This fraction-

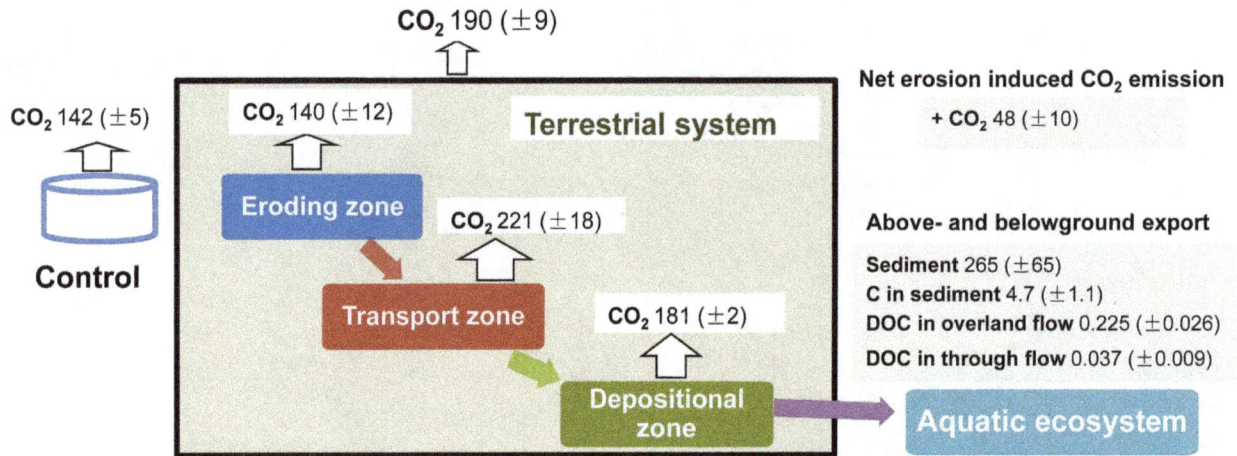

Figure 7. Conceptual diagram illustrating the total carbon budget as affected by soil erosion, transport and deposition in the four months rainfall simulation experiment. Fluxes were calculated on an annual base (interpolated from the 4-months experiments). The values were expressed as mean values and standard error of three replicates.

ation occurred between the different density fractions. Increasing C concentrations (MOC<oPOC<fPOC; Table 3) resulted in increasing CER of the sediments in the same order.

A significant portion of the eroded and transported C enriched sediment was not retained in the downslope parts of the depositional zone and was exported by overland flow and left the flume (Table 4). Particularly the weakly decomposed C of the fPOC should be a readily available C and nutrient source for aquatic organisms [16,42] contributing to CO_2 emission from aquatic ecosystems. This process linking terrestrial and aquatic systems cannot be neglected for modeling the C cycle and has to be studied in more detail in future.

Relationship between Erosion Rate and Carbon Enrichment

The inverse, non-linear relationship between the erosion rate and C enrichment of the sediments we found (Figure 3) is in agreement with previous studies [21,40,41,43]. This inverse relationship is the result of increasing sediment concentration in the overland flow during each single event and from the first to the fourth rainfall event. One of the most important reasons for this relationship should be the breakdown of macro-aggregates by the raindrops as already discussed to be the main reason for the preferential erosion of fPOC [8,10,35]. This process should be particularly important at the beginning of each rainfall event because rewetting of dry soils results in the disruption of aggregates and the release of organic matter [44]. It is also reasonable to assume that the importance of this process will decrease with increasing number of rainfall events. Heavy rainfall causes compaction, welding and crust formation resulting in reduced infiltration and increased erosion and suspended solid concentration with time [15]. The preferential removal of C enriched soil will result in C enriched sediments particularly at the beginning of the experiment where the erosion rate was still small. After removal of this soil enriched in C, the erosion rate increases because of decreasing infiltration and generation of more overland flow. That will result in even increasing erosion rates because soils are less protected by organic matter and aggregation. In field situations, the relationship between erosion rate and C enrichment might be weaker because of continuous above and belowground C input and its positive effect on aggregation [45].

Decreasing C enrichment with large erosion rates, i.e. increasing sediment concentration, indicated that an increasing erosion rate does not result in proportionally increasing C losses. However, this does not mean that more severe erosion events lead to less impact on soil C. Very strong erosion events will translocate large amounts of C. However, this C might be better protected against further mineralization after deposition because C is mostly deposited as mineral associated C (Table 4). The C loading of mineral surfaces should be low as well, resulting in a more efficient stabilization against microbial decay [46,47]. In addition, long-term erosion-induced C sequestration or depletion might depend on the precipitation frequency and intensity.

Soil CO_2 Effluxes

This study provides new data and insight on C decomposition under controlled conditions in an artificial landscape setting at eroding, transport and depositional positions allowing for a better process understanding. Although we did not scale up our results to the landscape level, it is important to know whether the fluxes measured do compare with observed field measurements and make any sense, also in comparison with previous indirect measurements on eroded sediments and soil profile investigations [5,12]. In the present study, CO_2 efflux rates measured (0.12 to 4.34 g C m^{-2} day^{-1}) were in the range of soil respiration rates from agricultural loess soils [5,19]. Initial increases of CO_2 emissions immediately after each of our rainfall events might be explained by the increase in microbiological activity after re-wetting the dry soil and/or increased bioavailability due to aggregate breakdown [44]. Aggregate breakdown and subsequent exposure of previously encapsulated SOC provide substrates for microbial decomposition [8]. The re-wetting effect was particularly important after the first event and decreased during the course of the experiment. This is in line with a decreasing capacity of soil to release C from aggregates over time [48,49].

Transport of topsoil and associated C influenced SOC decomposition rates at the different positions of the artificial slope. The small cumulative CO_2 emission from the eroding zone should be the result of the observed preferential removal of C enriched materials (i.e. higher CER of the sediments at the depositional zone in comparison to the eroding zone, Table 3), which was either preferentially deposited or left the gutter. This

Table 4. Soil organic C redistribution in three density fractions due to erosion (mass balance approach; C in free light fraction, fPOC; C in occluded light fraction, oPOC; C in heavy fraction, MOC).

Fraction	Source area (g C)[a]	Sink area (g C)			Relative value (% of SOC redistributed)			Erosion-induced fPOC[b] ΔC (g)	Aggregate Breakdown oPOC[c] ΔC (g)
		Depositional zone	Overland flow	CO2 emission	Depositional zone	Overland flow	CO2 emission	ΔC (g)	ΔC (g)
fPOC	0.5 (0.0)	0.6 (0.4)	0.14 (0.05)		4.5	1.0		+0.24	
oPOC	0.8 (0.0)	0.5 (0.1)	0.19 (0.04)		3.8	1.4			−0.11
MOC	12.3 (0.4)	9.8 (0.3)	2.13 (0.47)		72.1	15.7			
Total SOC/CO2	Σ 13.6	Σ 10.9 (0.6)	Σ 2.46 (0.60)	0.2 (0.0)	Σ 80.4	Σ 18.1	1.5		
SOC redistributed	Σ 13.6	Σ 13.6							

Results are given as mean and standard error of three replicates.
[a]. Original soil data were used to exclude any effect of soil erosion.
[b]. Erosion induced formation of fPOC (disruption of aggregates) ΔC (g) = fPOC in depositional zone + fPOC in overland flow- fPOC in source area.
[c]. Erosion-induced breakdown of aggregates (decline in oPOC) ΔC (g) = oPOC in depositional zone+oPOC in overland flow- oPOC in source area.

preferential removal of more labile C (POC: fPOC and oPOC) left behind less C, which was relatively more stable (Table 3). In turn, accumulation of labile C fractions in the depositional zone contributed to an increasing difference in cumulative CO_2 emission between the eroding and depositional zone over time (Figure 5).

The cumulative CO_2 emission was significantly and positively related to fPOC ($R^2 = 0.94$; $P = 0.03$) illustrating the more labile character of this SOC fraction. The large accumulation of fPOC probably explained that the transport zone had the largest cumulative CO_2 emissions (Table 3 and Figure 7). Larger CO_2 emissions from the depositional zone were anticipated because the deposited labile C (fPOC, oPOC) could be used as substrate and a source of energy for microbial respiration [50,51]. Although CO_2 emissions were large at the depositional zone, the SOC content increased by 6% in comparison to the control soils after four erosion events. Obviously, parts of the eroded and deposited SOC were preserved (Figure 5).

Considering all positions of the slope, mean DOC concentration in near-surface layers was positively correlated to median soil CO_2 efflux rate ($P = 0.02$). The largest CO_2 efflux was accompanied by largest DOC concentration in the transport zone – a second parameter (first fPOC) explaining the large CO_2 efflux in this zone. Creed et al. [52] found that substrates (i.e. DOC) in the near-surface soil were strongly related to median soil CO_2 efflux. Considering each position of the slope separately, median soil CO_2 efflux rates were not significantly related to mean DOC concentrations at the eroding ($P = 0.18$), transport ($P = 0.49$) and depositional zones ($P = 0.22$). However, DOC was significantly correlated to the median soil CO_2 efflux rate in the control soil ($P = 0.05$), which indicated DOC could be mineralized during the experimental period. Thus, DOC dynamics could not explain the observed additional C decomposition at the depositional zone. This might suggest a fast turnover of DOC or/and a direct use of POC by the microbial community.

Total Carbon Budget

We estimated an erosion-induced C loss of 53 g C m^{-2} yr^{-1} calculated as the sum of net erosion-induced CO_2 emission, C losses by overland flow (C in sediments, DOC) and through flow (cf. materials and methods, figure 7). During the entire experimental period, the averaged SOC fluxes leaving the flume with overland flow were 18 times larger than DOC fluxes including lateral fluxes by the through flow (Figure 7). Fluxes of DOC (0.26 g C m^{-2} yr^{-1}) were rather low particularly due to decreasing DOC concentration during the experiment, i.e. with increasing number of events. Fluxes of sediment associated C were equivalent to 8.9% of the erosion-induced C loss while DOC fluxes were equivalent to 0.5% of those C losses. Therefore, sediment associated C played a much larger role than DOC in the erosion-induced linking of terrestrial and aquatic ecosystems. However, erosion-induced DOC flux should not be neglected because DOC might be particularly important for aquatic food webs [42,53].

Erosion-induced CO_2 emission was the dominant form of C loss, representing 90.5% of erosion-induced C loss. Based on the assumption made (cf. material and methods), 1.5% of total C redistributed (deposited C at the depositional zone plus C exported to aquatic ecosystems) was mineralized to CO_2 (Table 4 and Figure 7). Previous estimates of decomposition of eroded SOC showed large variations, ranging from 0 to 100% (e.g. [7,14,15,23]). In modelling studies the assumption is often used that at least 20% of the eroded SOC is decomposed as a consequence of soil erosion [7,54]. Our measured values were much smaller than this conventional view of erosion effects on the C cycle. Palyakov and Lal [14] estimated 8% of SOC displaced by erosion had a potential to be mineralized and van Hemelryck et al. [23] estimated mineralization of 2% to 12% of the eroded SOC in a loess soils using laboratory rainfall simulation experiments. We propose three main reasons for the large difference. Firstly, the effects of the disruption of aggregates on extra CO_2 efflux were relatively short-lived [48]. Secondly, C stabilization as affected by soil erosion and deposition might be underestimated in the previous assumption [5,8]. Thirdly, the artificial slope was relatively short in our experimental setting in comparison to the field, which may result in an underestimation of transport effects on C mineralization.

Conclusions

Rainfall simulation experiments are a useful approach to determine the role of soil erosion for the C cycle. The data of our 4-months experiment were comparable to field situations, despite of well-known shortcomings of laboratory approaches. The erosion rate was estimated to be 2.1 mm yr^{-1} (26 t ha^{-1} yr^{-1}), which was comparable with estimations in this region based on field data [55]. Also C enrichment of exported sediments and soil CO_2 efflux were in the range of field measurements.

Erosion-induced CO_2 emission was the dominant form of C loss, representing about 90.5% of the erosion-induced C loss. In addition, a considerable amount of C rich sediments (265 g m^{-2} yr^{-1}) was laterally exported by overland flow. Carbon associated with sediments was the main form of erosion-induced lateral C loss and not DOC. This exported C plays an important role in the connection of terrestrial and aquatic ecosystems.

In our experiment, this redistribution of C rich materials resulted in a net additional CO_2 emission during transport and deposition. However, this enhanced CO_2 emission is much smaller than previously thought. Most of the redistributed C by overland flow was bound to soil minerals (heavy fraction), which might be one reason for the unexpected small mineralization. As a consequence, the induced C sink by deposition could be larger than assumed.

Our study clearly demonstrated a fractionation of SOC upon erosion, transport and deposition controlling C mineralization. Disruption of macro-aggregates was identified as the main process responsible for the observed preferential redistribution of labile particulate organic C. Future studies should determine the conditions and processes resulting in breakdown of the aggregates and related preferential erosion, transport and deposition of different sizes of aggregates and particles. Furthermore, the replacement of carbon at eroding zones has to be included in future studies determining the role of soil erosion as a potential C source or sink.

Acknowledgments

We gratefully acknowledge Leen de Lange, Dr. Chiara Cerli, Dr. Gillian Kopittke, Dr. Sebastiaan de Vet, Caridad Díaz López, and Bianca Pricope for their help during the rainfall simulation experiments. We also thank Leo Hoitinga and Bert de Leeuw for their lab support. The experimental station 'Proefboerderij Wijnandsrade' (The Netherlands) is acknowledged for providing soil material. Dr. John Parsons is also acknowledged for checking the language. Three anonymous reviewers are acknowledged for their useful and constructive suggestions.

Author Contributions

Conceived and designed the experiments: XW ELHC PR KK. Performed the experiments: XW PR. Analyzed the data: XW ELHC PR KK. Wrote the paper: XW ELHC KK.

References

1. Huxman TE, Snyder KA, Tissue D, Leffler AJ, Ogle K, et al. (2004) Precipitation pulses and carbon fluxes in semiarid and arid ecosystems. Oecologia 141: 254–268.

2. Chapin FS, McFarland J, McGuire AD, Euskirchen ES, Ruess RW, et al. (2009) The changing global carbon cycle: linking plant-soil carbon dynamics to global consequences. Journal of Ecology 97: 840–850.

3. IPCC (2007) Climate Change Fourth Assessment Report, Cambridge University Press, Cambridge, UK.

4. Stallard RF (1998) Terrestrial sedimentation and the carbon cycle: Coupling weathering and erosion to carbon burial. Global Biogeochemical Cycles 12: 231–257.

5. Van Oost K, Quine TA, Govers G, De Gryze S, Six J, et al. (2007) The impact of agricultural soil erosion on the global carbon cycle. Science 318: 626–629.

6. Harden JW, Sharpe JM, Parton WJ, Ojima DS, Fries TL, et al. (1999) Dynamic replacement and loss of soil carbon on eroding cropland. Global Biogeochemical Cycles 13: 885–901.

7. Lal R (2003) Soil erosion and the global carbon budget. Environment International 29: 437–450.

8. Berhe AA, Harden JW, Torn MS, Kleber M, Burton SD, et al. (2012) Persistence of soil organic matter in eroding versus depositional landform positions. Journal of Geophysical Research-Biogeosciences 117.

9. Zhang JH, Quine TA, Ni SJ, Ge FL (2006) Stocks and dynamics of SOC in relation to soil redistribution by water and tillage erosion. Global Change Biology 12: 1834–1841.

10. Gregorich EG, Greer KJ, Anderson DW, Liang BC (1998) Carbon distribution and losses: erosion and deposition effects. Soil & Tillage Research 47: 291–302.

11. Smith SV, Sleezer RO, Renwick WH, Buddemeier R (2005) Fates of eroded soil organic carbon: Mississippi basin case study. Ecological Applications 15: 1929–1940.

12. Wang X, Cammeraat LH, Wang Z, Zhou J, Govers G, et al. (2013) Stability of organic matter in soils of the Belgian Loess Belt upon erosion and deposition. European Journal of Soil Science 64: 219–228.

13. Smith SV, Renwick WH, Buddemeier RW, Crossland CJ (2001) Budgets of soil erosion and deposition for sediments and sedimentary organic carbon across the conterminous United States. Global Biogeochemical Cycles 15: 697–707.

14. Polyakov VO, Lal R (2008) Soil organic matter and CO_2 emission as affected by water erosion on field runoff plots. Geoderma 143: 216–222.

15. Jacinthe PA, Lal R, Kimble JM (2002) Carbon dioxide evolution in runoff from simulated rainfall on long-term no-till and plowed soils in southwestern Ohio. Soil & Tillage Research 66: 23–33.

16. Wan Y, El-Swaify SA (1998) Sediment enrichment mechanisms of organic carbon and phosphorus in a well-aggregated Oxisol. Journal of Environmental Quality 27: 132–138.

17. Kwaad FJPM, Mucher HJ (1994) Degradation of soil-structure by welding - a micromorphological study. Catena 23: 253–268.

18. Franzluebbers AJ (1999) Potential C and N mineralization and microbial biomass from intact and increasingly disturbed soils of varying texture. Soil Biology & Biochemistry 31: 1083–1090.

19. Bremenfeld S, Fiener P, Govers G (2013) Effects of interrill erosion, soil crusting and soil aggregate breakdown on in situ CO_2 effluxes. Catena 104: 14–20.

20. Strickland TC, Truman CC, Frauenfeld B (2005) Variable rainfall intensity effects on carbon characteristics of eroded sediments from two coastal plain ultisots in Georgia. Journal of Soil and Water Conservation 60: 142–148.

21. Truman CC, Strickland TC, Potter TL, Franklin DH, Bosch DD, et al. (2007) Variable rainfall intensity and tillage effects on runoff, sediment, and carbon losses from a loamy sand under simulated rainfall. Journal of Environmental Quality 36: 1495–1502.

22. Palis RG, Ghandiri H, Rose CW, Saffigna PG (1997) Soil erosion and nutrient loss. 3. Changes in the enrichment ratio of total nitrogen and organic carbon under rainfall detachment and entrainment. Australian Journal of Soil Research 35: 891–905.

23. Van Hemelryck H, Fiener P, Van Oost K, Govers G, Merckx R (2010) The effect of soil redistribution on soil organic carbon: an experimental study. Biogeosciences 7: 3971–3986.

24. Jacinthe PA, Lal R (2001) A mass balance approach to assess carbon dioxide evolution during erosional events. Land Degradation & Development 12: 329–339.

25. Koninklijk Nederlands Meteorologisch Instituut (KNMI) website. Available: http://www.klimaatatlas.nl/klimaatatlas.php?wel = neerslag. Assessed 2014 April 10.

26. WRB (2006) World reference base for soil resources. FAO, ISRIC, ISSS, Rome.

27. Blake GR HK (1986) Bulk density. In: Klute A (ed) Methods of soil analysis: Part 1 physical and mineralogical methods. American Society of Agronomy and Soil Science Society of America, Madison

28. Heimovaara TJ, Bouten W (1990) A computer-controlled 36-channel time domain reflectometry system for monitoring soil-water contents. Water Resources Research 26: 2311–2316.
29. Buishand TA, Velds CA (1980) Extreme neerslaghoeveelheden, Neerslag en Verdamping. KNMI, de Bild: pp. 104–118.
30. Borselli L, Torri D, Poesen J, Sanchis PS (2001) Effects of water quality on infiltration, runoff and interrill erosion processes during simulated rainfall. Earth Surface Processes and Landforms 26: 329–342.
31. Kuhn NJ (2007) Erodibility of soil and organic matter: independence of organic matter resistance to interrill erosion. Earth Surface Processes and Landforms 32: 794–802.
32. RIVM Landelijk meetnet luchtkwaliteit website (National measurement network air quality (includes rainwater quality). Available: http://www.lml.rivm.nl/data/gevalideerd/. Accessed 2014 April 10.
33. Golchin A, Oades JM, Skjemstad JO, Clarke P (1994) Study of free and occluded particulate organic matter in soils by solid state C13 Cp/MAS NMR spectroscopy and scanning electron microscopy. Australian Journal of Soil Research 32: 285–309.
34. Cerli C, Celi L, Kalbitz K, Guggenberger G, Kaiser K (2012) Separation of light and heavy organic matter fractions in soil - Testing for proper density cut-off and dispersion level. Geoderma 170: 403–416.
35. Ghadiri H, Rose CW (1993) Water eosion processes and the enrichment of sorbed pesticides. 2. Enrichment under rainfall dominated erosion process. Journal of Environmental Management 37: 37–50.
36. Chen FS, Zeng DH, Fahey TJ, Liao PF (2010) Organic carbon in soil physical fractions under different-aged plantations of Mongolian pine in semi-arid region of Northeast China. Applied Soil Ecology 44: 42–48.
37. Allison SD, Jastrow JD (2006) Activities of extracellular enzymes in physically isolated fractions of restored grassland soils. Soil Biology & Biochemistry 38: 3245–3256.
38. Denef K, Six J, Paustian K, Merckx R (2001) Importance of macroaggregate dynamics in controlling soil carbon stabilization: short-term effects of physical disturbance induced by dry-wet cycles. Soil Biology & Biochemistry 33: 2145–2153.
39. Nadeu E, Berhe AA, de Vente J, Boix-Fayos C (2012) Erosion, deposition and replacement of soil organic carbon in Mediterranean catchments: a geomorphological, isotopic and land use change approach. Biogeosciences 9: 1099–1111.
40. Wang ZG, Govers G, Steegen A, Clymans W, Van den Putte A, et al. (2010) Catchment-scale carbon redistribution and delivery by water erosion in an intensively cultivated area. Geomorphology 124: 65–74.
41. Schiettecatte W, Gabriels D, Cornelis WM, Hofman G (2008) Enrichment of organic carbon in sediment transport by interrill and rill erosion processes. Soil Science Society of America Journal 72: 50–55.
42. Cole JJ, Prairie YT, Caraco NF, McDowell WH, Tranvik LJ, et al. (2007) Plumbing the global carbon cycle: Integrating inland waters into the terrestrial carbon budget. Ecosystems 10: 171–184.
43. Ghadiri H, Rose CW (1991) Sorbed chemical-transport in overland-flow. 2. Enrichment ratio variation with erosion processes. Journal of Environmental Quality 20: 634–641.
44. Denef K, Six J, Bossuyt H, Frey SD, Elliott ET, et al. (2001) Influence of dry-wet cycles on the interrelationship between aggregate, particulate organic matter, and microbial community dynamics. Soil Biology & Biochemistry 33: 1599–1611.
45. Six J, Elliott ET, Paustian K, Doran JW (1998) Aggregation and soil organic matter accumulation in cultivated and native grassland soils. Soil Science Society of America Journal 62: 1367–1377.
46. Feng W, Plante AF, Aufdenkampe AK, Six J (2014) Soil organic matter stability in organo-mineral complexes as a function of increasing C loading. Soil Biology & Biochemistry 69: 398–405.
47. Kaiser K, Guggenberger G (2003) Mineral surfaces and soil organic matter. European Journal of Soil Science 54: 219–236.
48. Van Hemelryck H, Govers G, Van Oost K, Merckx R (2011) Evaluating the impact of soil redistribution on the in situ mineralization of soil organic carbon. Earth Surface Processes and Landforms 36: 427–438.
49. Casals P, Gimeno C, Carrara A, Lopez-Sangil L, Sanz MJ (2009) Soil CO_2 efflux and extractable organic carbon fractions under simulated precipitation events in a Mediterranean Dehesa. Soil Biology & Biochemistry 41: 1915–1922.
50. Doetterl S, Six J, Van Wesemael B, Van Oost K (2012) Carbon cycling in eroding landscapes: geomorphic controls on soil organic C pool composition and C stabilization. Global Change Biology 18: 2218–2232.
51. Fontaine S, Barot S, Barre P, Bdioui N, Mary B, et al. (2007) Stability of organic carbon in deep soil layers controlled by fresh carbon supply. Nature 450: 277–U210.
52. Creed IF, Webster KL, Braun GL, Bourbonniere RA, Beall FD (2013) Topographically regulated traps of dissolved organic carbon create hotspots of soil carbon dioxide efflux in forests. Biogeochemistry 112: 149–164.
53. Bianchi TS (2011) The role of terrestrially derived organic carbon in the coastal ocean: A changing paradigm and the priming effect. Proceedings of the National Academy of Sciences of the United States of America 108: 19473–19481.
54. Lal R (2004) Soil carbon sequestration impacts on global climate change and food security. Science 304: 1623–1627.
55. Kwaad FJPM, de Roo APJ, Jetten VG (2006) The Netherlands. In: Boardman J and Poesen J (eds) Soil Erosion in Europe. Wiley and Sons, Chichester.

16

Effect of Drying on Heavy Metal Fraction Distribution in Rice Paddy Soil

Yanbing Qi[1,2]*, Biao Huang[3], Jeremy Landon Darilek[3]

1 College of Resources and Environment, Northwest Agriculture and Forestry University, Yangling, Shaanxi, People's Republic of China, **2** Key Laboratory of Plant Nutrition and the Agri-environment in Northwest China, Ministry of Agriculture, Yangling, Shaanxi, People's Republic of China, **3** State Key Laboratory of Soil and Sustainable Agriculture, Institute of Soil Science, Chinese Academy of Sciences, Nanjing, Jiangsu, People's Republic of China

Abstract

An understanding of how redox conditions affect soil heavy metal fractions in rice paddies is important due to its implications for heavy metal mobility and plant uptake. Rice paddy soil samples routinely undergo oxidation prior to heavy metal analysis. Fraction distribution of Cu, Pb, Ni, and Cd from paddy soil with a wide pH range was investigated. Samples were both dried according to standard protocols and also preserved under anaerobic conditions through the sampling and analysis process and heavy metals were then sequentially extracted for the exchangeable and carbonate bound fraction (acid soluble fraction), iron and manganese oxide bound fraction (reducible fraction), organic bound fraction (oxidizable fraction), and residual fraction. Fractions were affected by redox conditions across all pH ranges. Drying decreased reducible fraction of all heavy metals. $Cu_{residual\ fraction}$, $Pb_{oxidizable\ fraction}$, $Cd_{residual\ fraction}$, and $Ni_{residual\ fraction}$ increased by 25%, 33%, 35%, and >60%, respectively. $Pb_{residual\ fraction}$, $Ni_{acid\ soluble\ fraction}$, and $Cd_{oxidizable\ fraction}$ decreased 33%, 25%, and 15%, respectively. Drying paddy soil prior to heavy metal analysis overestimated Pb and underestimated Cu, Ni, and Cd. In future studies, samples should be stored after injecting N_2 gas to maintain the redox potential of soil prior to heavy metal analysis, and investigate the correlation between heavy metal fraction distribution under field conditions and air-dried samples.

Editor: Wen-Xiong Lin, Agroecological Institute, China

Funding: This work was financially supported by the Natural Science Foundation of China (31100516), and the National "863" Plan Project of China (No. 2013AA102401). The funders had no role in study design, data collection and analysis, decision to publish, or preparation of the manuscript.

Competing Interests: The authors have declared that no competing interests exist.

* E-mail: yanbing_qi@sina.com

Introduction

Soil heavy metals Cu, Pb, Ni, and Cd, are regarded as "chemical time bombs" because of their propensity for accumulation in the soil and uptake by crops. This ultimately causes human toxicity in both the short- and long-term [1–2], making farmland ecosystems dangerous to health [3].

Morphological characteristics and processes of heavy metals have been studied to better understand heavy metal occurrence in various environments, transport pathways, and crop uptake. There are a wide range of soil inorganic and organic substances affected by redox conditions. Redox changes the valence of ions, subsequently affecting the forms of various elements and compounds and their transport processes. Furthermore, heavy metal behavior is strongly correlated with redox potential (Eh) [4].

Many studies of river and lake sediments have reported the effect of redox conditions on the distribution of heavy metals. For example, such changes in redox conditions affect heavy metal association with organic matter (OM) [5], and iron (Fe) and manganese (Mn) oxidized fractions are unstable in reduced environments [3,6]. Kelderman and Osman [2] reported that in river sediments, exchangeable and carbonate bound fractions of Cu, Zn, and Pb increased as Eh increased, and organic fractions, oxidizable Cu, and oxidizable Pb decreased. Lu et al. [7] found that in reduced Iraqi River sediment in Changchun City, Mn oxide fractions increased significantly, organic bound Cu de-creased by 40%, but decrease in organic bound Pb, Zn, and Ni was less than Cu.

Soil heavy metal studies which include plant available indices, total concentrations, and fraction distributions have also been conducted on many agricultural systems, including rice (*Oryza sativa*) production. Rice has a high water requirement compared to other crops and during the growing season, Eh can reach roughly 300–200 mV [4]. Soil samples, including from rice paddies under submerged conditions, are routinely air-dried prior to heavy metal analysis [8–12]. Zheng and Zhang [13] studied the effect of moisture regimes on paddy soil heavy metals and found that soil moisture did not affect the direction or pathways of fractionation distribution (from active to stable fractions), but did affect the transformation rate. Zheng and Zhang [13] dried rice paddy soil samples after collection and then reconstituted three moisture regimes under controlled conditions in the lab. It does not appear that sample preparation was conducted in an anaerobic environment, therefore this result does not represent in situ soil moisture regime that controls the distribution of heavy metals.

Redox conditions are known to have a significant effect on heavy metal speciation in sediments [14]. As indicated by Calmano et al. [5], if anoxic sediments are exposed to atmosphere, redox condition change and redistribution and transformation of heavy metal fractions in the sediments takes place. A few studies show the effect of redox conditions on heavy metal availability [15–17]. Paddy soil has the anoxic condition during rice growing season similar to the river sediment. Thus, we hypothesize that drying prior to analysis for paddy soil changes the anoxic condition to redox condition as a consequence mobility and uptake of heavy metal by crop are changes. To date, the effect of drying on soil heavy metal fraction distribution in rice paddies is largely unknown.

In order to fill this knowledge gap and work towards a better methodology for sampling and handling paddy soil prior to heavy metal analysis, we conducted a study in Zhangjiagang County, a typical rice production region of China, to 1) understand heavy metal distribution in paddy soil under various redox and pH conditions, and 2) determine the effect of soil air-drying on heavy metal fractions. This study intends to provide useful information for heavy metal assessment in situ. It is hoped that through a better understanding of heavy metal fractions in situ, a better understanding of correlation of heavy metal between soils and crops can be drawn, and a suitable set of corrective measures to prevent heavy metal pollution can be put in place.

Materials and Methods

Ethics Statement

All the sample sites were distributed in the private land and permission were approved by the land owner in each site (Zhigang Wang can be contacted for the future permissions). The field studies did not involve endangered or protected species because all the sample sites were in the farmland with rice planted. The coordinates of sample sites ranged from 120°35' to 120°42'E and from 31°44' to 31°51'N.

Sample Collection

Zhangjiagang County was selected as the research area because of two reasons, firstly, the soil is typical paddy soil with rice planting history of more than 100 years, secondly, heavy metals have been accumulated slightly in this area because of sparkly distributed chemical factories. Zhangjiagang County has a humid monsoon climate in the north subtropical zone, with four distinct seasons, and average annual temperature and precipitation is 15.0°C and 1045.9 mm, respectively.

An investigation of soil physical and chemical characteristics of Zhangjiagang County, Jiangsu, was conducted in 2004 [18–19]. In the current study, ten locations from the 2004 investigation, each with a long history of rice production, were selected to represent a wide pH range (4.81–8.16).

Topsoil (0–20 cm) was collected in October, 2007 from rice paddies before the harvest and after a full season of submersion. At the time of collection, there was still a 1–2 cm water layer covering the soil. Each soil sample was a composite of sub-samples taken from 6–8 points within 50 m^2 with a stainless steel soil probe. Half of the sample was quickly placed in polyethylene plastic tubes and the headspace was immediately purged with nitrogen (N) gas [20], pre-treated with acid, and weighed. Once transported to the laboratory, samples were stored under refrigeration at 4°C for the determination of heavy metals fractions under reduction conditions. The remaining sample was placed in polyethylene bags.

Chemical Analysis

Soil moisture was measured using soil from the polyethylene bags by drying at 105°C for 24 h. The remaining soil from the polyethylene bags were air-dried and sieved through a 2 mm sieve. Soil pH was measured with a glass electrode in a 1:2.5 soil: water suspension [21]. Soil OM was determined using the dichromate-wet combustion method [22]. The samples were digested using

HNO$_3$-HClO$_4$-HF for measurement of the total heavy metals [19].

The Community Bureau of Reference (BCR) sequential extraction procedure described by Ure et al. [23] was used to extract heavy metal fractions. For the exchangeable and carbonate bound fraction (acid soluble fraction), acetic acid (20 ml, 0.11 mol l^{-1}) was added to a 50 ml polypropylene centrifuge tube containing 0.5 g sample and shaken for 16 h at ambient temperature (20±1°C) on an end-over mechanical shaker operating at 40 rpm. The extract was centrifuged at 4000 rpm, decanted into a polyethylene container, and stored at −4°C until analysis. For the Fe and Mn oxide bound fraction (reducible fraction), the residue from the step above was shaken with hydroxylamine hydrochloride (20 ml, 0.1 mol l^{-1}, acidified to pH 2 with nitric acid). This was also centrifuged, decanted, and stored as described above. For the organic bound fraction (oxidizable fraction), the residue from reducible fraction was shaken with 30% hydrogen peroxide (20 ml, acidified to pH 2 with nitric acid) and then ammonium acetate (25 ml, 1 mol l^{-1}, acidified to pH 2 with nitric acid). This was also centrifuged, decanted, and stored as above. For the residual fraction, the residue from oxidizable fraction was digested using the method described above for the total digestion of heavy metals.

The entire extraction procedure for soil samples in polyethylene tubes was the same as described for the air-dried samples except that samples were handled in an anaerobic chamber, using N as the purge gas. All analysis of heavy metal concentrations was determined using inductively coupled plasma-atomic emission spectrometry (ICP-AES).

Statistical Analysis

Analysis of variance was performed using the GLM procedure. Mean separations were performed using Duncan's Multiple Range to test soil pH and redox on soil heavy metal fractions. Linear relationship (correlation) within fractions and regression within the percentage distribution of each metal fraction in oxidized and reduced condition were conducted. All statistical analysis was conducted using SPSS (version 13.0).

Results

Soil Physicochemical Properties

Soil physicochemical properties of rice paddies in Zhangjiagang are shown in Table 1. Average soil moisture of submerged soil was 30.2% and ranged from 24.6% to 33.9%. Soil OM ranged from 19.40 g kg^{-1} to 31.30 g kg^{-1} and had a mean of 25.47 g kg^{-1}. The most abundant metal was Pb, which had a mean concentration of 43.27 mg kg^{-1}. The least abundant metal was Cd, which had a mean concentration of 0.20 mg kg^{-1}. The coefficient of variation (CV) for Cd (41.3%) was higher than Cu, Pb, and Ni (25.6%, 8.8% and 9.3%, respectively).

Soil Heavy Metal Fraction Distribution

Heavy metal fraction distribution varied among elements (Fig. 1 and 2). Redox conditions had a significant effect on all Cu fractions (Fig. 1, Table 2), soil pH did not affect Cu fractions, and there was only a significant interaction effect of redox conditions and pH with reducible fraction. The predominant fractions were reducible fraction and oxidizable fraction (35–40%), followed by residual fraction (15–18%) and acid soluble fraction (<1%). Air-dried samples had significantly different Cu fraction distribution. Compared with reduced soil, acid soluble fraction was still the smallest fraction accounting for 5% of the total Cu. Reducible fraction and oxidizable fraction had significantly lower proportions

Table 1. Soil chemical properties of rice paddies in Zhangjiagang County, Jiangsu.

Samples	pH	Moisture (%)	Soil Organic Matter (g kg⁻¹)	Cu	Pb	Ni	Cd
				(mg kg⁻¹)			
1	4.81	33.93	31.30	34.87	48.96	28.82	0.236
2	5.54	30.75	26.50	28.35	38.10	26.41	0.087
3	5.87	26.74	26.80	28.57	41.64	24.66	0.160
4	6.12	31.21	28.70	26.64	40.10	25.70	0.190
5	6.45	24.62	23.70	46.71	42.92	30.15	0.107
6	6.94	28.10	29.30	25.73	39.13	24.61	0.135
7	7.17	29.77	19.40	32.56	43.46	31.74	0.180
8	7.61	31.79	21.10	30.99	44.81	28.80	0.327
9	8.02	32.66	26.70	53.48	44.26	28.54	0.300
10	8.16	32.53	21.20	30.26	49.34	24.62	0.281
Average	6.67	30.2	25.47	33.82	43.27	27.41	0.200
St dev	1.10	2.9	3.95	8.66	3.79	2.55	0.083
Cv (%)	16.49	9.7	15.53	25.62	8.76	9.31	41.28

while residual fraction had significantly higher proportion and was the predominant fraction.

Redox conditions affected all Pb fraction distributions except reducible fraction, and pH did not affect fraction distribution (p> 0.05) (Table 2). There was also no interaction effect of redox conditions and pH with Pb fractions. In reduced soil, the predominant fraction was residual fraction (60%), followed by reducible fraction (20%), oxidizable fraction and acid soluble fraction. In oxidized soil, acid soluble fraction was 3% higher than in the reduced soil, oxidizable fraction was 20% higher and residual fraction was 33% lower, respectively, than in reduced soil, and reducible fraction was not significantly different.

Redox conditions affected all Ni fraction distributions and pH affected all fraction distributions except oxidizable fraction (Table 2). In reduced soil with pH 6 to 7, $Ni_{oxidizable\ fraction}$ had the lowest proportion, and $Ni_{residual\ fraction}$ had the highest (Fig. 1). There was significant interaction of redox conditions and pH with acid soluble fraction. In oxidized soil, acid soluble fraction decreased from 30% to 5% and reducible fraction decreased from 35% to 15%. The residual fraction increased by more than 60% and oxidizable fraction increased less than 5%.

Redox conditions and pH affected $Cd_{oxidizable\ fraction}$ and $Cd_{residual\ fraction}$ (Table 2). In reduced soil the predominant fraction was reducible fraction (55%), followed by oxidizable fraction (25%), acid soluble fraction and residual fraction. In oxidized soil, residual fraction increased by 15% and oxidizable fraction decreased by 15%. Treatments had no significant effect on acid soluble fraction or reducible fraction.

Regression of Reduced and Oxidized Samples

Regression analysis did not reveal significant relationships between soil heavy metal distribution of oxidized and reduced soil. None of the Cu fractions had significant regression results. For Pb, only residual fraction was significant, and for Ni, only reducible fraction was significant (Fig. 3). For Cd fractions, regression of oxidized and reduced samples had R^2 values of 0.26, 0.69, 0.48, and 0.66 for acid soluble fraction, reducible fraction, oxidizable fraction, and residual fraction, respectively (Fig. 4). Regression analysis between relative concentration in oxidized soil and relative concentration in reduced heavy soil showed significant correlation for reducible fraction, oxidizable fraction and residual fraction (Fig. 4). Regression analysis separately by pH and OM did not improve correlation between the dependent and independent variable in Fig. 4.

Discussion

Total Heavy Metal Concentrations in Rice Paddies

Soil heavy metals originate from both the environment and anthropogenic sources. Although all heavy metal concentrations in this study, except three Cd samples, were higher than typical background concentrations [24], they are lower than the critical limits established by the Soil Environmental Quality Standard of China (GB 15618–1995) [25]. The results of our study are consistent with previous research in the area [25]. In rice paddies of China, there is some evidence that this accumulation is a result of heavy metal rich irrigation water [9]. However, low concentrations of heavy metals suggest that although heavy metal concentrations in Zhangjiagang have experienced accumulation from anthropogenic sources, it is not yet a pressing environmental concern.

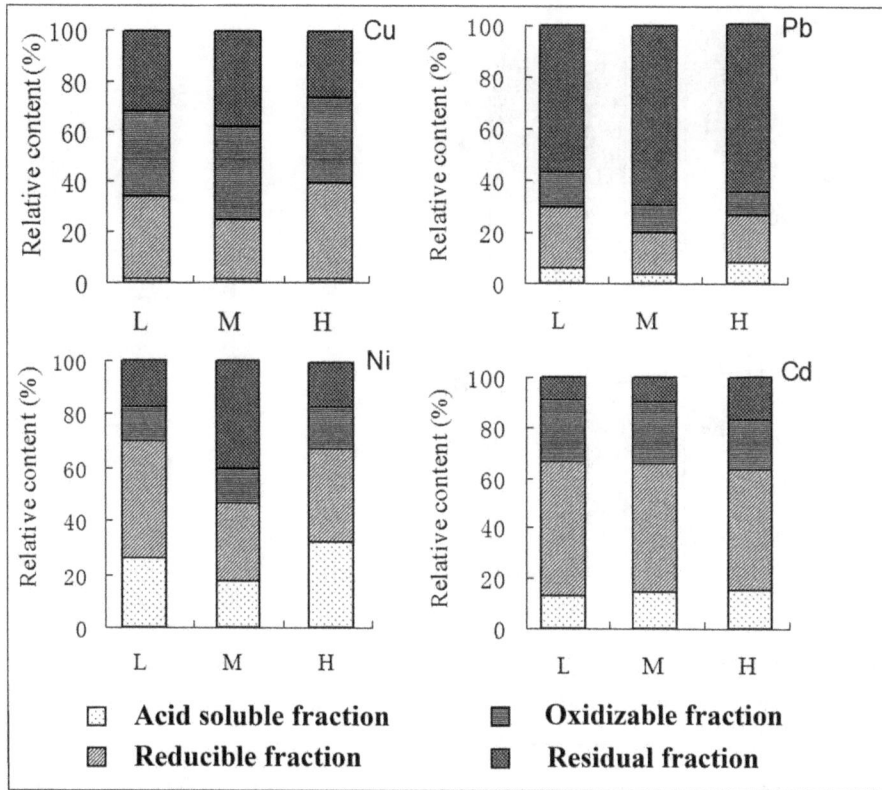

Figure 1. The percentage distribution of each metal fraction in reduced condition in paddy soil samples in Zhangjiagang County. L means lower pH (<6), M means moderate pH (6–7), H means higher pH (>7). Note: significant interactions within and between factors are presented in Table 2.

Heavy Metal Fractionation in Reduced Rice Paddies

Generally, rice paddies are submerged during rice cultivation and have lower Eh values during this part of the year. Reduced soil had lower $Cu_{acid\ soluble\ fraction}$ and $Pb_{acid\ soluble\ fraction}$ and higher $Ni_{acid\ soluble\ fraction}$, $Ni_{reducible\ fraction}$, and $Cd_{reducible\ fraction}$ than air-dried soil. This is similar to heavy metal fractionation changes that occur in various redox states of sewage sludge [26–28]. Förstner and Wittmann [3] explain that for Cu and Pb the most unstable fraction, acid soluble fraction, is replaced by reducible fraction and oxidizable fraction under reduced conditions.

For Ni and Cd, acid soluble fraction accounted for a relatively high proportion (15%–50%), which is not consistent with heavy metal fractionation ranges in sewage sludge. This may be due to Eh differences in rice paddies and sewage sludge, interactions in the rhizosphere of rice paddies, or both. Several studies address changes in soil heavy metal fraction distribution in the rhizosphere [29–31]. High $Ni_{acid\ soluble\ fraction}$ and $Ni_{reducible\ fraction}$ indicated that Ni is adsorbed to solid materials or combined with weak acid and Fe and Mn oxides in reduced soil [27]. High $Cd_{reducible\ fraction}$ was also found in dredged sediments [32].

High $Cu_{reducible\ fraction}$ and $Cu_{oxidizable\ fraction}$ in reduced soil indicates that Fe and Mn oxides play a crucial role in Cu processes, and that Fe and Mn oxides cause this fraction to be relatively stable [33–34]. This result is consistent with many studies which show dominant oxidizable fraction in reduced sediments [26,34–37].

We found the predominant fraction for Pb in rice paddies to be residual fraction. This fraction is generally considered to be the most stable heavy metal fraction in the soil and unavailable to plants because they are found in the lattice of clay mineral [35,38]. In addition to residual fraction, sediment Pb in many studies has been found to have large percentages of reducible fraction [2]. Lower reducible fraction in our study might be because of the higher Eh values, interactions in the rhizosphere of rice paddies, or both.

Effect of Drying on Heavy Metal Fractionation

When soils are routinely dried in the lab prior to chemical analysis, Eh rapidly increases and metals are oxidized. While this may not have a significant effect on heavy metal fractions in agriculture soil which is not submerged, this procedure can drastically change the heavy metal chemistry of reduced soils.

Generally, heavy metal acid soluble fraction increases as Eh increases, however, in our study, Ni and Cd did not respond accordingly. The reason for this is not clear and further study is needed. As Eh increases, reducible fraction generally decreases, most likely because of heavy metal bonding with Fe and Mn oxides [2,4–5,39]. Our study bears this out as well.

Previous research showed that the relationship of Eh with heavy metal oxidizable fraction and residual fraction is not clear [2–5,7,39]. In our study, increasing Eh led to increased $Pb_{oxidizable\ fraction}$ and decreased $Cu_{oxidizable\ fraction}$. Collavini et al. [40] suggested that decreased $Cu_{oxidizable\ fraction}$ as soil oxidizes occurs due to the interaction of copper sulfide oxidation and Fe and Mn oxides. Released Cu^{2+} is then redistributed to other fractions such as acid soluble fraction and residual fraction. Part of the obscurity may be because of the complex nature of bonding in copper sulfides, which

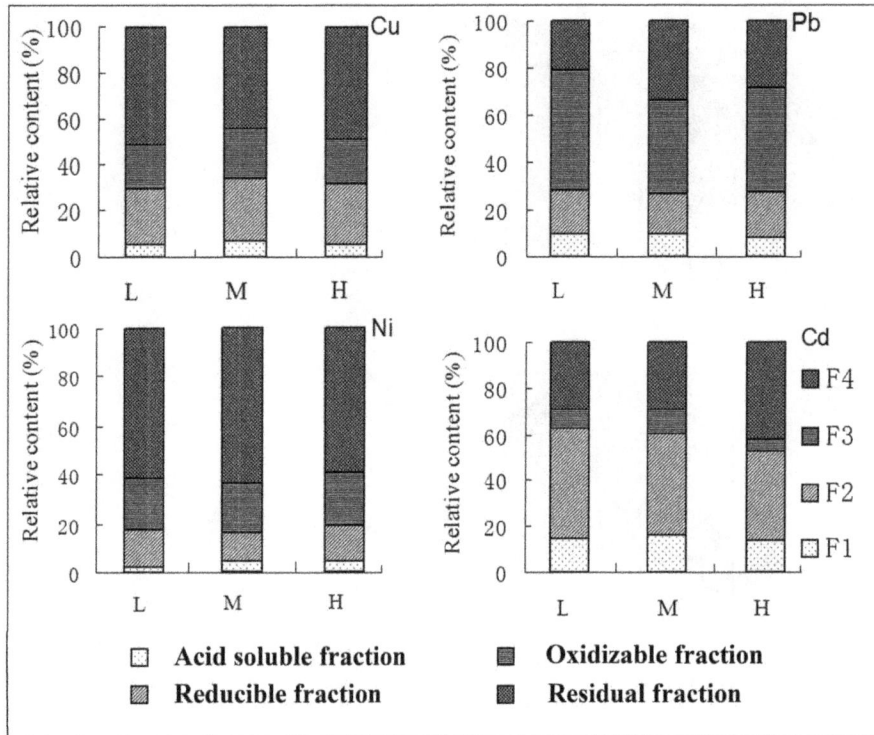

Figure 2. The percentage distribution of each metal fraction in oxidized condition in paddy soil samples in Zhangjiagang County. L means lower pH (<6), M means moderate pH (6–7), H means higher pH (>7). Note: significant interactions within and between factors are presented in Table 2.

have both covalent and ionic bonding characteristics and a high occurrence of delocalized electrons [41–42].

In this study, increasing Eh led to increased $Cu_{reducible\ fraction}$, which is in agreement with Kelderman and Osman [2], who reported a decrease in reducible fraction for river sediments under submerged conditions, and with Saeki et al. [14] who reported an increase in $Cu_{reducible\ fraction}$ after drying of lake sediments.

Drying caused an increase in $Pb_{oxidizable\ fraction}$, which contradicts previous literature [2,7]. As Eh increases, $Pb_{reducible\ fraction}$ slightly decreases, mainly due to bonding of Pb with Fe and Mn oxides [7]. Dried soil had lower $Ni_{reducible\ fraction}$ and $Cd_{oxidizable\ fraction}$ and higher $Ni_{residual\ fraction}$ and $Cd_{residual\ fraction}$ than reduced soil. This was consistent with the results of other studies [2,7,40].

Table 2. P-values from the analysis of variance of redox conditions and pH for heavy metal fraction distributions.

Heavy Metal	Factors	P value			
		Acid soluble fraction	Reducible fraction	Oxidizable fraction	Residual fraction
Cu	Redox	<0.001	0.050	<0.001	<0.001
	pH	0.685	0.109	0.325	0.555
	Interaction	0.308	0.034	0.873	0.111
Pb	Redox	0.001	0.901	<0.001	0.000
	pH	0.230	0.544	0.124	0.068
	Interaction	0.123	0.092	0.345	0.970
Ni	Redox	0.000	0.000	0.000	0.000
	pH	0.023	0.009	0.314	0.009
	Interaction	0.040	0.304	0.356	0.164
Cd	Redox	0.08	0.316	0.000	0.024
	pH	0.810	0.261	0.015	0.142
	Interaction	0.089	0.359	0.277	0.413

P-values <0.05 show significant difference within factors and significant interaction between factors for each heavy metal fraction, separately.

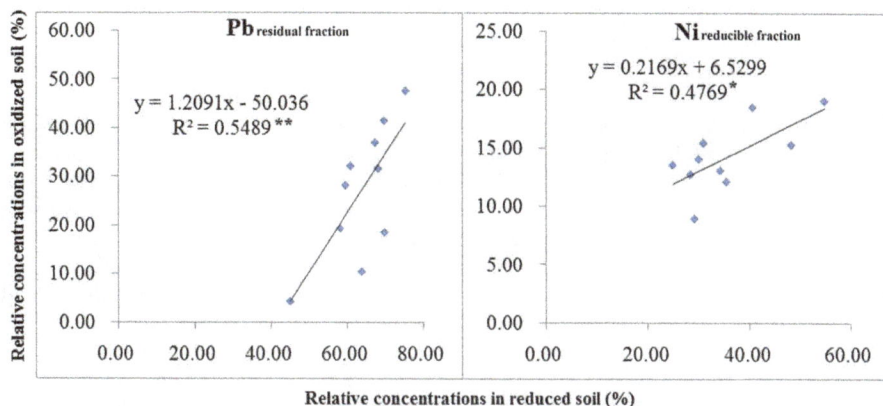

Figure 3. Correlation between the percentage distribution of iron and manganese bound nickel (Ni$_{reducible\ fraction}$) and residual lead (Pb$_{residual\ fraction}$) in oxidized and reduced paddy soil samples in Zhangjiagang County. *, and ** denote the significant correlation at $p \leq 0.05$ and 0.01, respectively.

There is strong evidence that drying significantly affects soil heavy metal fractionation of submerged rice paddy soil samples but effects are different than lake sediment, river sediment, and sewage sludge. One of the main reasons most likely is the differences in Eh of the sample sources. Generally, sediment Eh is < -100mv, while rice paddy soil Eh ranged from 200–300 mv [4]. Abundant roots in rice paddies may be another important factor. Root exudates can affect many soil properties such as pH and microbial activity. Changes in these properties can then indirectly affect soil heavy metal fraction distribution. Chen et al. [43]

reported that abundant rhizosphere can significantly increase Cu$_{acid\ soluble\ fraction}$ and significantly decrease Cd$_{acid\ soluble\ fraction}$. Shuman and Wang [44] reported that rice rhizosphere significantly increased Zn$_{reducible\ fraction}$ and Cu$_{oxidizable\ fraction}$. Similarly, Lin et al. [24] reported that rice rhizosphere significantly increased Cd$_{reducible\ fraction}$ and Cd$_{oxidizable\ fraction}$.

Effects of Drying on Heavy Metal Availability

The distribution, migration, and plant availability of soil heavy metals are reflected not in the total concentrations, but in specific

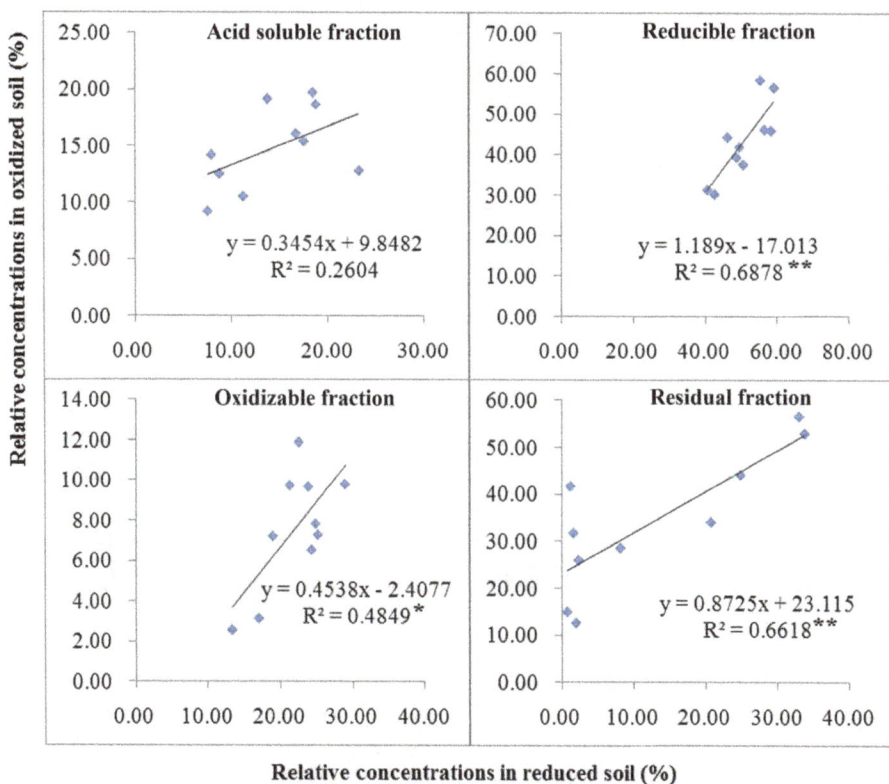

Figure 4. Correlation between the percentage distribution of cadmium in oxidized and reduced paddy soil samples in Zhangjiagang County. *, and ** denote the significant correlation at $p \leq 0.05$ and 0.01, respectively.

fractions. With the BCR sequential extraction method, heavy metal bioavailability decreases with each progressively stronger extraction [45]. acid soluble fraction has the highest bioavailability and residual fraction has the lowest. Acid soluble fraction, reducible fraction, and oxidizable fraction are considered active fractions and each contain some bioavailable heavy metals [46]. In this study, Ni and Cd have large active fractions (80%), followed by Cu (>60%), and Pb (<40%). Drying submerged paddy soil samples caused decreases in the active fractions of Cu, Ni, and Cd and increased active fractions of Pb. Zheng and Zhang [13] found an increased bioavailability with increased Eh across all heavy metals measured (Cu, Pb, Cd, and Hg). Our results suggest that when samples are dried prior to heavy metal bioavailability analysis, Pb is overestimated and Cu, Ni, and Cd are underestimated.

Regression analysis results showed that samples which have been dried prior to analysis are not suitable indicators of reduced soil conditions and cannot be used to predict field conditions. However, adding pH and OM did not improve the predictive power, particularly for Cd (data not shown).

Conclusions

Almost all heavy metal fractions were significantly affected by redox conditions, and few heavy metal fractions were affected by

pH. Drying the soil prior to analysis caused metal ions to change valence and a redistribution of heavy metal fractions. Some redistribution occurred as metals moving from active fractions to stable fractions. Therefore, drying decreases the representativeness of *in situ* conditions and the availability of some heavy metals are misestimated. We suggest that soil heavy metal fractionation procedures for rice paddies ensure anaerobic conditions from the time of sampling until analysis. Methodology studies to improve sampling and lab handling techniques would be beneficial, as would further investigation of possible correlation of non-dried sampling and analysis using bioavailable extractions with air-dried samples. For some heavy metals, the processes driving redistribution trends are still unclear. Because the relationship between soil Eh and heavy metal fractionation distribution directly affects heavy metal mobility and bioavailability, a better understanding of the soil heavy metal chemistry under various redox conditions is imperative.

Author Contributions

Conceived and designed the experiments: YQ. Performed the experiments: YQ JLD. Analyzed the data: YQ BH. Contributed reagents/materials/analysis tools: BH. Wrote the paper: YQ.

References

1. Gong ZT, Huang B (1998) Studies on potential "chemical timer bombs" and their igniting factors in soils. Adv. Earth Sci. 23: 184–191.
2. Kelderman P, Osman AA (2007) Effect of redox potential on heavy metal binding forms in polluted canal sediments in Delft (the Netherlands). Water Res. 41: 4251–4261.
3. Förstner U, Wittmann GTW (1983) Metal pollution in the aquatic environment, second ed. Springer, Berlin.
4. Li YG, Xue SG, Wu XY (2004) Transport and transformation of heavy metals in the soil-paddy plants system. Geo. China. 31(supp.): 87–92.
5. Calmano W, Hong J, Förstner U (1993) Binding and mobilization of heavy metals in contaminated sediments affected by pH and redox potential. Water Sci. Tech. 28: 223–235.
6. Stumm W, Morgan JJ (1996) Aquatic chemistry: an introduction emphasizing chemical equilibria in natural water, third ed. Wiley, New York.
7. Lu YZ, Dong DM, Fu Y (2006) Comparison of speciation patterns of heavy metals in Yitong River sediment under reduced and oxidized conditions. Chem. J. Chinese. Uni. 27: 449–453.
8. Andreu V, Gimeno-García E (1999) Evolution of heavy metals in marsh areas under rice farming. Environ. Pollution 104: 271–282.
9. Luo YM, Jiang XJ, Wu LH, Song J, Wu SC, Lu RH, Christie P (2003) Accumulation and chemical fractionation of Cu in a paddy soil irrigated with Cu-rich wastewater. Geoderma, 115: 113–120.
10. Li JX, Yang XE, He ZL, Jilani G, Sun CY, Chen SM (2007) Fractionation of lead in paddy soils and its bioavailability to rice plants. Geoderma, 141: 174–180.
11. Zhao KL, Liu XM, Xu JM, Selim HM (2010) Heavy metal contaminations in a soil-rice system: Identification of spatial dependence in relation to soil properties of paddy fields. J. Hazard. Mater. 181: 778–787.
12. Ngoc MN, Dultz S, Kasbohm J (2009) Simulation of retention and transport of copper, lead and zinc in a paddy soil of the Red River Delta, Vietnam, Agr. Ecosyst. Environ. 129: 8–16.
13. Zheng SA, Zhang MK (2011) Effect of moisture regime on the redistribution of heavy metals in paddy soil. J. Environ. Sci. 23: 434–443.
14. Saeki K, Okazaki M, Matsumoto S (1993) The chemical phase changes in heavy metals with drying and oxidation of the lake sediments. Water Res. 27: 1243–1251.
15. Reddy CN, Patrick WHJ, (1977) Effect of redox potential and pH on the uptake of cadmium and lead by rice plants. J. Environ. Qual. 6: 259–262.
16. Sajwan KS, Lindsay WL (1986) Effects of redox on zinc deficiency in paddy rice. Soil Sci. Soc. Am J. 50: 1264–1269.
17. Kashem MA, Singh BR (2001) Metal availability in contaminated soils: I. Effects of flooding and organic matter on changes in Eh, pH and solubility of Cd, Mi and Zn. Nutr. Cycl. Agroecosys. 61: 247–255.
18. Qi YB, Huang B, Gu ZQ, Sun WX, Zhao YC (2008) Spatial and temporal variation of C/N ratio of agricultural soils in typical area of Yangtze Delta region and its environmental significance. B. Miner. Petro. Geochem. 27: 50–56.
19. Shao XX, Huang B, Gu ZQ, Sun WX, Zhao YC (2006) Spatial-temporal variation of pH values of soil in a rapid economic developing area in the Yangtze Delta Region and their causing factors. B. Miner. Petro. Geochem. 25: 35–38.
20. Moore P, Coale F (2000) Phosphorus fractionation in flooded soils and sediments. In: G. M. Pierzenski, (Eds.), Methods of phosphorus analysis for soils, sediments, residuals, and waters. Available online at http://www.soil.ncsu.edu/sera17/publications/sera17-2/pm_cover.htm.
21. Lu RK (2004) Soil and agro-chemical analytical methods, China Agricultural Science and Technology Press, Beijing.
22. Nelson DW, Sommers LE (1996) Total carbon, organic carbon, and organic matter, In: D. L. Sparks (Eds.), Methods of soil analysis, part 3: Chemical methods, Madison, Wisc.: SSSA and ASA, 961–1010.
23. Ure AM, Quevauviller PH, Muntau H (1993) Speciation of heavy metals in soils and sediments-an account of the improvement and harmonization of extraction techniques undertaken under the auspices of the BCR of the Comission of European Communities. Int. J. Environ. Anal. Chem. 51: 135–151.
24. Lin Q, Zheng CR, Chen HM (1998) Transformation of cadmium species in rhizosphere. Acta Pedologica Sinica 35: 461–467.
25. Shao XX, Huang B, Zhao YC, Sun (2008) WX Pollution assessment of soil heavy metals in a representative area of the Yangtze River Delta region. Environ. Chem. 27: 218–221.
26. Tsai LJ, Yu KC, Chang JS (1998) Fractionation of heavy metals in sediment cores from the Ell-Ren River, Taiwan. Water Sci. Tech. 37: 217–224.
27. Tessier A, Campbell PGC, Bisson M (1979) Sequential extraction procedure for the speciation of particulate trace metals. Anal. Chem. 51: 844–851.
28. Argese E, Bettiol C (2001) Heavy metal partitioning in sediments from the lagoon of Venice (Italy). Toxicol. Environ. Chem. 79: 157–170.
29. Hinsinger P, Plassard C, Tang C, Jaillard B (2003) Origins of root-mediated pH changes in the rhizosphere and their responses to environmental constraints: a review. Plant Soil. 248: 43–59.
30. Tao S, Chen YJ, Xu FL, Cao J, Li BG (2003) Changes in copper speciation in maize rhizosphere soil. Environ Pollution. 122: 447–454.
31. Martínez-Alcalá I, Walker DJ, Bernal MP (2010) Chemical and biological properties in the rhizosphere of *Lupinus albus* alter soil heavy metal fractionation. Ecotox. Environ. Safe. 73: 595–602.
32. Singh SP, Ma LQ, Tack FMG (2000) Trace metal leach ability of land-disposed dredged sediments. J. Environ. Qual. 29: 1124–1132.
33. Wu HS, Zhang AQ, Wang LS (2004) Immobilization study of biosorption of heavy metal ions onto activated sludge. J. Environ. Sci. 16: 640–645.
34. Wang C, Li XC, Ma HT, Qian J, Zhai JB (2006) Distribution of extractable fractions of heavy metals in sludge during the wastewater treatment process. J. Hazard. Mater. 137: 1277–1283.
35. Šurija B, Branica M (1995) Distribution of Cd, Pb, Cu and Zn in carbonate sediments from the Krka River estuary obtained by sequential extraction. Sci. Total Environ. 170: 101–118.
36. Li RY, Yang H, Zhou ZG, Lv JJ, Shao XH, Jin F (2007) Fractionation of heavy metals in sediments from Dianchi lake, China. Pedosphere, 17: 265–272.
37. Shrivastaia SK, Banerjee DK (2004) Speciation of metals in sewage sludge and sludge-amended soils. Water, Air, Soil Pollut. 152: 219–232.
38. Teixeira E, Ortiz L, Alves M, Sanchez J (2001) Distribution of selected heavy metals in fluvial sediments of the coal mining region of Baixo Jacuí, RS, Brazil. Environ Geol. 41: 145–154.

39. Kazi TG, Jamali MK, Kazi GH, Arain MB, Afridi HI, et al. (2005) Evaluating the mobility of toxic metals in untreated industrial wastewater sludge using a BCR sequential extraction procedure and a leaching test. Anal. Bioanal. Chem. 383: 297–304.

40. Collavini F, Zonta R, Novelli AA, Zaggua L (2000) Heavy metals behavior during resuspension of the contaminated reduced sludge of the Venice canals. Toxicol. Environ. Chem. 77: 171–187.

41. Zheng SA, Zheng XQ, Chen C (2012) Leaching Behavior of Heavy Metals and Transformation of Their Speciation in Polluted Soil Receiving Simulated Acid Rain. PLoS ONE 7(11): e49664. doi:10.1371/journal.pone.0049664.

42. Goh SW, Buckley AN, Lamb RN, Rosenberg RA, Moran D (2006) The oxidation states of copper and iron in mineral sulfides, and the oxides formed on initial exposure of chalcopyrite and bornite to air. Geochim. Cosmochim. Ac. 70: 2210–2228.

43. Chen YJ, Tao S, Deng BS, Zhang XQ, Huang Y (2001) Effect of root system on metal fractionation in rhizosphere of contaminated soil. Acta Pedologica Sinica 38: 54–59.

44. Shuman LW, Wang J (1997) Effect of rice variety on zinc, cadmium, iron and manganese contents in rhizosphere and non-rhizosphere soil fractions. Commun. Soil Sci. Plant 28: 23–36.

45. Kashem MA, Singh BR, Kswai S (2007) Mobility and distribution of Cadmium, Nickel, and Zinc in contaminated soil profiles. Earth Environ. Sci. 77: 187–198.

46. Wang CQ, Dai TF, Li B (2007) The speciation and bioavailability of heavy metals in paddy soils under the rice-wheat cultivation rotation. Acta Ecologica Sinica 27: 889–897.

17

Arbuscular Mycorhizal Fungi Associated with the Olive Crop across the Andalusian Landscape: Factors Driving Community Differentiation

Miguel Montes-Borrego[1], Madis Metsis[2], Blanca B. Landa[1]*

1 Department of Crop Protection, Institute for Sustainable Agriculture (IAS-CSIC), Cordoba, Spain, **2** Tallinn University, Institute of Mathematics and Natural Sciences, Tallinn, Estonia

Abstract

Background: In the last years, many olive plantations in southern Spain have been mediated by the use of self-rooted planting stocks, which have incorporated commercial AMF during the nursery period to facilitate their establishment. However, this was practised without enough knowledge on the effect of cropping practices and environment on the biodiversity of AMF in olive orchards in Spain.

Methodology/Principal Findings: Two culture-independent molecular methods were used to study the AMF communities associated with olive in a wide-region analysis in southern Spain including 96 olive locations. The use of T-RFLP and pyrosequencing analysis of rDNA sequences provided the first evidence of an effect of agronomic and climatic characteristics, and soil physicochemical properties on AMF community composition associated with olive. Thus, the factors most strongly associated to AMF distribution varied according to the technique but included among the studied agronomic characteristics the cultivar genotype and age of plantation and the irrigation regimen but not the orchard management system or presence of a cover crop to prevent soil erosion. Soil physicochemical properties and climatic characteristics most strongly associated to the AMF community composition included pH, textural components and nutrient contents of soil, and average evapotranspiration, rainfall and minimum temperature of the sampled locations. Pyrosequencing analysis revealed 33 AMF OTUs belonging to five families, with *Archaeospora* spp., *Diversispora* spp. and *Paraglomus* spp., being first records in olive. Interestingly, two of the most frequent OTUs included a diverse group of Claroideoglomeraceae and Glomeraceae sequences, not assigned to any known AMF species commonly used as inoculants in olive during nursery propagation.

Conclusions/Significance: Our data suggests that AMF can exert higher host specificity in olive than previously thought, which may have important implications for redirecting the olive nursery process in the future as well as to take into consideration the specific soils and environments where the mycorrhized olive trees will be established.

Editor: Raffaella Balestrini, Institute for Plant Protection (IPP), CNR, Italy

Funding: This research was supported by grants from Projects AGL2008-00344/AGR from 'Ministerio de Ciencia e Innovación', Project P10-AGR-5908 from 'Consejería de Economía, Innovación y Ciencia' of Junta de Andalucía, Project AGL-2012-37521 from 'Ministerio de Economía y Competitividad' of Spain, and Fondo Europeo de desarrollo regional (FEDER) "Una manera de hacer Europa" from the European Union. The grant 219262 ArimNET_ERANET FP7 2012–2015 Project PESTOLIVE from Instituto Nacional de Investigación y Tecnología Agraria y Alimentaria (INIA), also provided partial financial support. M. Montes-Borrego enjoyed a contract from 'Consejería de Economía, Innovación y Ciencia' of Junta de Andalucía, Spain. The funders had no role in study design, data collection and analysis, decision to publish, or preparation of the manuscript.

Competing Interests: The authors have declared that no competing interests exist.

* E-mail: blanca.landa@csic.es

Introduction

Spain is the world's largest olive oil producer, accounting for more than one-third of global production [1,2]. In Andalusia (Southern Spain), olive orchards dominate the landscape in an impressive monoculture that covers approximately 17% of the total surface of the region (1.5 million ha) [1,3]. In this region, different olive farming systems can be found including: i) *conventional farming* with rain fed orchards of low plant density, intensive tillage, and low inputs in fertilizer, as well as intensive drip-irrigated orchards, grown with higher inputs of pesticides and fertilizers in order to push up olive yields, and ii) *organic farming* using no chemical inputs and mainly non or light-tillage and use of a vegetative cover to prevent soil erosion [3,4,5]. Additionally, in the last years, many new olive plantations have been mediated by the use of self-rooted planting stocks which have incorporated commercial arbuscular mycorrhizal fungi (AMF) in the potting mixture during the nursery period to facilitate establishment [6,7,8] due to their beneficial effects against biotic and abiotic stresses [9,10,11,12,13].

One of the critical steps for applying AMF to improve crop health is the appropriate selection of effective and well adapted-isolates to be used as inoculants. Although it is well known that olive tree is a mycotrophic plant [7] there is not enough knowledge concerning the effect of cropping practices, olive genotype and environment (soil type and climate) on the biodiversity of AMF in olive orchards in the Mediterranean Region. Knowledge of those factors may be essential to take advantage on the use of AMF in modern oliviculture.

In the present study we have examined the structure and diversity of AMF communities in the rhizosphere of cultivated and wild olives in Andalusia, southern Spain, by using two culture-independent molecular approaches: Fluorescent terminal restriction fragment length polymorphism (T-RFLP) analyses of amplified 28S rDNA sequences and SSU rDNA amplicon parallel 454 pyrosequencing. We also have determined which agronomic or environmental factors associated to the olive orchards sampled are the main drivers of the AMF structure.

Materials and Methods

Ethics Statement

No specific permits were required for the described field studies. Location of organic olive orchards was provided by the Andalusian Committee of Organic Farming (CAAE, Junta de Andalucía). Permission to sample the olive orchards were granted by the landowner. The samples from wild or feral forms of olive are located in public areas or degraded formations and abandoned groves. The specific location of all samples from the study is provided in Table S1. The 96 olive orchards sampled in this study were also included in previous studies [5,14] in which the bacterial communities and functional diversity of the olive rhizosphere was assessed. The sites are not protected in any way. The areas studied do not involve any species endangered or protected in Spain.

Location of Olive Orchards and Rhizosphere Sampling

Soil and roots samples were collected from 90 commercial orchards differing in management system [conventional (49 orchards) vs. organic (41 orchards)] located in the main olive-growing areas of Córdoba (41 orchards), Granada (3 orchards), Jaén (34 orchards), and Sevilla (12 orchards) provinces in Andalusia, southern Spain [5; Table S1]. In addition, six samples (LO, LOBA, BAETICA, MACO, LOMCO, EPCO) from three sites each in Córdoba and Cádiz provinces containing wild or feral forms of olive 'Acebuches' (i.e., secondary sexual derivatives of the cultivated clones or products of hybridization between cultivated trees and nearby oleasters) were included in the study [14; Table S1]. Orchards across all locations sampled differ in climate, soil texture and physicochemical characteristics, soil management system (use of cover crops vs. bare soil), and irrigation regimen (rain-fed vs. drip-irrigated). When possible, we tried to sample a representative distribution of the above considered factors within each orchard management system. Detailed description of soil physicochemical properties, and agronomic and climatic characteristics of the sampled orchards is provided in Table S1. Some of the soil physicochemical and climatic characteristics of the sampled locations were provided recently [5,14].

Root (only young and active) and soil samples were collected in May to July 2009 as described by Aranda et al. [14] and Montes-Borrego et al. [5] in the area of the canopy projection from the upper 5 to 30 cm of soil from three different points around each individual tree. Eight trees per orchard were sampled, and all roots from all trees were thoroughly mixed to obtain a single representative sample per orchard. Intact root systems were shaken gently by hand to remove all but the soil close- and naturally-adhering to the plant root and were kept at 5°C until processing.

Additionally the geographic location and altitude of the sampling sites were determined using a global positioning system (GPS), and climatic variables of each sampling site were obtained from SigMapa, Geographic Information System from the Spanish Ministry of "Medio Ambiente y Medio Rural y Marino" (http://sig.mapa.es/geoportal/) using ArcGIS 10 (ESRI, Redlands, California, EE.UU.) (Table S1).

DNA Extraction from Rhizosphere Samples

Pooled olive root samples were cut into 1-cm pieces with a sterile scissors to get a uniform sample per location. Rhizosphere suspensions (including rhizosphere soil and rhizoplane) were obtained by vigorously shaking 2 g of root segments (four independent replications) suspended in 20 ml of sterile distilled water in an orbital shaker for 10 min and sonicated (Ultrasons, JP Selecta SA, Barcelona, Spain) for 10 minutes. Then, 3 ml of those rhizosphere suspensions were subjected to two consecutive centrifugations at 11,000 rpm for 4 min and the pellet was kept at −20°C until processed. DNA from each of the four rhizosphere soil pellets (approximately 200 mg; four replication per each of the 96 olive orchards) was extracted using the PowerSoil DNA Isolation Kit (MO BIO Laboratories, Inc., Carlsbad, USA) and the FastPrep-24 (MP Biomedicals, Inc., Illkirch, France) instrument run at 6.0 m/s for 40 s as described elsewhere [14].

T-RFLP Analysis

For T-RFLP analysis PCR amplification of partial LSU of rDNA from mycorrhiza were performed using a nested-PCR approach, the first PCR round employing 20 ng of template and the primer pair LR1/FLR2 [15] and the second one the primer pair FLR3/FLR4 [16] following conditions described by Mummey and Rillig [17]. Primer FLR3 was 5′ end-labeled with the fluorescent dye FAM. T-RFLP analysis was performed for all samples using 5 μl of PCR products (about 1000 ng) and $TaqI$ restriction enzyme (Fast Digest, Fermentas, Germany) in a final volume of 10 μl. $TaqI$ restriction enzyme was selected from those ($AluI$, $MboI$ and $TaqI$) that were shown to discriminate more AMF groups in a previous study [17] after preliminary testing with a subset of our rhizosphere samples (data not shown). Terminal restriction fragments (TRF) were loaded and separated on a 3130XL genetic analyzer (Applied Biosystems, California, USA) at the SCAI-University of Córdoba sequencing facilities. Size of fragments were determined using a ROX500 size standard, and matrices containing incidence as well as peak area data of individual TRFs were generated for all samples with GeneMapper software (Applied Biosystems). Peaks of less than 100 fluorescence units (FU) and shorter than 50 bp were not included in the analysis to eliminate primer dimmers and other small charged molecules. Similarly, molecules that were not present in at least two of the four replicate profiles were disregarded. Also, TRFs that differed by less than 1 bp were clustered, unless individual peaks were detected in a reproducible manner. TRFs profiles were standardized based on methods described previously by Dunbar et al. [18]. The relative abundance of each TRF was calculated as the ratio of the peak area for that TRF to the sum of peak areas for all TRFs in the profile and was expressed as a percentage. Diversity statistics were calculated from standardized profiles of rhizosphere samples by using the number and area of peaks in each profile as representative of the number and relative abundance of OTUs, as defined by Dunbar et al. [19]. Phylotype richness was calculated as the total number of distinct TRF sizes (with length between 50 and 500 bp) in a profile and the Shannon-Wiener diversity index was calculated as described before [14]. Finally, a single standardized T-RFLP profile for each orchard was produced by averaging peak area for each TRF from four replicates.

Pyrosequencing Analysis

For 454-pyrosequencing SSU rRNA Glomeromycota sequences were amplified from a DNA mixture obtained from the four independent rhizosphere DNA extractions per olive orchard. This approach was taken to ideally cover as much biodiversity as possible and to ensure that representative AMF communities from each olive location were sampled [20]. The pyrosequencing was performed as described in Davison et al. [21] using a two-step PCR protocol with the primers NS31 and AML2, which target a ca. 560-bp central fragment of the SSU rRNA gene in Glomeromycota [22], the most widely used marker in AMF surveys to date [23,24]. These primers were linked to partial sequencing primers A and B, respectively. Bar-code sequences, 8 bp in length, were inserted between the A primer and NS31 primer sequences. Thus, the composite forward primer was: 5'-GTCTCCGACTCAG(NNNNNNNN) *TTGGAGGGCAAGTCTG-GTGCC*-3'; and the reverse primer was 5'-TTGGCAGTCT-CAG*GAACCCAAACACTTTGGTTTCC*-3, where partial sequences of A and B primers are underlined, barcode is indicated by N-s in parentheses and specific primers NS31 and AML2 are shown in italic. Then, a 10x dilution of the first PCR product was used in a second PCR where full sequencing adapters were added (Primer A 5'-CCATCTCATCCCTGCGTGTCTCCGAC-3' and Primer B 5'-CCTATCCCCTGTGTGCCTTGGCAGT -3'). The reactions contained 5 μl of Smart-Taq Hot Red 2x PCR Mix (Tartu, Naxo Ltd, Estonia), 1 μl of extracted DNA, and 0.2 μM of each primer in a final volume of 10 μl. The reactions were performed using a Thermal cycler 2720 (Applied Biosystems) under the following conditions: 95°C for 15 min; five cycles of 42°C for 30 s, 72°C for 90 s, 92°C for 45 s; 35 (first PCR) or 20 (second PCR) cycles of 65°C for 30 s, 72°C for 90 s, 92°C for 45 s; followed by 65°C for 30 s and 72°C for 10 min. PCR products were separated by electrophoresis using 1.5% agarose gels in 0.5 x TBE, and the PCR products were purified from the gel using the Qiagen QIAquick Gel Extraction kit (Qiagen Gmbh, Germany) and further purified with AgencourtH AMPureH XP PCR purification system (Agencourt Bioscience Co., Beverly, MA, USA). The 96 quantified samples were finally mixed at equimolar concentrations prior to sequencing. GATC Biotech (Constanz, Germany) performed sequencing procedures as custom service using a Genome Sequencer FLX System and Titanium Series reagents (Roche Applied Science, Indianapolis, IN, USA). Sequencing of 96 samples was performed as a part of a bigger dataset.

Processing of Pyrosequencing Data and phylogenetic analysis

Pyrosequencing data were processed as described by Fierer et al. [25] using the Quantitative Insights Into Microbial Ecology (QIIME) toolkit [26]. In brief, fungal sequences were quality trimmed, assigned to rhizosphere samples based on their barcodes and denoised using default parameters. Chimeras were identified with uclust_ref software [27] and removed, and the remaining sequences were binned into OTUs using a 97% identity threshold with uclust_ref software. Then, to take into account the different number of sequences obtained for each orchard sample in the pyrosequencing analysis we estimated the relative frequency of each OTU in each orchard. Next, the most abundant sequence from each OTU was selected as a representative sequence for that OTU and deposited in the Genbank database under accessions numbers KF831296-KF831328 and the entire dataset of reads in the Sequence Read Archive of Genbank under BioProject ID PRJNA237741. Alpha diversity statistics including Richness (numbers of OTUs) and the Shannon index were also determined for orchard samples with at least five sequences.

Taxonomy was assigned to OTUs by using the Basic Local Alignment Search Tool (BLAST) for each representative sequence against the Silva 108 database (http://www.arb-silva.de/documentation/release-108/) as well as by BLAST search against the MaarjAM database (http://maarjam.botany.ut.ee/, [23]). Sequences from the representative set of AMF OTUs obtained in this study, the reference AMF database from Redecker and Raab [28], the blast hits from Silva 108 and the MaarjAM databases, and those AMF sequences reported in olive from Calvente et al [7] and present in the GenBank database were aligned using ClustalW software [29] with default parameters. Sequence alignments were manually edited using BioEdit software [30]. Phylogenetic analysis of the sequence data sets was performed based on maximum likelihood (ML) and Bayesian inference (BI) using MrBayes version 3.1.2 software [31]. The best fitted model of DNA evolution was obtained using jModelTest v. 2.1.1 [32] with the Akaike information criterion (AIC). The Akaike-supported model, the base frequency, the proportion of invariable sites, and the gamma distribution shape parameters and substitution rates in the AIC were then used in phylogenetic analyses. BI analysis under a general time reversible of invariable sites and a gamma-shaped distribution (TIM2 +I+G) model for the SSU rRNA, were run with four chains for 1.0×10^6 generations.

Statistical Analysis

The rank-based Kruskall-Wallis test was used to determine differences in the Richness and Shannon diversity indexes in relation to the different agronomic factors of the olive orchard evaluated using the Statistical Analysis System software package (SAS version 9.2; SAS Institute, Cary, NC, USA). Non-metric multidimensional scaling (NMDS) analyses were performed using MetaMDS functions within the vegan package of R software (R Core Development Team, 2005) [33] based on dissimilarities calculated using the Bray–Curtis index obtained for T-RFLP and pyrosequencing results, using 1,000 runs with random starting configurations, and environmental variables (agronomic and climatic characteristics and soil physicochemical properties) were fitted using the envfit routine. For data derived from pyrosequencing analysis only the Glomeromycota sequences were used. Ordinations for the Bray–Curtis dissimilarity derived from relative frequency of OTUs in the pyrosequencing analysis did not reach acceptable [34] stress and stability levels and was not performed. Instead, a Multivariate Regression Tree (MRT) was calculated. MRT are a statistical technique that can be used to explore, describe, and predict relationships between multispecies data and environmental characteristics. MRT forms clusters of sites by repeated splitting of the data, with each split defined by a simple rule based on environmental variables. The splits are chosen to minimize the dissimilarity of sites within clusters [35]. The sums of squares multivariate regression tree was calculated within the mvpart package with the R software, using the one-standard error rule on the cross-validated relative error to determine the number of terminal nodes [35].

Results and Discussion

Diversity of olive AMF communities

i) **T-RFLP analysis.** A total of 36 unique TRFs profiles were consistently identified in the 384 rhizosphere samples analyzed by T-RFLP analysis, with 30 TRFs (83.3%) found in a reproducible manner in 93 of the 96 olive orchards sampled. Mean Richness values ranged from 1 to 15 depending of the rhizosphere sample with an average of 5 TRFs per olive orchard. This translates into an estimated total of 30 different OTUs present across all sampled orchards with 13 and 3 OTUs being common in at least 25% or 50% of olive orchards, respectively. Mean Shannon diversity index values ranged from 0.3 to 2.2 (Figure S1; Table S2). We did not find significant differences (*P*>0.120) in AMF richness or Shannon diversity indexes derived from T-RFLP analysis according to the orchard management system, use of irrigation, presence of vegetative cover, olive tree variety or olive age (Figure S1). It has been shown that although diversity indexes are useful in describing community characteristics, they do not provide information of important compositional features of biodiversity relating to the abundances of shared taxa [36] and statistical analyses that incorporate taxon abundance and identity are more appropriate to specifically assess changes in microbial community composition and to identify the existence, if any, of agronomic or environmental gradients.

ii) **Pyrosequencing analysis.** The pyrosequencing approach of the 96 composite samples yielded a total of 13,772 high-quality reads after denoising, with a length >100 bp and < 550 bp, a mean of 147 sequences per orchard field, and two samples with no sequences (Table S2). Of these, most of sequences 47.2% could not be assigned to any Eukaryota Phyla, 29.6% were assigned to the Phylum Metazoa, whereas 10.2% (1,108 reads) could be assigned to OTUs from fungal families (Figure 1), which indicates that primers NS31 and AML2 are not enough specific for amplifying AMF sequences. Other studies using pyrosequencing analysis have also shown that in spite of the supposed AMF primer specificity, 'contaminant' sequences belonging to taxa different from *Glomeromycota* are detected. For example, Öpik et al. [37] using same NS31 and AML2 primers and Ballestrini et al. [38] and Lumini et al. [20] using NS31 and AMmix primers also found about >55% of amplified sequences belonging to taxa from non-*Glomeromycota* fungi. In a recent work Kohout et al. [39] also demonstrated that a combination of up to five primer sets specifically designed to amplify *Glomeromycota*, including primer NS31/AML2, co-amplifies to a high extend non-target AMF sequences including plant, Asco- and Basidiomycota (ca. 20 to 50%, depending of the primer set used). In our study, due to the fact that we did not retrieve *Glomeromycota* sequences from some olive orchards our results could be somehow biased. Consequently, to improve the number of sequences and molecular species characterization of AMF, new designed specific AMF-primers should be tested in complex matrixes such as soil or rhizosphere [39,40], or deeper sequencing effort should be done in future studies to face this problem and to capture the total AMF diversity in olive.

The fact that we extracted DNA from the olive rhizosphere (i.e., soil tightly adhered to roots) might have accounted for the low presence of AMF sequences; probably extracting DNA from washed or entire roots may enhance the specificity of AMF amplification. In our study 59.84% of sequences assigned to fungi belonged to *Glomeromycota* (Figure 1), with 38 olive orchard samples retrieving no *Glomeromycota* sequences (Table S2). We did not find any pattern for the lack of *Glomeromycota* sequences in those samples with any of the agronomic, climatic or soil physicochemical properties of the sampled orchards (*data not shown*). When defining an OTU as belonging to AMF on the basis of having at least 97% similarity to sequences classified as AMF in the Silva and GenBank databases we identified 33 OTUs that could be unequivocally assigned to the *Glomeromycota* in 58 out of 94 olive rhizosphere

samples (with a mean of 11.4 sequences per orchard) (Figure 1; Table 1; Figure S2; Table S2). Mean richness values ranged from 1 to 9 with an average of 2.7 OTUs per olive orchard. Mean Shannon diversity index values ranged from 0 (1 single OTU) to 2.7 (Figures S1 and S2; Table S2). We did not find significant differences (*P*>0.230) in AMF richness or Shannon diversity indexes derived from pyrosequencing analysis according to the farm soil management system, use of irrigation, presence of vegetative cover, olive tree variety or olive age (Figure S1).

iii) **Comparison of T-RFLP and pyrosequencing analysis.** Although in the pyrosequencing analysis we got some orchard samples with no *Glomeromycota* sequences and this technique is costly, labour-intensive and allows lower number of samples to be processed, it provides some advantages over the T-RFLP analysis. For example, the latter technique does not provide any information on the taxa identified, different taxa (species) may share similar TRFs in electropherograms, and multiple TRFs profiles can exist within a single species [41]. Furthermore, the lack of specifity of the primers used for T-RFLP have also been shown in other studies and may also be a source of errors when PCR products serve as basis for those fingerprinting approaches [39,41]. Consequently, data dervived from T-RFLP analysis should be complemented with other techniques that provide taxa identity information such as library cloning or pyrosequencing as was the case of this study.

Species identity of olive AMF communities

In our study the 33 OTUs identified represented most of the major AMF lineages, including *Paraglomus* spp. (family Paraglomeraceae; two OTUs, comprising 4.20% of reads and 6.90% of fields), *Glomus* group A (family Glomeraceae; 22 OTUs, comprising 59.73% of reads and 81% of fields), *Glomus* group B (family Claroideoglomaceae; four OTUs, 26.40% of reads and 38% of fields), *Glomus* group C (family Diversisporaceae; three OTUs, 6.03% of reads and 14% of fields), and *Archaeospora* (two OTUs, 3.62% and 12% of fields) (Figures 1 and 2; Table 1); with *Archaeospora* spp., *Diversispora* spp. and *Paraglomus* spp. being first records in olive. It should be noted that OTU OAMF127 clustered with the virtual taxon sequence AF131054 that has been recently proposed as a potential new taxon (new family or even order) within Glomeromycota [37].

The fact that most sequences from our study belonged to the *Glomeraceae* family (*Glomus* group A) that contains several cryptic taxa with differences in ecological properties agrees with other studies that have found many isolates of this group in different locations through the world, including Mediterranean-type environments, on both natural woodlands to high input managed agro-ecosystems [20,42,43,44] suggesting that these taxa have a generalist ruderal style with tolerance to disturbance such as in agricultural ecosystems.

Interestingly, we found that those AMF species commonly used as olive inoculants or previously isolated from olive roots (i.e., *Claroideglomus claroideum*, *Funneliformis mosseae*, *Rhizophagus clarus*, *R. intraradices*, and *Septoglomus viscosum*; see [7,8,9,12,13]) showed low abundance since they were present in 1.7 to 18.97% of fields. This is in agreement with the fact that AMF belonging to Glomeraceae family colonize preferentially the roots and might be present in lower densities in the rhizosphere soil [45]. On the contrary, two AMF sequences including OAMF216 belonging to Claroideoglomeraceae (18.6% of sequences), and OAMF91246, a new

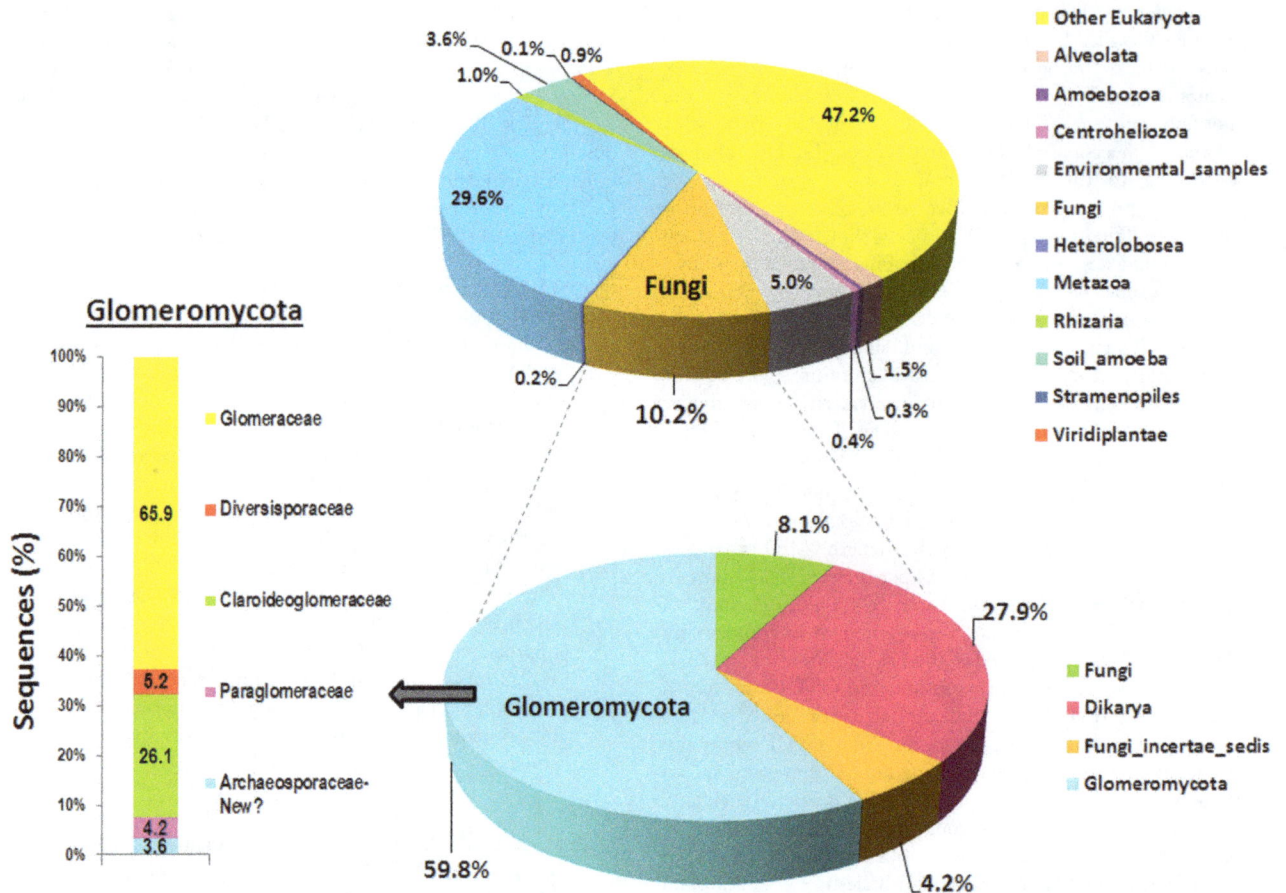

Figure 1. Proportion of overall phyla and disectioning of the fungal and Glomeromycota phyla detected by pyrosequencing analysis with primers NS31/AML2 from rhizosphere samples obtained from 96 olive orchards in Andalusia, southern Spain.

unidentified Glomeraceae (18.9% of sequences) that formed a separate cluster from other well-known AMF Glomeraceae taxa, were identified in 27.6% and 36.21% of olive orchards, respectively (Figure 2; Table 1). This might indicate that AMF can exert higher host specificity in olive than previously thought which may have implications for the olive nursery process. Thus, some authors have reported a differential growth response of olive cultivars to AMF inoculation where this responsiveness to mycorrhization has been found to depend on both the AMF species and the plant genotype [7,46]. In our study, we did not find any clear differences between the AMF sequences detected in the rhizosphere of wild olives and those found in the cultivated ones (Figure S2). This could be due to the small number of sequences that we sampled from wild olives which deserves further studies since wild olives have been shown to be a potential reservoir for discovering microbial species of diverse biotechnological and commercial interest [14,47].

Factors shaping the structure of AMF communities in olive rhizosphere

It has been shown that although diversity indexes (such as Richness and Shannon used in our study) are useful in describing community characteristics they do not provide information of important compositional features of microbial diversity related to the abundances of shared taxa [36] which migth explain that we did not find an effect of the environmental and agronomic

variables on the estimated alpha-diversity indexes. Consequently, in a second approach to specifically assess changes in AMF community composition (incorporating taxon abundance (frequency) and identity), we used NMDS ordination to represent, in two dimensions, the pairwise Bray-Curtis dissimilarities between AMF communities derived from T-RFLP analyses. Then, we projected each of the environmental and agronomic variables independently onto the NMDS ordination to identify hypothetical gradients likely related to the differentiation in AMF composition (Figure 3; Table 2). In relation to agronomic variables AMF communities were differentiated according to the cultivar genotype and age of plantation and the irrigation regimen of the olive orchard, in that order, whereas the grouping according to the orchard management system or presence of a vegetative cover was not significant (Table 2). Thus, there was a tendency to locate rhizosphere samples in the NMDS ordination from olive orchards <15 year old at the bottom quandrant of Y = 0 (with only two exceptions), whereas olive orchards of 15 to 30 year old were all located on the left cuadrant of X = 0 (Figure 3). The effect of olive gentoype in affecting soil biota has also been shown in a recent study [48] which demostrated that olive genotypes significantly influenced the nematode assemblages present in their rhizospheric soil.

We also identified C:N ratio, soil C and organic matter content and pH as the environmental variables better explaining ($P <$ 0.001; $0.2836 > r^2 > 0.1301$) the AMF community composition among the olive orchards, in that order (Figure 3, Table 2). Other

Table 1. Glomeromycota taxa detected in olive rhizosphere samples, taxonomic affiliation, number of sequences from each taxa, frequency of occurrence in the sampled olive orchards, and the closest related AMF sequence.

Family	OTU identification[a]		Acc. Number	Number of sequences	Frequency of sequences (%)	Frequency of orchards (%)	Closest taxa[b]	MaarjAM
	Phylogenetic group	Code					Silva 108	
Archaeosporaceae	Ia	OAMF127	KF831299	23	3.47	10.34	Glomeromycota sp.	Archaeospora sp. VT5
	Ib	OAMF64639	KF831323	1	0.15	1.72	Archaeospora trappei	Archaeospora trappei VT245
Paraglomeraceae	IIa	OAMF131	KF831300	1	4.07	1.72	Fungi	Paraglomus laccatum VT281
	IIb	OAMF60009	KF831322	27	0.15	6.90	Fungi	Paraglomus brasilianum VTX239
Claroideoglomeraceae	IIIa	OAMF26258	KF831310	52	7.84	18.97	Glomus etunicatum	Glomus sp. group B VT193
(Glomus group B)	IIIb	OAMF216	KF831306	121	18.25	27.59	Glomus etunicatum	Glomus geosporum VT65
		OAMF71	KF831324	1	0.15	1.72	Glomus etunicatum	Glomus sp. group B VT56
		OAMF333	KF831315	1	0.15	1.72	Glomus etunicatum	No match
	Total			123	18.55	27.59		
Diversisporaceae	IVa	OAMF79857	KF831327	19	2.87	3.45	Glomus eburneum	Glomus sp. MO-GC1 VT60
(Glomus group C)	IVb	OAMF264	KF831311	20	3.02	12.07	Glomus eburneum	Glomus sp. Wirsel OTU20 VT62
		OAMF156	KF831302	1	0.15	1.72	Glomus eburneum	Diversispora sp. VT62
	Total			21	3.31	13.79		
Glomeraceae	V	OAMF19034	KF831303	16	2.41	1.72	Glomeromycota	Glomus sp. Dictamnus2 VT163
(Glomus group A)		OAMF443	KF831318	11	1.66	5.17	Uncultured Glomus sp.	Glomus sp. Douhan6 VT143
	Total			27	4.07			
	VI	OAMF43	KF831317	23	3.47	5.17	Uncultured Glomus sp.	Glomus sp. VT145
	VIIa	OAMF359	KF831316	1	0.15	1.72	Glomeromycota	Glomus sp. VT153
		OAMF91246	KF831328	124	18.70	36.21	Uncultured Glomus sp.	Glomus sp. VT118
	Total			125	18.85	36.21		
	VIIb	OAMF73	KF831325	59	8.90	15.52	Uncultured Glomus sp.	Glomus Wirsel VT137
	VIIc	OAMF499	KF831320	9	1.36	1.72	Uncultured Glomus sp.	Glomus JP3 VT128
	VIIIa	OAMF213	KF831305	12	1.81	3.45	Glomeromycota	Glomus sp. Alguacil09b VT109
	VIIIb	OAMF452	KF831319	11	1.66	8.62	Uncultured Glomus sp.	Glomus sp. Glo45 VT109
	IXa	OAMF22696	KF831307	14	2.11	6.90	Uncultured Glomus sp.	Glomus sp. Glo24 VT105
		OAMF15	KF831301	2	0.30	3.45	Rhizophagus intraradices	Glomus intraradices VT113
		OAMF22729	KF831308	29	4.37	8.62	Rhizophagus intraradices	Glomus intraradices VT113
	Total			45	6.79	18.97		

Table. 1. Cont.

Family	OTU identification[a]		Acc. Number	Number of sequences	Frequency of sequences (%)	Frequency of orchards (%)	Closest taxa[b]	
	Phylogenetic group	Code					Silva 108	MaarjAM
	IXb	OAMF293	KF831312	8	1.21	6.90	Uncultured Glomus sp.	Glomus sp. VT94
	IXc	OAMF123	KF831297	2	0.30	3.45	Uncultured Glomus sp.	No match
	IXd	OAMF521	KF831321	17	2.56	8.62	Uncultured Glomus sp.	Glomus sp. Glo7 VT214
	Xa	OAMF102182	KF831296	4	0.60	1.72	Glomus sp.	Funneliformis caledonium VT65
	Xb	OAMF77556	KF831326	5	0.75	1.72	Glomus sp.	Funneliformis mosseae VT67
	Xc	OAMF311	KF831314	13	1.96	10.34	Uncultured Glomus sp.	Septoglomus viscosum VT63
	XI	OAMF3	KF831313	1	0.15	1.72	Uncultured Glomus sp.	Glomus sp. VT301
		OAMF20	KF831304	4	0.60	3.45	Uncultured Glomus sp.	Glomus sp. VT301
		Total		5	0.75	3.45		
	XII	OAMF25566	KF831309	30	4.52	18.97	Glomeromycota	Glomus Alguacil09c Glo4 VT166
		OAMF125	KF831298	1	0.15	1.72	Glomeromycota	Glomus Alguacil09c Glo4 VT166
		Total		31	4.68	18.97		

[a]The phylogenetic groups were arbitrarily named according to their position in the Bayesian analysis shown in Figure 2. Each AMF OTU sequence found in the study was assigned a Code (OAMF# S#) where OAMF refers to 'olive arbuscular mycorrhizal fungi' and # to the number assigned to each representative AMF OTUs derived from uclust_ref analysis with the QIIME software.
[b]Closest taxa assigned by BLAST analysis using the Silva 108 or MaarjAM data base. Numerical codes of 'virtual taxa' VT as appear in the MaarjAM database are shown as in Figure 2.

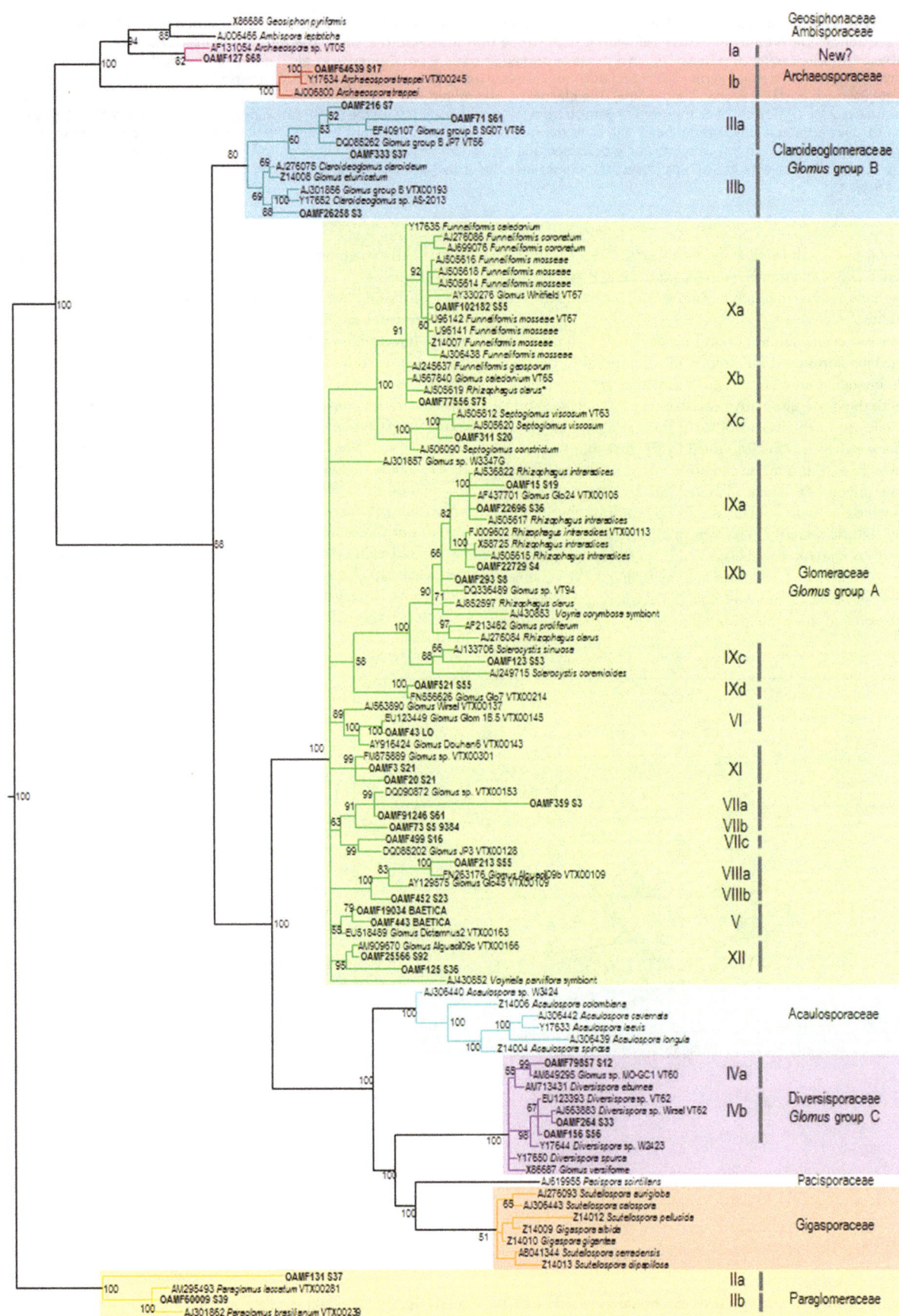

0.2

Figure 2. Phylogenetic relationships of nuclear small subunit ribosomal RNA (SSU rRNA) gene sequences of *Glomeromycota* reference sequences derived from uclust_ref search with those that matched the silva_108 and Maarj*AM* databases, the reference AMF database from Redecker and Raab [28] and those reported in olive from Calvente et al [7] and present in the GenBank. Bayesian 50% majority rule consensus tree as inferred from nSSU rRNA sequences alignments under the general time reversible + G + I model. Numbers on the nodes indicate Bayesian posterior probabilities (>50%). The phylogram was rooted with Paraglomeraceae sequences. Numerical codes in bold name each representative AMF OTUs from olive rhizosphere derived from uclust_ref analysis with the QIIME software and are labelled (OAMF# S#) where OAMF refers to 'olive arbuscular mycorrhizal fungi' and # to the number assigned and groups identified and the remaining code refers to the soil sample. Phylogenetic groups (I to Xd) were arbitrarily described and are shown in Table 1. (*) Although this sequence was originally identified as belonging to *R. clarus* by Calvente et al. [7] its closest taxonomic affiliation is to *Funneliformis* sp. and clearly differs from sequences AJ276084 and AJ852597 of *R. clarus*.

environmental factors showing a significant (P<0.034) but lower effect included clay, sand and N content, extractable P and annual evapotranspitation and minimum temperature of the sampled locatios (Table 2).

A multivariate regression tree was also calculated to summarize the relationships between AMF community composition derived from pyrosequencing analysis and environmental and agronomic variables with the most informative variables in each split shown in Figure 4. The tree explained >30% of the variability in AMF profiles, much of which were accounted by the first split based on exchangeable K and in a lower extend by altitude and sand and clay content (Figure 4). Then, climatic variables from sampled locations including total rainfall and evapotranspiration (ETP) followed by soil pH were the next best predictors for the second-order split. Two climatic variables (altitude and average rainfall), and nutrient contents of soil samples (including OM, C and extractable P and the C:N ratio) allowed differentiating five groups of soils that included three groups of soils showing high richness in

OTUs and two groups of soils characterized by two specific OTUs each (Figure 4).

In our study we were able to identify some agronomic and environmental gradients driving the AMF community differentiation; however, we found some differences when using data derived from T-RFLP or pyrosequencing analysis. This could be due to the unequal sample size included in each data set (93 vs. 58 olive orchards, respectively). Another factor that is likely to have contributed to those differences is the usage of two different rRNA regions and methods, SSU for pyrosequencing and LSU for T-RFLP. These regions differ in their phylogenetic resolutions, and the methods of T-RFLP versus full amplicon sequencing differ as well in this respect as found in previous studies [49]. Consequently, in our study results from the pyrosequencing analysis should be interpreted with caution due to the smaller data set analysed and the possibility of introducing some bias due to the fact that from some olive orchards we did not amplified any *Glomeromycota* sequences. Nevertheless, we retrieved consistent results with both techniques. Thus, soil pH, textural characteristics, nutrient

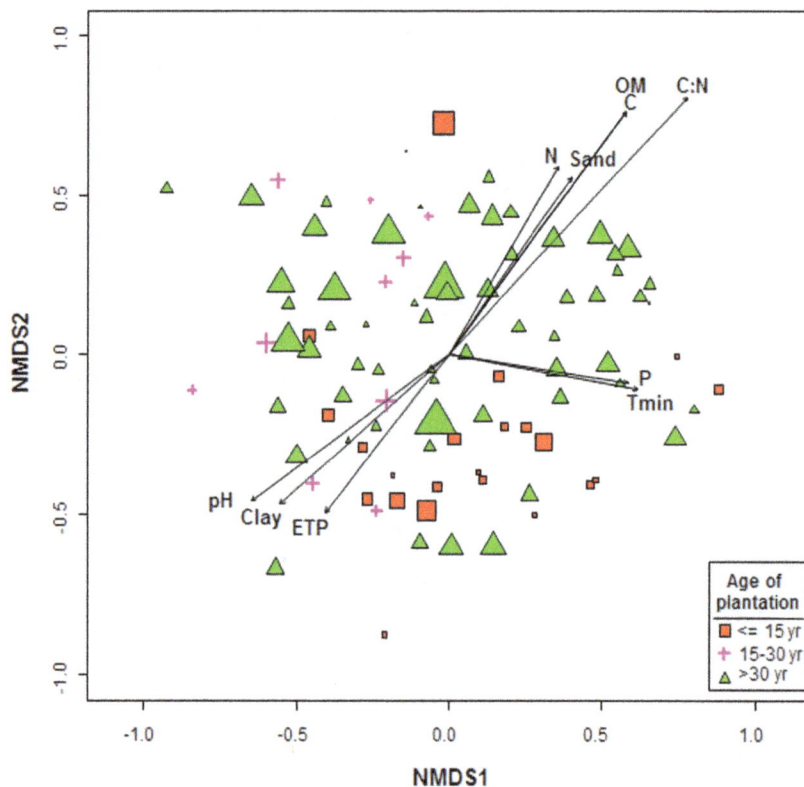

Figure 3. NMDS biplot of a Bray-Curtis dissimilarity matrix of T-RFLP analysis. The fitted vectors of environmental and physicochemical soil variables and the agronomic variable age of plantation (indicated with different symbols) most significantly and strongly associated (P<0.05) with the ordination and shown in Table 2 are also represented. Size of symbols is proportional to AMF richness in those olive orchards.

Table 2. Summary of relationships[a] between agronomic, soil and environmental factors and AMF communities assessed by T-RFLP analysis.

Factors [b]	r^2	P	
Soil physicochemical properties			
Clay (%)	**0.1087**	**0.003996**	**
Sand (%)	**0.0987**	**0.007992**	**
Organic C (%)	**0.1915**	**0.000999**	***
Organic N (%)	0.1006	0.010989	*
Extractable P (ppm)	0.0735	0.02997	*
Exchangeable K (ppm)	0.0315	0.228771	
CEC	0.0005	0.976024	
C:N ratio	**0.2623**	**0.000999**	***
pH(KCl)	**0.1301**	**0.000999**	***
SOM (%)	**0.1914**	**0.000999**	***
Climatic characteristics			
Total Rainfall	0.0183	0.438561	
Average Rainfall	0.0045	0.811189	
ETP	0.0849	0.026973	*
Tmax	0.0486	0.117882	
Tmin	0.0815	0.016983	*
Tmean	0.0412	0.154845	
Altitude	0.0274	0.306693	
Agronomic characteristics			
Olive variety	**0.2836**	**0.000999**	***
Presence of vegetative cover	0.0032	0.738262	
Age of plantation	**0.1233**	**0.001998**	**
Irrigation regimen	0.0401	0.033966	*
Orchard management system	0.002	0.986014	

[a]Correlations with all environmental variables (r^2) were obtained by fitting linear trends to the NMDS ordination obtained with each restriction enzyme and significance (P) was determined by permutation (nperm = 1000).
'***' = $P < 0.001$;
'**' = $P < 0.01$;
'*' = $P < 0.05$. Variables with highest significant weight are shown in bold.
[b]Orchard agronomic and climatic characteristics, and soil physicochemical properties are shown in Table S1 and some of them were reported before [5,14]. Climatic variables were obtained from SigMapa, Geographic Information System from the Spanish Ministry of "Medio Ambiente y Medio Rural y Marino" (http://sig.mapa.es/geoportal/) using ArcGIS 10 (ESRI, Redlands, California, EE.UU.).

contents, and some climatic variables appear as the most important vectors driving the AMF community differentiation in olive. Soil texture has typically not been identified as being of great importance on AMF community composition until recently [43,50]. Landis et al. [51] identified a texture effect in an oak savannah ecosystem, but the effects of texture could not be separated from plant community composition and soil N. A separate study [52] did find closely influences of clay, moisture, and pH in AMF composition in a maize agricultural system in Zimbabwe, along with strong effects of soil organic C and total N. On the other hand, Moebius-Clune et al. [51] studied AMF communities in an assemblage of maize fields across an eastern New York State landscape and found soil textural components as the most strongly related to AMF community differences followed by nutrient concentrations, particularly Mg, whereas soil P or pH, were less important.

These results demonstrate that when there are small differences in soil physicochemical characteristics, the composition of the AMF communities might be similar with an overlapping in the

AMF assemblages among different agronomic or soil use practises [20]. All the data obtained in this work, reinforce the concept that the general AMF assemblage structure and composition in olive might be influenced primarily by soil type and climate and at less extent by host plant features (age, vegetative stages, host genotype) or agricultural practices as it has been shown in other woody crop such as vineyards [20,53]. To our best knowledge this study provides the first evidence of a specific effect of such factors on AMF community composition in olive. Further research using a deeper pyrosequencing effort or more specific primers should be conducted to determine how this specific selection of AMF communities by the different olive varieties may be related to olive resilience to mycorrhization during the olive nursery process or to the successful establishment of those mycorrhized planting stocks when transplanted to soils in the different biogeographical areas (as identified by climatic and soil physicochemical properties) present in southern Spain.

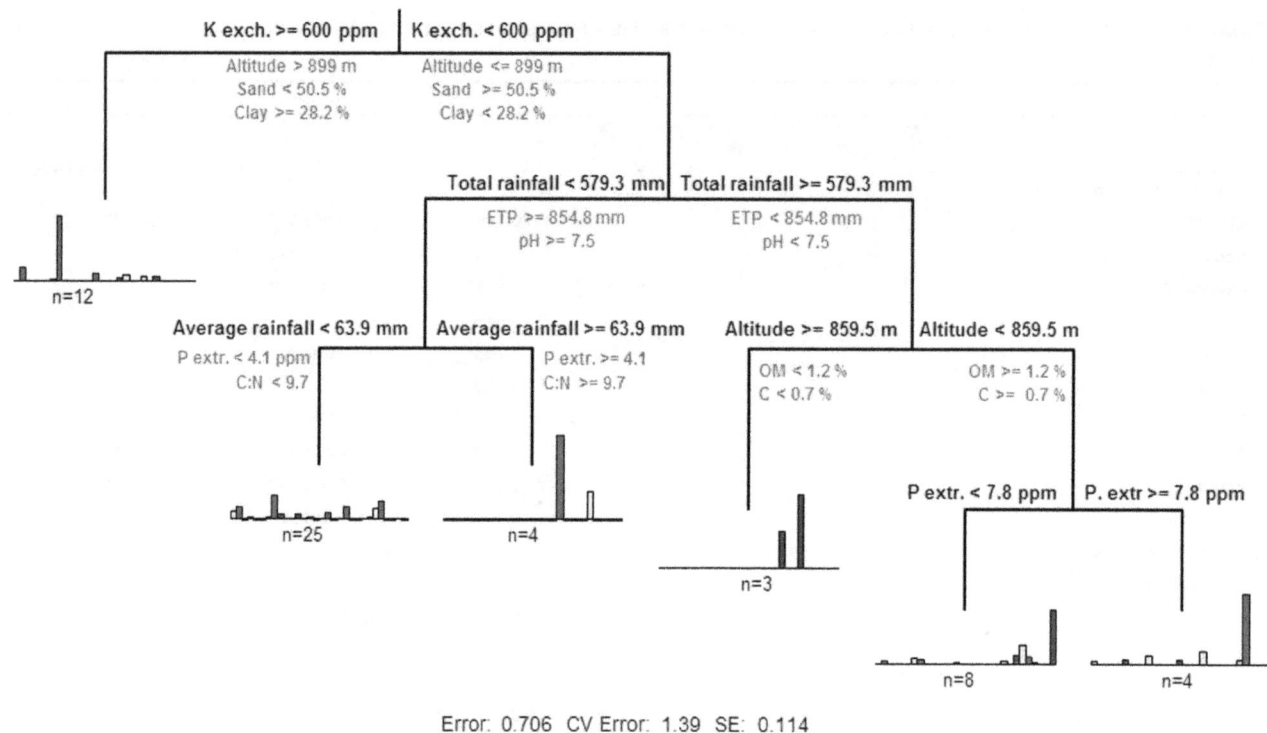

Figure 4. Sums of squares multivariate regression tree summarizing olive AMF community–agronomic, environmental and soil factors relationships. The tree was calculated using frequency of AMF OTUs derived from pyrosequencing analysis (Figure S1). For each split a rule is selected based on the predictors to minimize the dissimilarity within the AMF OTUs profiles in the resulting two nodes (main rule is shown above the node, and second rules are shown below the node). At each terminal node, the mean relative abundances of each AMF OTU are shown, together with the number of olive orchards for each group.

Supporting Information

Figure S1 Summary box-plots of Richness and Shannon diversity indexes derived from T-RFLP (93 olive orchards) and pyrosequencing analysis (43 olive orchards) grouped by the agronomic characteristics of the olive orchards sampled (Table S1).

Figure S2 Frequency of occurrence of the different *Glomeromycota* OTUs detected with primers NS31/AML2 and listed in Table 1 in 56 rhizosphere samples from 96 olive orchards sampled in Andalusia, southern Spain.

Table S1 Datasets, location and characteristics of the olive orchards sampled.

Table S2 Number of sequences and diversity indexes values obtained in the T-RFLP and pyrosequencing analysis in each orchard sampled.

Acknowledgments

We thank C. Cantalapiedra, F. Durán and G. Contreras for technical assistance and P. Castillo and J.A. Navas-Cortés for suggestions to improve the manuscript.

Author Contributions

Conceived and designed the experiments: MMB BBL. Performed the experiments: MMB BBL. Analyzed the data: BBL. Contributed reagents/materials/analysis tools: MM BBL MM. Wrote the paper: MM BBL MM.

References

1. CAP (2011) Estadísticas agrarias [Agricultural statistics]. Consejería de Agricultura y Pesca, Junta de Andalucía, Sevilla, Spain. Available at Web site http://www.juntadeandalucia.es/agriculturaypesca/portal/servicios/estadisticas/estadisticas/agrarias/index.html (verified October 21, 2013).
2. IOC (2011) World Olive Oil Figures. International Olive Council. Available at Web site http://www.internationaloliveoil.org/estaticos/view/131-world-olive-oil-figures (verified October 21, 2013).
3. Soriano MA, Álvarez S, Landa BB, Gómez JA (2013) Soil properties in organic olive orchards following different weed management in a rolling landscape of Andalusia, Spain. Renew Agr Food Syst: in press, doi:10.1017/S1742170512000361.
4. Milgroom J, Soriano MA, Garrido JM, Gómez JA, Fereres E (2007) The influence of a shift from conventional to organic olive farming on soil management and erosion risk in southern Spain. Renew Agr Food Syst 22: 1–10.

5. Montes-Borrego M, Navas-Cortés JA, Landa BB (2013) Linking microbial functional diversity of olive rhizosphere soil to management systems in commercial orchards in southern Spain. Agric Ecosyst Environ 181: 169–178.
6. Binet MN, Lemoine MC, Martin C, Chambon C, Gianinazzi S (2007) Micropropagation of olive (*Olea europaea* L.) and application of mycorrhiza to improve plantlet establishment. In Vitro Cell Dev Biol Plant 43: 473–478.
7. Calvente R, Cano C, Ferrol N, Azcon-Aguilar C, Barea JM (2004) Analysing natural diversity of arbuscular mycorrhizal fungi in olive tree (*Olea europaea* L.) plantations and assessment of the effectiveness of native fungal isolates as inoculants for commercial cultivars of olive plantlets. Appl Soil Ecol 26: 11–9.
8. Estaún V, Camprubi A, Calvet C, Pinochet J (2003) Nursery and field response of olive trees inoculated with two arbuscular mycorrhizal fungi, *Glomus intraradices* and *Glomus mosseae*. J Am Soc Horticult Sci 128:767–775.

9. Castillo P, Nico AI, Azcón-Aguilar C, Del Río Rincón C, Calvet C, et al. (2006). Protection of olive planting stocks against parasitism of root-knot nematodes by arbuscular mycorrhizal fungi. Plant Pathol 55: 705–713.

10. Castillo P, Nico A, Navas-Cortés JA, Landa BB, Jiménez-Díaz RM, et al. (2010) Plant-parasitic nematodes attacking olive trees and their management. Plant Dis 94: 148–162.

11. Dag A, Yermiyahu U, Ben-Gal A, Zipori I, Kapulnik Y (2009) Nursery and post-transplant field response of olive trees to arbuscular mycorrhizal fungi in an arid region. Crop Pasture Sci 60: 427–433.

12. Meddad-Hamza A, Beddiar A, Gollotte A, Lemoine MC, Kuszala C, et al. (2010) Arbuscular mycorrhizal fungi improve the growth of olive trees and their resistance to transplantation stress. Afr J Biotechnol 9: 1159–1167.

13. Porras-Soriano A, Soriano-Martín ML, Porras-Piedra A, Azcón R (2009). Arbuscular mycorrhizal fungi increased growth, nutrient uptake and tolerance to salinity in olive trees under nursery conditions. J Plant Physiol 166:1350–1359.

14. Aranda S, Montes-Borrego M, Jiménez-Díaz RM, Landa BB (2011) Microbial communities associated with the root system of wild olives (*Olea europaea* L. subsp. *europaea* var. *sylvestris*) are good reservoirs of bacteria with antagonistic potential against *Verticillium dahliae*. Plant Soil 343: 329–345.

15. Van Tuinen D, Jacquot E, Zhao B, Gollotte A, Gianinazzi-Pearson V (1998) Characterization of root colonization profiles by a microcosm community of arbuscular mycorrhizal fungi using 25S rDNA-targeted nested PCR. Mol Ecol 7: 879–887.

16. Gollotte A, van Tuinen D, Atkinson D (2004) Diversity of arbuscular mycorrhizal fungi colonizing roots of the grass species *Agrostis capillaries* and *Lolium perenne* in a field experiment. Mycorrhiza 14: 111–117.

17. Mummey DL, Rillig MC (2007) Evaluation of LSU rRNA-gene PCR primers for analysis of arbuscular mycorrhizal fungal communities via terminal restriction fragment length polymorphism analysis. J Microbiol Methods 70: 200–204.

18. Dunbar J, Ticknor LO, Kuske CR (2001) Phylogenetic specificity and reproducibility and new method for analysis of terminal restriction fragment profiles of 16S rRNA genes from bacterial communities. Appl Environ Microbiol 67: 190–197

19. Dunbar J, Ticknor LO, Kuske CR (2000) Assessment of microbial diversity in four southwestern United States soils by 16S rRNA gene terminal restriction fragment analysis. Appl Environ Microbiol 66: 2943–2950.

20. Lumini E, Orgiazzi A, Borriello R, Bonfante P, Bianciotto V (2010) Disclosing arbuscular mycorrhizal fungal biodiversity in soil through a land-use gradient using a pyrosequencing approach. Environ Microbiol 12: 2165–2179.

21. Davison J, Öpik M, Zobel M, Vasar M, Metsis M, et al. (2012) Communities of arbuscular mycorrhizal fungi detecting in forest soil are spatially heterogeneous but do not vary throughout the growing season. PLoS One 7:e41938.

22. Lee J, Lee S, Young JPW (2008) Improved PCR primers for the detection and identification of arbuscular mycorrhizal fungi. FEMS Microbiol Ecol 65: 339–349.

23. Öpik M, Vanatoa A, Vanatoa E, Moora M, Davison J, et al. (2010) The online database MaarjAM reveals global and ecosystemic distribution patterns in arbuscular mycorrhizal fungi (Glomeromycota). New Phytol 188:223–241.

24. Kivlin SN, Hawkes CV, Treseder KK (2011) Global diversity and distribution of arbuscular mycorrhizal fungi. Soil Biol Biochem 43: 2294–2303.

25. Fierer N, Hamady M, Lauber CL, Knight R (2008) The influence of sex, handedness, and washing on the diversity of hand surface bacteria. Proc Natl Acad Sci USA 105: 17994–17999.

26. Caporaso JG, Kuczynski J, Stombaugh J, Bittinger K, Bushman FD, et al. (2010). QIIME allows analysis of high-throughput community sequencing data. Nature Methods 7, 335 – 336.

27. Edgar RC (2010) Search and clustering orders of magnitude faster than BLAST, Bioinformatics 26(19): 2460-2461. doi: 10.1093/bioinformatics/btq461.

28. Redecker D, Raab P (2006) Phylogeny of the Glomeromycota (arbuscular mycorrhizal fungi): recent developments and new gene markers. Mycologia 98: 885–895.

29. Thompson JD, Gibson TJ, Plewniak F, Jeanmougin F, Higgins DG (1997) The CLUSTAL_X windows interface: flexible strategies for multiple sequence alignment aided by quality analysis tools. Nucleic Acids Res 25: 4876–4882.

30. Hall TA (1999). BioEdit: a user-friendly biological sequence alignment editor and analysis program for windows 95/98/NT. Nucleic Acids Symp Ser 41: 95–98.

31. Huelsenbeck JP, Ronquist F (2001) MrBAYES: Bayesian inference of phylogenetic trees. Bioinformatics 17: 754–755.

32. Darriba D, Taboada GL, Doallo R, Posada D (2012) jModelTest 2: more models, new heuristics and parallel computing. Nat Methods 9: 772.

33. Oksanen J, Blanchet FG, Kindt R, Legendre P, O'Hara RG, et al. (2011) Vegan: community ecology package. R package version 1.17-6. Available at: http://CRAN.R-project.org/package = vegan.; accessed 08/06/2013.

34. Clarke KR (1993) Non-parametric multivariate analyses of changes in community structure. Aust J Ecol 18: 117–143.

35. De'ath G (2002) Multivariate regression trees: a new technique for modeling species – environment relationships. Ecology 83: 1105–1117.

36. Griffiths RI, Thomson BC, James P, Bell T, Bailey MJ, et al. (2011) The bacterial biogeography of British soils. Environ Microbiol 13: 1642–1654.

37. Öpik M, Davison J, Moora M, Zobel M (2013) DNA-based detection and identification of Glomeromycota: the virtual taxonomy of environmental sequences. Botany 10.1139/cjb-2013-0110.

38. Balestrini R, Magurno F, Walker C, Lumini E, Bianciotto V (2013) Cohorts of arbuscular mycorrhizal fungi (AMF) in *Vitis vinifera*, a typical Mediterranean fruit crop. Environ Microbiol Rep 2: 594–604.

39. Kohout P, Sudováa R, Janoušková M, Čtvrtlíkovác M, Hejdaa M, et al. (2014) Comparison of commonly used primer sets for evaluating arbuscular mycorrhizal fungal communities: Is there a universal solution?. Soil Biol Biochem 68: 482–493.

40. Krüger M, Stockinger H, Krüger C, Schüßler A (2009) DNA-based species level detection of Glomeromycota: one PCR primer set for all arbuscular mycorrhizal fungi. New Phytol 183: 212–223.

41. Dickie IA, FitzJohn RG (2007) Using terminal restriction fragment length polymorphism (T-RFLP) to identify mycorrhizal fungi: a methods review. Mycorrhiza 17:259–270.42.

42. Öpik M, Zobel M, Cantero JJ, Davison J, Facelli JM, et al. (2013) Global sampling of plant roots expands the described molecular diversity of arbuscular mycorrhizal fungi. Mycorrhiza 23:411–430.

43. Oehl F, Laczko E, Bogenrieder A, Stahr K, Bosch R, et al. (2010) Soil type and land use intensity determine the composition of arbuscular mycorrhizal fungal communities. Soil Biol Biochem 42: 724–738.

44. Öpik M, Moora M, Liira J, Zobel M (2006) Composition of root-colonizing arbuscular mycorrhizal fungal communities in different ecosystems around the globe. J Ecol 94: 778–790.

45. Hart MH, Reader RJ (2002) Taxonomic basis for variation in the colonization strategy of arbuscular mycorrhizal fungi. New Phytol 153:335–344.

46. Estaún V, Calvet C, Campubrí A (2010) Effect of Differences among crop species and cultivars on the arbuscular mycorrhizal symbiosis. In: Arbuscular Mycorrhizas: Physiology and Function. Hinanit Koltai and Yoram Kapulnik, Editors. Springer pp. 279–295.

47. Aranda S, Montes-Borrego M, Landa BB (2011) Purple-pigmented violacein-producing *Duganella* spp. inhabit the rhizosphere of wild and cultivated olives in Southern Spain. Microb Ecol 62:446–459.

48. Palomares-Rius JE, Castillo P, Montes-Borrego M, Müller H, Landa BB (2012) Nematode community populations in the rhizosphere of cultivated olive differs according to the plant genotype. Soil Biol Biochem 45:168–171.

49. Verbruggen E, Kuramae EE, Hillekens R, de Hollander M, Kiers ET, et al. (2012). Testing potential effects of maize expressing the Bacillus thuringiensis Cry1Ab endotoxin (Bt maize) on mycorrhizal fungal communities via DNA- and RNA-based pyrosequencing and molecular fingerprinting. Appl Environ Microbiol 78: 7384–7392.

50. Moebius-Clune DJ, Moebius-Clune BN, van Es HM, Pawlowska TE (2013) Arbuscular mycorrhizal fungi associated with a single agronomic plant host across the landscape: Community differentiation along a soil textural gradient. Soil Biol Biochem 64: 191–199.

51. Landis FC, Gargas A, Givnish TJ (2004) Relationships among arbuscular mycorrhizal fungi, vascular plants and environmental conditions in oak savannas. New Phytol 164: 493–504.

52. Lekberg Y, Koide RT, Rohr JR, Aldrich-Wolfe L, Morton JB (2007) Role of niche restrictions and dispersal in the composition of arbuscular mycorrhizal fungal communities. J Ecol 95: 95–105.

53. Schreiner RP, Mihara K (2009) The diversity of arbuscular mycorrhizal fungi amplified from grapevine roots (*Vitis vinifera* L.) in Oregon vineyards is seasonally stable and influenced by soil and vine age. Mycologia 101: 599–611.

18

Linking Annual N$_2$O Emission in Organic Soils to Mineral Nitrogen Input as Estimated by Heterotrophic Respiration and Soil C/N Ratio

Zhijian Mu[1]*, Aiying Huang[2], Jiupai Ni[3], Deti Xie[3]

1 Chongqing Key Laboratory of Soil Multi-scale Interfacial Processes, College of Resources & Environment, Southwest University, Chongqing, China, **2** College of Agronomy & Biotechnology, Southwest University, Chongqing, China, **3** Chongqing Engineering Research Center for Agricultural Non-point Source Pollution Control in Three -Gorges Region, College of Resources & Environment, Southwest University, Chongqing, China

Abstract

Organic soils are an important source of N$_2$O, but global estimates of these fluxes remain uncertain because measurements are sparse. We tested the hypothesis that N$_2$O fluxes can be predicted from estimates of mineral nitrogen input, calculated from readily-available measurements of CO$_2$ flux and soil C/N ratio. From studies of organic soils throughout the world, we compiled a data set of annual CO$_2$ and N$_2$O fluxes which were measured concurrently. The input of soil mineral nitrogen in these studies was estimated from applied fertilizer nitrogen and organic nitrogen mineralization. The latter was calculated by dividing the rate of soil heterotrophic respiration by soil C/N ratio. This index of mineral nitrogen input explained up to 69% of the overall variability of N$_2$O fluxes, whereas CO$_2$ flux or soil C/N ratio alone explained only 49% and 36% of the variability, respectively. Including water table level in the model, along with mineral nitrogen input, further improved the model with the explanatory proportion of variability in N$_2$O flux increasing to 75%. Unlike grassland or cropland soils, forest soils were evidently nitrogen-limited, so water table level had no significant effect on N$_2$O flux. Our proposed approach, which uses the product of soil-derived CO$_2$ flux and the inverse of soil C/N ratio as a proxy for nitrogen mineralization, shows promise for estimating regional or global N$_2$O fluxes from organic soils, although some further enhancements may be warranted.

Editor: Shuijin Hu, North Carolina State University, United States of America

Funding: This research was supported by the National Natural Science Foundation of China (grant number 41371211) and the National Major Science and Technology Projects for Water Pollution Control and Management (grant number 2012ZX07104-003). The funders had no role in study design, data collection and analysis, decision to publish, or preparation of the manuscript.

Competing Interests: The authors have declared that no competing interests exist.

* E-mail: muzj01@gmail.com

Introduction

Although organic soils occupy only 3% of the Earth's land area, they contain approximately 40% (610 Pg) of the terrestrial soil organic carbon (SOC) [1]. Climate warming and human disturbance such as drainage and cultivation are expected to accelerate carbon decomposition in organic soils, and the decomposition of SOC can facilitate the release of mineral nitrogen which can then be utilized by denitrifying and nitrifying bacteria to produce the potent greenhouse gas N$_2$O [2,3]. N$_2$O emissions from organic soils under agricultural use in Nordic countries were on average four times higher than those from mineral soils, indicating that N$_2$O derived from SOC decomposition dominates overall fluxes [4]. However, no consistent and quantitative relationship has been reported for N$_2$O emission and organic carbon decomposition in organic soils.

Organic carbon and nitrogen in soils, plant and microbial biomass are usually covalently bonded at relatively constant ratios. It is thus logical to expect that N$_2$O and CO$_2$ originated from SOC decomposition should be closely linked. Some studies have indeed found a significant relationship between soil N$_2$O and CO$_2$ emissions at the site level [5,6]. This relationship, however, was weaker when data were pooled across sites or ecosystems[7,8]. The

variability of soil C/N ratio may be one of the important factors undermining the correlation for organic soils. The C/N ratio in organic soils ranges from 50~100 in weakly decomposed peat to 12~35 in highly decomposed peat [9]. The supply of mineral nitrogen from SOC decomposition is the outcome of two concurrent and oppositely directed microbial processes – nitrogen mineralization and immobilization [10]. Soils with a high C/N ratio may be characterized by rapid immobilization of nitrogen and soils with a low C/N ratio by higher net nitrogen mineralization and a surplus of available NH$_4^+$ and NO$_3^-$ [11]. A negative relationship has accordingly been shown for C/N ratio of soils and N$_2$O fluxes [9]. Similar to the relationship between N$_2$O and CO$_2$ emissions, the correlation of N$_2$O emission with soil C/N ratio tended to be weak when the data from different sites at larger scales were included [4,12], which makes it difficult to scale up N$_2$O fluxes by CO$_2$ emissions or C/N ratio alone from individual sites to regional scales. In view of the coupling of soil carbon and nitrogen processes and the bridging function of C/N ratio, we hypothesized that a combination of soil CO$_2$ emission and C/N ratio would likely provide better measurements of N$_2$O emission at larger scales. In fact, Mu et al. [13] have linked N$_2$O flux to soil mineral nitrogen as estimated by CO$_2$ emission and C/N ratio for agricultural mineral soils. To our knowledge, no such kind of attempt has ever been made for organic soils. The aim of this study was therefore to determine: 1) if N_2O flux from organic soils is related to soil mineral nitrogen input estimated from heterotrophic respiration divided by soil C/N ratio (a derived measure of soil nitrogen mineralization) plus fertilizer nitrogen; and 2) whether or not the relationship is sufficiently robust to serve as an approach for estimating N_2O flux from organic soils.

Materials and Methods

Data source

To test the hypothesis, we collected journal-published data of N_2O and CO_2 emissions measured simultaneously in the fields on peatlands or histosols for which the carbon and nitrogen content or ratio of the organic matter in the upper layers of the soil has been reported. Occasional and short-period flux measurements were not used and only data on annual emissions were considered. For long-term measurements, we used annual estimates rather than multi-year averages to reflect temporal variability. Annual emissions were directly reported by authors or estimated from points in the figures of publications. The final dataset comprised of 122 field measurements from 28 geographical sites (Table S1). Of all data, only 12 measurements at 9 sites were from the tropical regions and the rest were from the temperate regions. Most of the flux measurements were made using closed chamber technique with sampling frequency varying from 1–3 times per week to once per month. Other factors such as soil pH and water table level, if reported, were also recorded in the database. Readers should refer to the original papers for a more complete presentation of the data.

Estimation of soil mineral nitrogen input

The CO_2 emission measured in bare soils can be taken as the proxy of SOC decomposition or heterotrophic respiration [14]. There are limited studies in which CO_2 emission was measured in bare soils (Table S1). For the CO_2 emissions measured in soils with plants, the contribution of heterotrophic respiration or SOC decomposition was estimated using the following equation adapted from Bond-Lamberty and Thomson [15]:

$$R_h = 10e^{[0.22 + 0.87\ln(R_t/10)]} \qquad (1)$$

where R_h is heterotrophic respiration and R_t is total soil respiration (kg C ha^{-1} yr^{-1}).

The nitrogen mineralization rate from soil organic matter was then calculated using the following equation [13]:

$$N_m = R_h/S_{CN} \qquad (2)$$

where N_m is the gross nitrogen mineralization (kg N ha^{-1} yr^{-1}) and S_{CN} is soil C/N ratio.

The mineralized nitrogen from soil organic matter decomposition and the inorganic nitrogen from chemical fertilizers constitute the total input of soil mineral nitrogen (N_{mf}). Atmospheric nitrogen deposition, as another important external source of soil mineral nitrogen, was not considered for our study since there were few papers reporting it.

Statistical analysis

The dataset in the current study is of unbalanced nature with observations collected from peer-reviewed papers rather than from systematically designed experiments. Accordingly, the effects of soil mineral nitrogen input and other variables on N_2O flux were analyzed using the mixed model-REML estimation method of SAS/MIXED procedure (version 9.3), which is suitable for handling unbalanced data. The values of N_2O flux were first natural-log transformed to normalize their distribution and then analyzed by the following model:

$$\ln(f_{N2O}) = \text{constant} + \ln(N_{mf}) + pH + WT + NS_i + E\, cosys_j$$
$$+ NS_i \times \ln(N_{mf}) + E\, cosys_j \times \ln(N_{mf}) + E\, cos\, ys_j \times WT$$

where f_{N2O} is the N_2O flux; N_{mf}, pH, WT, NS_i and $Ecosys_j$ are the fixed effects of mineral nitrogen input, soil pH, water table level, nitrogen source (i is mineralized nitrogen only or a combination of mineralized nitrogen and inorganic nitrogen from chemical fertilizers), and ecosystem type (j is forest or non-forest type), respectively. A preliminary check of the data showed that the general trend of N_2O flux in forest system differed from grass and cropland, so the ecosystems were simply classified into two subclasses as forest and non-forest. Some two-factor interactions were also included in the model. A significant level of $p = 0.05$ was used to determine if a given variable or interactive effect was kept in the model to further seek solutions for fixed effects. Four negative values of N_2O flux reported by Inubushi et al. [16] and Mojeremane et al. [17] can not be subjected to log-transformation and were not included in the analysis. In addition to determination coefficient (i.e., R^2 value), concordance between observed N_2O fluxes and model fits was also analyzed using Lin's concordance correlation coefficient (CCC, Stata SE 12.0) to assess the goodness-of-fit of the finalized models. The resulting CCC was interpreted using the benchmarks described by Klevens et al. [18] as follows: <0.20 is considered virtually no agreement; 0.21–0.40 is considered slight; 0.41–0.60 is considered fair; 0.61–0.80 is considered moderate; and 0.81–0.99 is substantial.

Results

As shown in Table 1, soil pH, soil mineral nitrogen source (NS) and ecosystem type did not affect the annual N_2O flux ($p>0.05$), while the input of soil mineral nitrogen (N_{mf}) and water table level (WT) had significant effects on N_2O flux ($p<0.01$). The F value of N_{mf} was the biggest, indicating the input of soil mineral nitrogen was the main factor controlling N_2O emission in organic soils. The two-factor interactive effects between NS, N_{mf}, WT and ecosystem type on N_2O flux were not statistically significant ($p>0.05$).

Only the significant variables were then kept in the model to solve the estimates for their effects. Two models with different combinations of independent variables are shown in Table 2. The first model was the simplest one with N_{mf} as the single independent variable. The second model was expanded by adding the effect of water table level. The 95% confidence intervals of the estimated effect of N_{mf} were overlapped for different models. The models indicated that N_2O flux was positively correlated with N_{mf} and negatively with water table level. Using the estimated effects and the variables in the dataset allowed a comparison between predicted and observed annual N_2O fluxes from organic soils. The variable N_{mf} explained up to 69% of the variability in the overall data of observed N_2O fluxes (Fig. 1), while the addition of water table level increased the explanatory ability to 75% (Fig. 2). When the overall data were further divided by ecosystem types, the performance of models was somewhat different (Fig. 1 & 2). For forest, the determination coefficient (R^2) was nearly stable at the value of 0.63 for both models. In contrast, the introduction of water table level into models slightly improved the fitted results for

Table 1. Results of type III tests of fixed effects.

Effect	Numerator DF	Denominator DF	F Value	Pr>F
N_{mf}	1	96	13.16	0.0005
pH	1	96	1.43	0.2344
WT	1	96	5.15	0.0255
NS	1	96	0.10	0.7472
Ecosystem	1	96	0.70	0.4040
NS×N_{mf}	1	96	0.11	0.7426
WT×N_{mf}	1	96	3.20	0.0767
Ecosystem×N_{mf}	1	96	0.21	0.6506
Ecosystem×WT	1	96	2.17	0.1437

N_{mf}, the mineral nitrogen input to soil; WT, water table level; NS, the source of soil mineral nitrogen.

non-forest systems with R^2 values increasing from 0.59 to 0.69. This indicated that the input of mineral nitrogen was the most important predictor of N_2O flux, while water table level was a weak predictor of N_2O flux and appeared to be dependent on ecosystem type.

The slope of regression lines in Fig. 1 & 2 ranged from 0.50 to 0.75, indicating that the relationship strays from the ideal 1:1 line. Therefore the concordance correlation coefficient (CCC) between observed and predicted N_2O fluxes was calculated to measure robustness of the models. For the overall data with log-transformation, the concordance was substantial with the CCC ranging from 0.82 to 0.86 for the two models. When the log-transformed data were converted to actual N_2O fluxes, however, the cluster of fluxes greater than 15.0 kg N ha^{-1} yr^{-1} was found to be distinctly underestimated. The CCC for this cluster of data

ranged from -0.002 to 0.16 and showed virtually no agreement, suggesting that some important factors responsible for these high fluxes were not accounted for by the models. For the rest of the data (103 fluxes out of 118), the CCC (ranging from 0.63 to 0.68) still showed a moderate concordance.

The variable N_{mf} in the models can be decomposed into soil heterotrophic respiration (R_h), C/N ratio and inorganic nitrogen rate from chemical fertilizer (N_f). The mixed procedure analysis indicated that each of these components of N_{mf} had a significant influence on N_2O flux ($p<0.001$), with R_h and N_f being positively related to N_2O flux and C/N ratio negatively related to N_2O flux. Soil carbon and nitrogen contents, which could replace the variable of C/N ratio, were also significantly negatively or positively correlated with N_2O flux ($p<0.001$). The fitting efficiency between observed and predicted N_2O fluxes by models

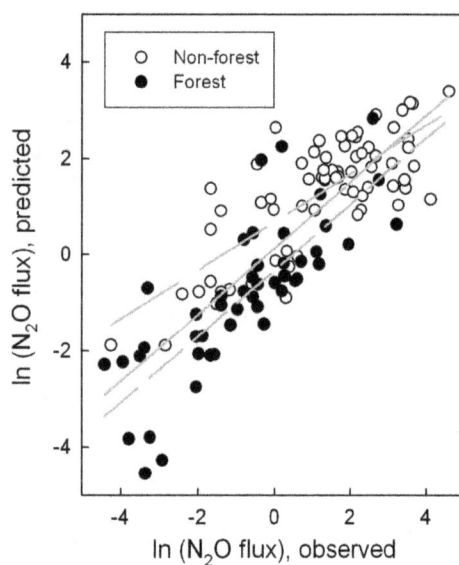

Figure 1. Correlation of observed fluxes of N_2O from organic soils and predicted values (kg N_2O-N ha^{-1}) by model 1 as presented in Table 2: ln(N_2O flux)=1.8685 ln(N_{mf})$-$9.0314. Solid line shows linear regression fit for the overall data: y=0.69x+0.13, R^2=0.69. Long-dashed line shows linear regression fit for the non-forest system: y=0.50x+0.68, R^2=0.59. Short-dashed line shows linear regression fit for the forest system: y=0.69x$-$0.33, R^2=0.63.

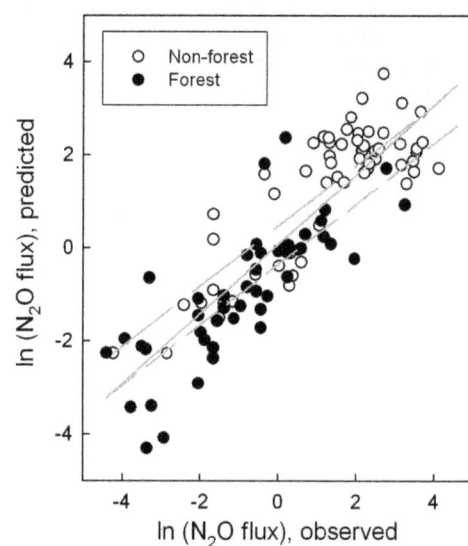

Figure 2. Correlation of observed fluxes of N_2O from organic soils and predicted values (kg N_2O-N ha^{-1}) by model 2 as presented in Table 2: ln(N_2O flux)=1.5374 ln(N_{mf})$-$0.0221 WT$-$8.2334. Solid line shows linear regression fit for the overall data: y=0.75x+0.08, R^2=0.75. Long-dashed line shows linear regression fit for the non-forest system: y=0.65x+0.48, R^2=0.69. Short-dashed line shows linear regression fit for the forest system: y=0.65x$-$0.35, R^2=0.63.

Table 2. Solutions for fixed effects of the models with log-transformed N_2O flux as dependent variable.

Model	Effect	Estimate	SE	DF	t Value	Pr>|t|	95% confidence	
							Lower	Upper
1	Intercept	−9.0314	0.5952	116	−15.17	<.0001	−10.2102	−7.8526
	ln (N_{mf})	1.8685	0.1157	116	16.15	<.0001	1.6393	2.0976
2	Intercept	−8.2334	0.6222	103	−13.23	<.0001	−9.4675	−6.9994
	ln (N_{mf})	1.5374	0.1479	103	10.39	<.0001	1.2441	1.8308
	WT	−0.0221	0.0053	103	−4.14	<.0001	−0.0326	−0.0115

N_{mf}, the mineral nitrogen input to soil (kg N ha^{-1} yr^{-1}); WT, water table level (cm).

using the above-mentioned components of N_{mf} as inputs were nearly the same as those of models using N_{mf} itself (data not shown).

Discussion

Previous studies have linked N_2O flux directly to either CO_2 flux or soil C/N ratio [5,8,9]. In this study, soil CO_2 emission and C/N ratio were combined to estimate mineral nitrogen input, and the latter accounted for up to 69% of the variability of N_2O fluxes from organic soils with various properties, land management practices and climates. Soil CO_2 flux or C/N ratio alone explained only 49% and 36% of the overall variability of N_2O fluxes, respectively (Fig. 3). This suggests the necessity of combining soil CO_2 flux and C/N ratio for predicting N_2O flux on a large scale. Of course, soil CO_2 flux and C/N ratio can be independently incorporated into the same models, but the interpretation of such models would be relatively complicated and evasive since there are various mechanisms which may explain the control of CO_2 flux and C/N ratio over N_2O flux [8,9,19]. In contrast, the quotient of soil CO_2 flux and C/N ratio can well represent in theory the gross nitrogen mineralization [20], and the implication of models using such a quotient as input is straightforward and self-evident in the importance of mineral nitrogen input for regulating soil N_2O flux. There is no significant difference in the influence of different sources of mineral nitrogen on N_2O flux (Table 1), suggesting that the simplified models might also be suitable for evaluating the effect of mineral nitrogen from other sources such as atmospheric deposition, though this idea needs further verification.

A negative relationship between N_2O flux and groundwater level has been observed for individual sites [21,22], and still holds at a large scale as shown in this study. This is logical simply because high moisture with increasing water table level can limit N_2O emission from soils due to the low availability of nitrate and/or efficient reduction of N_2O to N_2 through denitrification [16,23], while the lowering of water table increases oxygen penetration into the peat and enhances the decomposition of organic matter, as indicated by the negative relationship between heterotrophic respiration and water table level ($R^2 = 0.31$, $p<0.0001$). It has been reported that the control of soil water content or water table level over N_2O flux is important only when soil is not nitrogen limiting [24,25]. In this study, the percentage of observations with N_{mf} greater than 150 kg N ha^{-1} was only 19% for forest, but up to 87% for non-forest systems (Table S1). This suggests that forest soil is nitrogen limiting when compared with non-forest systems, which may be responsible for the insensitivity of N_2O flux to water table level for forest systems (Fig.1& 2). Besides the input of mineral nitrogen, forest differs from non-forest systems in many other factors, such as vegetation, below-/above-ground biomass, litter fall, soil compaction, and land management practices, all of which can influence N_2O flux but are not considered here due to limited and unsystematic information in literature sources of the current dataset. To fill the gap, ecosystem type was used as a proxy variable that we tried to incorporate into models; however, statistical analysis showed that its effect was not significant (Table 1).

It should be acknowledged that the models described here were dependent on simplifying assumptions that can introduce error. That is, the gross nitrogen mineralization was estimated from carbon mineralization and soil C/N ratio by assuming that the rate of carbon mineralization is the same as the rate of respiration and the C/N ratio of mineralized organic matter is the same as that of the bulk soil organic matter. In fact, carbon and nitrogen mineralization from soils originates from decomposable fractions

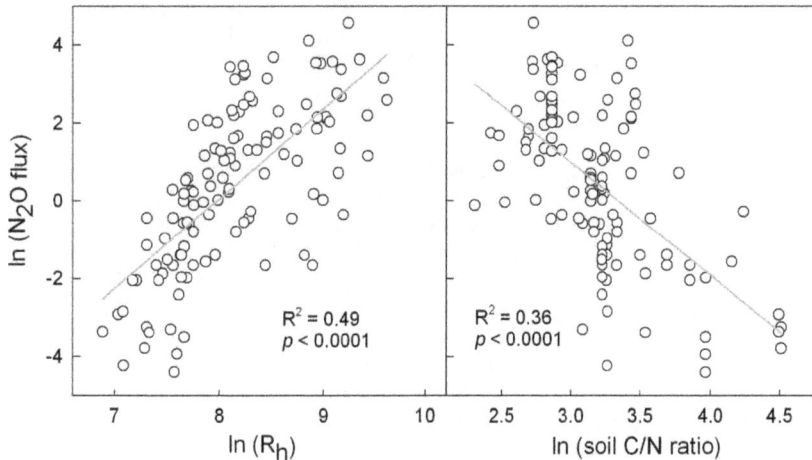

Figure 3. Correlation of observed fluxes of N₂O with estimated soil heterotrophic respiration (R$_h$, left panel) and C/N ratio (right panel).

of organic matter with different C/N ratios [26]. Most likely, the ratio of carbon evolved/nitrogen mineralized is much wider than the bulk soil carbon to nitrogen ratio [27,28]. This indicates that gross nitrogen mineralization might be over- or under-estimated if bulk soil C/N ratio was used in equation 2. The respiration process is also not exactly identical to carbon mineralization. The amount of carbon that is ultimately lost through respiration depends on how effectively the decomposer community converts mineralized carbon to biomass [29]. Similarly, the amount of nitrogen that is ultimately available to denitrifier or nitrifier for producing N₂O depends on how effectively the decomposer community converts mineralized nitrogen to biomass and plants compete with microbes for mineral nitrogen [10,20]. Empirical relationships have been established between nitrogen and carbon mineralization in studies performed usually under laboratory conditions [30]. Different organic matter fractions or their C/N ratios, and varying microbial use efficiency of carbon and nitrogen have also been proposed to predict nitrogen release [20,29]. However, these relationships are strongly dependent on the experimental conditions in which they have been established. Moreover, the current dataset is based on the in situ measurements in the field environment and contains only the basic information of respiratory carbon and bulk soil C/N ratio, thus necessitating the above-mentioned assumptions to estimate mineralized nitrogen. Such simplifications and assumptions may bring uncertainties, but it is necessary in some cases to understand the general trends and probabilistic nature of the environment [31].

N₂O emission from soils is of small magnitude and highly variable in space and time, and is thus very difficult to estimate. The measurement of soil N₂O flux also requires intricate techniques along with a lot of time and labor. In contrast, soil CO₂ emission is controlled primarily by soil temperature and moisture, and is relatively easy to measure or predict [32,33]. In addition, the estimates of soil respiration are currently more widely available than those of soil N₂O emission. The models developed in this study showed a promising approach to estimating N₂O emission from organic soils by using soil C/N ratio and CO₂ emission data derived from measurements or biogeochemical modeling. It should be mentioned, however, that several aspects of the information in the current dataset might impose uncertainties on these models. First, soil heterotrophic respiration was simply estimated from total soil respiration using a universal relationship

between them [15], but the relative contribution of organic matter decomposition or heterotrophic respiration would vary over time and depend on root respiration of the growing plants [8]. Second, the majority of the global organic soils are distributed in the boreal and sub-arctic regions and about 10%–15% in the tropical countries [1,3], but most of the current data came from northern Europe, indicating that the models developed in the present study might be biased to the temperate regions.

Conclusion

A fairly large number of data were collected to explore the relationship between annual N₂O emission and multiple variables for organic soil by a mixed-model analysis, and the input of soil mineral nitrogen was found to be the most useful predictor for N₂O flux. Soil mineral nitrogen was supposed to be composed of organic nitrogen mineralization as estimated by CO₂ emission and soil C/N ratio, thus providing a possibility for upscaling N₂O emission from organic soils by use of regional soil databases including information on C/N ratio and carbon storage change or CO₂ emission data. The approach proposed here may have validity as a whole, but needs further evaluation and advancement before practical application due to uncertainties associated with simplifying assumptions and a regionally unbalanced data source. A better understanding of the processes of carbon and nitrogen mineralization and their stoichiometric relationship as well as additional experimental data from organic soils outside of temperate Europe regions will help to improve the relationship established in this study.

Author Contributions

Conceived and designed the experiments: ZJM. Performed the experiments: ZJM AYH. Analyzed the data: ZJM. Contributed reagents/materials/analysis tools: JPN DTX. Wrote the paper: ZJM.

References

1. Page SE, Rieley J, Banks CJ (2011) Global and regional importance of the tropical peatland carbon pool. Global Change Biol 17: 798–818.
2. Goldberg SD, Knorr KH, Blodau C, Lischeid G, Gebauer G (2010) Impact of altering the water table height of an acidic fen on N_2O and NO fluxes and soil concentrations. Global Change Biol 16: 220–233.
3. Frolking S, Talbot J, Jones MC, Treat CC, Kauffman JB, et al. (2011) Peatlands in the Earth's 21st century climate system. Environ Rev 19: 371–396.
4. Maljanen M, Sigurdsson BD, Guðmundsson J, Óskarsson H, Huttunen JT, et al. (2010) Greenhouse gas balances of managed peatlands in the Nordic countries – present knowledge and gaps. Biogeosciences 7: 2711–2738.
5. Garcia-Montiel DC, Melillo JM, Steudler PA, Neill C, Feigl BJ, et al. (2002) Relationship between N_2O and CO_2 emissions from the Amazon Basin. Geophys Res Lett 29: Art.No. 1090.
6. Chatskikh D, Olesen JE (2007) Soil tillage enhanced CO_2 and N_2O emissions from loamy sand soil under spring barley. Soil Till Res 97: 5–18.
7. Keller M, Varner R, Dias JD, Silva H, Crill P, et al. (2005) Soil-atmosphere exchange of nitrous oxide, nitric oxide, methane, and carbon dioxide in logged and undisturbed forest in the Tapajos national forest, Brazil. Earth Interact 9: Art. No. 23.
8. Xu XF, Tian HQ, Hui DF (2008) Convergence in the relationship of CO_2 and N_2O exchanges between soil and atmosphere within terrestrial ecosystems. Global Change Biol 14: 1651–1660.
9. Klemedtsson L, Von Arnold K, Weslien P, Gundeersen P (2005) Soil CN ratio as a scalar parameter to predict nitrous oxide emissions. Global Change Biol 11: 1142–1147.
10. Luxhøi J, Bruun S, Stenberg B, Breland TA, Jensen LS (2006) Prediction of gross and net nitrogen mineralization-immobilization-turnover from respiration. Soil Sci Soc Am J 70: 1121–1128.
11. Bengtsson G, Bengtsson P, Månsson KF (2003) Gross nitrogen mineralization-, immobilization-, and nitrification rates as a function of soil C/N ratio and microbial activity. Soil Biol Biochem 35: 143–154.
12. Ojanen H, Minkkinen K, Alm J, Penttila T (2010) Soil – atmosphere CO_2, CH_4 and N_2O fluxes in boreal forestry-drained peatlands. Forest Ecol Manage 260: 411–421.
13. Mu ZJ, Huang AY, Kimura SD, Jin T, Wei SQ, et al. (2009) Linking N_2O emission to soil mineral N as estimated by CO_2 emission and soil C/N ratio. Soil Biol Biochem 41: 2593–2597.
14. Hanson PJ, Edwards NT, Garten CT, Andrews JA (2000) Separating root and soil microbial contributions to soil respiration: A review of methods and observations. Biogeochemistry 48: 115–146.
15. Bond-Lamberty B, Thomson A (2010) A global database of soil respiration data. Biogeosciences 7: 1915–1926.
16. Inubushi K, Furukawa Y, Hadi A, Purnomo E, Tsuruta H (2003) Seasonal changes of CO_2, CH_4 and N_2O fluxes in relation to land-use change in tropical peatlands located in coastal area of South Kalimantan. Chemosphere 52: 603–608.
17. Mojeremane W, Rees RM, Mencuccini M (2012) The effects of site preparation practices on carbon dioxide, methane and nitrous oxide fluxes from a peaty gley soil. Forestry 85: 1–15.
18. Klevens J, Trick WE, Kee R, Angulo F, Garcia D, et al. (2011) Concordance in the measurement of quality of life and health indicators between two methods of computer-assisted interviews: self-administered and by telephone. Qual Life Res 20: 1179–1186.
19. Rochette P, Tremblay N, Fallon E, Angers DA, Chantigny MH, et al. (2010) N_2O emissions from an irrigated and non-irrigated organic soil in eastern Canada as influenced by N fertilizer addition. Eur J Soil Sci 61: 186–196.
20. Murphy DV, Recous S, Stockdale EA, Fillery IRP, Jensen LS, et al. (2003) Gross nitrogen fluxes in soil: Theory, measurement and application of ^{15}N pool dilution techniques. Adv Agron 79: 69–118.
21. Regina K, Silvola J, Martikainen PJ (1999) Short-term effects of changing water table on N_2O fluxes from peat monoliths from natural and drained boreal peatlands. Global Change Biol 5: 183–189.
22. Danevcic T, Mandic-Mulec I, Stres B, Stopar D, Hacin J (2010) Emissions of CO_2, CH_4 and N_2O from southern European peatlands. Soil Biol Biochem 42: 1437–1446.
23. Maljanen M, Shurpali N, Hytönen J, Mäkiranta P, Aro L, et al. (2012) Afforestation does not necessarily reduce nitrous oxide emissions from managed boreal peat soils. Biogeochemistry 108: 199–218.
24. Smith KA, Thomson PE, Clayton PE, McTaggart IP, Conen F (1998) Effects of temperature, water content and nitrogen fertilization on emissions of nitrous oxide by soil. Atmos Environ 32: 3301–3309.
25. Weslien P, Klemedtsson AK, Borjesson G, Klemedtsson L (2009) Strong pH influence on N_2O and CH_4 fluxes from forested organic soils. Eur J Soil Sci 60: 311–320.
26. Springob G, Kirchmann H (2003) Bulk soil C to N ratio as a simple measure of net N mineralization from stabilized soil organic matter in sandy arable soils. Soil Biol Biochem 35: 629–632.
27. Sollins P, Spycher G, Glassman CA (1984) Net nitrogen mineralization from light- and heavy-fraction forest soil organic matter. Soil Biol Biochem 16: 31–37.
28. Kader MA, Sleutel S, Begum SA, D'Haene K, Jegajeevagan K, et al. (2010) Soil organic matter fractionation as a tool for predicting nitrogen mineralization in silty arable soils. Soil Use Manage 26: 494–507.
29. Manzoni S, Taylor P, Richter A, Porporato A, Ågren GI (2012) Environmental and stoichiometric controls on microbial carbon-use efficiency in soils. New Phytol 196: 79–91.
30. Nicolardot B, Recous S, Mary B (2001) Simulation of C and N mineralization during crop residue decomposition: a simple dynamic model based on the C:N ratio of the residues. Plant Soil 228: 83–103.
31. Yan XY, Yagi K, Akiyama H, Akimoto H (2005) Statisical analysis of the major variables controlling methane emission from rice fields. Global Change Biol 11: 1131–1141.
32. Lloyd J, Taylor JA (1994) On the temperature-dependence of soil respiration. Funct Ecol 8: 315–323.
33. Raich JW, Potter CS, Bhagawati D (2002) Interannual variability in global soil respiration, 1984-94. Global Change Biol 8: 800–812.

19

Biochemical and Molecular Characterization of Potential Phosphate-Solubilizing Bacteria in Acid Sulfate Soils and Their Beneficial Effects on Rice Growth

Qurban Ali Panhwar[1], Umme Aminun Naher[2,3], Shamshuddin Jusop[1]*, Radziah Othman[1,2], Md Abdul Latif[2,3], Mohd Razi Ismail[2]

1 Department of Land Management, Faculty of Agriculture, Universiti Putra Malaysia (UPM), Serdang, Selangor, Malaysia, 2 Institute of Tropical Agriculture, Universiti Putra Malaysia (UPM), Serdang, Selangor, Malaysia, 3 Bangladesh Rice Research Institute, Gazipur, Bangladesh

Abstract

A study was conducted to determine the total microbial population, the occurrence of growth promoting bacteria and their beneficial traits in acid sulfate soils. The mechanisms by which the bacteria enhance rice seedlings grown under high Al and low pH stress were investigated. Soils and rice root samples were randomly collected from four sites in the study area (Kelantan, Malaysia). The topsoil pH and exchangeable Al ranged from 3.3 to 4.7 and 1.24 to 4.25 $cmol_c$ kg^{-1}, respectively, which are considered unsuitable for rice production. Total bacterial and actinomycetes population in the acidic soils were found to be higher than fungal populations. A total of 21 phosphate-solubilizing bacteria (PSB) including 19 N_2-fixing strains were isolated from the acid sulfate soil. Using 16S rRNA gene sequence analysis, three potential PSB strains based on their beneficial characteristics were identified (*Burkholderia thailandensis*, *Sphingomonas pituitosa* and *Burkholderia seminalis*). The isolated strains were capable of producing indoleacetic acid (IAA) and organic acids that were able to reduce Al availability via a chelation process. These PSB isolates solubilized P (43.65%) existing in the growth media within 72 hours of incubation. Seedling of rice variety, MR 219, grown at pH 4, and with different concentrations of Al (0, 50 and 100 µM) was inoculated with these PSB strains. Results showed that the bacteria increased the pH with a concomitant reduction in Al concentration, which translated into better rice growth. The improved root volume and seedling dry weight of the inoculated plants indicated the potential of these isolates to be used in a bio-fertilizer formulation for rice cultivation on acid sulfate soils.

Editor: Luis Herrera-Estrella, Centro de Investigación y de Estudios Avanzados del IPN, Mexico

Funding: The authors acknowledge the financial and technical support given by Universiti Putra Malaysia and the Ministry of Education, Malaysia under the Long-term Research Grant Scheme (LRGS) fund for Food Security. The funders had no role in study design, data collection and analysis, decision to publish, or preparation of the manuscript.

Competing Interests: The authors have declared that no competing interests exist.

* Email: shamshud@upm.edu.my

Introduction

Acid sulfate soils are sporadically spread throughout the coastal plains around the globe. The estimated worldwide extent of acid sulfate soils is about 24 million hectares [1]. In Malaysia, acid sulfate soils mainly occur in the west coastal plains of Peninsular Malaysia and Sarawak (Borneo Island). The soils are characterized by the presence of pyrite (FeS_2), which is associated with high acidity and Al content, resulting from its oxidation upon exposure to the atmosphere [2]. Under acidic soil conditions, Al^{3+} restricts the growth of roots either by inhibiting cell division, cell elongation or both [3]. It is also known that rice grown on acid sulfate soils are subjected to Fe^{2+} toxicity [4].

Acid sulfate soils are normally not suitable for crop production unless they are properly ameliorated and their fertility improved [5]. They usually contain high concentration of Al and Fe which influence the biochemical properties of the soils. There are only a few acid-tolerant plant species and microbes that are able to survive in these soils. The soils contain low amount of total microorganisms, which vary considerably depending on the vegetation type and soil management. Microorganisms perform a major role in nutritional chains that are an important part of the biological balance in soils [6]. Bacteria are important for closing nutrient and geochemical cycles, involving carbon, nitrogen, sulfur and phosphorous. Beside chemical treatment, addition of microbes may improve nutrient availability in soils (especially phosphorus) and reduce Al toxicity.

There are some species of bacteria which have the potential to mineralize and solubilize organic and inorganic phosphorus in soil [7]. Bacterial strains such as *Pseudomonas*, *Bacillus* and *Rhizobium* are the dominant phosphate solubilizers [8] and single novel genus *Sphingomonas* spp. has the capability to fix nitrogen [9]. Moreover, in recent years, a growing number of *Burkholderia* strains and species have also been reported as plant-associated bacteria. *Burkholderia* spp. can be free-living in the rhizosphere as well as epiphytic or endophytic, obligate endosymbionts or phytopathogens [10]. Among the beneficial properties of these bacteria is the production of plant growth promoting phytohormone, polysaccharides and organic acids that are important for rice growth. Indoleacetic acid produced by the bacteria is known to promote an extensive root architecture capable of absorbing nutrient elements efficiently from the surroundings which ultimately improves rice growth [11–12].

PSB produce large amounts of organic acids [13] resulting in Al binding via a chelation process which is a plausible mechanism for reducing Al toxicity to rice roots in acid sulfate soils. Few attempts have been made to isolate and characterize these potential microorganisms living in the rhizosphere of rice plants grown in acid sulfate soils. Hence, the present study was undertaken to isolate and enumerate microbial occurrences in acid sulfate soils, and to identify beneficial isolates such as N_2-fixing and P-solubilizing bacteria that have important role in nutrient cycling, reducing Al toxicity and producing IAA in an acid sulfate soil environment.

Materials and Methods

Sampling location and isolation of microbes

Experimental site and conditions. This experiment was conducted at the Soil Microbiology Laboratory, Department of Land Management, Faculty of Agriculture, Universiti Putra Malaysia, Serdang, Malaysia. Soil samples were randomly collected from four sites in an acid sulfate soil area at Semerak, Kelantan, Malaysia which is located at a latitude of 30.01°N and longitude of 101°.70E.

We are working in a Government agency (University) so there is no need to take any permission for research activities from any one as we are working in a Government Project" Long term Research Grant Scheme" (LRGS) by the Ministry of Higher Education Malaysia. Hence, no specific permissions were required for these locations/activities. It has no any conflict on this issue." The methods to determine the soil physico-chemical analyses are given in Table 1.

Enumeration of the total microbial population from rice cultivated on acid sulfate soil. A series of 10-fold dilutions were prepared up to 10^{-8} for soil and rhizosphere microbial population determinations using the spread plate count method. Total fungal, bacterial and actinomycetes populations were determined using potato dextrose agar (PDA), nutrient agar (NA), and actinomycetes agar plates, respectively in five replicates.

Enumeration of PSB from rice cultivated on acid sulfate soil. Phosphate-solubilizing bacterial (PSB) population was determined from the soil, rhizosphere and endosphere using selective media plates at various pH levels: (i) the National Botanical Research Institute's phosphate growth medium (NBRIP) [17] at pH 5.0 and 6.7; and (ii) PDYA-AlPO$_4$ at pH 3.5 and 5. For the determination of the rhizosphere population, approximately 3 g of rice plant roots with its adhering soil were transferred into conical flask containing 99 mL of sterile distilled water and the contents were vigorously shaken. A 10-fold dilution series was prepared up to 10^{-8} and 0.1 mL aliquots were spread on selective media and incubated at $28\pm2°C$ in an incubator. For the determination of the endophytic population, fresh roots were taken and surface sterilized with 70% ethanol for 5 minutes and treated with 3% Clorox for 30 seconds [11]. Roots were cut into small pieces and surface sterilized by dipping into 95% ethanol and a flame. Surface sterilized roots were homogenized using a sterilized mortar and pestle. Endophytic PSB populations were determined using the total plate count method in five replicates.

Estimation of diazotrophs from rice cultivated on acid sulfate soil. A series of 10-fold dilutions were prepared up to 10^{-8} using rhizosphere and non-rhizosphere soil and the diazotroph populations were determined using the most-probable number (MPN) method in nitrogen free (Nfb) semi-solid medium in five replicates. The Nfb semi-solid medium [18] contained 5 g malic acid, 0.5 g K_2HPO_4, 0.2 g $MgSO_4$. 7 H_2O, 0.1 g NaCl, 0.02 g $CaCl_2$ and 0.5% bromothymol blue in 0.2 N KOH (2 mL), 1.64% Fe-EDTA solution (4 mL) and 2 g agar.

Determination of the beneficial traits of isolated bacteria

Determination of indoleacetic acid (IAA). The isolates were inoculated in nutrient agar (NB) broth with the addition of tryptophan (2 mg L^{-1}) and incubated at $28\pm2°C$ for 48 hours. The culture was centrifuged at 7000 rpm for 7 minutes and 1.0 mL of the supernatant was mixed with 2 mL of Salkowsky's reagent [19]. The IAA concentration was determined using a spectrophotometer at 535 μm in five replicates.

Phosphate solubilization

Phosphate solubilization on media plates. The phosphate solubilizing activities of the isolates were assayed by spotting 10 μL of the cultures on different phosphate containing media such as NBRIP [17], Pikovaskaya, Christmas Island Rock Phosphate (CIRP) and PDYA-AlP in five replicates. The plates were incubated at 30°C for one week and the phosphate solubilization efficiency was measured [20]:

$$Solubilization\,efficiency = \frac{Solubilization\,diameter}{Growth\,diameter} \times 100 \quad (1)$$

Phosphate solubilization in broth culture. The isolated PSB strains were evaluated for their phosphate-solubilizing activity in broth culture. Three different media containing insoluble

Table 1. Analytical methods used in the study.

S No.	Analysis	Procedure
1	Soil pH and EC	Soil: water (1:2.5) extract using PHM210 standard pH meter and EC meter by glass electrode [14]
2	CEC	1 M NH₄OAc solution buffered at pH 7.0 was used [14]
3	Soil texture	Pipette method [15]
4	Total N	Kjeldahl digestion method [16]
5	Exchangeable cations (Ca, Mg, K)	Extracted by 1M NH₄OAc solution at pH 7 [14]
6	Exchangeable Al	Extracted by 1 M KCl solution
7	Total carbon	CN analyzer (LECO CR-412)
8	Micronutrients (Cu, Mn, Zn, Fe)	Inductivity coupled plasma - atomic emission spectroscopy (ICP-AES)

ICP-AES = inductively coupled plasma atomic emission spectroscopy, CEC = cation exchange capacity.

phosphate: i) calcium phosphate in NBRIP broth [17] containing (g L^{-1}) MgCl$_2$.6H$_2$O$_5$ g, MgSO$_4$.H$_2$O 0.25 g, KCl 0.2 g, (NH$_4$)$_2$SO$_4$ 0.1 g, Ca$_3$(PO$_4$)$_2$ amended with glucose 10 g; ii) CIRP broth modified from NBRIP supplemented with phosphate rock (CIRP) instead of calcium phosphate; and iii) PDYA-AlP broth [21] containing (g L^1): PDA agar 39 g, yeast extract 2 g, sterilized 10% K$_2$HPO$_4$, and 10% AlCl$_3$ (100 mL). Each media (200 mL) were inoculated with PSB inoculum and were incubated at 30°C on a Kottermann 4020 shaker at 80 rpm for 3 days. Each treatment was replicated five times.

Determination of the population in broth culture. One milliliter of broth was taken from the respective flasks at different periods (initial & after 72 h) for determination of the bacterial population. A series of 10-fold dilutions were prepared up to 10^{-10}. The population was determined using the drop plate count method according to the method of Somasegaran and Hoben [22].

Determination of phosphorus solubilization in broth cultures. Exactly 2 mL of samples were taken for P determination. The samples were first allowed to sediment for 15 minutes and were then centrifuged at 4000× g for 5 minutes. The supernatant was filtered through 0.2 μm filter paper and kept at − 20°C until analysis. The available P was determined according to the published methods [23].

Nitrogen-fixing activities. The N$_2$-fixing activity was determined using the Nfb semi-solid liquid medium [18]. The presence of pellicle formation below the surface of media indicated the N$_2$-fixing activity.

Molecular identification of PSB strains

DNA extraction and primers. The identification of PSB was carried out on the basis of 16S rRNA gene sequencing. The genomic DNA of PSB isolates was extracted by the Qiagen DNeasy Plant Mini Kit (Qiagen, Valencia, CA). Forward primers D1 (5-AGAGTTTGATCCTGGCTCAG-3) and reverse P2 (3-ACGGCTACCTTGTTACGACTT-5) were used for amplification of 16S rRNA gene [24]. Each sample was replicated five times.

PCR protocols and gel electrophoresis. The total PCR reaction mixture was 50.0 μL, comprising 200 μM dNTPs, 50 μM of each primer, 1× PCR buffer, 3 U *Taq* polymerase and 100 μg genomic DNA. The thermal cycler (MJ Mini personal Thermal Cycler, Bio-Rad, Model- PTC-1148) conditions were as follows: 95°C for 3 minutes, followed by 30 cycles of denaturation at 95°C for 1 minute, annealing at 48°C for 1 minute and primer extension at 72°C for 2 minutes. This was followed by a final extension at 72°C for 10 minutes. The reaction products were separated by running 5 μL of the PCR reaction mixture in a 1.0% (w/v) agarose gel and staining the bands with ethidium bromide.

Strain identification using gene sequencing. Three potential isolates with greater beneficial characteristics were selected for 16S rRNA gene sequencing analysis. Sequence data were aligned and compared with the available standard sequences of bacterial lineage in the National Center for Biotechnology Information GenBank (http://www.ncbi.nlm.nih.gov/) using BLAST [25]. A phylogenetic tree was constructed by the neighbor-joining method using the software MEGA 4 [26]. The obtained sequences were deposited in the European Molecular Biology Laboratory data (accession number NR 074312.1, NR 042635.1 and NR 25363.1).

Principal Components Analysis. Principal Components Analysis (PCA) is an ordination technique. Here we performed an eigen analysis of the covariance matrix. The eigen value is a measure of the strength of an axis, the amount of variation along an axis, and ideally the importance of an ecological gradient. The precise meaning depends on the ordination method used. Two-dimensional graph (two axes of PCA) was constructed using co-variance matrix and find out the variation among the characters of 21 bacterial isolates.

Efficiency of isolates to improve rice seedling growth with high Al and low pH. Three potential phosphate-solubilizing bacterial strains (PSB7, PSB17 and PSB21) were selected on the basis of their beneficial characteristics. Seven-days-old MR 219 rice seedlings were grown in Hoagland solution containing different concentrations of Al (0, 50 and 100 μM) with five replicates. The initial pH of the solution was adjusted to 4.0. Rice seedlings were harvested 21 days after sowing. The bacterial population, plant dry biomass, solution pH and organic acids were determined soon thereafter.

Determination of the microbial population at different Al concentrations. One milliliter of broth was taken from the respective flasks at different time periods (6, 12, 24, and 48 h) for the determination of bacterial growth. A series of 10-fold dilution were prepared up to 10^{-10}. The population was determined using the drop plate count method [22].

Determination of organic acids. About 20 μL of the samples from each treatment were injected into HPLC with a UV detector set at 210 nm. A Rezex ROA-organic acid "H$^+$" (8%) column (250×4.6 mm) from Phenomenex Co. was used, the mobile phase was 0.005 N H$_2$SO$_4$ with a flow rate of 0.17 mL min^{-1}.

Determination of root morphology. The root morphology of the rice plants was determined using a root scanner (model Epson Expression 1680 equipped with root scanning analysis software). Total root length (cm), total surface area (cm^2) and total volume (cm^3) were quantified using a scanner (Expression 1680, Epson) equipped with a 2 cm deep plexiglass tank (20.30 cm) filled with up H$_2$O [27]. The scanner was connected to a computer and scanned data were processed by Win-Rhizo software (Regent Instruments Inc., Québec, Canada).

Statistical analysis

All data were statistically analyzed using SAS Software (Version 9.2), and treatment means were separated using Tukey's test (P< 0.05).

Results

Properties of the acid sulfate soils

The pH of the acid sulfate soils under study ranged from 3.3 to 4.7 pH with values decreasing with depth (Table 2). The soils were taxonomically classified as Typic Sulfaquepts. Soils of this nature normally contain pyrite in the subsoil, having pH<3.5 below a depth of 50 cm [2]. It is the oxidation of this pyrite that produces acidity and the subsequent release of Al and/or Fe into the environment. The low soil pH is consistent with the presence of high exchangeable Al in the soils. Exchangeable Al in the topsoil at all sampling sites ranged from 1.24 to 4.25 cmol$_c$ kg^{-1}, occurring at a toxic level for rice growth. Total N and exchangeable K were found to be at sufficient levels for rice growth, while available P, exchangeable Ca and Mg were insufficient. The Zn, Cu and Mn contents were low, while extractable Fe was high (124 to 181 mg kg^{-1}) (Table 2). Total C was high in the topsoil with values above 2% and it decreased consistently with depth. The high organic matter in the topsoil would have a profound effect on the availability of Al and Fe, which can be partly fixed (chelated) by it and hence deactivated. Chelated Al and Fe are non-toxic to rice plants in the field.

Table 2. Chemical characteristics of the acid sulfate soils.

Site	Soil depth	Soil pH	Total C	Total N	Avail. P	K	Al	Ca	Mg	Fe	Zn	Cu	Mn
							Exchangeable				Extractable		
	(cm)		(%)		(mg kg^{-1})		(cmol$_c$ kg^{-1})				(mg kg^{-1})		
1	0-15	4.0b	2.1c	0.18b	26.3a	0.05b	1.7cd	0.43c	1.0bc	174abc	1.6ab	3.2a	8.1ab
	15-30	4.0b	1.6d	0.14c	19.1b	0.05b	2.1c	0.40c	0.9cd	170abc	1.6ab	2.5b	7.8ab
	30-45	3.8bc	1.2e	0.10d	13.1def	0.04c	4.5b	0.30d	0.6ef	129bcd	0.9d	2.3bc	7.4abc
	45-60	3.6c	0.9f	0.09d	12.9def	0.04c	5.5a	0.25d	0.5ef	124d	0.7d	2.0bcd	6.4bcd
2	0-15	4.7a	2.9a	0.17b	25.2a	0.06a	1.24d	0.57b	0.7de	181a	2.3a	2.4b	8.8a
	15-30	3.6c	1.1e	0.12c	16.6bc	0.04c	1.5cd	0.43c	0.5ef	176ab	1.8a	1.9cde	8.2ab
	30-45	3.4de	1.1e	0.09de	15.5cd	0.04c	1.7cd	0.40c	0.4f	167abcd	1.8a	1.7def	7.9ab
	45-60	3.3e	0.9f	0.08de	11.4f	0.03d	1.9c	0.12e	0.4f	163bcd	1.5b	1.5ef	6.0bcd
3	0-15	3.8bc	2.3b	0.21a	19.2b	0.05b	1.8cd	0.80a	1.3a	180a	2.0a	1.4ef	7.5ab
	15-30	3.5d	1.1e	0.13c	14.8cde	0.04c	1.9c	0.73a	1.2b	178a	1.3bc	1.5def	6.1cd
	30-45	3.4de	0.9f	0.09d	12.2ef	0.04c	1.9c	0.63b	0.94bc	145cd	1.2c	1.5ef	5.3cd
	45-60	3.3e	0.6g	0.06e	11.6f	0.04c	4.3b	0.60b	0.7de	124d	1.1c	1.1f	4.5d
4	0-15	3.8bc	2.2b	0.20a	19.1b	0.05b	1.7cd	0.72a	1.2b	179a	1.97a	1.3ef	7.2ab
	15-30	3.6c	1.15e	0.11c	14.3cde	0.04c	1.8cd	0.70a	1.0bc	170abc	1.2c	1.4ef	6.0cd
	30-45	3.5d	1.01ef	0.08de	11.5f	0.03d	2.03c	0.60b	0.94bc	154c	1.1c	1.4ef	5.2cd
	45-60	3.4de	0.5g	0.05e	10.7fg	0.03d	4.03b	0.56b	0.8de	130d	1.0c	1.2f	4.3d

Data values are means of five replicates. Means followed by the same letter within a column are not significantly different (P<0.05).

Total population of microorganisms in the rhizosphere, non-rhizosphere and endosphere

Higher microbial populations were found in the rice rhizosphere compared to the non-rhizosphere. The total bacterial and actinomycetes population were higher than fungal populations (Table 3) and higher PSB populations were observed on NBRIP plates compared to PDYA-AlP media plates (Table 4). However, there were no significant population differences of PSB present in the NBRIP media plates at media pH of 5.0 and 6.7. The highest diazotrophic population was found in the endosphere (25×10^6 cfu root g^{-1}) and the lowest was in the non-rhizosphere (13×10^3 cfu soil g^{-1}) (Table 4).

Biochemical properties of the strains isolated from rice cultivated on acid sulfate soils

The indigenous bacterial isolates were capable of producing indoleacetic acid (IAA). Among all the isolates, PSB21 (14.96 mg L^{-1}), followed by PSB7 (13.16 mg L^{-1}) produced high IAA levels as compared to the other isolates. The lowest amount of IAA was produced by PSB20 (Table 5). A total of 21 phosphate-solubilizing bacterial (PSB) isolates were identified from the acid sulfate soils. However, only 19 strains were able to grow in nitrogen-free medium (Table 5).

Phosphate solubilization in media plates

All the selected 21 PSB isolates were able to solubilize P on NBRIP and Pikovskaya media plates, with P-solubilizing activity indicated by a clear halo zone (Plate S1). The highest P solubilizing activity in NBRIP media plate was contributed by PSB17 (70.23%), followed by PSB7 (57.5%), while the lowest activity (23.23%) was contributed by PSB20 strain (Table 5). On the other hand, on Pikovskaya media plates, the highest activity was contributed by PSB17 (76.03%), followed by PSB21 (70.12%).

Phosphate solubilization in broth culture

The isolates from the soils were able to solubilize P from different forms of inorganic phosphate incorporated into the broth culture after 72 hours of incubation (Table 5). Comparatively, the highest P solubilization activity was found in NBRIP broth with PSB7 (43.65%), followed by PSB21 (43.34%). Most of the strains were able to solubilize P in CIRP broth, but higher P solubilization was observed in samples containing PSB1 and PSB5 compared to other isolates, and lower amounts of soluble P were observed in the PDYA-Al broth.

Principal component analysis among 21 bacterial isolates based on biochemical properties

In order to assess the patterns of variation, PCA was done by considering seven biochemical characters. The first three principal components (PCs) explained 99% of the total variation in 21 bacterial isolates and showed 73.27, 21.63, 4.14% variations, respectively (Table 6). The PC1 and PC2 were loaded with all seven characters such as IAA production, Nfb activity, P solubilization in different sources of P media and culture (NBRIP, Pikovskaya, PDYA-AlP and CIRP). All the characters either in PC1 or PC2 showed positive contribution for the differences between the bacterial isolates. The PC1 was strongly responsible for variation of the 21 bacterial isolates.

The PCA scatter plot is shown in Fig. 1. All the bacterial isolates were grouped into seven clusters. The highest number of bacterial isolates from acid sulfate soil was in cluster III (PSB3, PSB4, PSB6, PSB8, PSB9, PSB16 and PSB19), followed by cluster II (PSB10, PSB11, PSB12 and PSB13), cluster I (PSB1, PSB2 and PSB6), cluster VII (PSB7, PSB17 and PSB21) and cluster VII (PSB15 and PSB18). Clusters IV and V each contained one strain. The results show that cluster VII with bacterial isolates PSB17, PSB21 and PSB7 was the most diverged group.

Identification of the potential strains

Molecular analysis by the 16S rDNA identification technique was adopted in this study. These excellent markers for the clarification of bacterial phylogeny are ribosomal ribonucleic acids. In this study, we used gene sequences from the β-subclass of Proteobacteria to determine the phylogenetic relationships among the tested isolates. The neighbor-joining tree was subjected to the numerical re-sampling by bootstrapping, and the resulting bootstrap values were observed at the tree branch nodes. Each value represents the number of times (out of 1000 replicates) that the represented groupings occurred in the re-samplings. The consensus tree showed 98–100% confidence levels between 3 potential isolates from the β-subclass of Proteobacteria PSB7 [*Burkholderia thailandensis*] and PSB21 [*Burkholderia seminalis*], whereas PSB17 [*Sphingomonas pituitosa*] had a 100% confidence level (Fig. 2 & 3).

All of the sequences have higher than 98% identity with the queried sequence. Two PSB7 and PSB21 sequences were identified as *Burkholderia* spp. with accession numbers NR 074312.1 and NR 042635.1, respectively, while one belonged to *Sphingomonas* sp. (NR 25363.1).

Table 3. Total microbial population of microorganisms isolated from the acid sulfate soils.

Site	Bacterial		Fungal		Actinomycetes	
	------------------------------- (*cfu soil g^{-1}) -------------------------------					
	Rhizosphere	Non-rhizosphere	Rhizosphere	Non-rhizosphere	Rhizosphere	Non-rhizosphere
1	1.5×10^{7b}	7.1×10^{5a}	7.0×10^{3ab}	6.0×10^{3ab}	3.0×10^{4c}	2.1×10^{5a}
2	9.0×10^{6c}	3.5×10^{5b}	10.0×10^{3ab}	3.0×10^{3b}	8.0×10^{4b}	3.0×10^{4c}
3	2.8×10^{7a}	1.8×10^{5c}	7.0×10^{3a}	9.0×10^{3a}	1.2×10^{5ab}	10.0×10^{4ab}
4	1.9×10^{7ab}	1.5×10^{5c}	2.0×10^{3b}	5.0×10^{3ab}	1.2×10^{5ab}	4.0×10^{4c}

*cfu = colony forming unit, Data values are means of five replicates.
Means followed by the same letter within a column are not significantly different (P<0.05).

Table 4. Phosphate-solubilizing bacteria and diazotroph population from rice cultivated on acid sulfate soils.

Site	Phosphate-solubilizing bacterial population												Diazotrophs population		
	Rhizosphere (cfu soil g^{-1})				Non-rhizosphere (cfu soil g^{-1})				Endosphere (cfu root g^{-1})				Rhizosphere (cfu soil g^{-1})	Non-rhizosphere (cfu soil g^{-1})	Endosphere (cfu root g^{-1})
	NBRIP		PDYA-AIP		NBRIP		PDYA-AIP		NBRIP		PDYA-AIP		N-free media		
	pH 5.0	pH 6.7	pH 3.5	pH 5.0	pH 5.0	pH 6.7	pH 3.5	pH 5.0	pH 5.0	pH 6.7	pH 3.5	pH 5.0			
1	2×10^{7a}	4×10^{4b}	-	-	5.7×10^{4a}	3×10^{4a}	-	-	28×10^{4a}	19×10^{4b}	-	-	23×10^{5b}	24×10^{3b}	25×10^{6a}
2	70×10^{4b}	41×10^{4b}	-	-	2×10^{3b}	3×10^{3b}	-	-	8×10^4	4×10^{4a}	-	-	15×10^{6a}	3×10^{4a}	11×10^{5b}
3	57×10^{4b}	60×10^{4b}	-	5×10^{4a}	3×10^{4a}	21×10^{4a}	-	6×10^{4a}	9×10^4	9×10^{4a}	-	-	16×10^{5b}	13×10^{3b}	38×10^{5b}
4	24×10^{4b}	5×10^{4b}	-	2×10^{4a}	2×10^{4a}	21×10^{4a}	-	5×10^{4a}	50×10^{4a}	71×10^{4a}	-	5×10^{4a}	24×10^{3b}	27×10^{3b}	3×10^{6a}

*cfu = colony forming unit,
†NBRIP = National Botanical Research Institute's phosphate growth medium,
‡PDYA-AIP = Potato dextrose Yeast Agar- Aluminium Phosphate respectively.
Data values are means of five replicates. Means followed by the same letter within a column are not significantly different (P<0.05).

The efficacy of PSB strains to reduce Al toxicity

It was observed that high Al concentration severely affected the growth of the rice seedlings. Plant height and dry biomass significantly decreased with increased Al concentrations (Table 7). In comparison, the bacterial inoculated plants were less affected by Al toxicity. High plant height (18 cm) and dry biomass (0.76 g) were observed in the *Burkholderia seminalis* inoculated treatments at 0 μM Al (Table 7). The PSB associated with the plant roots produced organic acids that provided P and chelated the Al toxicity. As shown by the SEM and TEM micrographs (Plate S2), the bacteria existed in association with the rice seedlings.

The morphology of the PSB-inoculated and untreated rice roots

This study found that root length, root surface area and volume varied with the Al concentration and bacterial inoculation. Presence of Al affected rice root development, especially at high concentrations (Table 7). Generally, greater root length, surface area and volume were found in inoculated compared to non-inoculated rice plants. Significantly greatest root length (58.51 cm) was observed at 0 μM Al by PSB21 inoculated rice plant compared those at high Al concentration. The greatest root surface area of 74.12 cm^2 was recorded in PSB17 inoculated rice at 0 μM Al, while the highest root volume was observed in the *Sphingomonas pituitosa* and *Burkholderia seminalis* (6.9 and 6.88 cm^3, respectively) inoculated rice plants at the same Al concentration.

Effects of Al and PSB inoculation on the release of organic acids

It was observed that organic acids released by the rice roots varied with the Al concentration and bacterial inoculation (Table 8). Plants with highest Al concentration was found to secrete the highest amounts of organic acids. The release of organic acids was affected by the Al concentration. Higher amounts of organic acids were released by PSB-inoculated plants compared to the non-inoculated rice seedlings. Among the organic acids released, higher amounts of oxalic and citric acids were observed at 100 μM Al compared to malic acid, while the amount of malic acid was found to be low, particularly at 0 and 50 μM Al concentration.

Effect of Al on the population of PSB strains with or without rice seedlings

It was found that the population of PSB was affected by the Al concentrations in both of the plant and without plant systems. It seemed that a higher Al concentration lowered the population of PSB (Fig. 4a &b). Hence, at 0 μM Al concentration, higher PSB population was observed compared to that observed at higher Al concentrations. Among the inoculated PSB strains, the highest population was recorded by PSB21 at all the Al concentrations without plant system, while in plant system all strains were showing the same response. The population decreased with the increasing Al concentration, and this trend was observed for all the PSB strains in both culture systems. It was clear from the current study that some bacteria strains survived under the condition of low pH and high Al concentration; these strains have the potential to be used for rice cultivated on acid sulfate soil containing high Al concentration.

Table 5. Biochemical properties of the isolated PSB strains.

Isolates	IAA production (mg L^{-1})	Nfb activity	P solubilization (%) in different media plates		P solubilization in broth culture after 72 hours of incubation					
			NBRIP	Pikovskaya	*NBRIP (ppm)	(%)	†CIRP (ppm)	(%)	‡PDYA-AlP (ppm)	(%)
PSB1	10.00bc	+ve	50.56c	45.26d	104c	31.30	8.43bc	2.54	3.21c	0.97
PSB 2	9.60c	+ve	45.23d	55.23c	103c	31.00	5.67de	1.71	4.32b	1.30
PSB 3	9.00c	+ve	40.00e	40.24e	108c	32.51	7.23c	2.18	1.34e	0.40
PSB 4	5.28e	+ve	36.67f	45.65d	104c	31.30	4.32f	1.30	2.87d	0.86
PSB 5	7.32d	+ve	45.23d	50.13c	106c	31.91	10.12a	3.05	4.89ab	1.47
PSB 6	4.32f	+ve	30.22f	45.81d	109c	32.81	3.21g	0.97	3.23c	0.97
PSB 7	13.16a	+ve	57.5b	60.12b	145a	43.65	11.3a	3.40	5.43a	1.63
PSB 8	3.84fg	–ve	42.2e	45.36d	107c	32.21	6.32d	1.90	3.45c	1.04
PSB 9	4.40f	+ve	34.28f	35.93f	106c	31.91	5.64de	1.70	1.76e	0.53
PSB 10	5.44e	+ve	45.65d	50.22c	79de	23.78	6.54d	1.97	2.65d	0.80
PSB 11	9.20c	+ve	51.11c	52.16c	65f	19.57	5.21e	1.57	3.43c	1.03
PSB 12	10.00bc	+ve	55.73b	47.27d	89d	26.79	4.21f	1.27	3.28c	0.99
PSB 13	11.60b	+ve	50.32c	34.44fg	73e	21.97	4.02f	1.21	2.69d	0.81
PSB 14	5.60e	+ve	40.24e	35.38f	85d	25.59	3.41g	1.03	3.38c	1.02
PSB 15	8.16d	+ve	49.41cd	45.42d	142a	42.74	6.03d	1.82	2.76d	0.83
PSB 16	9.63c	+ve	31.11f	40.83e	112bc	33.71	4.07f	1.23	3.48c	1.05
PSB 17	12.16b	+ve	70.23a	76.03a	142a	42.74	9.34b	2.81	5.73a	1.72
PSB 18	11.36b	–ve	45.09d	44.1d	122b	36.72	5.74de	1.73	4.21b	1.27
PSB 19	10.56b	+ve	50.21c	32.5g	107c	32.21	6.24d	1.88	4.89ab	1.47
PSB 20	1.56h	+ve	23.23g	31.02g	124b	37.32	2.02h	0.61	4.1b	1.23
PSB 21	14.96a	+ve	56.65b	70.12a	144a	43.34	9.57b	2.88	5.23a	1.57

*NBRIP = National Botanical Research Institute's phosphate growth medium,

†CIRP = Christmas Island Rock Phosphate and

‡PDYA-AlP = Potato dextrose Yeast Agar- Aluminium Phosphate respectively.

Data values are means of five replicates. Means in each column followed by the same letters are not significantly different according to Tukey's HSD at P≤0.05. Note: (+ve) for N$_2$ fixing and (–) for not N$_2$ fixing activities.

Table 6. Principal component analysis of seven biochemical characters and proportion of variation for each component.

Biochemical characters	[a]PC1	PC2	PC3
% Variation	73.27	21.63	4.14
[b]IAA production (mg L^{-1})	0.49485	−0.04136	0.46886
Nfb activity	0.71264	0.4016	−0.03145
P solubilization (%) on [c]NBRIP	0.5499	−2.0133	1.8433
P solubilization (%) on Pikovskaya media plates	0.60905	−1.5349	−2.3093
P solubilization (ppm) in NBRIP broth culture	2.4185	1.2426	0.19441
P solubilization (ppm) in [d]CIRP broth culture	−0.55919	0.18271	−0.0435
P solubilization (ppm) in [e]PDYA-AIP broth culture	−0.63265	0.33844	−0.06201

[a]PC = Principle component,
[b]IAA = Indoleacetic acid,
[c]NBRIP = National Botanical Research Institute's phosphate growth medium,
[d]CIRP = Christmas Island Rock Phosphate and
[e]PDYA-AIP = Potato dextrose Yeast Agar- Aluminium Phosphate respectively.

Effect of PSB inoculation on nutrient solution pH at different Al concentration

There was a clear effect of PSB inoculation found in the nutrient solution pH at different Al concentrations in both of the plant and without plant systems (Fig. 5a &b). The initial solution pH was 4.0 and after 24 h of inoculation, the pH increased to 7.00 for all PSB inoculated treatments (with and without plant system). However, for the non-inoculated treatment, it remained almost the same as the initial pH (3 to 4.0) in both of the conditions regardless of Al concentrations. Without plant, the highest pH of 7.69 was obtained for the 100 μM Al treatment with PSB 21(Fig. 5a). For the inoculated plant, the solution pH was a bit higher in 0 μM Al compared to the other treatments (Fig. 5b). The lowest solution pH of 2.80 was found in the non-inoculated control at 100 μM of

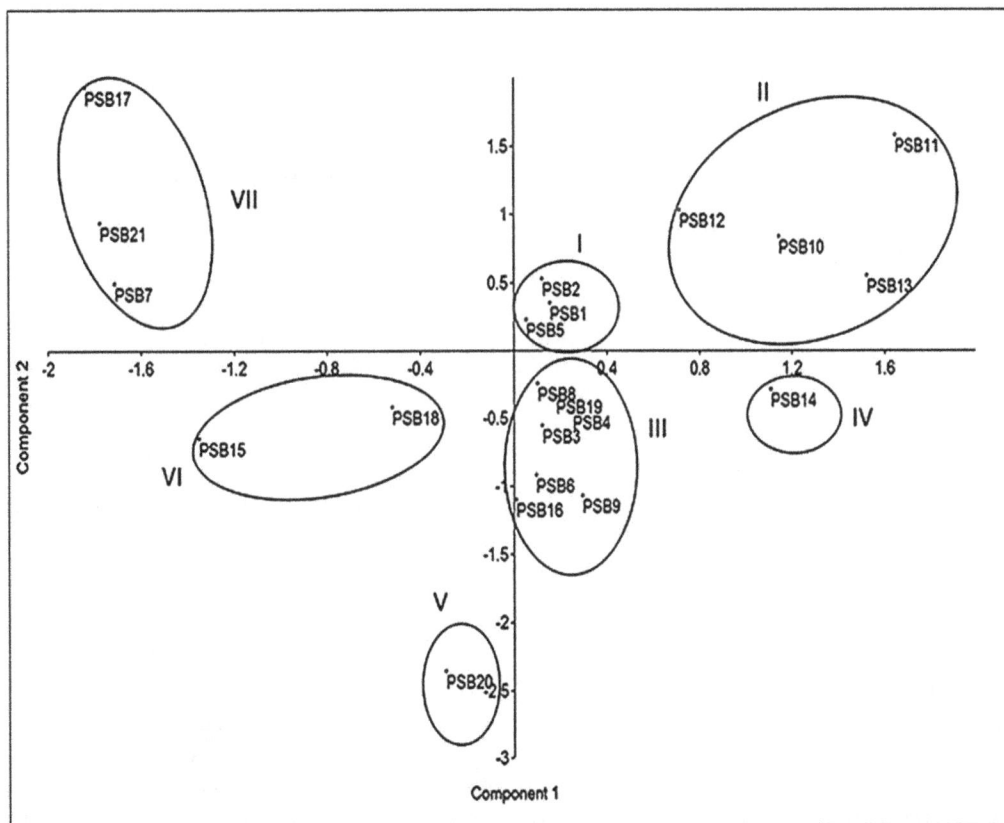

Figure 1. Plotting of two principal axes in principal component analysis showed the variation among 21 bacterial isolates based on seven biochemical characters using co-variance matrix.

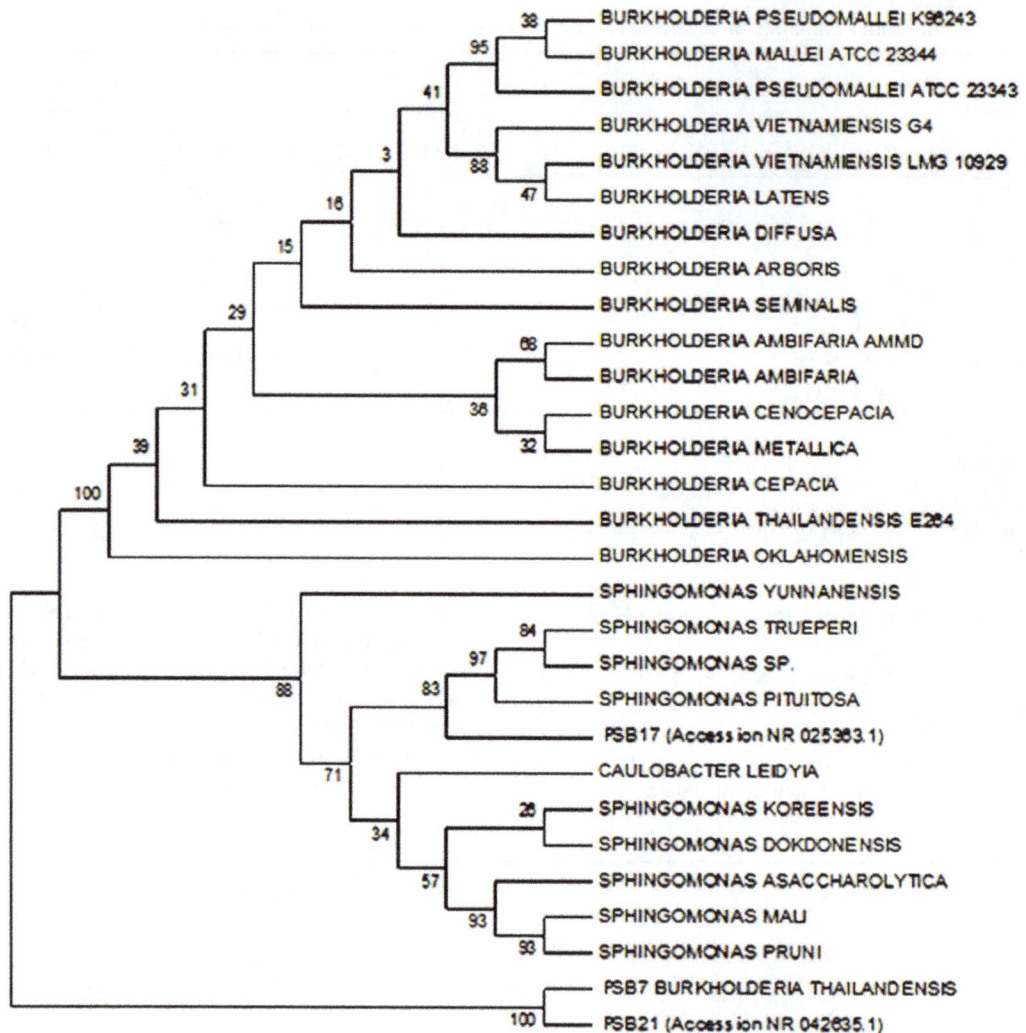

Figure 2. Phylogenetic tree with bootstrap values. Tree constructed using Neighbor-Joining (NJ) method. PSB7 accession NR 074312.1, PSB17 accession NR025363.1, and PSB21 accession NR 042635.1.

Al concentration (with rice seedling). Rice seedlings were affected severely by the low pH. However, PSB inoculation increased the pH of solution that resulted in enhanced growth of rice seedlings (Fig. 6).

Discussion

Acid sulfate soils are dominated by sulfur or oxidizing-reducing bacteria [28]. These soils may also contain some beneficial microorganisms that help improve rice growth. In this study, it was found that the pH of the acid sulfate soils was very low and Al concentration was very high, especially at the depth below 50 cm, proving that these were true acid sulfate soils [29]. Although the soils were very acidic there were still some microorganisms living in them. These microorganisms (bacteria, actinomycetes and fungus) can be potentially beneficial for rice production. It was found that the population of the bacteria changed from place to place and varied according to the native vegetation. Furthermore, the bacterial population was found to be higher at the rhizosphere compared to the non-rhizosphere, indicating bacterial synergism with plant roots [11].

A total of 21 PSB were isolated from the acid sulfate soils in the present study. These PSB isolates have the potential to be used in rice production. Higher PSB populations were observed in the calcium phosphate (NBRIP) medium compared to other P media sources. In this medium, the PSB population was not affected by pH. Panhwar *et al.* [13] reported significantly higher PSB population observed in tricalcium phosphate medium. Furthermore, it was reported that most of the strains after an incubation period formed pellicles in N free semisolid-malate media, pointing to their N_2 fixing abilities [30].

The isolated strains have the capability to produce growth hormones (IAA) that could help to the plants to enhance their root and shoot growth. This shows the potential of these bacteria for use in crop production [31]. It is known that rhizosphere microorganisms mediate many soil processes, such as decomposition, nutrient mineralization and nitrogen fixation [32]. Other researchers have reported that bacteria in rice fields have the potential to produce IAA and are able to fix N [11–33]. In the current study, it was found that the isolated strains were able to solubilize phosphate in NBRIP and Pikovskaya media, with P-solubilizing activity indicated by clear halo zone around colonies [34–35]. A similar trend was observed for liquid broth culture

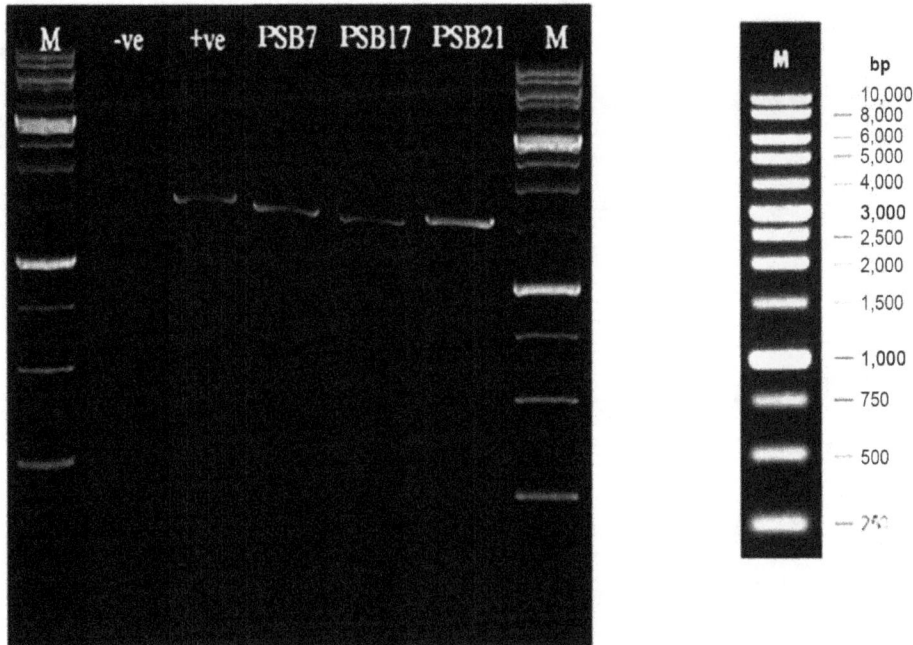

Figure 3. SDS-PAGE of the PSB7 (*Burkholderia thailandensis*), PSB17 (*Sphingomonas pituitosa*) and PSB21 (*Burkholderia seminalis*). (M: DNA ladder; −ve: negative control; +ve: positive control).

using calcium phosphate (NBRIP) and rock phosphate (CIRP). The highest P solubilization occurred in NBRIP broth, while the lowest was in the CIRP and PDYA-AlP broth cultures. This is similar to the findings of Chakraborty *et al.* [36] who reported that isolated PSB strains solubilized P from calcium phosphate to a greater amount than rock phosphate, aluminum phosphate and iron phosphate.

The bacterial population was significantly affected by the different forms of inorganic phosphate incorporated into the broth culture. NBRIP and CIRP broths were found to have the highest bacterial population, while lower bacterial growth was found in the PDYA-AlP broth after 72 hours of inoculation. The low content of solubilized P in the PDYA-A broth may be due to its insolubility although the isolates had the ability to solubilise it. Al in the soils may fix P, even further reducing solubilization. It has been established that phosphate rock has a lower content of soluble P compared to calcium phosphate [37].

The three principal components showed 99% variation of the total variation in 21 bacterial isolates. Similarly, Naher *et al.* [38] found distinct variations in different soil bacterial isolates using principal component analysis. Based on the production of indoleacetic acid and phosphate solubilizing activities, three isolates PSB7, PSB21 and PSB21 were found to have great potential. This noteworthy finding is supported by the PCA scatter plot based on the biochemical properties and these three bacterial isolates which were grouped distantly from the other strains.

The assessment of the bacterial 16S rRNA gene sequence has emerged as a preferred genetic technique as it can better identify weakly described, rarely isolated, or phenotypically aberrant strains [39–40]. In our study, we employed a molecular phylogenetic approach based on 16S rRNA sequences to identify pure potential isolates. Three potential isolates, namely PSB7 (*Burkholderia thailandensis*), PSB21 (*Burkholderia seminalis*), and PSB17 (*Sphingomonas pituitosa*) were identified.

The present isolates demonstrated beneficial traits like N_2 fixation, phosphate solubilization and phytohormones production (IAA). Several *Burkholderia* spp. (*Burkholderia unamae*, *Burkholderia xenovorans*, *Burkholderia silvatlantica*, *Burkholderia tropica*, *Burkholderia tuberum*, *Burkholderia phymatum*, *Burkholderia mimosarum* and *Burkholderia nodosa*) have been known to fix N_2 [41] and are commonly find in rice, maize, sugar cane, sorghum, coffee and tomato [42]. The *Sphingomonas* sp. has beneficial attributes like N_2 fixing [43] and high polysaccharide production. These species are able to synthesize the bacterial exopolysaccharide gellan and related polymers and were shown to possess constitutive gellanase activity [44]. Exopolysaccharides have been known to perform a major role in providing protection to the cell as a boundary layer [45], as well as by chelating heavy metals due to the presence of several active functional groups [46].

The isolated PSB were able to grow well under low pH conditions. This might be due to the deposition of lipopolysaccharide in bacterial cultures at a pH below 4.5. The possibility of lipopolysaccharide incorporation into the cell wall is episodic at acidic pH that would argue against the involvement of these polysaccharides in the acid pH toterant bacteria (*R. tropici* UMR1899) as other outer-membrane components may confer greater pH tolerance to the strain [47]. Exopolysaccharides were partially distinguished by their FTIR spectra indicating the presence of carboxylic acid group and H^- bonded group that might have been the reason for the increase of pH in broth culture [44]. However, in the present study, increase in the pH of the nutrient solution from 4 to 7 (both plant and without plant system) might be the result of bacterial exopolysaccharide production. An increase in the pH of bacterial culture medium was the main factor that governs cell growth and exopolysaccharides production [48]. Moreover, bacteria produce polysaccharides in the presence of carbon source and when the pH of the broth is increased from 5.0 to 7.0 [49].

Table 7. Effects of Al and PSB inoculation on the growth of rice seedlings.

| Treatments | Plant height (cm) | | | Dry weight (g) | | | Root length (cm) | | | Root surface area (cm²) | | | Root volume (cm³) | | |
| | Al conc. (µM) | | | | | | | | | | | | | | |
	0	50	100	0	50	100	0	50	100	0	50	100	0	50	100
Control	14.5c	13.3c	11.6c	0.61c	0.59c	0.58c	24.31c	23.69c	18.99d	42.63d	28.29d	25.47c	3.51c	2.62d	1.03c
Burkholderia thailandensis	17.7b	15.6b	14.1a	0.71b	0.69a	0.67a	57.69b	36.32b	33.78c	69.08c	58.43a	42.36a	4.80b	4.19b	3.42a
Burkholderia seminalis	18a	16.4a	14.1a	0.76a	0.69a	0.65b	58.02a	47.28a	36.85b	74.12a	53.02b	43.40a	6.88a	3.76c	2.03b
Sphingomonas pituitosa	17b	15b	13b	0.71b	0.66b	0.65b	58.51a	49.66a	38.01a	72.59b	45.45c	38.52b	6.90a	5.18a	3.81a

Data values are means of five replicates. Means within the same column followed by the same letters are not significantly different at $P<0.05$.

Table 8. Effects of Al and PSB inoculation on the release of organic acids.

| Treatments | Oxalic acid (µM) | | | Citric acid (µM) | | | Malic acid (µM) | | |
| | Al conc. (µM) | | | | | | | | |
	0	50	100	0	50	100	0	50	100
Control	72c	82d	94d	13d	40d	52d	3d	4d	55d
Burkholderia thailandensis	78c	131a	265 a	45b	196a	257b	40a	121a	336a
Burkholderia seminalis	98a	123b	114 c	146a	170b	268a	7b	86b	315b
Sphingomonas pituitosa	95ab	94c	185 b	38c	137c	241c	5c	54c	151c

Data values are means of five replicates. Means within the same column followed by the same letters are not significantly different at $P<0.05$.

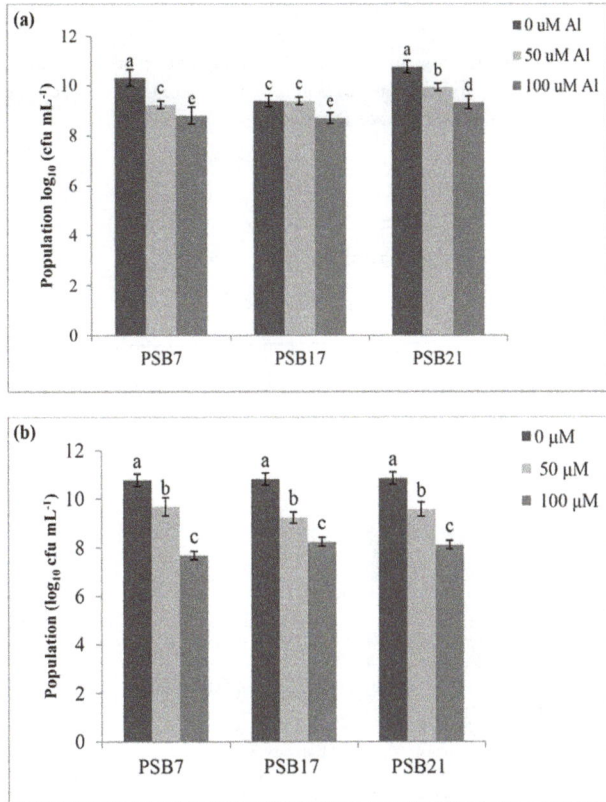

Figure 4. Effect of different Al concentrations on the PSB population (a) without plant, (b) with plant system.

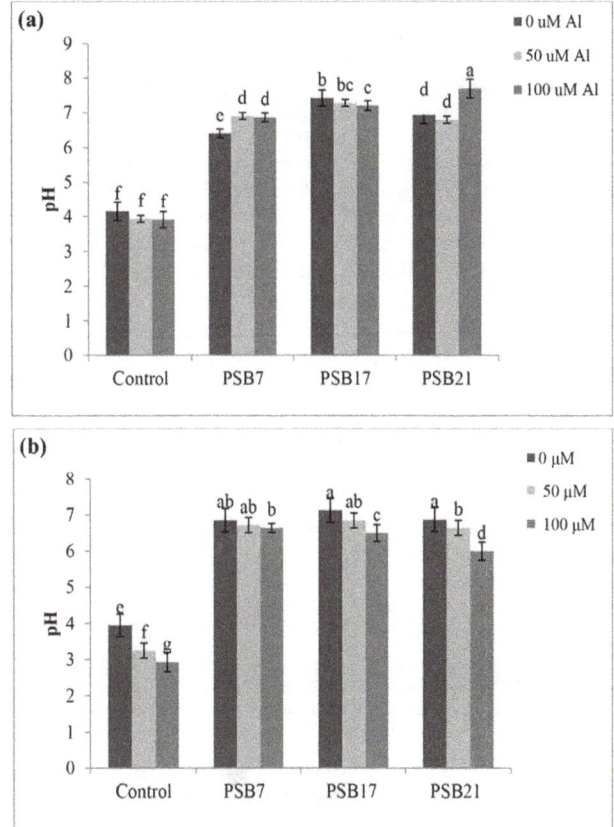

Figure 5. Effect of PSB inoculation on solution pH at different Al concentration a) without plant, b) with plant.

Al toxicity is one of the major constraints to rice root development. In the present study, PSB-inoculated rice seedlings showed better root growth compared to non-inoculated seedlings. The pH of the solution changed from acidic to neutral, providing a favourable environment for plant growth. The better root growth was also promoted by the release of organic acids produced by PSB that chelated Al, rendering it inactive [4]. It is known that rice roots also secrete organic acids (citric, oxalic, and malic acids) and the secretion of these organic acids is localized to the root apex [50]. The inoculated rice seedlings without the presence of Al produced higher plant biomass. This might have been due to the production of phytohormone by the bacteria (such as IAA) that stimulated rice plant growth.

In the present study, multiple reasons might have contributed to a reasonable seedling growth under high Al and low pH conditions. First, the strains were able to produce some polysaccharides that increased the solution pH. Moreover, the strains produced organic acids that helped chelate Al in the solution and consequently reduced Al toxicity. The organic acids performed two main important functions: 1) to detoxify Al; and 2) to make P available [51]. There has been evidence of aluminum resistance in some plant species that has been ascribed to organic acid exudation from roots and by microorganisms that have the ability to chelate Al [52]. However, the most effective organic acids to chelate Al, in decreasing order, are citric, oxalic and tartaric acids [53].

In this study, it is proven that these isolates have the potential to reduce Al toxicity, fix nitrogen, solubilize phosphate, increase soil

pH and produce phytohormone. Therefore, these three bacteria

Figure 6. Effect of Al concentration (100 μM) at pH 4.0 on rice seedlings.

strains can be exploited to enhance the productivity of rice planted on acid sulfate soils either in Malaysia or other part of the tropics.

Conclusions

Rice growing on acid sulfate soils suffers from Al^{3+} toxicity and H^+ which affects its growth and eventually the yield. To some extent, rice with the help of PSB is able to reduce Al^{3+} by excreting organic acids (oxalic, citric and malic acids) via its roots at the appropriate time, and these acids are able to detoxify Al^{3+} via a chelation process. The growth of rice can be improved further with the help of PSB. In this study, three locally isolated bacteria; *Burkholderia thailandensis* (PSB7), *Sphingomonas pituitosa* (PSB17) and *Burkholderia seminalis* (PSB21) of these essential microbes were found to exist under adverse conditions (low pH and high Al concentration). These PSB not only helped to solubilize P in the soils, but their activities resulted in increased pH due to the release of exo-polysaccharides. If the water pH increased above 5, Al would have precipitated as inert Al-hydroxides, making it unavailable to the growing rice in the field.

These PSB could also produce growth enhancement phytohormones such as IAA. The potential of these PSB should be exploited further for the sustainable management of acid sulfate soils, especially for rice production. For instance, they can be used as bio-fertilizers to improve the fertility of acid sulfate soils.

Author Contributions

Conceived and designed the experiments: QAP UAN SJ RO. Performed the experiments: QAP UAN. Analyzed the data: QAP UAN MAL. Contributed reagents/materials/analysis tools: RO MRI. Wrote the paper: QAP UAN SJ RO.

References

1. Sullivan S (2004) Sustainable management of acid sulfate soils. Aust J Soil Res 42(5/6): 595–602.
2. Shamsuddin J, Elisa Azura A, Shazana MARS, Fauziah CI, Panhwar QA, et al. (2014) Properties and management of acid sulfate soils in Southeast Asia for sustainable cultivation of rice, oil palm and cocoa. Adv Agron 124: 91–142.
3. Marschner H (1991) Mechanisms of adaptation of plants to acid soils. Plant Soil 134: 1–20.
4. Shamsuddin J, Elisa AA, Shazana MARS, Che Fauziah I (2013) Rice defense mechanisms against the presence of excess amount of Al^{3+} and Fe^{2+} in the water. Aust J Crop Sci 7(3): 314–320.
5. Shamsuddin J, Che Fauziah I (2010) Alleviating acid soil infertility constraints using basalt, ground magnesium limestone and gypsum in a tropical environment. Malaysian J Soil Sci 14: 1–13.
6. Arias ME, Gonzalez-Perez JA, Gonzalez-Vila FJ, Ball AS (2005) Soil health-a new challenge for microbiologists and chemists. Inter Microbiol 8: 13–21.
7. Khiari L, Parent LE (2005) Phosphorus transformations in acid light-textured soils treated with dry swine manure. Can J Soil Sci Soc 85: 75–87.
8. Rodriguez H, Fraga R (1999) Phosphate solubilizing bacteria and their role in plant growth promotion. Biotech Adv 17: 319–339.
9. Takeuchi M, Hamana K, Hiraishi A (2001) Proposal of the genus *Sphingomonas sensu stricto* and three new genera, *Sphingobium, Novosphingobium* and *Sphingopyxis*, on the basis of phylogenetic and chemotaxonomic analyses. Inter J Syst Evo Microbiol 51: 1405–1417.
10. Janssen PH (2006) Identifying the dominant soil bacteria taxa in libraries of 16S rRNA and 16S rRNA genes. App Environ Microbiol 72: 1719–1728.
11. Naher UA, Radziah O, Shamsuddin ZH, Halimi MS, Mohd Razi I (2009) Isolation of diazotrophs from different soils of Tanjong Karang Rice growing area in Malaysia. Inter J Agri Biol 11(5): 547–552.
12. Naher UA, Radziah O, Shamsuddin ZH, Halimi MS, Mohd Razi I, et al. (2011) Effect of root exuded specific sugars on biological nitrogen fixation and growth promotion in rice (*Oryza sativa*). Aust J Crop Sci 5(10): 1210–1217.
13. Panhwar QA, Radziah O, Zaharah AR, Sariah M, Mohd Razi I (2012) Isolation and characterization of phosphorus solubilizing bacteria from aerobic rice. African J Biotech 11(11): 2711–2719.
14. Benton J Jr (2001) Laboratory guide for conducting soil tests and plant analysis. CRC Press LLC, New York.
15. Teh SC, Talib J (2006) Soil physics analyses volume 1. Universiti Putra Malaysia Press, Kuala Lumpur, Malaysia.
16. Bremner JM, Mulvaney CS (1982) Nitrogen-Total. In: Methods of soil analysis. Agron. No. 9, Part 2: Chemical and microbiological properties, 2nd edn. Madison, WI, USA, 595–624 pp.
17. Nautiyal CS (1999) An efficient microbiological growth medium for screening phosphate-solubilizing microorganisms. FEMS Microbiology Letters 170: 265–270.
18. Prasad G, James EK, Mathan N, Reddy PM, Reinhold-Hurek B, et al. (2001) Endophytic Colonization of Rice by a Diazotrophic Strain of *Serratia marcescens*. J Bacteriol 183(8): 2634–2645.
19. Gordon AS, Weber RP (1950) Colorometric estimation of indoleacetic acid. Plant Physiol 26: 192–195.
20. Nguyen C, Yan W, Le Tacon F, Lapeyrie F (1992) Genetic variability of phosphate solubilizing activity by monocaryotic and dicaryotic mycelia of the ectomycorrhizal fungus *Laccaria bicolor* (Maire) P.D. Orton. Plant Soil 143: 193–199.
21. Katzenelson H, Bose B (1959) Metabolic activity and phosphate dissolving ability of bacterial isolates from wheat roots rhizosphere and non-rhizosphere soil. Can J Microbiol 5: 79–85.
22. Somasegaran P, Hoben HJ (1985) General Microbiology of Rhizobium. Methods in Legume- *Rhizobium* Technology 39–53.
23. Murphy J, Riley JP (1962) A modified single solution method for the determination of phosphate in natural waters. Anal Chim Acta 27: 31–36.
24. Weisburg WG, Barns SM, Pelletier DA (1991) 16S ribosomal DNA amplification for phylogenetic study. J Bacteriol 173: 697–703.
25. Chen YP, Rekha PD, Arun AB (2006) Phosphate solubilizing bacteria from subtropical soil and their tricalcium phosphate solubilizing abilities. App Soil Ecol 34: 33–41.
26. Chung H, Park M, Madhaiyan M (2005) Isolation and characterization of phosphate solubilizing bacteria from the rhizosphere of crop plants of Korea. Soil Biol Biochem 37: 1970–1974.
27. Hamdy EL Z, Czarnes S, Hallett PD, Alamercery S, Bally R, et al. (2007) Early changes in root characteristics of maize (*Zea mays*) following seed inoculation with the PGPR *Azospirillum lipoferum* CRT1. Plant Soil 291: 109–118.
28. Mathew EK, Panda RK, Nair M (2001) Influence of subsurface drainage on crop production and soil quality in a low-lying acid sulphate soil. Agriculture Water Management 47: 191–209.
29. Shamsuddin J, Muhrizal S, Fauziah I, Van Ranst E (2004) A laboratory study of pyrite oxidation in an acid sulfate soils. Comm Soil Sci Plant Anal 35: 117–129.
30. Azlin CO, Amir HG, Chan LK (2005) Isolation and characterization of diazotrophic rhizobacteria of Oil Palm roots. Malaysian J Microbiol 1: 31–35.
31. Glick BR (2005) Modulation of plant ethylene levels by the enzyme ACC deaminase. FEMS Microbiology Letters 251: 1–7.
32. Pradhan N, Sukla LB (2005) Solubilization of inorganic phosphate by fungi isolated from agriculture soil. African J Biotechnol 5: 850–854.
33. Woo SM, Lee MK, Hong IS, Poonguzhali S, Sa TM (2010) Isolation and characterization of phosphate solubilizing bacteria from Chinese cabbage. 19th World Congress of Soil Science, Soil Solutions for a Changing World 1–6 August 2010, Brisbane, Australia.
34. Kumar A, Bhargava P, Rai LC (2010) Isolation and molecular characterization of phosphate solubilizing Enterobacter and Exiguobacterium species from paddy fields of Eastern Uttar Pradesh, India. African J Microbiol Res 4(9): 820–829.
35. Parasanna A, Deepa V, Balakirashna Murthy P, Deecaraman M, Sridhar R, et al. (2011) Insoluble phosphate solubilization by bacterial strains isolated form rice rhizosphere soils from southern India. Inter J Soil Sci 6(2): 134–141.
36. Chakraborty BN, Chakraborty U, Saha A, Sunar K, Dey PL (2010) Evaluation of Phosphate Solubilizers from Soils of North Bengal and Their Diversity Analysis. World J Agri Sci 6(2): 195–200.
37. Nahas E (1996) Factors determining rock phosphate solubilization by microorganisms isolated from soil. World J Microbiol Biotechnol 12: 567–572.
38. Naher UA, Radziah O, Latif MA, Panhwar QA, Puteri AMA, et al. (2013) Biomolecular Characterization of Diazotrophs Isolated from the Tropical Soil in Malaysia. Inter J Mol Sci 14: 17812–17829.
39. Clarridge JE (2004) Impact of 16S rRNA Gene Sequence Analysis for Identification of Bacteria on Clinical Microbiology and Infectious Diseases. Clinical Microbiol Rev 17(4): 840–862.
40. Ludwig W, Amann R, Martinez-Romero E, Schönhuber W, Bauer S, et al. (1998) rRNA based identification and detection systems for rhizobia and other bacteria. Plant Soil 204: 1–19.

41. Suárez-Moreno ZR, Caballero-Mellado J, Venturi V (2008) The new group of non-pathogenic plant-associated nitrogen-fixing *Burkholderia* spp. shares a conserved quorum-sensing system, which is tightly regulated by the RsaL repressor. Microbiology 154(7): 2048–2059.

42. Caballero-Mellado J, Onofre-Lemus J, Estrada-de los Santos P, Martinez-Aguilar L (2007) The tomato rhizosphere, an environment rich in nitrogen-fixing Burkholderia species with capabilities of interest for agriculture and bioremediation. App Environ Microbiol 73: 5308–5319.

43. Zhang JY, Liu XY, Liu SJ (2010) *Sphingomonas changbaiensis* sp. nov., isolated from forest soil. Inter J Syst Evo Microbiol 60: 790–795.

44. Sunil PT, Amarsinh Bhosale A, Trishala Gawade B, Tejswini Nale R (2013) Isolation, screening and optimization of exopolysaccharide producing bacterium from saline soil. J Microbiol Biotechnol Res 3(3): 24–31.

45. Caiola MG, Billi D, Friedmann EI (1996) Effect of desiccation on envelopes of the *cyanobacterium Chroococcidiopsis* sp. (Chroococcales). European J Phycol 31: 97–105.

46. Kaplan D, Christiaen D, Arad SM (1987) Chelating Properties of Extracellular Polysaccharides from *Chlorella* spp. App Environ Microbiol 53(12): 2953–2956.

47. Graham PH, Draeger KJ, Ferrey ML, Conroy MJ, Hammer BE, et al. (1994) Acid pH tolerance in strains of *Rhizobium, Bradyrhizobium*, and initial studies on the basis for acid tolerance of *Rhizobium tropici* UMR 1899. Can J Microbiol 40: 198–207.

48. Singh S, Das A (2011) Screening, production, optimization and characterization of cyanobacterial polysaccharide World J Microbiol Biotechnol 27: 1971–1980.

49. Bueno SM, Garcia-Cruz CH (2006) Optimization of polysaccharides production by bacteria isolated from soil. Brazilian J Microbiol 37(3): 296–301.

50. Ma JF (2000) Role of organic acids in detoxification of aluminum in higher plants. Plant Cell Physiol 41: 383–390.

51. Haynes RJ, Mokolobate MS (2001) Amelioration of Al toxicity and P deficiency in acid soils by additions of organic residues: a critical review of the phenomenon and the mechanisms involved. Nutr Cycl Agroecosyst 59: 47–63.

52. Klugh KR, Cumming JR (2007) Variations in organic acid exudation and aluminum resistance among arbuscular mycorrhizal species colonizing *Liriodendron tulipifera*. Tree Physiol 27: 1103–1112.

53. Hue NV, Craddock GR, Adams F (1986) Effect of organic acids on aluminium toxicity in subsoils. Soil Sci Soc Am J 50: 28–34.

20

Spatial and Temporal Variations of Ecosystem Service Values in Relation to Land Use Pattern in the Loess Plateau of China at Town Scale

Xuan Fang[1], Guoan Tang[1]*, Bicheng Li[2], Ruiming Han[3]

1 Key Laboratory of Virtual Geographic Environment, Ministry of Education, School of Geography Science, Nanjing Normal University, Nanjing, China, 2 Research Center of Soil and Water Conservation and Ecological Environment, Chinese Academy of Sciences, Yangling, Shaanxi, China, 3 School of Geography Science, Nanjing Normal University, Nanjing, China

Abstract

Understanding the relationship between land use change and ecosystem service values (ESVs) is the key for improving ecosystem health and sustainability. This study estimated the spatial and temporal variations of ESVs at town scale in relation to land use change in the Loess Plateau which is characterized by its environmental vulnerability, then analyzed and discussed the relationship between ESVs and land use pattern. The result showed that ESVs increased with land use change from 1982 to 2008. The total ESVs increased by 16.17% from US$ 6.315 million at 1982 to US$ 7.336 million at 2002 before the start of the Grain to Green project, while increased significantly thereafter by 67.61% to US$ 11.275 million at 2008 along with the project progressed. Areas with high ESVs appeared mainly in the center and the east where largely distributing orchard and forestland, while those with low ESVs occurred mainly in the north and the south where largely distributing cropland. Correlation and regression analysis showed that land use pattern was significantly positively related with ESVs. The proportion of forestland had a positive effect on ESVs, however, that of cropland had a negative effect. Diversification, fragmentation and interspersion of landscape positively affected ESVs, while land use intensity showed a negative effect. It is concluded that continuing the Grain to Green project and encouraging diversified agriculture benefit to improve the ecosystem service.

Editor: Ricardo Bomfim Machado, University of Brasilia, Brazil

Funding: This study was sponsored by the Jiangsu Planned Projects for Postdoctoral Research Funds (No. 1401033C), the National Natural Science Foundation of China (No. 41401441), and the Priority Academic Program Development of Jiangsu Higher Education Institutions (PAPD) (No. 164320H101). The funders had no role in study design, data collection and analysis, decision to publish, or preparation of the manuscript.

Competing Interests: The authors have declared that no competing interests exist.

* Email: tangguoan@njnu.edu.cn

Introduction

Ecosystem contributes to human welfare by providing goods and services directly and indirectly [1–2]. With widely spreading of environmental problems, ecosystem service received increasing attention. Many studies showed human factors, such as urban sprawl [3,4,5], socioeconomic changes [6], agricultural policies [7,8], could affect natural or artificial ecosystems. Land use, an original and foundational human activity and represents the most substantial human alteration to systems on the planet of earth for long-term study [9], plays an important role in providing ecosystem services, including biodiversity, water filtration, retention of soil, etc. [10] Inappropriate land use may lead to significant degradation of local and regional ecological services [11]. Moreover, there were studies showed that ecosystem service trade-offs could successful apply to land use planning [12,13]. Understanding the relationship between ecosystem services and land use change is essential for maintaining a healthy ecosystem and getting sustainable services.

The growing body of literatures focused on how ecosystem service changes in response to land use change of different regions [14,15,16,17,18]. However, these studied focus on the impact of land use type on ecosystem service, while the spatial pattern of land that reflects ecological processed and functions [19] get less attention. Monitoring the characteristic of landscape patterns including area, shape, diversity, etc., is helpful to deeply understand the relationship between ecosystem service and land use change and then to provide complete references for land use planning.

The Loess Plateau is the area suffered from the most severe soil erosion in the world, and it is also a major agricultural production region in China [20]. Long-term poor land use has resulted in vegetation destruction and accelerated soil erosion [21]. To control soil erosion and restore the ecosystem, the Grain for Green project converting slope cropland to grassland or forestland was implemented in 1999 by the Chinese Government [22]. The land use on the plateau under the project has changed significantly. Studying the ecosystem service in relation to land use change before and after the Grain to Green project was crucial for ecosystem protection and agricultural sustainability for the area. Researchers have analyzed ecosystem service at different scales within the Loess Plateau [17,18,23]. However, town is a basic administrative area in China. Exploring the characteristic of ecosystem services change at town scale is of practical significance to provide operable land use planning.

Ecosystem service values (ESVs) is monetary assessment of ecosystem services. This paper examined the characteristics of ESVs at Hechuan town, a typical town in the hilly and gully region of the Loess Plateau. The objectives of this study were: 1) to analyze the changes in land use pattern from 1982 to 2008; 2) to access the spatial and temporal variation in ESVs in response to land use during this period; 3) to quantitively analysis the relationship between ESVs and land use pattern; and 4) to discuss how land use management is favorable for ecosystem service supply and the ecological and economic sustainable development.

Data and Methods

2.1 Ethics statement

No specific permits were required for the described studies, and the work did not involve any endangered or protected species.

2.2 Study area

The study area, Hechuan town ($106°18'43''{\sim}106°32'16''$E, $35°54'59''{\sim}36°06'05''$N), is located in Guyuan city of the Ningxia Hui Autonomous Region of northwest China (Fig. 1), consisting 12 villages with 16,524 people. The reasons that Hechuan Town was chosen as the study area were, on the one hand, Hechuan town has the typical characteristics of Loess Plateau including the

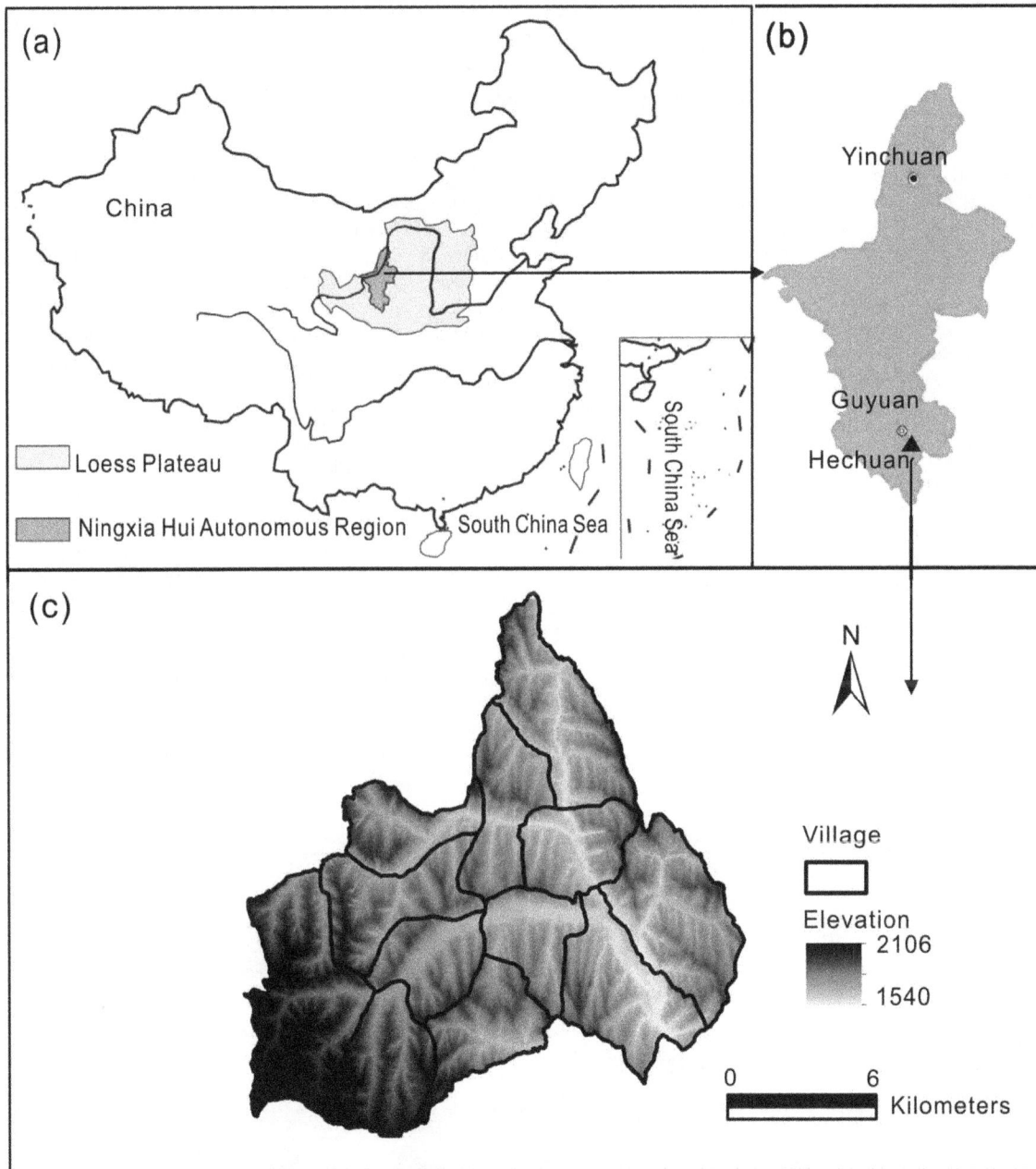

Figure 1. Location of the study area. Ningxia Province and the Loess Plateau, China (a), the location of Hechuan Town in the Loess area of Ningxia Province (b) and, the village boundary and the digital elevation model (DEM) map of Hechuan Town (c).

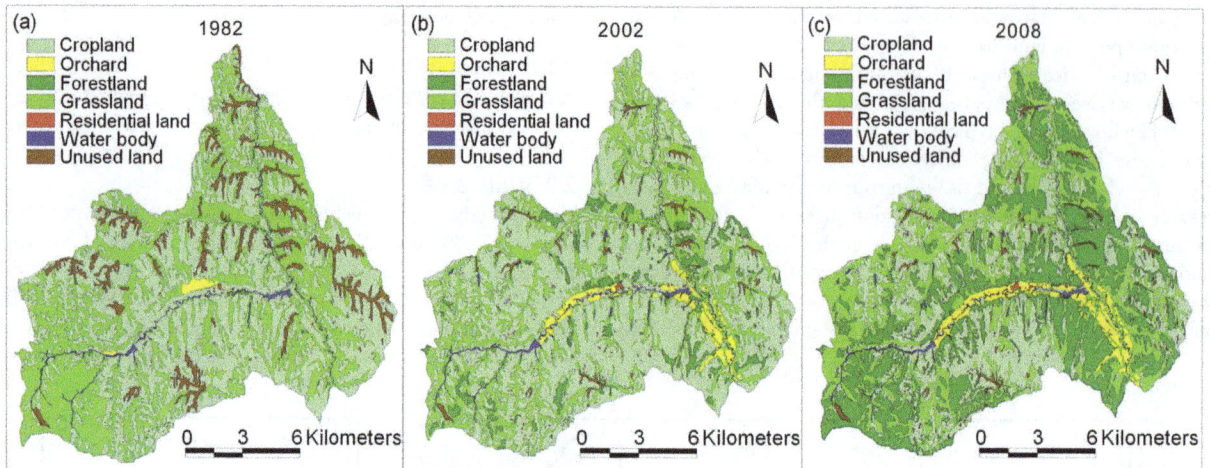

Figure 2. Land use maps of Hechuan town in 1982 (a), 2002 (b) and 2008 (c).

terrain of hill and gull, the fragile ecosystem and the backward economy; on the other hand, there was a long term ecological observation and experiment station in the study area, which facilitated the survey of land use and ecosystems. This town has an altitude ranging from 1540 to 2106 m, covering an area of 215.58 km^2. There exist the topographic differences in the town. The central area with river terrace stretches smoothly with a low elevation. The terrain in the northern area is fragmented while that of southern area is relatively simple. Hechuan town has a semi-arid continental temperate climate with the average annual temperature of 6.9°C and precipitation of 419 mm (1982–2002). Most of the annual precipitation is concentrated between June to September in the form of heavy storms that can cause severe soil erosion. The soil is composed of loessial soil and Dark loessial soils, which is erodible due to its weak cohesion and high infiltrability.

The ecosystem in Hechuan town is fragile with serious soil erosion and frequent natural disasters. Human disturbances of excessive land use, such as deforestation, overgrazing and over-reclamation further destructed the native natural grassland. Therefore, this area has long been in a vicious circle, endless cultivation and poverty. Since the early 1980s, a variety of comprehensive investigation of soil erosion was practiced by Chinese Academy of Sciences. Shanghuang watershed, located in the east of Hechuan town, was taken as a key test area. The

ecological restoration covering the whole town was started from implementing the Grain for Green project after 2002 (launched in 1999 by China government). Since then, abandoned cropland, shrubland (*Caragana korshinskii*, *Hippophae rhamnoides*) and artificial grassland (*Medicago sativa*) was generated, which made a significant change on landscape pattern and ecosystem components providing a variety of ecosystem services. Meanwhile, farming and grazing, the traditional way of living, had to be changed, and raising livestock, orchards, and migrant working diversified their incomes.

2.3 Data acquisition and preprocessing

Land use data was the key data for evaluating landscape pattern and ecosystem service. The land use data of 1982 was obtained by digitizing the land use patches from the 1:10,000 scale topographic maps of 1982, in which the information of land use types and its boundary are clearly shown. The 10 m resolution of remote sensing image could be considered to be corresponding with the scale of 1:50000 [24,25]. The land use data of 1982 acquired from 1:10000 topographic maps was therefore generalized to be at 1:50000 scale [26]. The land use data of 2002 and 2008 were respectively extracted from the 10 m resolution multispectral Spot-5 image of 2002 and 2008 by updating the land use patches of

Table 1. Equivalent weight factor of ecosystem service values (ESVs) per hectare of terrestrial ecosystem in China [30].

	Cropland	Forestland	Grass land	Water body	Barren land
Gas regulation	0.72	4.32	1.5	0.51	0.06
Climate regulation	0.97	4.07	1.56	2.06	0.13
Water supply	0.77	4.09	1.52	18.77	0.07
Soil formation and retention	1.47	4.02	2.24	0.41	0.17
Waste treatment	1.39	1.72	1.32	14.85	0.26
Biodiversity protection	1.02	4.51	1.87	3.43	0.40
Food production	1.00	0.33	0.43	0.53	0.02
Raw material	0.39	2.98	0.36	0.35	0.04
Recreation and culture	0.17	2.08	0.87	4.44	0.24
Total	7.90	28.12	11.67	45.35	1.39

Table 2. The ecosystem service values (ESVs) per hectare of different land use types in Hechuan town (US$·ha-1·yr-1).

	Cropland	Orchard	Forestland	Grass land	Water body	Unused land
Gas regulation	22.570	91.222	135.422	47.022	15.987	1.881
Climate regulation	30.407	88.244	127.585	48.902	64.576	4.075
Water supply	24.138	87.930	128.212	47.649	588.397	2.194
Soil formation and retention	46.081	98.118	126.018	70.219	12.853	5.329
Waste treatment	43.573	47.649	53.918	41.379	465.514	8.150
Biodiversity protection	31.975	99.999	141.378	58.620	107.523	12.539
Food production	31.348	11.912	10.345	13.480	16.614	0.627
Raw material	12.226	52.351	93.416	11.285	10.972	1.254
Recreation and culture	5.329	46.238	65.203	27.272	139.184	7.523
Total	247.647	623.663	881.498	365.828	1421.619	43.573

1982 one by one in visual interpretation method. The interpretation sign was established by understanding the Spot image characteristics and carrying out field surveys in order to further determine the relationship between the true ground and the image. The kappa accuracy index [27] was used to assess the accuracy of the interpretation. The stratified random sampling method was used to generate the reference points on the classified image for the accuracy test. These reference points were located in the field with a GPS with 5-m precision for ground truth. The total kappa indexes are all higher than 0.85, which are higher than the minimum acceptable (0.7) [28]. Considering the characteristic of the land use in study area and the interpretation level of the data and to facilitate the calculation of ESVs, the land use was classified into seven types: cropland, orchard, forestland, grassland, residential area, water area, and unused land (Fig. 2).

To acquire accurate area data of the land use for ESVs estimation and facilitate analyzing the spatial distribution of ESVs, the topographic maps and Spot images were transformed to the same projection and coordinate system (the Albers-Conical-Equal-Area projection system and Krasovsky 1940 coordinate system) before the extraction of land use data, and all acquired land use data were transformed to Arc-grid formats with the same grid size (10 m×10 m). The above data processing was completed using ERDAS and ArcGIS software.

2.4 Analysis on land use pattern

The transfer matrix analysis of land use was produced to understand how land use changed. Landscape metrics analysis was used for spatial pattern analysis of land use. Landscape metrics has been adopted widely; meanwhile, its abilities to indicate ecological process gained increasing attention [29,30,31]. Conceptual flaws in landscape pattern analysis, limitations inherent in landscape metrics and the improper use of pattern analysis may lead to the misuse of landscape metrics [32]. For better explanations and predictions of ecological phenomena from ecological pattern, the landscape metrics in this study was therefore selected by two steps. Firstly, the diversity, the fragmentation and the dominance of landscape were all considered, and then 34 metrics was selected, by understanding the knowledge of the landscape pattern and the ecological services indication of landscape metrics [33,34] and referring to the previous studies on landscape pattern [4,31,35,36]. Secondly, a correlation analysis for the 34 metrics was employed to ensure the low redundancy among landscape metrics. If the coefficient between two metrics was significant at 0.05 level, only one metric of them could be eventually selected.

Landscape-level metrics providing general landscape information and class-level metrics providing more specific information about variations at the local level and spatial patterns of land use classes [37] were used to monitor the characteristics of landscape pattern. The selected landscape-level metrics were patch density (PD), area-weighted mean shape index (SHAPE_AM), Interspersion and Justaposition Index (IJI), and Shannon's diversity index (SHDI). The selected class-level metrics were PD, the percentage of landscape (PLAND), SHAPE_AM and IJI. PD and SHAPE_AM could show the fragmentation of landscape. SHDI and PLAND reflect the dominance of some land use type and the diversity of landscape, respectively. IJI reflects whether the patches or classes are contiguous. Landscape metrics analysis was conducted with above metrics by FRAGSTATS 3.3, in which the eight-neighbor rule was used to derive the patch number. Besides these metrics, the land use intensity index (LUII) was also used to describe the landscape pattern. It was calculated by the following equation [31]:

$$LUII = \sum_{i=1}^{n} A_i \times C_i \qquad (1)$$

where $LUII$ is the land use intensity index, A_i is the percentage of for a give land use type i, and C_i is the coefficient value of intensity for a give land use type i, that is assigned 4 for build-ups, 3 for farmland and 2 for forest, orchard, grassland and water bodies, and 1 for unused land.

2.5 Estimation of ESVs

Costanza et al.'s model of ESVs estimation was adopted in this study [1,2]. The model classified ecosystem service into 17 types of service functions and estimated the ESVs by placing an economic value on different biomes [34]. For the defects of this model, such as overestimating the agriculture ESVs and underestimating the wetland ESVs, Xie et al. proposed refined coefficients for ESVs assessment both solving the above problem and making it apply to China [33,34]. Based on this model, the total ESVs in the study area was calculated using the following formulas:

$$ESV_k = \sum_f A_k VC_{kf} \qquad (2)$$

Table 3. Land use transition matrix from 1982 to 2002 and from 2002 to 2008 (%).

1982	2002								
	Cropland	Orchard	Forestland	Grassland	Residential land	Water body	Unused land	Total	Loss
Cropland	44.42	2.33	1.18	2.50	0.35	0.04	0.01	50.83	6.41
Orchard	0.17	0.05	0.26	0.09	0.00	0.00	0.00	0.57	0.53
Forestland	0.10	0.01	0.15	0.02	0.00	0.00	0.00	0.28	0.13
Grass land	12.63	0.39	4.74	22.15	0.01	0.08	0.00	40.01	17.86
Residential land	0.05	0.01	0.00	0.01	0.14	0.00	0.00	0.22	0.07
Water body	0.01	0.00	0.01	0.00	0.00	0.80	0.00	0.81	0.02
Unused land	1.39	0.00	0.16	3.98	0.00	0.07	1.66	7.27	5.61
Total	58.76	2.78	6.51	28.77	0.51	0.99	1.68	100.00	
Gain	14.35	2.73	6.36	6.62	0.37	0.19	0.01		

2002	2008								
	Cropland	Orchard	Forestland	Grass land	Residential land	Water body	Unused land	Total	Loss
Cropland	27.09	1.00	20.75	9.84	0.07	0.00	0.00	58.76	31.68
Orchard	0.00	2.75	0.01	0.00	0.01	0.00	0.00	2.78	0.03
Forestland	0.05	0.04	5.84	0.57	0.01	0.01	0.00	6.51	0.67
Grass land	0.12	0.00	7.17	21.42	0.02	0.04	0.00	28.77	7.35
Residential land	0.00	0.00	0.00	0.00	0.51	0.00	0.00	0.51	0.00
Water body	0.02	0.02	0.00	0.01	0.00	0.94	0.00	0.99	0.05
Unused land	0.00	0.00	0.27	0.08	0.00	0.00	1.33	1.68	0.35
Total	27.27	3.82	34.05	31.91	0.62	1.00	1.33	100.00	
Gain	0.18	1.07	28.21	10.50	0.11	0.05	0.00		

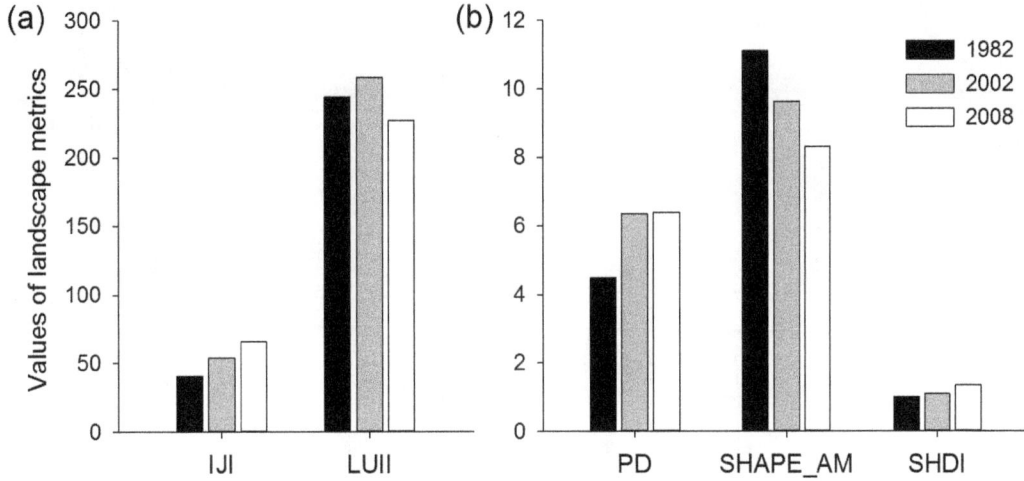

Figure 3. Landscape metrics at the landscape level in Hechuan Town in 1982, 2002 and 2008. IJI: Interspersion and Justaposition Index; LUII: land use intensity index; PD: patch density; SHAPE_AM: area-weighted mean shape index; SHDI: Shannon's diversity index.

$$ESV_f = \sum_k A_k VC_{kf} \qquad (3)$$

$$ESV = \sum_k \sum_f A_k VC_{kf} \qquad (4)$$

where ESV_k, ESV_f, ESV are the ESVs of land use type k, the ESVs of ecosystem service function type f, and the total ESVs respectively. A_k is the area (ha) for land use types. VC_{kf} is the value coefficient (US\$·ha-1·yr-1) for land use type k and ecosystem service function type f, which is the key for ESVs estimating. Xie et al.'s model was used to determine VC_{kf}, which can be expressed as follows:

$$VC_{kf} = R_{kf} \times V_f \qquad (5)$$

where R_{kf} is the equivalent weight factor of ecosystem service, V_f is food production values of agriculture land per area per year.

The equivalent weight factor was presented for customizing Chinese terrestrial ecosystem based on Costanza et al.'s model by surveying 500 Chinese ecologists (Table 1) [34]. It is the ratio of the ESVs to the economic value of average natural food production provided by agricultural land per hectare per year. The factors of land use types in our study were basically assigned based on the nearest ecosystems in Xie et al.'s model. However, minor adjustments were made. The equivalent weight factor of orchard which was not put forward clearly in Xie et al.'s model was determined by the mean of grassland and forestland by referring some researches [5,18]. The factor of unused land equates to that of barren land, and that of residential land was determined to zero.

The value of food production service of agriculture land per area per year was considered to be 1/7 of the actual price of food production in Xie et al.'s model. With the average actual food production of cropland in Hechuan town from 1982 to 2008 of 901.77 kg/ha which was get from *Statistic yearbook of the Yuanzhou District, Guyuan City, Ningxia Hui Autonomous Region* and the average grain price of US\$ 0.243 per kilogram (i.e. an equivalent of RMB Yuan 1.69 according to the average exchange

rate of 2008) in 2008, the value of food production service of cropland per area per year was calculated to be US\$ 31.348 (i.e. an equivalent of RMB Yuan 217.713 according to the average exchange rate of 2008). ESVs of one unit area of each land use types were then assigned as shown in Table 2.

After the ESVs were calculated by above processing, a sensitivity analysis was conducted to test the land use type's representative for ecosystem types and the certainty of the coefficients value for ecosystem service. A coefficient of sensitivity (CS) was used to indicate the degree of sensitivity of ESVs to a coefficients value, calculated by the following formula [5]:

$$CS = \left| \frac{(ESV_j - ESV_i)/ESV_i}{(VC_{jk} - VC_{ik})/VC_{ik}} \right| \qquad (6)$$

where ESV_j an ESV_i are the total ESVs of the initial status j and the adjusted status i, and VC_{jk} and VC_{ik} are the initial and adjusted coefficients. A 50% adjustment in the coefficients was made in the study. The greater the CS responded to the adjustment, the more critical is the use of an accurate coefficient [38]. A CS lower than 1 indicates the ESVs is inelastic to the coefficient and the estimation of ESVS is reliable. Otherwise, a CS greater than 1 indicates the estimation of ESVs is sensitive to the coefficient.

2.6 Correlation and regression analysis

The data of ESVs and landscape metrics was used to analysis the relationship between ecosystem service and land use pattern change. Because the spatial variation of landscape pattern exist among 12 villages in Hechuan town, the land use data of the three years (1982, 2002 and 2008) for the 12 villages can be considered as representing different landscape pattern on a time-for-space perspective [39]. Therefore, there were totally 36 sample data. Correlation and regression was employed for the relationship analysis, in which Multiple stepwise regression was specifically chosen considering the multicollinearity among landscape metrics. The dependents were the nine categories and total ESVs, while the corresponding independents were the landscape-level and class-level landscape metrics.

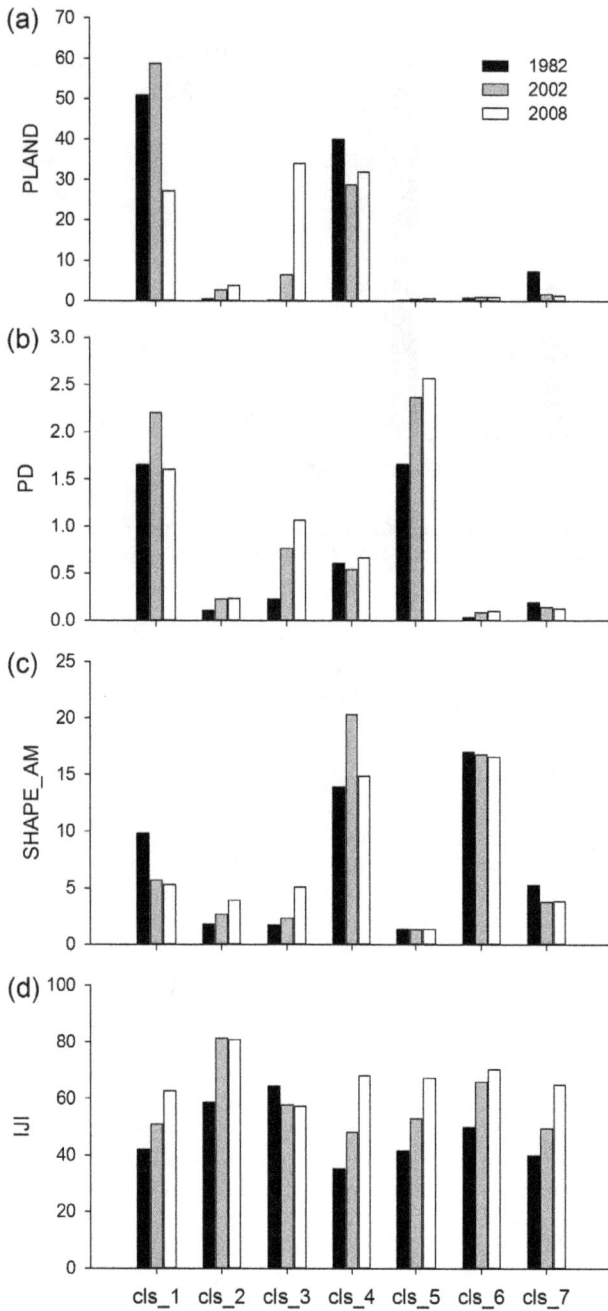

Figure 4. Landscape metrics at the class-level in Hechuan Town in 1982, 2002 and 2008. cls_1, cls_2, cls_3, cls_4, cls_5, cls_6, and cls_7 represent cropland, orchard, forestland, grassland, residential land, water body and unused land. PLAND: the percentage of landscape; PD: patch density; SHAPE_AM: area-weighted mean shape index; IJI: Interspersion and Justaposition Index.

Results

3.1 Changes of land use pattern

Table 3 showed the land use transition matrix. From 1982 to 2002, cropland as the dominant land use type increased from 50.83% to 58.76%. Grassland was the land use type with the largest change in area, decreasing from 40.01% to 28.77%. Orchard increased by 6.24% of total area, indicating the economic driving force of fruit trees on land use change. Forestland

Table 4. The change of ecosystem service values (ESVs) in Hechuan Town from 1982 to 2008.

		Cropland	Orchard	Forestland	Grass land	Water body	Unused land	Total
ESVs (10^6 US$ yr^-1)	1982	2.714	0.077	0.051	3.155	0.249	0.068	6.315
	2002	3.137	0.374	1.237	2.269	0.304	0.016	7.336
	2008	1.456	0.514	6.470	2.517	0.305	0.013	11.275
Change of ESVs (10^6 US$ yr^-1)	1982–2002	0.423	0.297	1.186	-0.887	0.054	-0.053	1.021
	2002–2008	-1.681	0.140	5.234	0.248	0.002	-0.003	3.939
	1982–2008	-1.258	0.437	6.419	-0.638	0.056	-0.056	4.960
Change of ESVs (%)	1982–2002	2.248	55.387	333.431	-4.045	3.145	-11.081	2.328
	2002–2008	-7.716	5.404	60.934	1.574	0.072	-2.992	7.731
	1982–2008	-6.674	81.578	1805.410	-2.913	3.232	-11.771	11.309
Average annual Change (%yr^-1)	1982–2002	0.112	2.769	16.672	-0.202	0.157	-0.554	0.117
	2002–2008	-1.286	0.901	10.155	0.262	0.012	-0.498	1.289
	1982–2008	-0.256	3.137	69.439	-0.112	0.124	-0.452	0.435

Table 5. Values of different ecosystem service functions in 1982, 2002, and 2008.

	1982			2002			2008		
	ESVs (10⁶ US$·yr⁻¹)	%	Rank	ESVs (10⁶ US$·yr⁻¹)	%	Rank	ESVs (10⁶ US$·yr⁻¹)	%	Rank
Gas regulation	0.678	10.73	6	0.826	11.26	6	1.529	13.56	5
Climate regulation	0.791	12.53	5	0.936	12.75	5	1.540	13.65	4
Water supply	0.800	12.67	4	0.960	13.09	4	1.610	14.28	3
Soil formation and retention	1.141	18.06	1	1.260	17.17	1	1.764	15.65	1
Waste treatment	0.938	14.85	2	1.015	13.84	3	1.078	9.56	6
Biodiversity protection	0.915	14.49	3	1.054	14.37	2	1.738	15.42	2
Food production	0.466	7.38	7	0.506	6.90	7	0.367	3.25	9
Raw material	0.247	3.91	9	0.390	5.32	8	0.881	7.81	7
Recreation and culture	0.339	5.37	8	0.388	5.29	9	0.768	6.81	8
Total	6.315	100.00		7.336	100.00		11.275	100.00	

increased from 0.57% to 2.78%, reflecting that ecological restoration began to gain attention. From 2002 to 2008, cropland and forestland changed significantly, decreasing from 58.76% to 27.27% and increasing from 6.51% to 34.05% respectively. Land use structure was transferred from cropland dominated (58.76%) to cultivated land (27.27%), forestland (34.05%) and grassland (31.91%) relatively balanced distributed.

The most notable change of land use from 1982 to 2002 was the conversion from grassland to cropland and forestland with 12.63% and 4.74% of the total area respectively. The conversions from cropland (2.50%) and unused land (3.98%) to grassland were not adequate to compensate for the grass loss. From 2002 to 2008, the notable changes of land use were cropland to forestland, cropland to grassland, and grassland to forestland, with the rates of 20.75%, 9.84%, and 7.17% respectively. It was found that the conversion among land use types was more outstanding and concentrated than that before 2002, reflecting that the Grain for Green project as an ecological policy had great influence on land use change.

The results of landscape-level metric analysis were exhibited in Fig. 3. The significant increased PD from 1982 to 2002 reflected the landscape fragmentation. It was relative to the increase of patches on the land use types with intense human disturbance, such as cropland, residential land and artificial reservoir. Oppositely, the slight change of PD from 2002 to 2008 reflected that human disturbance became stable. The change of human disturbance was also demonstrated by the change of LUII which increased before 2002 and decreased after 2002. SHAPE_AM decreased in the study period, showing the landscape became more regular in shape. The increase of IJI suggested that the landscape became more contiguous and the ecological connectivity among land use types increased. SHDI increase obviously from 2002 to 2008, which related to that the land use structure became even.

Fig. 4 showed the change of class-level metrics. The PLAND of land use types indicated that cropland, forestland, and grassland had significantly influence on land use pattern. PD in orchard, forestland, and residential land increased obviously, attributing to the increasing area of these land use types and the fragmental terrain. SHAPE_AM showed that cropland and unused land became more regular in shape, while orchard and forestland more complicated. IJI increased generally in land use types. Orchard was the most contiguous with high IJI, which was relative to its concentrated distribution across the river terrace.

3.2 ESVs from 1982 to 2008

The ESVs of each land use type and the total ESVs was shown in Table 4. The total ESVs of Hechuan town was US$ 6.315, US$ 7.336 and US$ 11.275 million in 1982, 2002 and 2008, respectively. From 1982 to 2002, the decline of ESVs caused by the decrease of grassland was offset by the increase of forestland, orchard and cropland, resulting that the total ESVs increased by US$ 1.021 million. From 2002 to 2008, the total ESVs increased by US$ 3.939 million, mainly due to the increase of forestland. The average annual change rate of total ESVs before and after 2002 was quite different, that is 0.81% and 8.95% respectively. It indicated the Grain to Green project implemented since 2002 had a significant effect on the ecosystem service. It was also shown from the value of ESVs produced by forestland occupying 57.39% of the total ESVs. Overall, the total ESVs increased US$ 4.960 million during the study period, mainly due to the increase of ESVs by the increase of forestland and orchard far beyond the decrease of ESVs by the decrease of cropland and grassland. It was essentially because of the higher coefficient value of forestland and orchard than that of cropland and grassland.

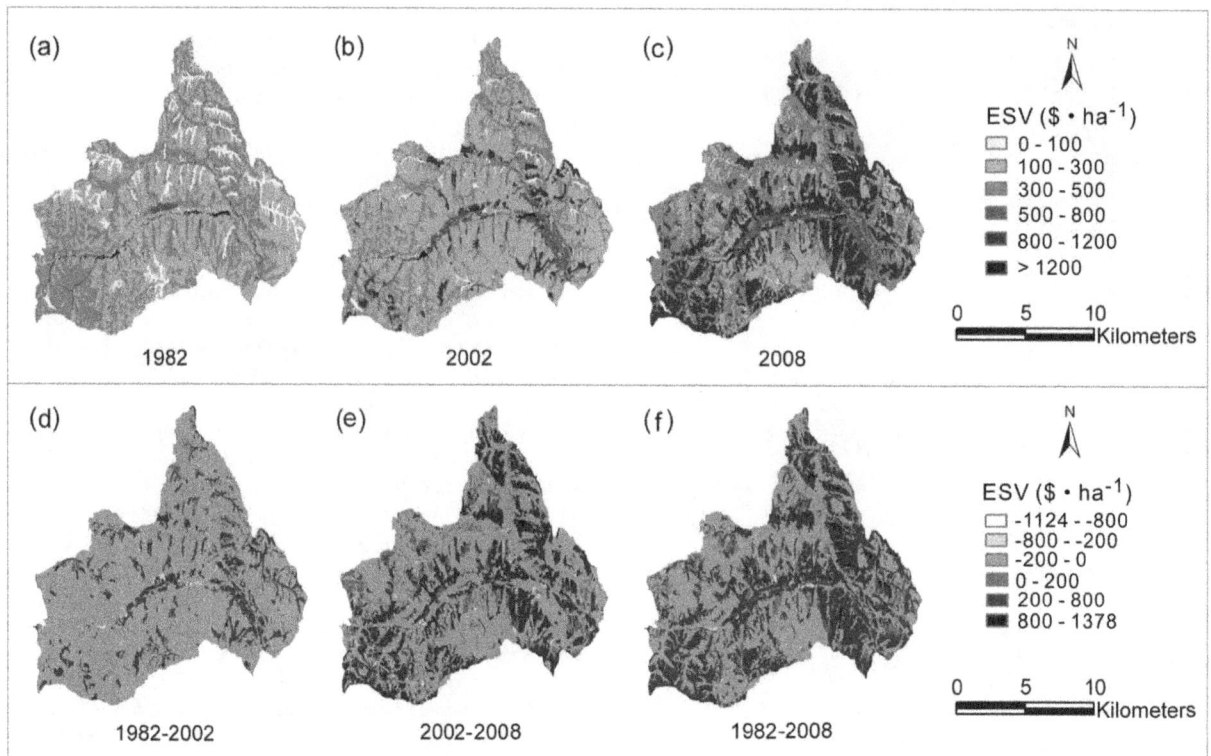

Figure 5. Spatial and temporal distribution of ecosystem service values (ESVs) in Hechuan Town from 1982 to 2008. The spatial distribution of ESVs in 1982 (a), 2002 (b) and 2008 (c), and the spatial-temporal changes of ESVs between time intervals from 1982 to 2002 (d), 2002 to 2008 (e) and 1982 to 2008 (f).

The ESVs of each ecosystem function type was shown in Table 5. Expect for food production, the values of ecosystem service functions increased especially after 2002. The decrease of food production was due to the great decline of cropland in the Grain to Green project. The ESVs proportion of each ecosystem function type to the total ESVs represented the contribution of each ecosystem function to the total ESVs. It was found that the functions of soil formation and retention, waste treatment, and food production were decline during 1982 to 2008, while other functions were improved. The rank of the contribution by each ecosystem service function was also estimated. It was basically stable except for relatively obvious decline in the rank of waste treatment and food production. In 2008, the rank order for each ecosystem service was as follows from high to low, soil formation and retention, biodiversity protection, water supply, climate regulation, gas regulation, waste treatment, raw material, recreation and culture, and food production. Soil formation and retention was the highest during the study period.

3.3 Spatial distribution of ESVs

Maps of ESVs in different periods (Fig. 5) showed the spatial distribution of ESVs of unit area in Hechuan town, directly reflecting the difference of ESVs among land use types. In 1982, the ESVs>4000 mostly appeared in the center of the town where river and river terrace located. It was because water body and orchard which intensely distributed in river terrace for its high water demand both had high ESVs. Therefore, due to the orchard increasing intensely and the forest increasing scatteredly, the increase of ESVs also mainly happened across the river terrace in 2002. Since 2008, the ESVs>4000 spread widely with the increase of forestland transformed from cropland. The lowest ESVs mostly

occurred in the gully where unused land was distributed in 1982. With vegetation recovery in the gully, the low ESVs happened from gully to terraced hillside where cropland with low ESVs was distributed in 2008. Fig. 5d–f showed the temporal change of ESVs spatial distribution. The change characteristic of 2002 to 2008 was adjacent to that during the total study period, reflecting that the change of ESVs mainly occurred after 2002, just after the Grain to Green project.

3.4 Relationship between ESVs and land use pattern

From the above analysis on the change of land use and ESVs in quantity and spatial distribution, we could infer there was some relationship between land use change and ecosystem service. To quantitively understand the relationship, the correlation analysis and regression analysis between ESVs and landscape pattern metrics was conducted.

Table 6 showed there existed significant correlations between ESVs and many landscape metrics (p<0.01), which explained that landscape pattern affected ESVs significantly. For example, the correlation coefficients between total ESVs and landscape metrics showed that there existed significantly positive relationship between SHDI (0.433), PLAND_3 (0.677), SHAPE_AM_3 (0.744), IJI_4 (0.513) and ESVs, and negative relationship between LUII (−0.634), PLAND_1 (−0.752) and ESVs. It reflected that the diversity and intensity of land use had important effects on total ESVs. It also reflected that cropland, forestland and grassland were the land use types which had significant effects on total ESVs. On quantity, the less the cropland and the more the forestland, the higher the total ESVs were. As to the landscape shape, the more regular the cropland and the more complex the forestland, the higher the total ESVs were. The higher the IJI of grassland, the

Table 6. Correlation coefficients between ecosystem service values (ESVs) and landscape pattern metrics.

	TESVs	ESVs_1	ESVs_2	ESVs_3	ESVs_4	ESVs_5	ESVs_6	ESVs_7	ESVs_8	ESVs_9
PD	0.035	0.497*	0.509*	0.539*	0.477*	0.516*	0.499*	−0.221	0.547	0.478*
SHAPE_AM	0.326	−0.216	−0.220	−0.188	−0.177	−0.088	−0.205	0.026	−0.290	−0.166
IJI	0.292	0.624*	0.635*	0.639*	0.597*	0.534*	0.621*	−0.293	0.687	0.586*
SHDI	0.433*	0.763*	0.764*	0.766*	0.741*	0.507*	0.765*	−0.636*	0.775*	0.765*
LUII	−0.634*	−0.681*	−0.658*	−0.618*	−0.675*	−0.113	−0.684*	0.977*	−0.599*	−0.734*
PLAND_1	−0.752*	−0.810*	−0.795*	−0.772*	−0.811*	−0.334	−0.815*	0.952*	−0.742*	−0.853*
PD_1	0.045	−0.055	−0.063	−0.056	−0.054	−0.113	−0.051	−0.134	−0.091	−0.022
SHAPE_AM_1	−0.476*	−0.369	−0.358	−0.330	−0.368	−0.063	−0.369	0.495	−0.328	−0.390
IJI_1	0.189	0.542*	0.552*	0.530*	0.510*	0.405	0.533*	−0.199	0.619*	0.485*
PLAND_2	0.323	0.527*	0.541*	0.590*	0.520*	0.595*	0.534*	−0.246	0.558	0.525*
PD_2	0.420	0.457*	0.471*	0.515*	0.449	0.529*	0.463*	−0.193	0.491	0.452
SHAPE_AM_2	0.159	0.450	0.468*	0.541*	0.439	0.629*	0.460*	−0.149	0.493	0.452
IJI_2	0.207	0.392	0.409	0.483*	0.376	0.583*	0.402	−0.115	0.438	0.396
PLAND_3	0.677*	0.984*	0.983*	0.941*	0.975	0.558*	0.980*	−0.770*	0.988*	0.961*
PD_3	0.276	0.631*	0.637*	0.629	0.625	0.477*	0.629*	−0.383	0.653	0.606*
SHAPE_AM_3	0.744*	0.828*	0.827*	0.780*	0.820	0.449*	0.821*	−0.623*	0.836*	0.799*
IJI_3	0.040	0.231	0.241	0.291	0.208	0.356	0.236	−0.069	0.276	0.231
PLAND_4	0.224	−0.192	−0.212	−0.203	−0.159	−0.298	−0.181	−0.264	−0.311	−0.110
PD_4	−0.294	0.199	0.201	0.190	0.160	0.101	0.192	−0.086	0.257	0.171
SHAPE_AM_4	0.455	−0.061	−0.065	−0.049	−0.028	−0.029	−0.053	−0.086	−0.125	−0.022
IJI_4	0.513*	0.717*	0.719*	0.705*	0.697*	0.457*	0.715*	−0.539*	0.739*	0.701*
PLAND_5	−0.035	0.290	0.313	0.381	0.279	0.578*	0.297	0.081	0.357	0.271
PD_5	−0.047	0.244	0.269	0.322	0.244	0.548*	0.248	0.188	0.314	0.208
SHAPE_AM_5	0.118	0.081	0.082	0.053	0.072	−0.009	0.075	−0.004	0.105	0.055
IJI_5	0.307	0.461*	0.470*	0.525*	0.446*	0.512*	0.471*	−0.312	0.482	0.477*
PLAND_6	0.047	0.139	0.167	0.360	0.148	0.852	0.170	0.064	0.153	0.207
PD_6	−0.160	0.122	0.137	0.198	0.105	0.371	0.128	0.080	0.172	0.118
SHAPE_AM_6	0.378	0.088	0.086	0.140	0.080	0.141	0.099	−0.218	0.067	0.137
IJI_6	0.020	0.201	0.217	0.276	0.197	0.438	0.208	0.028	0.239	0.197
PLAND_7	−0.385	−0.447	−0.474*	−0.525*	−0.507*	−0.769*	−0.456*	−0.041	−0.442	−0.422
PD_7	−0.270	−0.105	−0.129	−0.182	−0.154	−0.494*	−0.114	−0.245	−0.108	−0.089
SHAPE_AM_7	−0.236	−0.313	−0.329	−0.434	−0.348	−0.650*	−0.333	0.150	−0.279	−0.358
IJI_7	0.210	0.416	0.408	0.287	0.402	−0.083	0.394	−0.254	0.443	0.341

TESVs: the total ecosystem service values (ESVs); ESVs_1: the ESVs of gas regulation; ESVs_2 climate regulation; ESVs_3: the ESVs of water supply; ESVs_4: the ESVs of soil formation and retention; ESVs_5: the ESVs of waste treatment, ESVs_6: the ESVs of biodiversity protection; ESVs_7 the ESVs of food production; ESVs_8: the ESVs of raw material; ESVs_9: the ESVs of recreation and culture. PD: patch density; SHAPE_AM: area-weighted mean shape index; IJI: Interspersion and Justaposition Index; SHDI: Shannon's diversity index; LUII: land use intensity index; PLAND: percentage of landscape. The 1, 2, 3, 4, 5, 6, 7 after the above landscape metrics respects different landscape, that is cropland, orchard, forestland, grassland, residential land, water body and unused land, respectively. *significant at 0.01 level.

Table 7. Regression analysis between ecosystem service values (ESVs) and landscape patterns (n = 36).

Dependent	Standardized coefficients regression	R^2	Sig.
Gas regulation	0.878×PLAND_3+0.166×PLAND_2-0.099×PLAND_1-0.068×IJI_1	0.990	*
Climate regulation	0.790×PLAND_3-0.197×PLAND_7-0.190×LUII+0.081×PLAND_2	0.998	*
Water supply	0.665×PLAND_3-0.317×PLAND_7-0.254×LUII+0.106×PLAND_2	0.955	*
Soil formation and retention	0.684×PLAND_3-0.301×PLAND_7-0.284×LUII+0.066×PLAND_2	0.998	*
Waste treatment	0.672×PLAND_6+0.365×PLAND_3-0.352×PLAND_7+0.051×PLAND_2+0.049×PLAND_5	0.993	*
Biodiversity protection	0.059×SHDI +0.861×PLAND_3-0.033×SHAPE_AM_3+0.133×PLAND_1	0.967	*
Food production	0.742×LUII-0.173×PLAND_3-0.052×SHDI+0.106×PLAND_1	0.991	*
Raw material	0.964×PLAND_3+0.091×SHDI+0.068×LUII	0.981	*
Recreation and culture	−0.747×PLAND_1+0.380×IJI_1	0.853	*
Total	-0.588×PLAND_1+0.569× SHAPE_AM_3-0.303×SHDI	0.709	*

*significant at 0.01 level.
PLAND_1: the percentage of cropland; PLAND_2: the percentage of orchard; PLAND_3: the percentage of forestland; PLAND_5: the percentage of residential land; PLAND_6: the percentage of water body; PLAND_7: the percentage of unused land; SHAPE_AM_3: the area-weighted mean shape index of forestland; IJI_1: the Interspersion and Justaposition Index of cropland; LUII: land use intensity index; SHDI: Shannon's diversity index.

higher the total ESVs were. This indicated that the connectivity of grassland was important for ecosystem service.

Correlation also occurred between ESVs of all the functions and landscape metrics (Table 6). However, the relationships between ESVs of different functions and landscape pattern were different. For example, the correlation between food production and landscape pattern was almost opposite from that between other ecosystem functions and landscape pattern. For example, PLAND_1 had a positive effect on food production; SHDI, PLAND_3, SHAPE_3, and IJI_4 had a negative effect on food production. It could infer that there were contradictions between food production and other ecosystem functions.

As shown in Table 7, the result of regression analysis further explained that the ESVs was correlated significantly with landscape pattern. The total ESVs could be predicted by PLAND on cropland, SHAPE_AM on grassland, and SHDI. ESVs of all kinds of ecosystem functions also could be explained by landscape metrics. These regression equations indicated that landscape-level metrics (such as SHDI and LUII) and class_level metrics (such as PLAND of forestland, orchard, and cropland, unused land, SHAPE of forestland, IJI of cropland) acted as predictors for categories of ecosystem services. Specifically, the proportion of forest (PLAND_3) accounted for almost all of the categories of ecosystem services.

Discussion

4.1 Reliability of ESVs

This study estimated ESVs by multiplying the area for each land use types by the corresponding value coefficients. As discussed in the previous researches, estimations using this method was coarse

with high variation and uncertainty for the following reasons, limitations on the economic evaluation [1], problems of double counting and scales [40,41,42], the complex, dynamic and nonlinear ecosystems [43], the imperfect matches of land use categories as proxies [38] and the accuracy of the ecosystem value coefficients [5]. This study also existed such uncertainty on ESVs estimation. For example, the value coefficient of orchard, determined by the average of forest and grassland, was an approximate estimation and need a further exploration. However, the estimation of temporal variation on ESVs was considered to be more reliable than that of cross-sectional analysis [5]. In addition, the sensitivity analysis of the estimated ESVs with 50% adjustment in the value coefficients was conducted. The result showed that the sensitivity coefficients of all land use categories were lower than 1 (Table 8), which suggested that despite of the above limitations, the estimated ESVs are reliable and useful for subsequent study.

4.2 Relationship between ESVs and landscape pattern

It is usually assumed that land use can affect the ecosystem service. Moreover, a few studies showed that there was a correlation between landscape pattern and ESVs [41,44]. This study signified this statement at town scale on the Loess Plateau. Land use configuration, land use intensity, landscape diversity, fragmentation and connectivity all affected ecosystem service.

The correlation analysis between ESVs and PLAND implied land use structure had significant impact on ecosystem service. Especially, the increase of forestland and the decrease of cropland played an important part in improving the ESVs in the past twenty years. It is closely related to the Grain to Green project comprehensively started in study area since 2002. In the project,

Table 8. The coefficient of sensitivity (CS) resulting from adjustment of ecosystem valuation coefficients.

	Cropland	Orchard	Forestland	Grass land	Water body	Unused land
1982	−0.430	−0.012	−0.008	−0.500	−0.039	−0.011
2002	−0.428	−0.052	−0.169	−0.309	−0.041	−0.002
2008	−0.129	−0.048	−0.574	−0.223	−0.027	−0.001

measures for optimizing land use structure were implemented, including restoring slope cropland into forest and grassland, banning grazing, transforming slopes into terraces, and building reservoirs, etc. Forestland and grassland increased by 423.19% (27.54% of the study area) and 10.93% (3.15% of the study area), and cropland decreased by 53.59%(31.49% of the study area) (Table 3). The increase of ESVs due to the increase of forestland occupied 46.28% of the total ESVs in 2008 (Table 4). The result of the correlation analysis between ESVs and PLAND reflected that vegetation recovery could strongly enhance ecosystem service, and it was coincident with many other studies on the Loess Plateau [17,18,23,45]. LUII, which also related to the proportion of land use types, implied the intensity of human activities. This study showed land use intensity had a negative effect on ecosystem service with negative correlation coefficients (-0.634) (Table 6). It was coincident with some studies on ESVs change under urbanization [5,31]. These studies showed that urbanization which means the increase of land use intensity led to considerable declines in ESVs.

Landscape diversity always presents high positive relevance with biodiversity [46]. Our results were coincident to previous statements given the positive relationships between SHDI and biodiversity conservation. However, there were studies reporting the negative relationships between them, in which the increase of SHDI was the result of rapid urban sprawl [31]. In our study, the increase of SHDI was because land use structure became more balanced, which was the result of the increase of forestland. In addition, landscape diversity could also promote agricultural production [47]. Our study disagreed with this statement, and showed that food production was weakened with landscape diversification. It was because that the increase of SHDI was the result of a larger number of conversion from cropland to forestland. Therefore, the relationship between landscape diversity and biodiversity conservation as well as food production should not be treat as the same but be understood considering the driving force of SHDI change.

Fragmentation could lead to declining habitat quality, lower wildlife survival, and limited movement of soil microorganisms [48], and subsequently cause the decrease of ecosystem service [30]. Our study disagreed with this statement. For example, PD of the total landscape, PD_Forest, PD_orchard and shape_ Forest revealed significantly positive impacts on most categories of ESVs (Table 6-7). The increase of PD and the decrease of connectivity of landscape were usually simultaneous, which is disagreed in our study (Fig. 3 and Fig. 4). The landscape became more contiguous as IJI shown. Table 6 showed the IJI had significantly positive impacts on ESVs. Especially, the increase of IJI of grassland promoted the total ESVs and all categories of ESVs. This maybe because the connectivity of landscape has contribution to habitat corridors [49] and forest production [50].

Based on the relationship between ESVs and landscape pattern, we could improve the ecosystem service by the adjustment of land use policy. On the one hand, continuing to implement the Grain to Green project is helpful for improving ESVS, because it could increase the vegetation coverage, decline the intensity of land use, and make cropland become regular by canceling the slope cropland. On the other hand, diversified agriculture gathering planing fruit trees, planting crops and breeding, which could promote the diversification of land use, should be encouraged to increase both ESVs and farmer's incomes.

Conclusion

ESVs at town scale in the Loess Plateau were estimated in Hechuan town of Ningxia Hui Autonomous Region from 1982 to 2008. It was concluded that ESVs varied with land use change. ESVs in 1982, 2002, and 2008 were US$ 6.315, US$ 7.336 and US$ 11.275 million respectively. Among all the land use types, forestland, grassland and cropland had important contribution (> 90%) on ESVs. The total ESVs increased slowly by 16.17% due to the decrease of grassland from 1982 to 2002, while the total ESVS increased significantly by 67.61% due to the increase of forestland from 2002 to 2008. Areas with high services level were mainly located in the center due to orchard and east due to forestland, while areas with low services level mainly located in the north and south sides due to cropland.

Land use pattern had a significant effect on ecosystem service in our study by analyzing and discussing the relationship between landscape pattern and ESVs. The proportion of forestland had a positive effect on ecosystem service while that of cropland had a negative effect on ESVs. The diversity and interspersion of landscape both had a positive effect on ESVs. Land use intensity which reflects the intensity of human activities had a negative effect on ESVs. Fragmentation had positive effect on ESVs, which was disagreed with the previous studies because the fragmentation in study area was related to the increased patch of such land use types as forestland, water body, orchard.

Based on the results of this study, it was conclude that land use pattern was important for ecosystem service. Therefore, we could improve the ecosystem service by the adjustment of land use policy. Continuing the Grain to Green project is reasonable and significant because it could increase the vegetation coverage and decline land use intensity. Diversified agriculture collecting planing fruit trees, growing food and breeding should be encouraged, because it could not only promote ecosystem service by increasing landscape diversification but also improve people's incomes.

Author Contributions

Conceived and designed the experiments: XF. Analyzed the data: XF. Contributed reagents/materials/analysis tools: GAT BCL. Contributed to the writing of the manuscript: XF GAT RMH.

References

1. Costanza R, Arge DR, Groot DR, Farber S, Grasso M, et al. (1997) The value of the world's ecosystem services and natural capital. Nature 387: 253−260.
2. Costanza R, Cumberland J, Daly H, Goodland R, Norgaard R (1997) An Introduction to ecological economics. Delray Beach Fla USA: St Lucie Press.
3. Kreuter UP, Harris HG, Matlock MD, Lacey RE (2001) Change in ESVs in the San Antonio area, Texas. Ecological Economics 39: 333−346.
4. Ronald CE, Yuji M (2013) Landscape pattern and ESV changes: Implications for environmental sustainability planning for the rapidly urbanizing summer capital of the Philippines. Landscape and Urban Planning 116: 60−72.
5. Li TH, Li WK, Qian ZH (2010) Variations in ESV in response to land use changes in Shenzhen. Ecological Economics 69: 1427−1435.
6. Cai YB, Zhang H, Pan WB, Chen YH, Wang XR (2013) Land use pattern, socio-economic development, and assessment of their impacts on ESV: study on natural wetlands distribution area (NWDA) in Fuzhou city, southeastern China. Environ Monit Assess 185: 5111−5123.
7. Zaehle S, Bondeau A, Carter RT, Cramer W, Erhard M, et al. (2007) Projected changes in terrestrial carbon storage in europe under climate and land-use change, 1990–2100. Ecosystems 10: 380−401.
8. Eliska L, Jana F, Edward N, David V (2013) Past and future impacts of land use and climate change on agricultural ecosystem services in the Czech Republic. Land Use Policy, 33: 183−194.
9. Vitousek PM, Mooney HA, Lubchenco J, Melillo JM (1997) Human domination of earth's ecosystems. Science 277: 494−499.
10. Nasiri F, Huang GH (2007) Ecological viability assessment: A fuzzy multi-pleattribute analysis with respect to three classes of ordering techniques. Ecol Inform 2: 128−137.

11. Collin ML, Melloul AJ (2001) Combined land-use and environmental factors for sustainable groundwater management. Urban Water 3: 229–237.
12. Schmidta JP, Mooreb R, Alber M (2014) Integrating ecosystem services and local government finances into land use planning: A case study from coastal Georgia. Landscape and Urban Planning 122: 56–67.
13. Ernesto FV, Federico CF (2006) Land-use options for Del Plata Basin in South America: Tradeoffs analysis based on ecosystem service provision. Ecological Economics 57: 140–151.
14. Christine F, Susanne F, Anke W, Lars K, Franz M (2013) Assessment of the effects of forestland use strategies on the provision of ecosystem services at regional scale. Journal of Environmental Management 127: 96–116.
15. Ignacio P, Berta M, Pedro Z, David GDA, Carlos M (2014) Deliberative mapping of ecosystem services within and around Donana National Park (SW Spain) in relation to land use change. Reg Environ Change14: 237–251.
16. Mendoza-Gonzalez G, Martinez ML, Lithgow D, Perez-Maqueo O, Simonin P (2012) Land use change and its effects on the value of ecosystem services along the coast of the Gulf of Mexico. Ecological Economics 82: 23–32.
17. Su CH, Fu BJ (2013) Evolution of ecosystem services in the Chinese Loess Plateau under climatic and land use changes. Global and Planetary Change 101: 119–128.
18. Si J, Nasiri FZ, Han P, Li TH (2014) Variation in ESVs in response to land use changes in Zhifanggou watershed of Loess plateau: a comparative study. Environmental Systems Research 3: 2.
19. Turner MG, Gardner RH, O'Neill RV (2001) Landscape Ecology in theory and practice. New York: Springer-Verlag.
20. Ritsema CJ (2003) Introduction: soil erosion and participatory land use planning on the Loess Plateau in China. Catena 54: 1–5.
21. Fu BJ, Wang YF, Lu YH, He CS, Chen LD, et al. (2009) The effects of land-use combinations on soil erosion: a case study in the Loess Plateau of China. Progress in Physical Geography 33: 793–804.
22. Fu BJ, Chen DX, Qiu Y, Wang J, Meng QH (2002) Land Use Structure and Ecological Processes in the LoessHilly Area, China. Beijing: Commercial Press. (in Chinese).
23. Jing L, Zhiyuan R (2011) Variations in ESV in Response to Land use Changes in the Loess Plateau in Northern Shaanxi Province, China. Int. J. Environ. Res 5: 109–118.
24. Zhang TB, Tang JX, Liu DZ (2006) Feasibility of Satellite Remote Sensing Image About Spatial Resolution. Journal of Earth Sciences and Environment 28: 79–83.
25. Chu YF, Li ES, Lu J, Zhang KK (2007) The Adaptability Analysis to the Satellite Image Spatial Resolution and Mapping Scale. Hydrographic Surveying and Charting 27: 47–50.
26. Li Q, Liu C, Xi CY, Liu ML (2002) Cartographic Generalization of Digital Land Use Current Situation Map. Bulletin of Surveying and Mapping 9: 59–63.
27. Congalton RG (1991) A review of assessing the accuracy of classifications of remotely sensed data. Remote Sensing of Environment 37: 35–46.
28. Wang Y, Gao JX, Wang JS, Qiu J (2014) Value Assessment of Ecosystem Services in Nature Reserves in Ningxia, China: A Response to Ecological Restoration. PloS One 9: e89174. doi:10.1371/journal.Pone.0089174.
29. Ribeiro SC, Lovett A (2009) Associations between forest characteristics and socio-economic development: a case study from Portugal. Journal of Environmental Management 90: 2873–2881.
30. Su S, Jiang Z, Zhang Q, Zhang Y (2011) Transformation of agricultural landscapes under rapid urbanization: a threat to sustainability in Hang-Jia-Hu region, China. Applied Geography 31: 439–449.
31. Su SL, Xiao R, Jiang ZL, Zhang Y (2012) Characterizing landscape pattern and ESV changes for urbanization impacts at an eco-regional scale. Applied Geography, 34: 295–305.
32. Li H, Wu J (2004) Use and misuse of landscape indices. Landscape Ecology 19: 389–399.
33. Xie GD, Lu CX, Xiao Y, Zheng D (2003) The Economic Evaluation of Grassland Ecosystem Services in Qinghai Tibet Plateau, Journal of Mountain Science 21: 50–55. (in Chinese).
34. Xie GD, LU CX, Leng YF, Zheng D, Li SC (2008) Ecological assets valuation of Tibetan Plateau. Journal of Natural Resources 18: 190–196. (in Chinese).
35. Liu DL, Li BC, Liu Xianzhao Z, Warrington DN (2011) Monitoring land use change at a small watershed scale on the Loess Plateau, China: applications of landscape metrics, remote sensing and GIS. Environmental Earth Sciences 64: 2229–2239.
36. Pan WKY, Walsh SJ, Bilsborrow RE, Frizzelle BG, Erlien CM, et al. (2004) Farm-level models of spatial patterns of land use and land cover dynamics in the Ecuadorian Amazon. Agriculture, Ecosystems and Environment 101: 117–134.
37. de Groot RS, Wilson MA, Boumans RMJ (2002) A typology for the classification, description and valuation of ecosystem functions, goods and services. Ecological Economics 41: 393–408.
38. Kreuter UP, Harris HG, Matlock MD, Lacey RE (2001) Change in ESVs in the San Antonio area, Texas. Ecological Economics 39: 333–346.
39. Wu J, Jenerette GD, Buyantuyev A, Redman CL (2011) Quantifying spatiotemporal patterns of urbanization: the case of the two fastest growing metropolitan regions in the United States. Ecological Complexity 8: 1–8.
40. Turner RK, Paavola J, Coopera P, Farber S, Jessamya V, et al. (2003) Valuing nature: lessons learned and future research directions. Ecological Economics 46: 493–510.
41. Hein L, Koppen VK, de Groot RS, van Ierland EC (2006) Spatial scales, stakeholders and the valuation of ecosystem services. Ecological Economics 57: 209–228.
42. Konarska KM, Sutton PC, Castellon M (2002) Evaluating scale dependence of ecosystem service valuation: a comparison of NOAA-AVHRR and Landsat TM datasets. Ecological Economics 41: 491–507.
43. Limburg KE, O' Neill RV, Costanza R, Farber S (2002) Complex systems and valuation. Ecological Economics 41: 409–420.
44. Zhang MY, Wang KL, Liu HY, Zhang CH (2011) Responses of Spatial-temporal Variation of Karst Ecosystem Service Values to Landscape Pattern in Northwest of Guangxi, China. Chin. Geogra. Sci. 21: 446–453.
45. Deng L, Shangguan ZP, Li R (2012) Effects of the grain-for-green program on soil erosion in China. International Journal of Sediment Research 27: 120–127.
46. Nagendra H (2002) Opposite trends in response for the Shannon and Simpson indices of landscape diversity. Applied Geography, 22: 175–186.
47. Shrestha RP, Schmidt-Vogt D, Gnanavelrajah N (2010) Relating plant diversity to biomass and soil erosion in a cultivated landscape of the eastern seaboard region of Thailand. Applied Geography 30: 606–617.
48. Sherrouse BC, Clement JM, Semmens DJ (2011) A GIS application for assessing, mapping, and quantifying the social values of ecosystem services. Applied Geography 31: 748–760.
49. Li M, Zhu Z, Vogelmann JE, Xu D, Wen W, et al. (2011) Characterizing fragmentation of the collective forests in southern China from multitemporal Landsat imagery: a case study from Kecheng district of Zhejiang province.Applied Geography 31: 1026–1035.
50. Long JA, Nelson TA, Wulder MA (2010) Characterizing forest fragmentation: distinguishing change in composition from configuration. Applied Geography 30: 426–435.

21

Global Agricultural Land Resources – A High Resolution Suitability Evaluation and Its Perspectives until 2100 under Climate Change Conditions

Florian Zabel*, Birgitta Putzenlechner, Wolfram Mauser

Department of Geography, Ludwig Maximilians University, Munich, Germany

Abstract

Changing natural conditions determine the land's suitability for agriculture. The growing demand for food, feed, fiber and bioenergy increases pressure on land and causes trade-offs between different uses of land and ecosystem services. Accordingly, an inventory is required on the changing potentially suitable areas for agriculture under changing climate conditions. We applied a fuzzy logic approach to compute global agricultural suitability to grow the 16 most important food and energy crops according to the climatic, soil and topographic conditions at a spatial resolution of 30 arc seconds. We present our results for current climate conditions (1981–2010), considering today's irrigated areas and separately investigate the suitability of densely forested as well as protected areas, in order to investigate their potentials for agriculture. The impact of climate change under SRES A1B conditions, as simulated by the global climate model ECHAM5, on agricultural suitability is shown by comparing the time-period 2071–2100 with 1981–2010. Our results show that climate change will expand suitable cropland by additionally 5.6 million km^2, particularly in the Northern high latitudes (mainly in Canada, China and Russia). Most sensitive regions with decreasing suitability are found in the Global South, mainly in tropical regions, where also the suitability for multiple cropping decreases.

Editor: Juergen P. Kropp, Potsdam Institute for Climate Impact Research, Germany

Funding: This research was carried out within the framework of the GLUES (Global Assessment of Land Use Dynamics, Greenhouse Gas Emissions and Ecosystem Services) Project, which has been supported by the German Ministry of Education and Research (BMBF) program on sustainable land management (FKZ 01LL0901E). (http://modul-a.nachhaltiges-landmanagement.de/en/). The funders had no role in study design, data collection and analysis, decision to publish, or preparation of the manuscript.

Competing Interests: The authors have declared that no competing interests exist.

* Email: f.zabel@lmu.de

Introduction

Natural constraints are limiting the land's suitability for agriculture and cultivation practices. They consist of prevailing local climatic, soil and topographic conditions determining the available energy, water and nutrient supply for agricultural crops. Besides natural conditions, complex interactions of social, economic, political, and cultural aspects determine whether and how land is used for agriculture. Agricultural land has become one of the largest terrestrial biomes on the planet, occupying approx. 40% of the land surface [1]. Thereby, a variety of different land use types and intensities determine heterogeneously distributed patterns, including e.g. the choice of crop varieties, irrigation practices, fertilization, terracing and the level of technological input [2]. Thus, natural constraints are to a limited extent suspended by human actions [3].

The demand for agricultural products is expected to increase by 70–110% by 2050, driven by a projected world population of 9 billion people, increasing meat consumption and a growing use for bio-based materials and biofuel [4–15].

An increase in agricultural production can be accomplished by agricultural intensification and expansion, while considering social and environmental externalities and changing climate conditions

[5,16]. Bruinsma [16] concluded that additionally 1.2 million km^2 of converted land are projected to be necessary until 2030 and another 5% up to 2050 with most land expected to be transformed in South America and Sub Saharan Africa, while latest studies project an increase of cropland between 10-25% by 2050 compared to 2005 for different socio-economic and climate scenarios [17]. Nonetheless, the expansion of agricultural land into forested or protected areas must be viewed critically, in order to conserve valuable ecosystem services e.g. for regulating climate or conserving biodiversity [5–8].

Changing patterns of temperature and precipitation and man-made degradation affect the suitability of land for agricultural use. For example, 19-23 ha of suitable land are lost per minute due to soil erosion and desertification [18,19]. Additionally, the area of suitable land is decreasing due to urbanization, with an estimate of 1.5 million km^2 until 2030 [20,21].

When focusing on the natural potentials of land for agricultural use, suitability analyses give local evidence on todays and future availability and quality. Thus, they help answering questions for managing a transition towards a more environmentally efficient and sustainable land use and involve better information on the global scale impacts of land use decisions [1].

Table 1. List of investigated food, feed and energy crops.

Crop name
Barley (*hordeum vulgare*)
Cassava (*manihot esculenta*)
Groundnut (*arachis hypogaea*)
Maize (*zea mays*)
Millet (*pennisetum americanum*)
Oil palm (*elaeis guineensis*)
Potato (*solanum tuberosum*)
Rapeseed (*brassica napus*)
Paddy rice (*oryza sativa*)
Rye (*secale cereale*)
Sorghum (*sorghum bicolor*)
Soy (*glycine maximum*)
Sugarcane (*saccharum officinarum*)
Sunflower (*helianthus annus*)
Summer wheat (*triticum aestivum*)
Winter wheat (*triticum gestivum*)

The relationship between climate, soil, topography and agricultural suitability has long been recognized. As such, suitability analysis combine heterogeneous soil, terrain and climate information and determine whether specific crop requirements are fulfilled under the given local conditions and assumptions. A variety of regional suitability studies for specific crops exist [22–28], while only a few exist on a global scale and for a broad variety of crops [3,29–31].

In the meantime, global soil and topography data are available at high spatial resolution and global climate models have improved their capabilities and spatial resolution. Previous analysis showed that questions of scale play a major role in suitability analysis as coarse data affect the validity of results [32]. In this context, we present our results in modelling global crop-suitability using a fuzzy logic approach at a spatial resolution of 30 arc seconds. The results of this approach include the potentially suitable area for agriculture differentiated for 16 crops for rainfed and irrigated conditions, the start of the growing cycles and the number of crop cycles. We analyze global distribution of agricultural suitability and changes until 2100 considering the numbers of crop cycles.

Thereby, we identify changes, opportunities and challenges in global agriculture related to the expansion of agricultural land competing with protected and forested areas as ecosystem services.

Material and Methods

Local climate, soil and topography determine the natural suitability of land for agricultural use. Thereby, the climatic, soil and topographic requirements may vary over a wide range of different agricultural crops. This analysis investigates the suitability for the following 16 crops that are most important for the global economy, food security and biofuel issues (see Table 1).

We aggregated the world into 23 regions in order to regionally analyse the results (see Fig. 1). We applied a fuzzy-logic approach [33,34] in order to calculate the crops' suitability on the globe at a spatial resolution of 30 arc seconds (0.00833°, approx. 1 km^2 at the equator). The length of the growing cycle (*lgc*) and the 'membership functions' that describe the crop-specific requirements for each of the crops during the growing period (Fig. 2) are derived from [35].

The membership functions representing climate constraints describe the degree of membership of each selected crop with regard to mean temperature and total precipitation during its

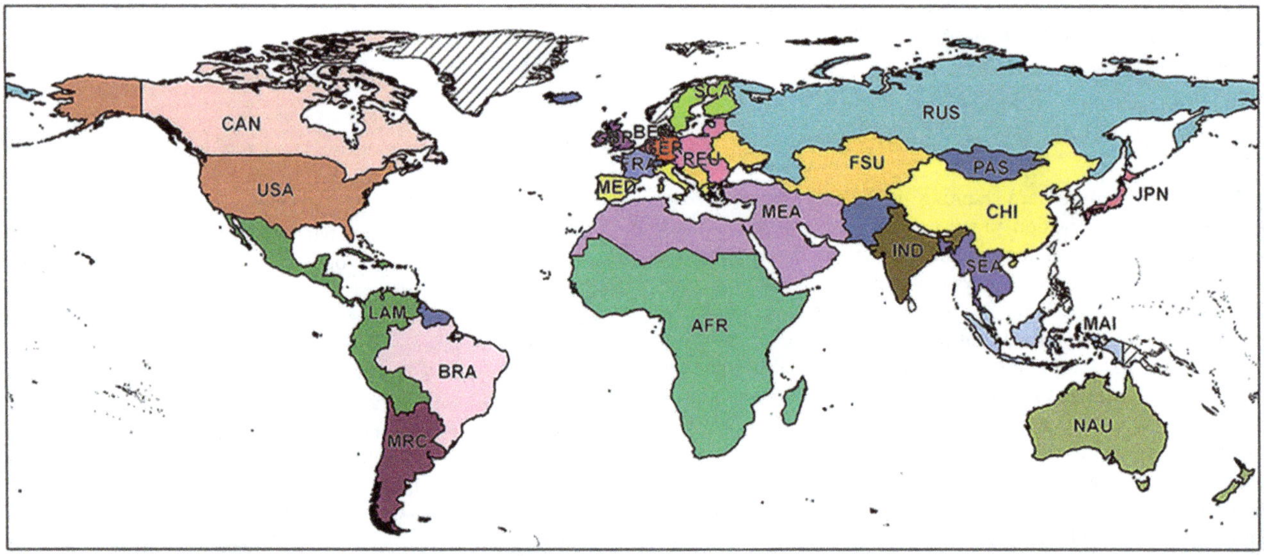

Figure 1. Map of the 23 world regions: AFR (Sub Saharan Africa), ANZ (Australia, New Zealand), BEN (Belgium, Netherlands, Luxemburg), BRA (Brazil), CAN (Canada), CHN (China), FRA (France), FSU (Rest of Former Soviet Union and Rest of Europe), GBR (Great Britain), GER (Germany), IND (India), JPN (Japan), LAM (Rest of Latin America), MAI (Malaysia, Indonesia), MEA (Middle East, North Africa), MED (Italy, Spain, Portugal, Greece, Malta, Cyprus), PAC (Paraguay, Argentina, Chile, Uruguay), ROW (Rest of the World), REU (Austria, Estonia, Latvia, Lithuania, Poland, Hungary, Slovakia, Slovenia, Czech Republic, Romania, Bulgaria), RUS (Russia), SCA (Finland, Denmark, Sweden), SEA (Cambodia, Laos, Thailand, Vietnam, Myanmar, Bangladesh), USA (United States of America).

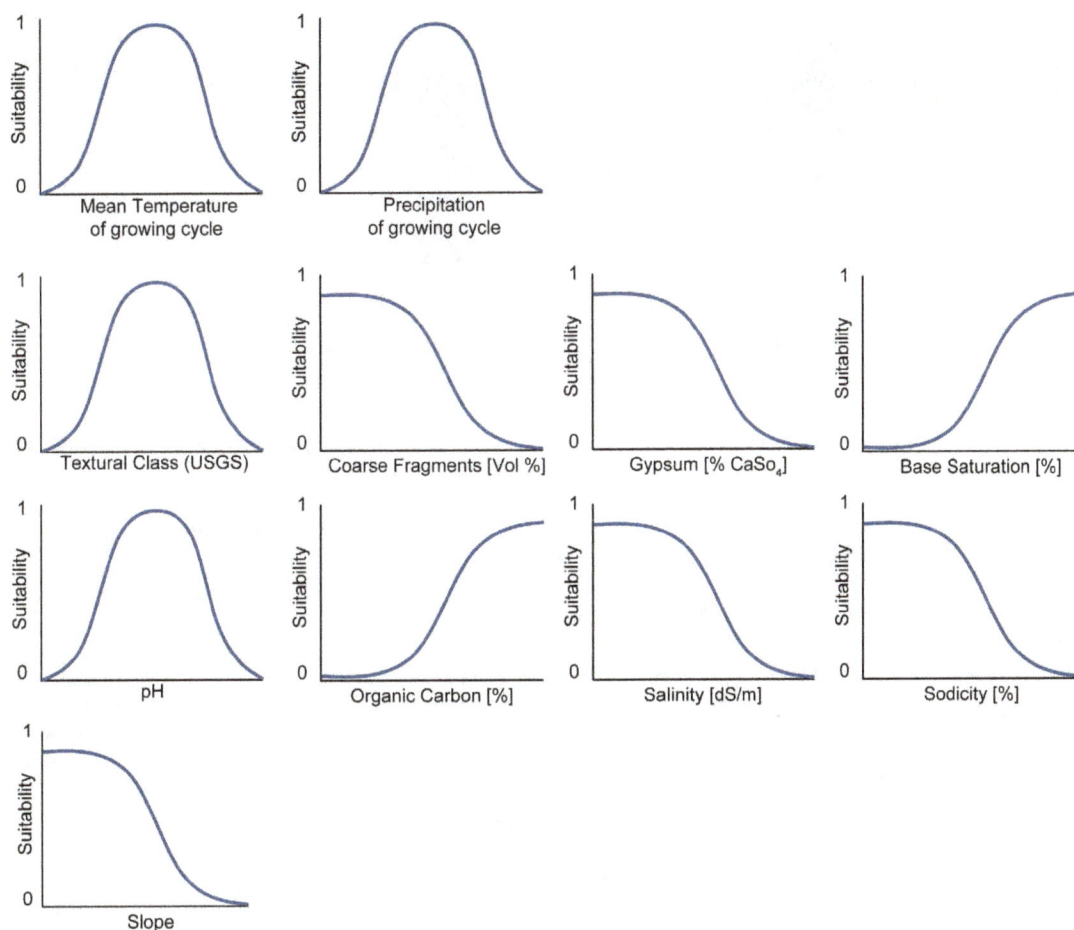

Figure 2. Membership functions for climatic, soil and topographic conditions.

respective growing cycle. Depending on the crop, membership functions have different curves according to [35]. Three shapes are in principle possible: 'more is better', 'less is better' and 'optimum'. For temperature e.g., the suitability is increasing from a minimum towards an optimal temperature and again decreasing until a maximum temperature is reached (Fig. 2). Eight soil parameters are considered: texture, proportion of coarse fragments and gypsum, base saturation, pH content, organic carbon content, salinity, and sodicity. Terrain is considered by the slope. The fuzzy-logic approach calculates *fuzzy values* based on the ecological rules (between 0 and 1), which determine the crops'

suitability on a specific location by the lowest membership value of all parameters.

An overview of the applied global datasets is given in Table 2. The climate data applied in this study are outputs from the global circulation model ECHAM5 of the Max-Planck Institute for Meteorology (MPI-M) [36,37]. It uses radiative forcing, sea surface temperature and sea ice concentrations from a 20th century/SRES A1B scenario simulation. The 6-hourly dataset (temperature, precipitation) are converted to daily values for the climate period of 1981–2010 and 2071–2100. The daily data is spatially downscaled from its original resolution of 0.56° to 0.00833° (30

Table 2. Applied global datasets.

Parameter	Source	Detailed Description
Climate	ECHAM5	[37]
Soil	Harmonized World Soil Database (HWSD)	[41]
Topography	Space Shuttle Topography Mission (SRTM)	[39]
Crop-requirements	FAO Land Evaluation Part III: Crop Requirements	[35]
Irrigation	Global Map of Irrigation Areas (GMIA) v5.0	[44]
Protected Areas	International Union for Conservation of Nature (IUCN) Protected Areas	[45]
Forested Areas	GlobCover 2009	[46]

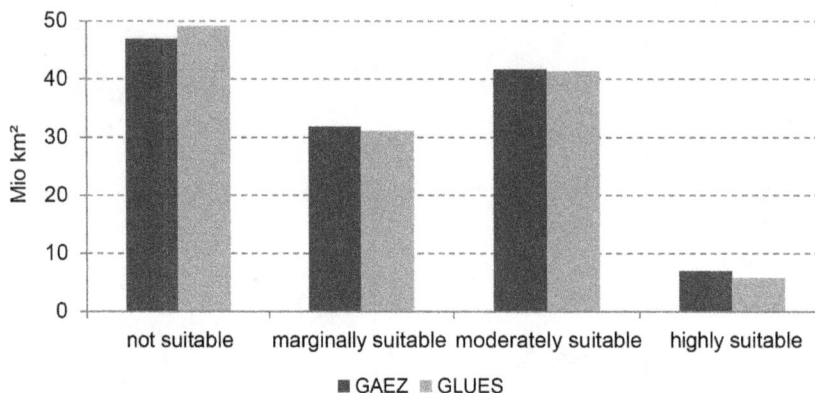

Figure 3. Global comparison of agriculturally suitable area between GAEZ (Baseline period 1961–1990) and GLUES (1961–1990).

arc seconds), based on an approach by [38], using sub-grid terrain information provided by the SRTM-dataset [39]. A bias-correction is executed during the downscaling procedure for temperature and precipitation based on monthly derived factors from the WorldClim dataset [40].

Mean temperature $(\bar{T})$ and total precipitation $(\bar{P})$ are calculated over the length of the growing cycle for each day of the year (doy) (see eq. 1 and eq. 2). Starting on the 1st of January $(doy = 1)$, the growing cycle is shifted day by day until the 31st of December $(doy = 365)$. The suitability value (S) is calculated for each doy as in eq. 3 for $\bar{T}$ and $\bar{P}$ according to the membership function (mf).

$$\bar{T}_{doy} = \overline{T_{doy}, \ldots, T_{doy+lgc}} \qquad (eq.1)$$

$$\bar{P}_{doy} = \sum_{doy}^{doy+lgc} P \qquad (eq.2)$$

$$S(\bar{T}_{doy}) = mf\left(\bar{T}_{doy}\right) ; \ S(\bar{P}_{doy}) = mf\left(\bar{P}_{doy}\right) \qquad (eq.3)$$

Since the natural suitability of crop growth is limited by the minimum value, the smaller value of the temperature and precipitation fuzzy value determines the climate suitability $S(C)$

which is calculated for each doy (eq. 4).

$$S(C_{doy}) = min \left\{ S(\bar{T}_{doy}), S(\bar{P}_{doy}) \right\} \qquad (eq.4)$$

Among all daily fuzzy values of $S(C)$ within the year, the maximum of $S(C)$ determines the climate suitability over the growing cycle and thus, the optimal start of the growing cycle (eq. 5) for cultivation of a single crop within the entire growing season.

$$S(C_{start \ of \ the \ growing \ cycle}) = max \left\{ S(C_{doy \ 1}), \ldots, S(C_{doy \ 365}) \right\} \quad (eq.5)$$

In order to allow for the calculation of multiple cropping, the fuzzy values for each possible combination of days for the start of the growing cycle are tested as to how often they would fit within one year. The number of multiple cropping is selected that generates the highest accumulated value. Multiple cropping and the start of the growing cycle(s) are obtained for single, double and triple cropping. Hereby, the start of the growing cycle(s) in the context of this paper describes an optimal time for cultivation of a crop to reach the maximum suitability within a year. Crop mixing is not considered. Regarding temporal demands for technical field work, we assume a break of two weeks between crop cycles.

Moreover, the following assumptions are made: At least 20 mm of precipitation are required within the first two weeks of the growing season in order to provide enough soil moisture for germination. No day within the growing period must be below 5°C and below 1°C for winter crops. Vernalisation requirements are considered separately from the growing period for winter

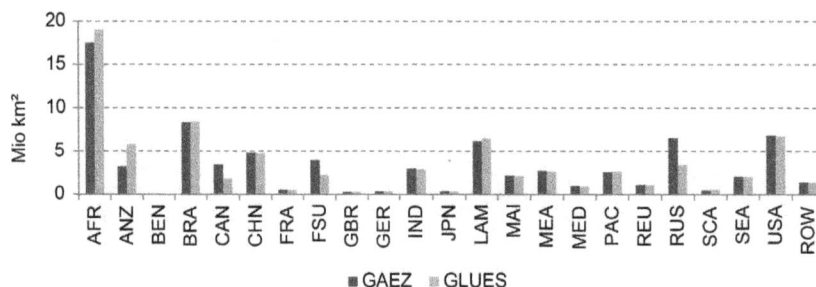

Figure 4. Comparison of total agriculturally suitable area of GAEZ (Baseline period 1961–1990) and GLUES (1961–1990) for different regions.

Table 3. Classified suitability considering rainfed conditions (1981–2010).

Not Suitable	Marginally Suitable	Moderately Suitable	Highly Suitable
49.8 million km²	30.6 million km²	41.3 million km²	5.8 million km²

crops: Vernalisation period starts 150 days before the start of the growing period. At least 20 days below 5°C must exist during the vernalisation period and there must not exist more than 3 days below −30°C. In order to consider permafrost conditions that exclude agricultural use, mean annual temperature must not be below 0°C. Mean daily incoming solar radiation must exceed 60 W/m² to provide enough energy for crop growth.

Thus, suitability values, number of crop cycles and the start of the growing cycle are calculated on each land surface pixel for both rainfed and irrigated conditions. For irrigated conditions, fuzzy values for precipitation are neglected during the calculation process. Due to a lack of global information on irrigation practices, we assume perennial irrigation on irrigated areas.

Besides climatic constraints, soil properties are limiting agricultural suitability. According to the membership functions (Fig. 2), the fuzzy values representing each of the soil properties are calculated. The minimum of the eight values represents the value of the soil suitability. Soil information was taken from the Harmonized World Soil Database (HWSD) [41], considering the topsoil (0–30 cm) of the dominant and all (up to 8) component soils at a spatial resolution of 30 arc seconds [42]. Within the calculation of soil suitability, fuzzy values of each of the component soils are calculated and weighted according to their share.

The suitability for crops to be cultivated is decreasing with increasing slope (see Fig. 2). The slope must not exceed 16% for the considered crops, except for oil palm and paddy rice. The slope was calculated and resampled to 30 arc seconds from Shuttle Radar Topography Mission (SRTM) data [39].

Across all climate, soil and topography fuzzy values, the lowest fuzzy value quantifies the crops' suitability at a certain location. The highest value across all crops determines the suitability for agriculture at a certain location.

This methodology does not allow for yield estimations, in which socio-economic and bio-physical aspects, which our approach does not consider, play an important role. However, this approach is well suited to draw conclusions about where areas are agriculturally suitable and how these areas may change with future climate conditions.

Results

The Earth surface consists of 510 million km² of which 149 million km² are land surface. Up to 60°S, excluding Antarctica, and considering a lack of input data, in total 127.5 million km² of land surface remain to be analyzed regarding their suitability for agriculture. We classified the results of the suitability analysis into

four categories: not suitable (0), marginally suitable (>0.0), moderately suitable (>0.33) and highly suitable (>0.75).

Comparison

Our results (further named GLUES in the Figures) highly correlate with existing studies, such as the GAEZ approach [29], when comparing the area of each of the four classified categories in each of the 23 World Regions ($R^2 = 0.99$).

The global aggregation of the classified areas and the regional distribution of not suitable and suitable areas show a high level of agreement (Fig. 3 and 4). Compared to the distribution of global cropland in the year 2000 [43], our approach identifies 95.5% of current cropland as suitable.

Rainfed

For the period 1981–2010, our suitability analysis shows that in total 77.7 million km² are potentially suitable for purely rainfed agricultural cultivation, while 49.8 million km² are not suitable for rainfed conditions (Table 3). Further, 30.6 million km² are marginally suitable, 41.3 million km² are moderately suitable and 5.8 million km² are highly suitable (Table 3).

Irrigation

Irrigated agriculture produces 40% of the world's food (FAO) on 3.1 million km² [44]. When considering irrigation, suitability is area weighted according to the fraction of rainfed and irrigated agricultural area (given by GMIA Version 5.0 [44]). Thereby, irrigation increases suitability on irrigated areas in global average by 0.13, adds 1.8 million km² of suitable land (Table 4) and allows for multiple cropping on 1.2 million km² (assuming sufficient water available for irrigation). Accordingly, huge areas e.g. in the Nile and Ganges delta are only becoming suitable due to irrigation. Overall, 79.6 million km² are suitable with spatially varying patterns (Fig. 5).

Figure 5 represents the global distribution of agricultural suitability as a result of local climate, soil and terrain conditions. In boreal regions, the growing season over all stages of phenology usually is too short for cultivation. The temperate zones seasonally have adequate temperatures and enough precipitation and often sufficient soil, while in subtropical regions, the annual distribution of precipitation strongly determines crop growth and soils often are alkaline. In inner tropics have adequate temperature and moisture throughout the year, but soil quality often restricts cultivation due to low organic content and acidity [3].

Table 4. Classified suitability considering rainfed and irrigated conditions (1981–2010).

Not Suitable	Marginally Suitable	Moderately Suitable	Highly Suitable
48.0 million km²	31.8 million km²	41.8 million km²	5.9 million km²

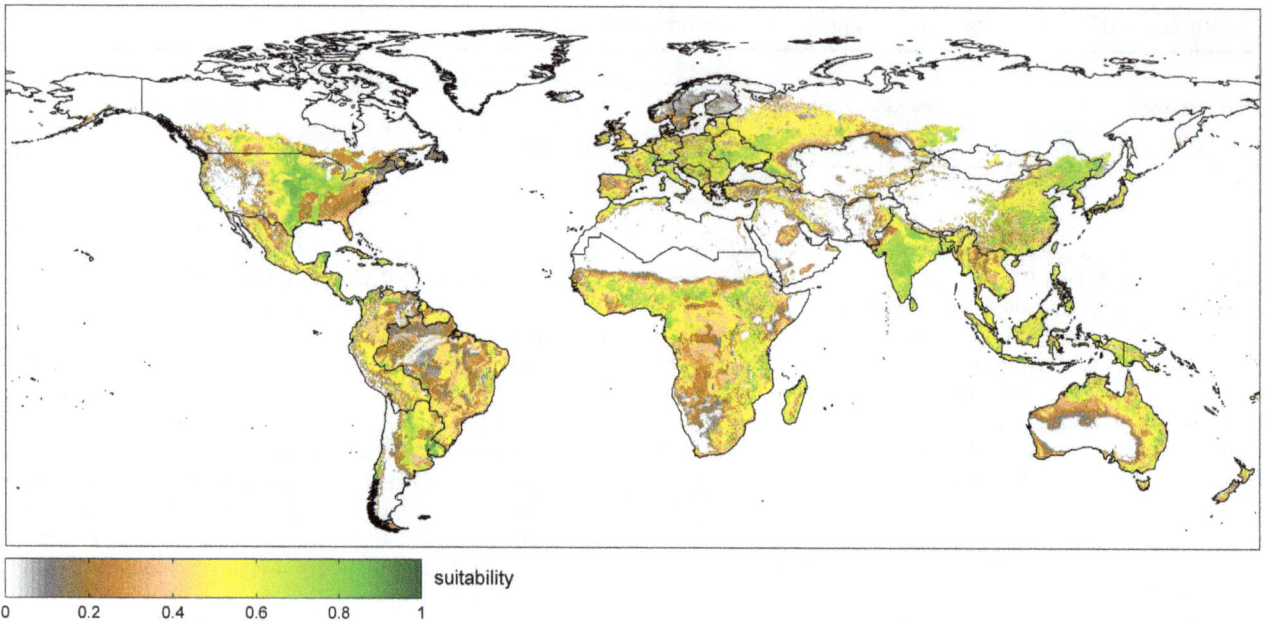

Figure 5. Agricultural suitability considering rainfed conditions and irrigated areas (1981–2010).

Protected Areas

Protected areas globally account for 8.3 million km². Information on actual protected areas is gathered from IUCN [45]. When excluding protected areas from the suitability calculation, 74.8 million km² remain suitable for cultivation. Thereby, protected areas are mainly situated in not suitable or marginally suitable areas (Table 5). Only 2% (0.2 million km²) of the global protected area are located on land highly suitable for agriculture, 25% (2.1 million km²) are on moderately suitable land, 30% (2.4 million km²) on marginally suitable land while 43% (3.6 million km²) are situated on unsuitable land. Overall, only 57% of global protected areas are suitable for agriculture.

Forested Areas

Dense forests are highly important to provide numerous ecosystem services. Densely forested areas account for 23.3 million km² according to GlobCover [46] and 23.5 million km² according to [47]. GlobCover defines forests as being dense when 75% of the pixel is forest [46]. Only 1.5 million km² or 6.2% of the global densely forested areas are currently protected.

4.9% (1.1 million km²) of the densely forested areas (excluding forests within protected areas) are located in highly suitable land, 49.4% (11.1 million km²) in moderately suitable land, 37.5% (8.4 million km²) in marginally suitable land and only 8.2% (1.9 million km²) are situated on unsuitable land. Overall, 92% of densely forested areas are potentially suitable for agriculture which indicates that global forests are subject to increasing societal stress.

Current Use of Suitable Land and Trade-Offs

When excluding both, protected areas and dense forests from the suitability calculation, 54.1 million km² remain suitable (Table 6). In comparison, currently used agricultural land (including pasture) today covers 49.1 million km², of which 15.5 million km² (status for 2011) are arable land (land under temporary and permanent crops; double-cropped areas are counted only once) [48]. Accordingly, 91% of all suitable land is already occupied by agriculture when today's protected and densely forested areas are preserved in the future. This illustrates that agricultural expansion is only possible by substituting other uses/covers of land which causes high social and ecological externalities. Figure 6 gives an overview of the current use/cover of suitable areas in the different regions of the world.

Figure 6 shows, that the current fraction of suitable area, which is not protected or dense forest is highly variable across regions. The most efficient use of current agriculturally suitable land is obvious in the USA, where only 2% of currently suitable land is not yet used or protected/dense forest.

In Africa, about 20% of the agriculturally suitable area is currently not used for agriculture or is statistically not recorded in the data of currently used agricultural land (Ramankutty *et al.*, 2008). This shows the extraordinary potentials of Africa for future expansion of agricultural land. However, agricultural expansion would always take place at ecological costs (e.g. conversion of tropical rainforest, grassland and savannah). In Latin America large suitable areas are protected or covered with dense forest and

Table 5. Classified suitability for 1981–2010 considering rainfed and irrigated conditions, excluding protected areas.

Not Suitable	Marginally Suitable	Moderately Suitable	Highly Suitable
44.4 million km²	29.4 million km²	39.7 million km²	5.7 million km²

Table 6. Classified suitability for 1981–2010 considering rainfed and irrigated conditions, excluding protected and densely forested areas.

Not Suitable	Marginally Suitable	Moderately Suitable	Highly Suitable
42.6 million km²	21.0 million km²	28.6 million km²	4.6 million km²

Table 7. Classified suitability for 2071–2100 considering rainfed and irrigated conditions, excluding protected and densely forested areas.

Not Suitable	Marginally Suitable	Moderately Suitable	Highly Suitable
37.8 million km²	24.8 million km²	30.2 million km²	3.9 million km²

the current fraction of remaining suitable area is smaller than in Africa, India is the prototype of a country, which is already using very large parts of its suitable agricultural land - and by for using the largest proportion (58%) of current cropland. Australia and larger parts of Asia still have reasonable land resources left for future expansion (Fig 6).

Future Change

For the investigation of future agricultural suitability for the time-period 2071–2100 as determined by the simulated climate effects of the SRES A1B emission scenario, we assume no changes in irrigated areas, soil properties, terrain or any adaptations, such as crop breeding. As result, when again excluding protected and densely forested areas, the global area being highly suitable for agriculture decreases from 4.6 to 3.9 million km², while marginally and moderately suitable areas increase (Table 7). In total, agriculturally suitable areas increase by 4.8 million km² due to the selected climate change scenario. However, most of the additional area is only marginally suitable for agricultural use.

Without excluding any areas, the impact of climate change increases the potentially suitable areas on the globe by 5.6 million km². Marginally suitable areas increase by 4.2 million km², moderately suitable areas increase by 2.3 million km², while highly suitable areas decrease by 0.8 million km² (Fig. 7).

A more regional analysis shows that the world is divided into regions that receive additional suitable land and regions where land that used to be suitable turns into not suitable land (Fig. 8). Regions in the northern hemisphere, such as Canada (+2.1 million km² of suitable land), Russia (+3.1 million km²) and China (+0.9 million km²), benefit most.

On the global scale, suitability improves on 18.7 million km² and worsens on 22.2 million km². In total, the area with decreasing suitability is 3.5 million km² more than the area with increasing suitability (Fig. 9). The highest absolute net loss of suitable areas is found in Sub-Saharan Africa.

Thereby, the globally averaged suitability value (averaged over all suitable areas), decreases from 0.41 to 0.39. The greatest losses of suitability are simulated for France and the Mediterranean (Fig. 10). The changing suitability is mapped in Fig. 11.

Growing Cycle and Multiple Cropping

The seasonal development of temperature and precipitation determines the length of the growing season, the start of the growing cycle and the potential number of annual cropping. Thus, the option of multiple cropping represents an important measure for farmers to increase production. Figure 12 shows the spatial distribution of the start of the growing cycle for the time period 1981–2010, exemplarily for maize.

Changing climate does not only affect the suitability of land, but also the start and length of the growing cycle. As an example, the start of the growing cycle for maize in Germany shifts in average 23 days earlier in time, when comparing the period of 2071–2100 with 1981–2010. The shift of growing cycles again influences the possibility for multiple cropping. Today's maximal achievable multiple cropping according to the course of temperature and precipitation is shown in Fig. 13.

Our results suggest that climate change has huge impacts on the areas suitable for multiple cropping under the assumed climate scenario. Until 2100, 6.0 million km² are globally lost for triple cropping until 2100, while the area which is suitable for double cropping increases by 2.3 million km². Multiplying the area with

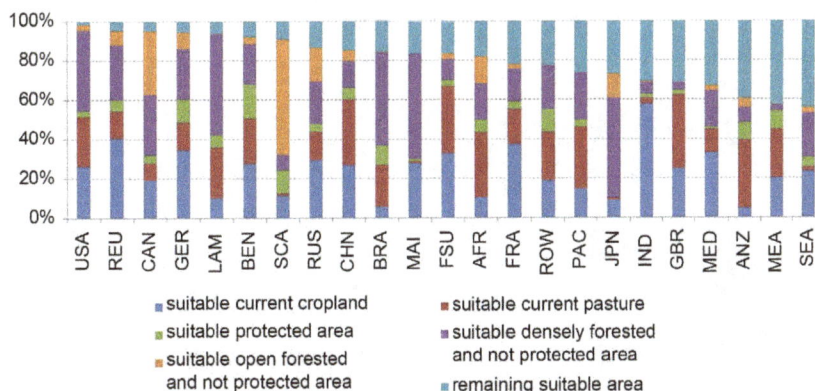

Figure 6. Current use of suitable areas (1981–2010), considering forest cover [46], **protected areas** [45] **and current pasture and cropland (Ramankutty et al., 2008).** If forested areas are agriculturally suitable and protected, they are attributed to 'suitable protected area'.

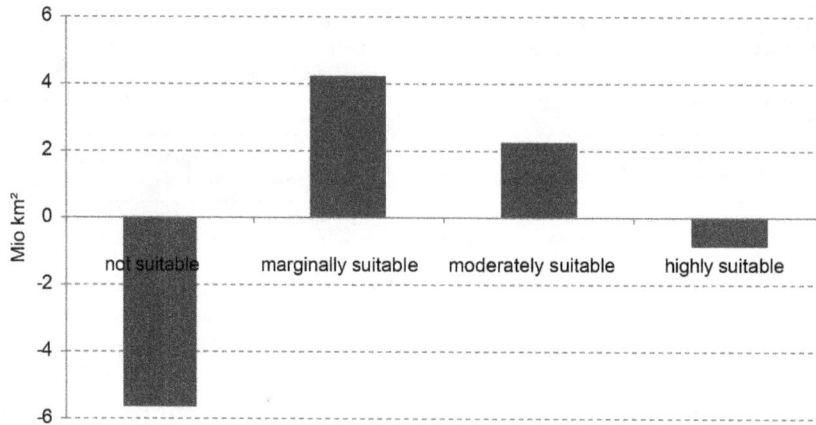

Figure 7. Global changes in agricultural suitability categories (million km²) between 1981–2010 and 2071–2100.

the number of cycles, this means a global decrease of 13.4 million km². Most of the increase in double cropping areas results from the transformation from triple to double cropping. Again, no change in irrigation is assumed in this calculation.

The largest decrease in multiple copping area can be found in Brazil (BRA) and in Sub-Saharan Africa (AFR), where areas suitable for triple cropping decrease by 1.7 (AFR) and 2.9 million km² (BRA) (Fig. 14), while the area for double cropping increases by 0.2 and 1.3 million km², respectively. In total, this means a decrease of multiple cropping area by 1.5 (AFR) and 1.6 million km² (BRA). This is equivalent to the amount of 4.7 and 6.1 million km² respectively, which are lost for agriculture, when multiplying the area with the number of possible crop cycles. This corresponds to 20.2 (AFR) and 28.8% (BRA) of today's potentially suitable area for multiple cropping. In the same manner, France (FRA) and the Mediterranean (MED) lose 24.1 (FRA) and 13.2% (MED) of their total equivalent area when considering the change of multiple cropping, which means a decrease by 93 (FRA) and 55% (MED) according to the multiple cropping area of 1981–2010. Regions where areas that potentially allow for more than one crop cycle increase due to climate change are CHI, IND, JPN, MEA, REU, RUS and USA, while the total area considerably increases mainly in the USA for both, double (0.35 million km²) and triple (0.12 million km²) cropping (Fig. 14).

Conclusions

The analyses of the present situation demonstrats that there is extraordinary potential e.g. for Sub Saharan Africa for future expansion of agricultural land without expanding into protected or forested areas. Further research is necessary to identify the environmental and social costs and consequences of agricultural expansion in these regions. Also further investigation is needed to give answers on how this land could be managed sustainable with benefit to local food systems and socio-economy.

Our results show at high spatial resolution how agricultural suitability may change until 2100 due to changing climate under the chosen scenario (SRES A1B), assuming no adaptation measurements by farmers. First, suitable areas increase especially in the northern regions such as Canada, China and Russia, where new land will be available for agricultural use. The increase in suitable areas mainly takes place in sparsely populated areas, which could imply a lack of labor for open up new agricultural land and prepare soils. Certainly, it will be related with high investment costs and it will take a long time to extend agriculture here. Secondly, global average suitability decreases under the chosen climate scenario. Especially the extend of highly suitable areas is reduced by the effect of climate change. Finally, suitable areas indirectly are reduced due to a substantial global reduction of the suitability for multiple cropping, especially in Sub Saharan Africa, and Brazil.

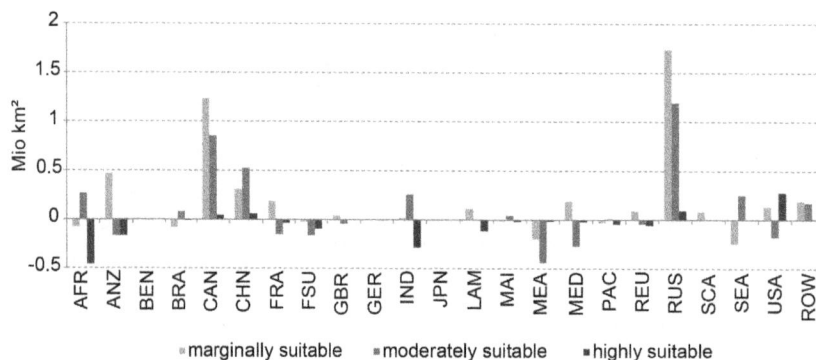

Figure 8. Regional change in agricultural suitability categories (million km²) between 1981–2010 and 2071–2100.

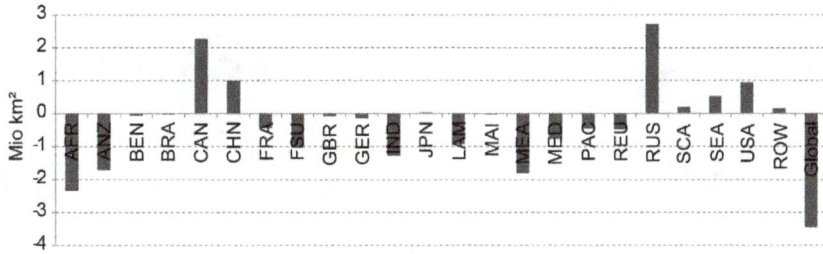

Figure 9. Regional change of agriculturally suitable area due to A1B climate change scenario between 1981–2010 and 2071–2100.

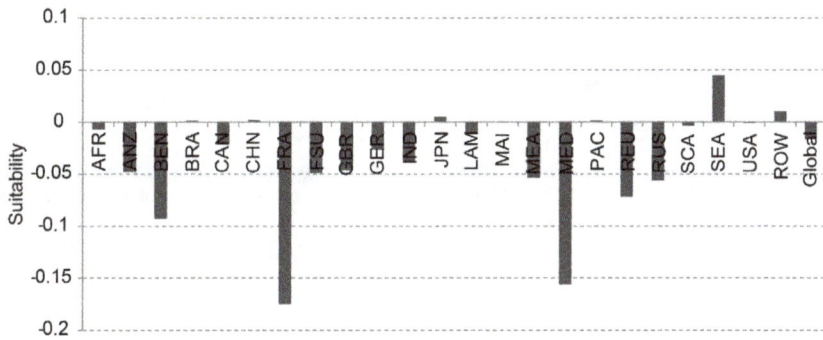

Figure 10. Regional changes in the average suitability between 1981–2010 and 2071–2100.

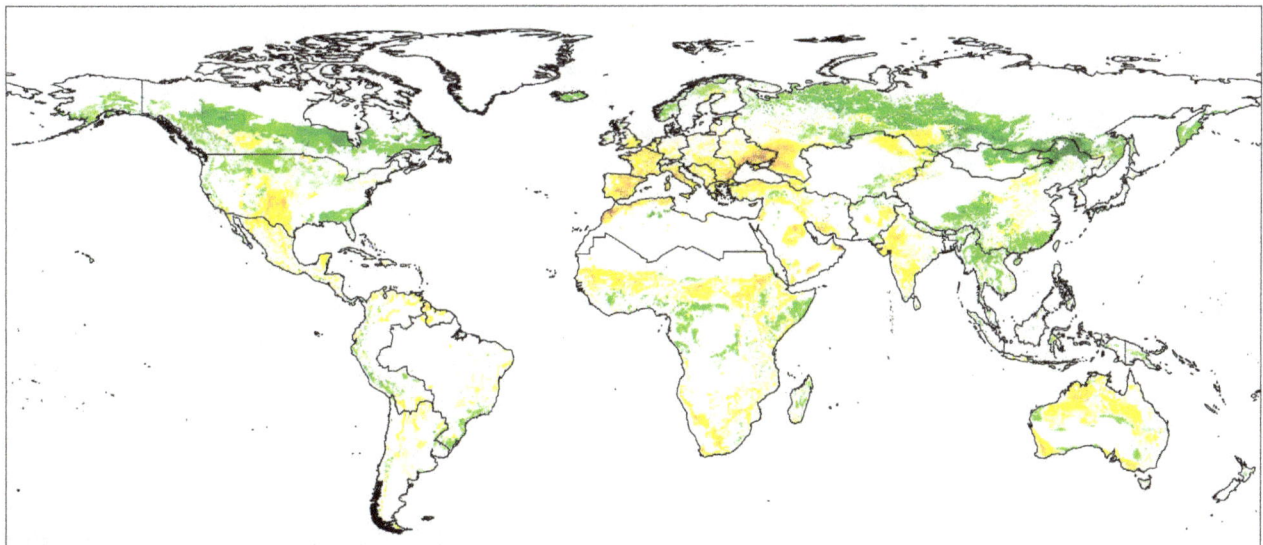

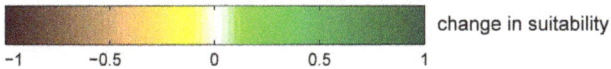

change in suitability

Figure 11. Change in agricultural suitability between 1981–2010 and 2071–2100. Green areas indicate an increase in suitability while brown areas show a decreasing suitability.

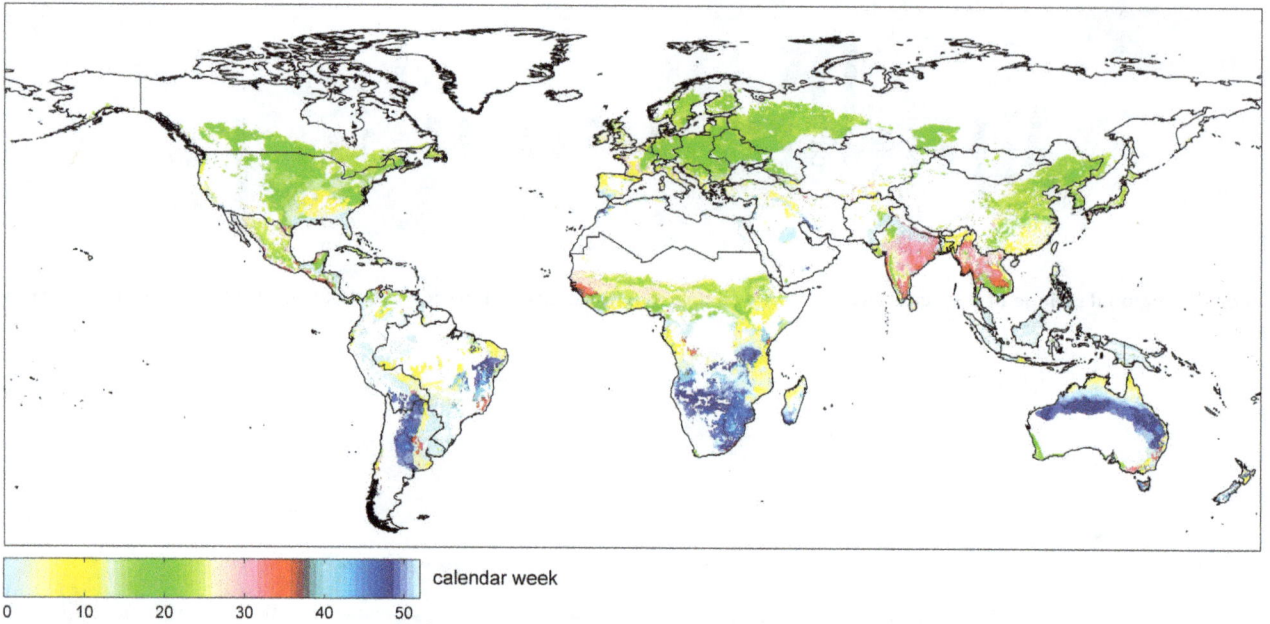

Figure 12. Start of the growing cycle for maize (1981–2010). The start of the growing cycle is illustrated for rainfed conditions and for irrigated conditions on predominantly irrigated areas (irrigated area > 50%). In case of multiple cropping, the map shows the start of the first growing cycle.

Overall, the Global North regionally increases suitability and the number of crop cycles, while the Global South and the Mediterranean area lose agriculturally suitable land without adaptations. This will decisively affect smallholder farmers as their options for adaptations through e.g. irrigation are limited.

Scientific knowledge on the geographical distribution has decisively being increased with the availability of global data sets, also based on remote sensing. The tensions between both limits of land expansion and intensification within the context of sustainable agricultural intensification stresses the ongoing debate on

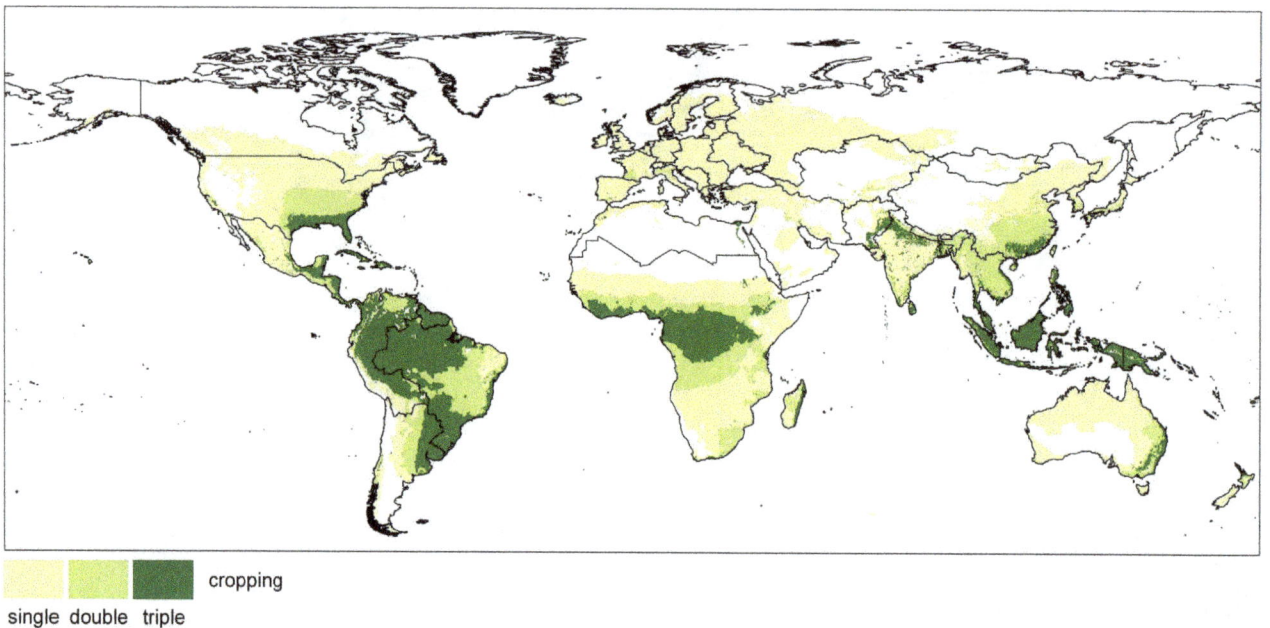

Figure 13. Suitable areas for single, double and triple cropping (1981–2010). Multiple cropping is illustrated for rainfed conditions and for irrigated conditions on predominantly irrigated areas (irrigated area > 50%).

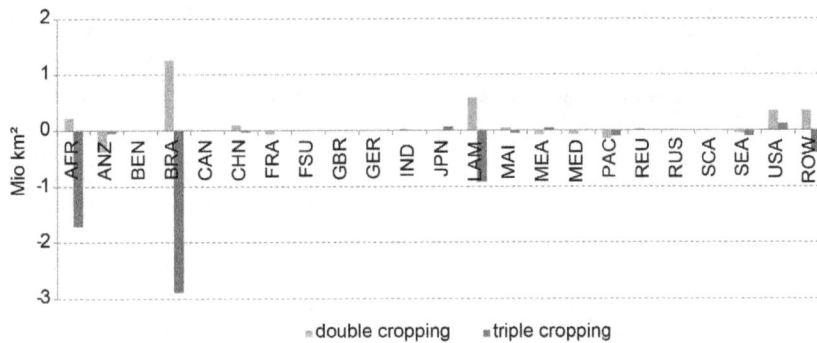

Figure 14. Change in area suitable for double and triple cropping (million km²) between 1981–2010 and 2071–2100.

global land management, considering the complex interplay and trade-offs between different uses of land and ecosystem services.

Acknowledgments

This research was carried out within the framework of the GLUES (Global Assessment of Land Use Dynamics, Greenhouse Gas Emissions and Ecosystem Services) Project. Thanks to all project members and to Jonas Maier who contributed to this study as a student assistant.

Author Contributions

Conceived and designed the experiments: FZ BP WM. Performed the experiments: FZ. Analyzed the data: FZ. Contributed reagents/materials/analysis tools: FZ. Contributed to the writing of the manuscript: FZ.

References

1. Foley JA, DeFries R, Asner GP, Barford C, Bonan G, et al. (2005) Global Consequences of Land Use. Science 309: 570–574.
2. Václavík T, Lautenbach S, Kuemmerle T and Seppelt R (2013) Mapping global land system archetypes. Global Environmental Change 23: 1637–1647.
3. Ramankutty N, Foley JA, Norman J, McSweeney K (2002) The global distribution of cultivable lands: current patterns and sensitivity to possible climate change. Global Ecology and Biogeography 11: 377–392.
4. Alexandratos N, Bruinsma J (2012) World agriculture towards 2030/2050: the 2012 revision. ESA Working Paper No 12-03. Rome: FAO.
5. Tilman D, Balzer C, Hill J, Befort BL (2011) Global food demand and the sustainable intensification of agriculture. Proceedings of the National Academy of Sciences 108: 20260–20264.
6. Foley JA, Ramankutty N, Brauman KA, Cassidy ES, Gerber JS, et al. (2011) Solutions for a cultivated planet. Nature 478: 337–342.
7. Godfray HCJ, Beddington JR, Crute IR, Haddad L, Lawrence D, et al. (2010) Food Security: The Challenge of Feeding 9 Billion People. Science 327: 812–818.
8. Pretty J, Sutherland WJ, Ashby J, Auburn J, Baulcombe D, et al. (2010) The top 100 questions of importance to the future of global agriculture. International Journal of Agricultural Sustainability 8: 219–236.
9. Gregory PJ and George TS (2011) Feeding nine billion: the challenge to sustainable crop production. Journal of Experimental Botany.
10. Ray DK, Mueller ND, West PC, Foley JA (2013) Yield Trends Are Insufficient to Double Global Crop Production by 2050. PLoS ONE 8: e66428.
11. Vermeulen S, Zougmoré R, Wollenberg E, Thornton P, Nelson G, et al. (2012) Climate change, agriculture and food security: a global partnership to link research and action for low-income agricultural producers and consumers. Current Opinion in Environmental Sustainability 4: 128–133.
12. Kastner T, Rivas MJI, Koch W, Nonhebel S (2012) Global changes in diets and the consequences for land requirements for food. Proceedings of the National Academy of Sciences 109: 6868–6872.
13. Cassidy ES, West PC, Gerber JS, Foley JA (2013) Redefining agricultural yields: from tonnes to people nourished per hectare. Environmental Research Letters 8: 034015.
14. Erb K-H, Haberl H, Krausmann F, Lauk C, Plutzar C, et al. (2009) Eating the Planet: Feeding and fuelling the world sustainably, fairly and humanely – a scoping study. Social Ecology Working Paper. Alpen-Adria Universität Klagenfurt.
15. Spiertz JHJ, Ewert F (2009) Crop production and resource use to meet the growing demand for food, feed and fuel: opportunities and constraints. NJAS - Wageningen Journal of Life Sciences 56: 281–300.
16. Bruinsma J (2011) The resources outlook: by how much do land, water and crop yields need to increase by 2050? In: P. Conforti , editor. Looking ahead in world food and agriculture: Perspectives to 2050. Rome: FAO.
17. Schmitz C, van Meijl H, Kyle P, Nelson GC, Fujimori S, et al. (2014) Land-use change trajectories up to 2050: insights from a global agro-economic model comparison. Agricultural Economics 45: 69–84.
18. Pimentel D, Harvey C, Resosudarmo P, Sinclair K, Kurz D, et al. (1995) Environmental and Economic Costs of Soil Erosion and Conservation Benefits. Science 267: 1117–1123.

19. UNCCD (2014) Desertification - The Invisible Frontline. Bonn, Germany: United Nations Convention to Combat Desertification.
20. Seto KC, Fragkias M, Güneralp B, Reilly MK (2011) A Meta-Analysis of Global Urban Land Expansion. PLoS ONE 6: e23777.
21. Avellan T, Meier J, Mauser W (2012) Are urban areas endangering the availability of rainfed crop suitable land? Remote Sensing Letters 3: 631–638.
22. Teka K, Haftu M (2012) Land Suitability Characterization for Crop and Fruit Production in Midlands of Tigray, Ethiopia. Momona Ethiopian Journal of Science 4: 12.
23. Kalogirou S (2002) Expert systems and GIS: an application of land suitability evaluation. Computers, Environment and Urban Systems 26: 89–112.
24. Baja S, Chapman DM, Dragovich D (2002) A Conceptual Model for Defining and Assessing Land Management Units Using a Fuzzy Modeling Approach in GIS Environment. Environmental Management 29: 647–661.
25. Braimoh AK, Vlek PLG, Stein A (2004) Land Evaluation for Maize Based on Fuzzy Set and Interpolation. Environmental Management 33: 226–238.
26. Kurtener D, Torbert HA, Krueger E (2008) Evaluation of Agricultural Land Suitability: Application of Fuzzy Indicators. In: O . Gervasi, B . Murgante, A . Laganûâ, D . Taniar, Y . Mun and M . Gavrilova, editors. Computational Science and Its Applications – ICCSA 2008. Springer Berlin Heidelberg. pp. 475–490.
27. Nisar Ahamed TR, Gopal Rao K, Murthy JSR (2000) GIS-based fuzzy membership model for crop-land suitability analysis. Agricultural Systems 63: 75–95.
28. Van Ranst E, Tang H, Groenemam R, Sinthurahat S (1996) Application of fuzzy logic to land suitability for rubber production in peninsular Thailand. Geoderma 70: 1–19.
29. IIASA/FAO (2012) Global Agro-ecological Zones (GAEZ v3.0) - Model Documentation. IIASA, Laxenburg, Austria and FAO, Rome, Italy: IIASA, FAO.
30. Fischer G, Hizsnyik E, Prieler S, Wiberg D (2011) Scarcity and abundance of land resources: competing uses and the shrinking land resource base. SOLAW Background Thematic Report. FAO.
31. Lane A, Jarvis A (2007) Changes in Climate will modify the Geography of Crop Suitability: Agricultural Biodiversity can help with Adaptation. Journal of SAT Agricultural Research 4: 12.
32. Avellan T, Zabel F, Mauser W (2012) The influence of input data quality in determining areas suitable for crop growth at the global scale – a comparative analysis of two soil and climate datasets. Soil Use and Management 28: 249–265.
33. Burrough PA (1989) Fuzzy mathematical methods for soil survey and land evaluation. Journal of Soil Science 40: 477–492.
34. Burrough PA, Macmillan RA, van Deursen W (1992) Fuzzy classification methods for determining land suitability from soil profile observations and topography. Journal of Soil Science 43: 193–210.
35. Sys CO, van Ranst E, Debaveye J, Beernaert F (1993) Land evaluation: Part III Crop requirements. Agricultural Publications. Brussels: G.A.D.C.
36. Bengtsson L, Hodges KI, Keenlyside N (2009) Will Extratropical Storms Intensify in a Warmer Climate? Journal of Climate 22: 2276–2301.

37. Jungclaus JH, Keenlyside N, Botzet M, Haak H, Luo JJ, et al. (2006) Ocean Circulation and Tropical Variability in the Coupled Model ECHAM5/MPI-OM. Journal of Climate 19: 3952–3972.

38. Marke T, Mauser W, Pfeiffer A, Zängl G, Jacob D, et al. (2013) Application of a hydrometeorological model chain to investigate the effect of global boundaries and downscaling on simulated river discharge. Environ Earth Sci: 1–20.

39. Farr TG, Rosen PA, Caro E, Crippen R, Duren R, et al. (2007) The Shuttle Radar Topography Mission. Reviews of Geophysics 45: RG2004.

40. Hijmans RJ, Cameron SE, Parra JL, Jones PG, Jarvis A (2005) Very high resolution interpolated climate surfaces for global land areas. International Journal of Climatology 25: 1965–1978.

41. FAO/IIASA/ISRIC/ISSCAS/JRC (2012) Harmonized World Soil Database (version 1.2). FAO, Rome, Italy and IIASA, Laxenburg, Austria.

42. Avellan T, Zabel F, Putzenlechner B, Mauser W (2013) A Comparison of Using Dominant Soil and Weighted Average of the Component Soils in Determining Global Crop Growth Suitability. Environment and Pollution 2: 11.

43. Ramankutty N, Evan AT, Monfreda C, Foley JA (2008) Farming the planet: 1. Geographic distribution of global agricultural lands in the year 2000. Global Biogeochemical Cycles 22: GB1003.

44. Siebert S, Henrich V, Frenken K, Burke J (2013) Global Map of Irrigation Areas version 5. Rheinische Friedrich-Wilhelms-University, Bonn, Germany/Food and Agriculture Organization of the United Nations, Rome, Italy.

45. IUCN (2008) Guidelines for Applying Protected Area Management Categories In: N. Dudley, editor. Gland, Switzerland: IUCN.

46. Bontemps S, Defourny P, Bogaert Ev, Arino O, Kalogirou V (2009) GLOBCOVER 2009, Products Description and Validation Report. ESA, Univ. catholique de Louvain.

47. Hansen MC, Potapov PV, Moore R, Hancher M, Turubanova SA, et al. (2013) High-Resolution Global Maps of 21st-Century Forest Cover Change. Science 342: 850–853.

48. FAOSTAT (2014) FAOSTAT Land USE module. Retrieved 24 February 2014. Available at: http://faostat.fao.org/site/377/DesktopDefault.aspx?PageID=377#ancor.

Permissions

The contributors of this book come from diverse backgrounds, making this book a truly international effort. This book will bring forth new frontiers with its revolutionizing research information and detailed analysis of the nascent developments around the world.

We would like to thank all the contributing authors for lending their expertise to make the book truly unique. They have played a crucial role in the development of this book. Without their invaluable contributions this book wouldn't have been possible. They have made vital efforts to compile up to date information on the varied aspects of this subject to make this book a valuable addition to the collection of many professionals and students.

This book was conceptualized with the vision of imparting up-to-date information and advanced data in this field. To ensure the same, a matchless editorial board was set up. Every individual on the board went through rigorous rounds of assessment to prove their worth. After which they invested a large part of their time researching and compiling the most relevant data for our readers.

The editorial board has been involved in producing this book since its inception. They have spent rigorous hours researching and exploring the diverse topics which have resulted in the successful publishing of this book. They have passed on their knowledge of decades through this book. To expedite this challenging task, the publisher supported the team at every step. A small team of assistant editors was also appointed to further simplify the editing procedure and attain best results for the readers.

Apart from the editorial board, the designing team has also invested a significant amount of their time in understanding the subject and creating the most relevant covers. They scrutinized every image to scout for the most suitable representation of the subject and create an appropriate cover for the book.

The publishing team has been an ardent support to the editorial, designing and production team. Their endless efforts to recruit the best for this project, has resulted in the accomplishment of this book. They are a veteran in the field of academics and their pool of knowledge is as vast as their experience in printing. Their expertise and guidance has proved useful at every step. Their uncompromising quality standards have made this book an exceptional effort. Their encouragement from time to time has been an inspiration for everyone.

The publisher and the editorial board hope that this book will prove to be a valuable piece of knowledge for researchers, students, practitioners and scholars across the globe.

List of Contributors

Jorge Torres-Sánchez, Francisca López-Granados, Ana Isabel De Castro and José Manuel Peña-Barragán
Department of Crop Protection, Institute for Sustainable Agriculture (IAS) Spanish National Research Council (CSIC), Córdoba, Spain

Innocent Nzeyimana and Violette Geissen
Soil Physics and Land Management Group, Wageningen University, Wageningen, The Netherlands

Alfred E. Hartemink
Department of Soil Science, FD Hole Soils Lab, University of Wisconsin, Madison, Madison, Wisconsin, United States of America

Yuan Liu, Yongzhuo Liu, Yuanjun Ding, Jinwei Zheng, Tong Zhou, Lianqing Li, Jufeng Zheng, Xuhui Zhang, Xinyan Yu and Jiafang Wang
Institute of Resource, Ecosystem and Environment of Agriculture, Nanjing Agricultural University, Nanjing, China

Genxing Pan
nstitute of Resource, Ecosystem and Environment of Agriculture, Nanjing Agricultural University, Nanjing, China
Center of Ecosystem Carbon Sink and Environment Remediation, Zhejiang Agricultural and Forestry University, Linan, Hangzhou, China

David Crowley
Department of Environment Sciences, University of California Riverside, Riverside California, United States of America

Srivathsa Nallanchakravarthula, Shahid Mahmood, Sadhna Alström and Roger D. Finlay
Uppsala BioCenter, Department of Forest Mycology and Plant Pathology, Swedish University of Agricultural Sciences, Uppsala, Sweden

Richard V. Scholtz III and Allen R. Overman
Agricultural & Biological Engineering Department, University of Florida, Gainesville, Florida, United States of America

Shiqing Li
State Key Laboratory of Soil Erosion and Dryland Farming on the Loess Plateau, Institute of Soil and Water Conservation, Northwest A&F University, Yangling, Shaanxi, China

Yaai Dang
State Key Laboratory of Soil Erosion and Dryland Farming on the Loess Plateau, Institute of Soil and Water Conservation, Northwest A&F University, Yangling, Shaanxi, China
International Center for Climate and Global Change Research, School of Forestry & Wildlife Sciences, Auburn University, Auburn, Alabama, United States of America
College of Science, Northwest A&F University, Yangling, Shaanxi, China

Wei Ren, Bo Tao, Guangsheng Chen, Chaoqun Lu, Jia Yang, Shufen Pan and Hanqin Tian
International Center for Climate and Global Change Research, School of Forestry & Wildlife Sciences, Auburn University, Auburn, Alabama, United States of America

Guodong Wang
College of Science, Northwest A&F University, Yangling, Shaanxi, China

Huhe, Shinchilelt Borjigin, Nobukiko Nomura, Toshiaki Nakajima,
Toru Nakamura and Hiroo Uchiyama
Graduate School of Life and Environmental Sciences, University of Tsukuba, Tsukuba, Ibaraki, Japan

Yunxiang Cheng
State Key Laboratory of Grassland Agro-Ecosystems, College of Pastoral Agriculture Science and Technology, Lanzhou University, Lanzhou, China

Chao Chai, Hongzhen Cheng, Dong Ma, Yanxi Shi
College of Resources and Environment, Qingdao Agricultural University, Qingdao, China

Wei Ge
College of Life Sciences, Qingdao Agricultural University, Qingdao, China

Basit Yousuf, Raghawendra Kumar, Avinash Mishra, Bhavanath Jha
Discipline of Marine Biotechnology and Ecology, CSIR-Central Salt and Marine Chemicals Research Institute, Bhavnagar, Gujarat, India
Academy of Scientific and Innovative Research, CSIR, New Delhi, India

JoséA. Siles and Inmaculada García-Romera
Department of Soil Microbiology and Symbiotic Systems, Estación Experimental del Zaidín, Consejo Superior de Investigaciones Científicas (CSIC), Granada, Spain

Inmaculada Sampedro
Department of Soil Microbiology and Symbiotic Systems, Estación Experimental del Zaidín, Consejo Superior de Investigaciones Científicas (CSIC), Granada, Spain
Thayer School of Engineering, Dartmouth College, Hanover, New Hampshire, United States of America

Caio T. C. C. Rachid
Institute of Microbiology Paulo de Góes, Federal University of Rio de Janeiro, Rio de Janeiro, Brazil
Center for Microbial Ecology, Michigan State University, East Lansing, Michigan, United States of America

James M. Tiedje
Center for Microbial Ecology, Michigan State University, East Lansing, Michigan, United States of America

Biao Liu
Nanjing Institute of Environmental Sciences, Ministry of Environmental Protection of China, Nanjing, Jiangsu Province, China

Xiaogang Li
Nanjing Institute of Environmental Sciences, Ministry of Environmental Protection of China, Nanjing, Jiangsu Province, China
Key Laboratory of Soil Environment and Pollution Remediation, Institute of Soil Science, Chinese Academy of Sciences, Nanjing, Jiangsu Province, China

Weihua Li, Qinbin Zhang, Long Wang, Xiaoxiang Zhang, Guiliang Song, Zhiyuan Fu, Dong Ding, Zonghua Liu and Jihua Tang
1 National Key Laboratory of Wheat and Maize Crops Science, Collaborative Innovation Center of Henan Grain Crops, College of Agronomy, Henan Agricultural University, Zhengzhou, China

Zhongjun Fu
Maize Research Institute, Chongqing Academy of Agricultural Sciences, Chongqing, Chin

Franziska Lauer, Katharina Prost, Stefan Pätzold, Mareike Wolf
Sarah Urmersbach, Eva Lehndorff and Wulf Amelung
Institute of Crop Science and Resource Conservation – Soil Science and Soil Ecology, University of Bonn, Bonn, Germany

Renate Gerlach
Archaeological Heritage Management Rhineland (LVR-Amt für Bodendenkmalpflege im Rheinland), Bonn, Germany

Eileen Eckmeier
Department of Geography, RWTH Aachen University, Aachen, Germany

Risheng Ding, Shaozhong Kang, Taisheng Du and Xinmei Hao
Center for Agricultural Water Research in China, China Agricultural University, Beijing, China

Yanqun Zhang
National Center of Efficient Irrigation Engineering and Technology
Research-Beijing, China Institute of Water Resources and Hydropower Research, Beijing, China

Xiang Wang, Erik L. H. Cammeraat, Paul Romeijn and Karsten Kalbitz
Earth Surface Science, Institute for Biodiversity and Ecosystem Dynamics, University of Amsterdam, Amsterdam, The Netherlands

Yanbing Qi
College of Resources and Environment, Northwest Agriculture and ForestryUniversity, Yangling, Shaanxi, Peoplés Republic of China
Key Laboratory of Plant Nutrition and the Agri-environment in Northwest China, Ministry of Agriculture, Yangling, Shaanxi, People's Republic of China

Biao Huang and Jeremy Landon Darilek
State Key Laboratory of Soil and Sustainable Agriculture, Institute of Soil Science, Chinese Academy of Sciences, Nanjing, Jiangsu, People's Republic of China

Miguel Montes-Borrego and Blanca B. Landa
Department of Crop Protection, Institute for Sustainable Agriculture (IAS-CSIC), Cordoba, Spain

Madis Metsis
Tallinn University, Institute of Mathematics and Natural Sciences, Tallinn, Estonia

Zhijian Mu
Chongqing Key Laboratory of Soil Multi-scale Interfacial Processes, College of Resources & Environment, Southwest University, Chongqing, China

Aiying Huang
College of Agronomy & Biotechnology, Southwest University, Chongqing, China

Jiupai Ni and Deti Xie
Chongqing Engineering Research Center for Agricultural Non-point Source Pollution Control in Three -Gorges Region, College of Resources & Environment, Southwest University, Chongqing, China

Qurban Ali Panhwar and Shamshuddin Jusop
Department of Land Management, Faculty of Agriculture, Universiti Putra Malaysia (UPM), Serdang, Selangor, Malaysia

Radziah Othman
Department of Land Management, Faculty of Agriculture, Universiti Putra Malaysia (UPM), Serdang, Selangor, Malaysia

Institute of Tropical Agriculture, Universiti Putra Malaysia (UPM), Serdang, Selangor, Malaysia

Mohd Razi Ismail
Institute of Tropical Agriculture, Universiti Putra Malaysia (UPM), Serdang, Selangor, Malaysia

Umme Aminun Naher and Md Abdul Latif
Institute of Tropical Agriculture, Universiti Putra Malaysia (UPM), Serdang, Selangor, Malaysia
Bangladesh Rice Research Institute, Gazipur, Bangladesh

Xuan Fang and Guoan Tang
Key Laboratory of Virtual Geographic Environment, Ministry of Education, School of Geography Science, Nanjing Normal University, Nanjing, China

Bicheng Li
Research Center of Soil and Water Conservation and Ecological Environment, Chinese Academy of Sciences, Yangling, Shaanxi, China

Ruiming Han
School of Geography Science, Nanjing Normal University, Nanjing, China

Florian Zabel, Birgitta Putzenlechner and Wolfram Mauser
Department of Geography, Ludwig Maximilians University, Munich, Germany

Index